OXFORD STATISTICAL SCIENCE SERIES

Series Editors

OXFORD STATISTICAL SCIENCE SERIES

Books in the series

1. A.C. Atkinson: *Plots, Transformations, and Regression*
2. M. Stone: *Coordinate-free Multivariable Statistics*
3. W.J. Krzanowski: *Principles of Multivariate Analysis: A User's Perspective*
4. M. Aitkin, D. Anderson, B. Francis, and J. Hinde: *Statistical Modelling in GLIM*
5. Peter J. Diggle: *Time Series: A Biostatistical Introduction*
6. Howell Tong: *Non-linear Time Series: A Dynamical System Approach*
7. V.P. Godambe: *Estimating Functions*
8. A.C. Atkinson and A.N. Donev: *Optimum Experimental Designs*
9. U.N. Bhat and I.V. Basawa: *Queuing and Related Models*
10. J.K. Lindsey: *Models for Repeated Measurements*
11. N.T. Longford: *Random Coefficient Models*
12. P.J. Brown: *Measurement, Regression, and Calibration*
13. Peter J. Diggle, Kung-Yee Liang, and Scott L. Zeger: *Analysis of Longitudinal Data*
14. J.I. Ansell and M.J. Phillips: *Practical Methods for Reliability Data Analysis*
15. J.K. Lindsey: *Modelling Frequency and Count Data*
16. J.L. Jensen: *Saddlepoint Approximations*
17. Steffen L. Lauritzen: *Graphical Models*
18. A.W. Bowman and A. Azzalini: *Applied Smoothing Methods for Data Analysis*
19. J.K. Lindsey: *Models for Repeated Measurements, Second Edition*
20. Michael Evans and Tim Swartz: *Approximating Integrals via Monte Carlo and Deterministic Methods*
21. D.F. Andrews and J.E. Stafford: *Symbolic Computation for Statistical Inference*
22. T.A. Severini: *Likelihood Methods in Statistics*
23. W.J. Krzanowski: *Principles of Multivariate Analysis: A User's Perspective, Revised Edition*
24. J. Durbin and S.J. Koopman: *Time Series Analysis by State Space Models*
25. Peter J. Diggle, Patrick Heagerty, Kung-Yee Liang, and Scott L. Zeger: *Analysis of Longitudinal Data, Second Edition*
26. J.K. Lindsey: *Nonlinear Models in Medical Statistics*
27. Peter J. Green, Nils L. Hjort, and Sylvia Richardson: *Highly Structured Stochastic Systems*
28. Margaret S. Pepe: *The Statistical Evaluation of Medical Tests for Classification and Prediction*
29. Christopher G. Small and Jinfang Wang: *Numerical Methods for Nonlinear Estimating Equations*
30. John C. Gower and Garmt B. Dijksterhuis: *Procrustes Problems*
31. Margaret S. Pepe: *The Statistical Evaluation of Medical Tests for Classification and Prediction, Paperback*
32. Murray Aitkin, Brian Francis, and John Hinde: *Generalized Linear Models: Statistical Modelling with GLIM4*
33. Anthony C. Davison, Yadolah Dodge, and N. Wermuth: Celebrating Statistics: *Papers in honour of Sir David Cox on his 80th birthday*
34. Anthony Atkinson, Alexander Donev, and Randall Tobias: *Optimum Experimental Designs, with SAS*

Optimum Experimental Designs, with SAS

A. C. Atkinson

London School of Economics

A. N. Donev

AstraZeneca

R. D. Tobias

SAS Institute Inc

OXFORD
UNIVERSITY PRESS

Great Clarendon Street, Oxford OX2 6DP

Oxford University Press is a department of the University of Oxford.
It furthers the University's objective of excellence in research, scholarship,
and education by publishing worldwide in

Oxford New York

Auckland Cape Town Dar es Salaam Hong Kong Karachi
Kuala Lumpur Madrid Melbourne Mexico City Nairobi
New Delhi Shanghai Taipei Toronto

With offices in

Argentina Austria Brazil Chile Czech Republic France Greece
Guatemala Hungary Italy Japan Poland Portugal Singapore
South Korea Switzerland Thailand Turkey Ukraine Vietnam

Oxford is a registered trade mark of Oxford University Press
in the UK and in certain other countries

Published in the United States
by Oxford University Press Inc., New York

First published 2007

British Library Cataloguing in Publication Data
Data available

Library of Congress Cataloging in Publication Data
Data available

Typeset by Newgen Imaging Systems (P) Ltd., Chennai, India
Printed in Great Britain
by
Biddles Ltd., King's Lynn, Norfolk

ISBN 978–0–19–929659–0 (Hbk.)
ISBN 978–0–19–929660–6 (Pbk.)

1 3 5 7 9 10 8 6 4 2

In gratitude for 966 and all that

To Lubov, Nina and Lora

To my father Thurman, for encouragement always to be curious

PREFACE

A well-designed experiment is an efficient method of learning about the world. Because experiments in the world, and even in the carefully controlled conditions of laboratories, cannot avoid random error, statistical methods are essential for their efficient design and analysis.

The fundamental idea behind this book is the importance of the model relating the responses observed in the experiment to the experimental factors. The purpose of the experiment is to find out about the model, including its adequacy. The model can be very general; in several chapters we explore designs for response surfaces in which the response is a smoothly varying function of the settings of the experimental variables. The typical model would be a second-order polynomial. Experiments can then be designed to answer a variety of questions. Often estimates of the parameters are of interest together with predictions of the responses from the fitted model. The variances of the parameter estimates and predictions depend on the particular experimental design used and should be as small as possible. Poorly designed experiments waste resources by yielding unnecessarily large variances and imprecise predictions.

To design experiments we use results from the theory of optimum experimental design. The great power of this theory is that it leads to algorithms for the construction of designs; as we show these can be applied in a wide variety of circumstances. To implement these algorithms we use the powerful and general computational tools available in SAS.

One purpose of the book is to describe enough of the theory to make apparent the overall pattern of optimum designs. Our second purpose is to provide a thorough grounding in the use of SAS for the design of optimum experiments. We do this through examples and through a series of 'SAS tasks' that allow the reader to extend the examples to further design problems. As we hope the title makes clear, our emphasis is on the designs themselves and their construction, rather than on the underlying general theory.

The material has been divided into two parts. The first eight chapters, 'Background', discuss the advantages of a statistical approach to the design of experiments and introduce many of the models and examples which are used in later chapters. The examples are mainly drawn from scientific, engineering, and pharmaceutical experiments, with rather fewer from agriculture. Whatever the area of experimentation, the ideas of Part I are fundamental. These include an introduction to the ideas of models, random variation, and least squares fitting. The principles of optimum experimental design are introduced by comparing the variances of parameter estimates and the variances of the predicted responses from a variety of designs and models. In the much longer Part II the relationship between these two sets of variances leads to the General Equivalence Theorem. In turn this leads to efficient algorithms for the construction of designs. As well as these ideas, we include, in Chapter 7 a description of many 'standard' designs and demonstrate how to generate them in SAS. In order to keep the book to a reasonable length, there is rather little on the analysis of experimental results. However, Chapter 8 discusses the use of SAS in the analysis of data with linear models. The analysis of data from non-linear models is treated in §17.10.

Part II opens with a general discussion of the theory of optimum design followed, in Chapter 10, by a description of a wide variety of optimality criteria that may be appropriate for designing an experiment. Of these the most often used is D-optimality, which is the subject of Chapter 11. Algorithms for the construction of D-optimum designs are described in the following chapter. The SAS implementation of algorithms for a variety of criteria are covered in Chapter 13. These are applied in Chapter 14 to extensions to response surface designs in which there are both qualitative and quantitative factors. The blocking of response surface designs is the subject of Chapter 15 with mixture experiments discussed in Chapter 16.

Each chapter is intended to cover a self-contained topic. As a result the chapters are of varying lengths. The longest is Chapter 17 which describes the extension of the methods to non-linear regression models, including those defined by sets of differential equations. Designs for non-linear models require prior information about parameter values. The formal use of prior information in design is the subject of Chapter 18 which describes Bayesian procedures. Design augmentation is covered in Chapter 19 and designs for model checking and for discriminating between models are handled in Chapter 20. In Chapter 21 we explore the use of compound designs that are simultaneously efficient for more than one of the criteria of Chapter 10.

In Chapter 22 we move beyond regression to generalized linear models, appropriate, for example, for when the outcome is a count measurement with a binomial distribution. The transformation of observations provides a second extension of the customary linear model. Designs for response transformation and for structured variances are the subject of Chapter 23. Chapter 24 contains material on experimental design when the observations are from a time series and so are correlated.

In the last chapter we gather together a number of further topics. These include crossover designs used in clinical trials to reduce the effect of patient to patient variation whilst accommodating the persistent effects of treatments. A second medical application is the use of optimum design methods in sequential clinical trials. In some cases the purpose of the design is to provide a mixture of randomization and treatment balance. But, in §25.4, we extend this to adaptive designs where earlier responses are used to reduce the number of patients receiving inferior treatments. Other topics include the design of experiments for training neural networks, designs for models when some of the parameters are random, and designs for computer simulation experiments.

A feature of many of these chapters is that we include SAS examples and tasks that provide the ability to construct the designs. These descriptions also reinforce understanding of the design process. Further understanding will be obtained from the exercises of Chapter 26. Some of these are intended to develop an insight into the principles of experimental design; others are more specifically focused on optimum design. Most require pencil and paper for solution, in contrast with the computer-based SAS tasks.

It is hard to overrate the importance of the statistical input to the design of experiments. Sloppily designed experiments may not only waste resources, but may completely fail to provide answers, either in time or at all. This is particularly true of experiments in which there are several interacting factors. In a competitive world, an experiment that takes too long to contribute to the development of a competitive product is valueless, as is a clinical trial in which balance and randomization have not been properly attended to; such trials are often the cause of the irreproducible results that plague drug development.

Despite the importance of statistical methods in the design of experiments, the subject is seriously neglected in the training of many statisticians, scientists, and technologists. Obviously, in writing this book, we have had in mind students and practitioners of statistics. But there is also much here of importance for anyone who has to perform experiments in the laboratory or factory. So in writing we have also had in mind experimenters, the statisticians who sometimes advise them, and anyone who will be training experimenters in universities or in industrial short courses. There are also those who need a more structured account of the subject than can be obtained from browsing the web. The material of Part I, which is at a relatively low mathematical level, should be accessible to members of all these groups. The mathematical level of Part II is slightly higher, but we have avoided derivations of mathematical results—these can be found in the references and in the suggestions for further reading at the ends of most chapters. Little previous statistical knowledge is assumed; although a first course in statistics with an introduction to regression would be helpful for Part I, such knowledge is not essential.

This book is derived from the well-received 1992 book '*Optimum Experimental Designs*' by Atkinson and Donev. This was written almost exactly 30 years after Jack Kiefer read a paper to the Royal Statistical Society in London that marked the beginning of the systematic study of the properties and construction of optimum experimental designs. In the 14 years since the appearance of that book there has been a steady increase in statistical work on the theory and applications of optimum designs. There has also, since 1992, been an explosion in computer power. The combination of these two factors means that most of the text of Part II of this book is new, not just that about SAS. Although we give much SAS code, the detailed code for all our examples is available on the web at `http://www.oup.com/uk/companion/atkinson`. Solutions to the exercises will also be found there.

Although we give many new references, further references continue to appear on the web. We have called our book 'Optimum Experimental Designs, with SAS', rather than 'Optimal ...' because this is the slightly older form in English and avoids the construction 'optim(um) + al'—there is no 'optimalis' in Latin, although there is, for example, in Hungarian. None of this would matter except for searching the web. In September 2006 'optimal design' in Google returned 1.58 million entries, whereas 'optimum design' returned almost exactly a third of that number. Of course, most of the items were not about experimental design. But 'D optimal design' returned 25,000 entries.

London, Macclesfield and Cary
October 2006

Anthony Atkinson
`a.c.atkinson@lse.ac.uk`
`http://stats.lse.ac.uk/atkinson/`

Alexander Donev
`alex.donev@astrazeneca.com`

Randall Tobias
`randy.tobias@sas.com`

CONTENTS

PART I

BACKGROUND

1

INTRODUCTION

1.1 Some Examples

This book is concerned with the design of experiments when random variation in the measured responses is appreciable compared with the effects to be investigated. Under such conditions, statistical methods are essential for experiments to provide unambiguous answers with a minimum of effort and expense. This is particularly so if the effects of several experimental factors are to be studied. The emphasis will be on designs derived using the theory of optimum experimental design. Two main contributions result from such methods. One is the provision of algorithms for the construction of designs, which is of particular importance for non-standard problems. The other is the availability of quantitative methods for the comparison of proposed experiments. As we shall see, many of the widely used standard designs are optimum in ways that are defined in later chapters. The algorithms associated with the theory in addition often allow incorporation of extra features into designs, such as the blocking required in Examples 1.3 and 1.4. This is often achieved with little loss of efficiency relative to simpler designs.

The book is in two parts. In Part I the ideas of the statistical design of experiments are introduced. In Part II, beginning with Chapter 9, the theory of optimum experimental design is developed and numerous applications are described. To begin we describe six examples of experimental design of increasing statistical complexity. The first four, three from technology and one from agriculture, can be modelled using the normal theory linear model. The fifth, from pharmacokinetics, requires a non-linear model with additive errors. In the final example the response is binary, survival or death as a function of dose; a generalized linear model is required. Comments on the scope and limits of the statistical contribution to experimental design are given in the second section of this chapter. We then introduce some aspects of SAS that will be important in later chapters, before concluding with a guide to the literature.

Example 1.1 The Desorption of Carbon Monoxide During the nineteenth century gas works, in which coal was converted to coke and town gas, were

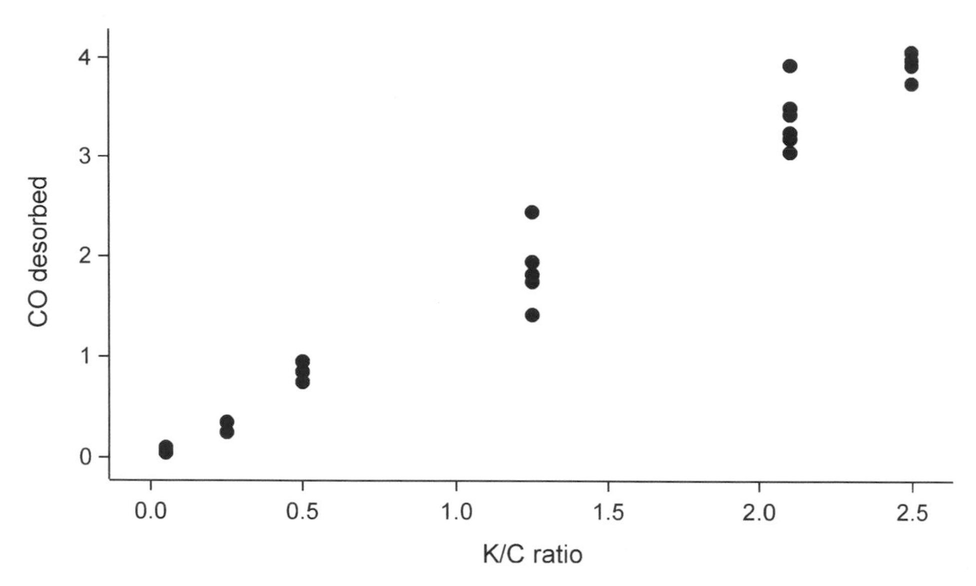

FIG. 1.1. Example 1.1 the desorption of carbon monoxide. Yield (carbon monoxide desorbed) against K/C ratio.

a major source of chemicals and fuel. At the beginning of the twenty-first century, as the reality of a future with reduced supplies of oil approaches, the gasification of coal is again being studied. Typical of this renewed interest is the series of experiments described by Sams and Shadman (1986) on the potassium catalyzed production of carbon monoxide from carbon dioxide and carbon.

In the experiment, graphitized carbon was impregnated with potassium carbonate. This material was then heated in a stream of 15% carbon dioxide in nitrogen. The full experiment used two complicated temperature–time profiles and several responses were measured. The part of the experiment of interest here consisted in measuring the total amount of carbon monoxide desorbed. The results are given in Table 1.1, together with the initial potassium/carbon (K/C) ratio, and are plotted in Figure 1.1. They show a clear linear relationship between the amount of carbon monoxide desorbed and the initial K/C ratio.

Experimental design questions raised by this experiment include:

1. Six levels of K/C ratio were used. Why six levels, and why these six?

2. The numbers of replications at the various K/C ratios vary from 2 to 6. Again why?

TABLE 1.1. Example 1.1: the desorption of carbon monoxide

Observation number	Initial K/C atomic ratio (%)	CO desorbed (mole/mole C %)
1	0.05	0.05
2	0.05	0.10
3	0.25	0.25
4	0.25	0.35
5	0.50	0.75
6	0.50	0.85
7	0.50	0.95
8	1.25	1.42
9	1.25	1.75
10	1.25	1.82
11	1.25	1.95
12	1.25	2.45
13	2.10	3.05
14	2.10	3.19
15	2.10	3.25
16	2.10	3.43
17	2.10	3.50
18	2.10	3.93
19	2.50	3.75
20	2.50	3.93
21	2.50	3.99
22	2.50	4.07

The purpose of questions such as these is to find out whether it is possible to do better by a different selection of numbers of levels and replicates. Better, in this context, means obtaining more precise answers for less experimental effort. We shall be concerned with the need to define the questions that an experiment such as this is intended to answer. Once these are established, designs can be compared for answering a variety of questions and efficient experimental designs can be specified. For example, if it is known that the relationship between desorption and the K/C ratio is linear and passes through the origin and that the magnitude of experimental errors does not depend on the value of the response, then only one value of the K/C ratio needs to be used in the experiment. This should be as large as possible, but not so large that the assumption of linearity is challenged. If the relationship is thought to be more complicated, further levels of the ratio would need to be included in the experiment. The more that is known, the more specific the design can be. While these recommendations may make intuitive sense, the theory we discuss in this book will give an objective and quantitative basis for them. ■

Example 1.1 is an experiment with one continuous factor. In principle, an experimental run or trial could be performed for any non-negative value of the K/C ratio, although there will always be limits imposed by technical considerations, such as the the strength of an apparatus, as well as the values of the factors that are of interest to the experimenter. But often the factors in an experiment are qualitative, having a few fixed levels. The next example presents an experiment with a factor of this type.

Example 1.2 The Viscosity of Elastomer Blends Derringer (1974) reports the results of an experiment on the viscosity of styrene butadiene rubber (SBR) blends. He introduces the experiment as follows:

> Most commercial elastomer formulations contain various types and amounts of fillers and/or plasticizers, all of which exert major effects on the viscosity of the system. A means of predicting the viscosity of a proposed formulation is obviously highly desirable since viscosity control is crucial to processing operations. To date, considerable work has been done on the viscosity of elastomer-filler systems, considerably less on elastomer-plasticizer systems, and virtually none on the complete elastomer-filler-plasticizer systems. The purpose of this work was the development of a viscosity model for the elastomer-filler-plasticizer system which could be used for prediction.

Some of his experimental results are given in Table 1.2. The response is the viscosity of the elastomer blend. There are two continuously variable quantitative factors, the levels of fillers and of naphthenic oils, both of which are measured in parts per hundred (phr) of the pure elastomer. The single qualitative factor is the kind of filler, of which there are three. This factor therefore has three levels.

For each of the three fillers, the experimental design is a 4×6 factorial, that is, measurements of viscosity are taken at all combinations of four levels of naphthenic oil and of the six levels of filler. Some of the questions that this design raises are extensions of those asked about Example 1.1.

1. Why four equally spaced levels of naphthenic oil?
2. Why six equally spaced levels of filler?
3. To investigate filler-plasticizer systems it is necessary to vary the levels of both factors together; experiments in which one factor is varied while the other is held constant will fail to describe the dependency of the response on the factors, unless the two factors act independently. But,

TABLE 1.2. Example 1.2: viscosity of elastomer blends (Mooney viscosity MS_4 at 100°C as a function of filler and oil levels in SBR-1500)

Napththenic		Filler level (phr)					
oil (phr)	Filler	0	12	24	36	48	60
0	A	26	28	30	32	34	37
	B	26	38	50	76	108	157
	C	25	30	35	40	50	60
10	A	18	19	20	21	24	24
	B	17	26	37	53	83	124
	C	18	21	24	28	33	41
20	A	12	14	14	16	17	17
	B	13	20	27	37	57	87
	C	13	15	17	20	24	29
30	A	-	12	12	13	14	14
	B	-	15	22	27	41	63
	C	11	14	15	17	18	25

The fillers are as follows: A, N900, Cabot Corporation; B, Silica A, Hi-Sil 223, PPG Industries; C, Silica B, Hi-Sil EP, PPG Industries.

is a complete factorial with 24 trials necessary? If the viscosity varies smoothly with the levels of the two factors, a simple polynomial model will explain the relationship. As we shall see in §11.5, we do not need so many design points for such models.

4. If there is a common structure for the different fillers, can this be used to provide an improved design?

5. Measurements of viscosity are non-negative and often include some very high values. As we show in §8.3 it is sensible to take the logarithms of the viscosities before analysing the data. The effect of such transformations on good experimental design is the subject of Chapter 23.

The purpose of asking such questions is to find experimental designs that provide sufficiently accurate answers with a minimum number of trials, that is a minumum of experimental effort. To extend this example slightly, if there were two factors at four levels and two at 6, the complete $4^2 \times 6^2$ factorial design would require $16 \times 36 = 576$ trials. The use of a fractional factorial design, or an optimum design for a smooth response surface would lead to an appreciable saving of resources. ■

TABLE 1.3. Example 1.3: breaking strength of cotton fibres

	Treatments				
	T_1	T_2	T_3	T_4	T_5
Block 1	7.62	8.14	7.76	7.17	7.46
Block 2	8.00	8.15	7.73	7.57	7.68
Block 3	7.93	7.87	7.74	7.80	7.21

These two examples come from the technological literature and are typical of the earlier applications of optimum design theory, in that the underlying models are linear with independent errors of constant variance. In addition the experimental factors are continuously variable over a defined range. The next example shows a kind of experiment that arose in agriculture, where the factors are often discrete and the experimental material highly variable, so that blocking is important.

Example 1.3 The Breaking Strength of Cotton Fibres Cox (1958, p. 26) discusses an agricultural example taken from Cochran and Cox (1957, §4.23). The data, given in Table 1.3 are from an experiment in which five fertilizer treatments $T_1, \ldots, T_5$ are applied to cotton plants. These treatments are the levels of a continuous factor, the amount of potash per acre. The response is the breaking strength of the cotton fibres. Although it is well known that fertilizer helps plants to grow, this experiment is concerned with the quality of the product.

In such agricultural experiments there is often great variation in the yields of experimental plots of land even when they receive the same treatment combination, in this case fertilizer. In order to reduce the effect of this variation, which can mask the presence of treatment effects, the experimental plots are gathered into blocks; plots in the same block are expected to have more in common than plots in different blocks. A block might consist of plots from a single field or from one farm. In Table 1.3 there are three blocks of five plots each. A different treatment is given to each plot within the block. The primary interest in the experiment is the differences between the yields for the various treatments. The differences between blocks are usually of lesser interest. A distinction between this experiment and Example 1.2 is that the differences between fillers in Example 1.2 were of the same interest as the effect of plasticizer and filler levels. ■

Example 1.4 Valve Wear Experiment Experiments may include several blocking factors. Goos and Donev (2006*a*) describe an experiment on the wear of valves in an internal combustion engine in which some of the experimental variables were the conditions under which the engine was run, while others were related to the properties of the valves themselves such as materials, dimensions, and coatings. It was anticipated that a quadratic model in these variables was needed.

Six cylinder engines were used so there were six valve positions. However, it was known from experience of similar experiments that the wear characteristics of the cylinders were different in a consistent way. In addition, the wear of a valve in one cylinder had a negligible effect on the wear of the valves in the other cylinders. Valve position could therefore be used as a blocking variable giving blocks similar in structure to those of Table 1.3.

The experiment also used several engines. A benefit from carrying out the experiment in this way is that if the engines are selected at random from a class of engines, then the results from the study apply to that class and it is possible to estimate the between engine variability. This introduces a second blocking variable acting at as many levels as the number of engines used, a number decided by practical constraints. The two blocking variables are different in nature. The valve position is a fixed effect, whereas the block effects for engines are random parameters from which the variability between engines can be estimated. ■

Examples of blocking variables in other statistical investigations include the effect of operators of equipment or apparatus, the batches of raw material used in a chemical plant and the centres of a multi-centre clinical trial.

The SAS tools for constructing randomized block experiments such as Example 1.3, often elaborated to allow for several blocking factors, are described in §7.7. Designs with one or more quantitative factors combined with blocking factors are the subject of Chapter 15.

These first four examples are of data for which simple linear models are appropriate: means for Example 1.3 and linear regression for Example 1.1. But many important applications require models in which the expected response is a non-linear function of parameters and experimental variables, even though the observational errors can often still be assumed independent and additive.

Example 1.5 The Concentration of Theophylline in the Blood of a Horse Fresen (1984) presents the results of an experiment in which six horses each received 15 mg/kg of theophylline. Table 1.4 gives the results of 18 measurements of the concentration of theophylline for one of the horses at a series of times. The data are plotted in Figure 1.2. Initially the concentration is zero. It then rises to a peak before gradually returning to zero as the material is eliminated from the horses's system.

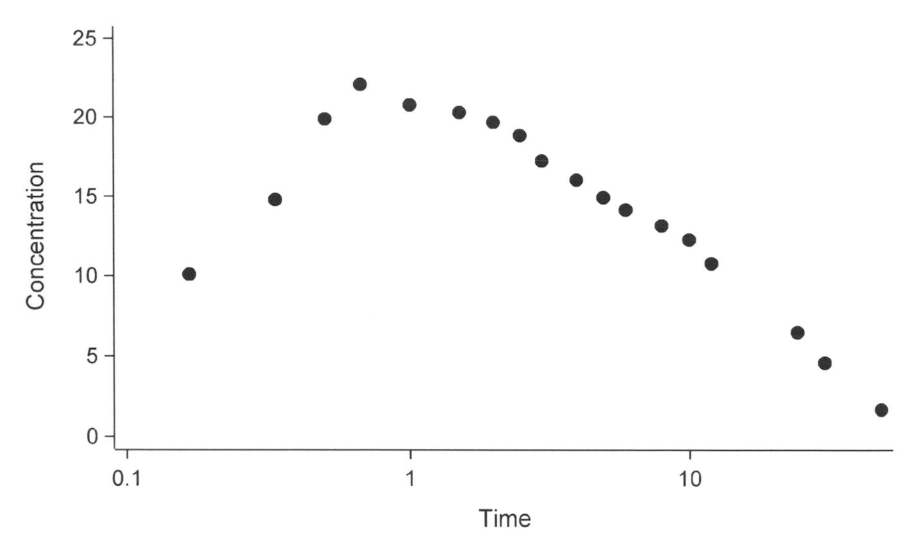

FIG. 1.2. Example 1.5: the concentration of theophylline against time.

TABLE 1.4. Example 1.5: concentration of theophylline in the blood of a horse

x: time (min)	0.166	0.333	0.5	0.666	1	1.5
y: concentration	10.1	14.8	19.9	22.1	20.8	20.3
x (continued)	2	2.5	3	4	5	6
y (continued)	19.7	18.9	17.3	16.1	15.0	14.2
x (continued)	8	10	12	24	30	48
y (continued)	13.2	12.3	10.8	6.5	4.6	1.7

Such data can frequently be represented by non-linear models that are a series of exponential terms. These arise either from an understanding of the chemical kinetics involved or an approximation to it using simple compartmental models with assumed first-order reactions. As in the earlier examples, questions include why were 18 time points chosen and why these 18?

As we shall see in Chapter 17, because the model is non-linear, the design depends upon the values of the unknown parameters; if the parameters are large, the reaction will be fast and there is a danger that measurements will be taken when the theophylline has already been eliminated. On the other hand, if the parameters are small and the reaction slow, a series of uninformative almost constant measurements may result. Another thing we shall see is that designs for non-linear models depend not only on the values of the parameters but also on the aspect of the model that is of importance. For example, efficient estimation of the maximum concentration of theophylline may require a different design from that for efficient estimation of all the parameters in the model. ■

As a final introductory example we consider an experiment with a binomial response, the number of successes or failures in a fixed number of trials.

Example 1.6 Bliss's Beetle Data The data in Table 1.5 result from subjecting eight groups of around 60 beetles to eight different doses of insecticide. The number of beetles killed was recorded. (The data were originally given by Bliss (1935) and are reported in many text books, for example Flury 1997, p. 526). The resulting data are binomial with variables:

x_i: dose of the insecticide;
n_i: number of insects exposed to dose x_i;
R_i: number of insects dying at dose x_i.

At dose level x_i the model is that the observations are binomially distributed, with parameter θ_i. Interest is in whether there is a relationship between the probability of success θ_i and the dose level. The plot of the proportion of success R_i/n_i against x_i in Figure 1.3 clearly shows some relationship. Models for the data have θ_i increasing from zero to one with x_i. One family of functions with this property are cumulative distribution functions of univariate distributions. If, for example, the cumulative normal distribution is used, probit analysis of binomial data results. Optimum design for such generalized linear models is developed in Chapter 22. One design question is what dose levels are best for estimating the parameters of the model relating x_i to θ_i? ■

1.2 Scope and Limitations

The common structure to all six examples is the allocation of treatments, or factor combinations, to experimental units. In Example 1.2, the unit would be a specimen of pure elastomer which is then blended with specified amounts of filler and naphthenic oil, which are the treatments. In Example 1.3, the unit is the plot of land receiving a unique treatment combination, here a level of fertilizer. In Example 1.6, the unit is a set of around 60 beetles, all of which receive the same dose of insecticide.

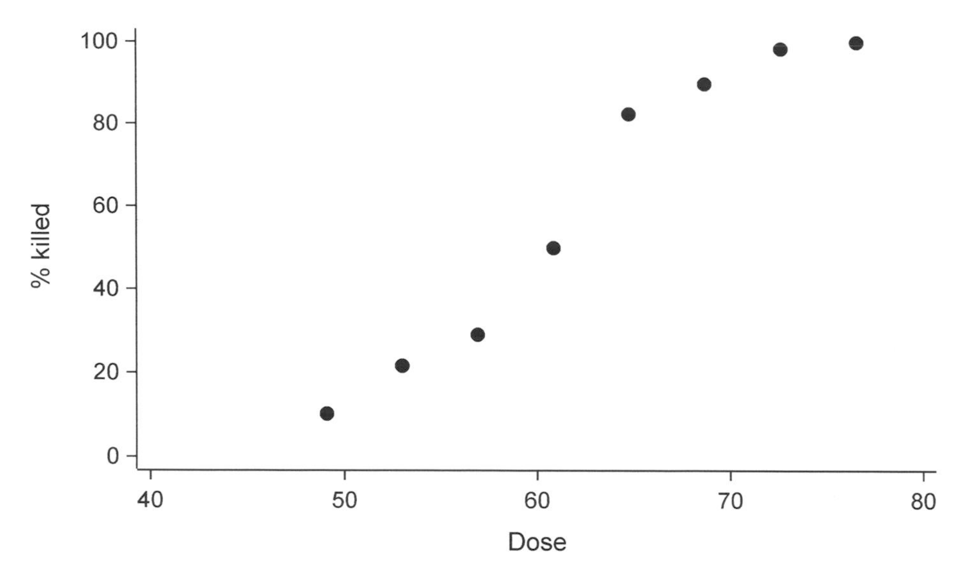

FIG. 1.3. Bliss's beetle data: proportion of deaths increasing with dose.

TABLE 1.5. Bliss's beetle data on the effect of an insecticide

Number	Dose	Killed	Total
1	49.09	6	59
2	52.99	13	60
3	56.91	18	62
4	60.84	28	56
5	64.76	52	63
6	68.69	53	59
7	72.61	61	62
8	76.54	60	60

In optimum experimental design the allocation of treatments to units often depends upon the model or models that are expected to be used to explain the data and the questions that are asked about the models. In Example 1.1 the question might be whether the relationship between carbon monoxide and potassium carbonate was linear, or whether some curvature was present, requiring a second-order term in the model. The optimum design for answering this question is different from that for the efficient fitting of a first-order model, or of a second-order model. The theory could be used to find the best design for any one of these purposes. Another possibility is to find designs efficient for a specified set of purposes, with weightings formalizing the relative importance of the various aspects of the design.

Whatever the procedure followed, the resulting optimum design is a list of treatments to be applied to units. Other procedures for generating designs include custom ('it has always been done this way') and the knowledge and hunches of the experimenter. These procedures also lead to a list of treatments to be applied. A powerful use of the methods described in this book is the principled assessment of proposed designs for all the purposes for which the results might be used.

However, there are many aspects of the design that cannot be determined by optimum design methods. The purpose of the experiment and the design of the apparatus are outside the scope of statistics, as usually are the size of the units and the responses to be measured. In Example 1.1 the amount of carbon needed for a single experiment at a specified carbonate level depends upon the design of the apparatus. However, in Example 1.6 the number of insects forming a unit might be subject to statistical considerations of the power of the experiment, that is, the ability of an experiment involving a given number of beetles to detect an effect of a given size. The experimental region also depends upon the knowledge and intentions of the experimenter. Once the region has been defined, the techniques of this book are concerned with the choice of treatment combinations, which may be from specified levels of qualitative factors, or values from ranges of quantitative variables. The total size of the experiment will depend upon the resources, both of money and time, that are available. The statistical contribution at the planning stage of an investigation is often to calculate the size of effects which can be detected, with reasonable certainty, in the presence of errors of the size anticipated in the particular experiment.

This is a book about an important, powerful, and very general method for the design of experiments. Given the list of treatments to be applied to the experimental units, which units receive which treatments must be arranged in such a way as to avoid systematic bias. Usually this is achieved by randomizing the application of treatments to units, to avoid the confounding of treatment effects with those due to omitted variables which are nevertheless of importance. In many technological experiments the time of day is an important factor if an apparatus is switched off overnight. Randomization of treatment allocation over time provides insurance against the confounding of observed treatment effects with time of day.

Many books on the design of experiments contain almost as much material on the analysis of data as on the design of experiments. The focus here is on design. Analysis is mentioned specifically only in Chapter 8. In Part I of the book, of which Chapter 8 is the last chapter, we give the background to our approach. Chapters 9 onwards are concerned with the central theory of optimum experimental design, illustrated with numerous examples.

1.2.1 SAS Software

SAS software is one of the world's best established packages for data processing and statistical analysis. We will demonstrate the ideas and techniques that we discuss in this book with SAS® tools. Most often, these tools take the form of SAS programs, but we will also touch on SAS point-and-click interfaces.

While most of the SAS facilities that we employ are available in any recent version of the software, the code that we will present was developed using SAS 9.1, released in 2004. SAS software tools comprise a number of different products. Base SAS is the foundation product, and SAS/STAT® provides tools for statistical analysis. SAS facilities for construction of experimental designs, including optimal designs, are located in SAS/QC® software. SAS/IML® provides a language for matrix programming and facilities for optimization.

You will need all of these products to run all the SAS code presented in this book. Note that most universities and business organizations license a bundle of SAS products that includes these tools.

1.3 Background Reading

There is a vast statistical literature on the design of experiments. Cox (1958) remains a relatively short non-mathematical introduction to the basic ideas that is still in print as we write. Cox and Reid (2000) provide an introduction to the theory. Box, Hunter, and Hunter (2005) is a stimulating introduction to statistics and experimental design which reflects the authors' experience in the process industries. The essays in the bravely titled collection Box (2006) extend Box's ideas on design to problems outside these industries. Many books, such as Cobb (1998) and Montgomery (2000), place more emphasis on the analysis of experimental data than on the choice and construction of the designs. Dean and Voss (2003), like us, stresses the use of SAS. Such books typically do not dwell on optimum design. But see Chapter 7 of Cox and Reid (2000).

The pioneering book, in English, on optimum experimental design is Fedorov (1972). Silvey (1980) provides a concise introduction to the

® SAS® and all other SAS Institute Inc. product or service names are registered trademarks or trademarks of SAS Institute Inc. in the USA and other countries.

central theory of the General Equivalence Theorem which we introduce in Chapter 9. From the mathematical point of view, our book can be considered as a series of special cases of the theorem with computer code for the numerical construction of optimum designs. More theoretical treatments are given by Pázman (1986), Pukelsheim (1993), Schwabe (1996), Melas (2006) and, again very concisely, by Fedorov and Hackl (1997). Chapter 6 of Walter and Pronzato (1997) provides an introduction to the theory with engineering examples. Uciński (2005) is devoted to the design of experiments where measurements are taken over time, perhaps from measuring devices, or sensors, whose optimum spatial trajectories are to be determined. Rafajłowicz (2005) (in Polish) emphasizes applications in control, the process industries and image sampling.

Developments in the theory and practice of optimum experimental design can be followed in the proceedings volumes of the MODA conferences. The three most recent volumes are Atkinson, Pronzato, and Wynn (1998), Atkinson, Hackl, and Müller (2001) and Di Bucchianico, Läuter, and Wynn (2004). Two other collections of papers are Atkinson, Bogacka, and Zhigljavsky (2001) and Berger and Wong (2005).

Several books on optimum experimental design were written in the German Democratic Republic. These include the brief introduction of Bandemer *et al.* (1973), which is at the opposite extreme to the two-volume handbook of Bandemer *et al.* (1977) and Bandemer and Näther (1980). Another book from the former German Democratic Republic is Rasch and Herrendörfer (1982). Russian books include Ermakov (1983) and Ermakov and Zhiglijavsky (1987).

The optimum design of experiments is based on a theory which, like any general theory, provides a unification of many separate results and a way of generating new results in novel situations. Part of this flexibility results from the algorithms derived from the General Equivalence Theorem, combined with computer-intensive search methods. Of all these books, only Atkinson and Donev (1992) provides computer code, in the form of a Fortran program, for the generation of optimum designs. Our book, derived from Atkinson and Donev, provides SAS code for the generation of designs in a wide variety of applications.

As well as optimum design there are other well-established approaches to some areas of experimental design. The modern statistical approach was founded by Fisher (1960) (first edition 1935). The methods of Box and Draper (1987) for response surface problems lead to widely used designs that can be evaluated using the methods of our book. In addition, the algorithms of optimum design theory are not, in general, the best way of constructing 'treatment' designs in which the factors are either all qualitative, or quantitative with specified levels to be used a specified number of times. For such problems combinatorial methods of construction, as described by Street and Street (1987) and by Bailey (2004) are preferable. The classic book on analysis of treatment designs is Cochran and Cox (1957). Calinski and Kageyama (2000) present the analysis of block designs, with design itself in a second volume (Calinski and Kageyama 2003). Bailey (2006) gives a careful introduction to comparative experiments. Optimality for treatment designs is given booklength treatment in Shah and Sinha (1980).

2

SOME KEY IDEAS

2.1 Scaled Variables

Experiments are conducted in order to study how the settings of certain variables affect the observed values of other variables. Figure 2.1 is one schematic representation of an experiment. A single trial consists of measuring the values of the h response, or output, variables $y_1, \ldots, y_h$. These values are believed to depend upon the values of the m factors or explanatory variables $u_i, \ldots, u_m$, the values of which will be prescribed by the experimental design. The values of the responses may also depend on t concomitant variables $z_1, \ldots, z_t$ which may, or may not, be known to the experimenter. In clinical trials the patients' responses to drugs frequently depend on body

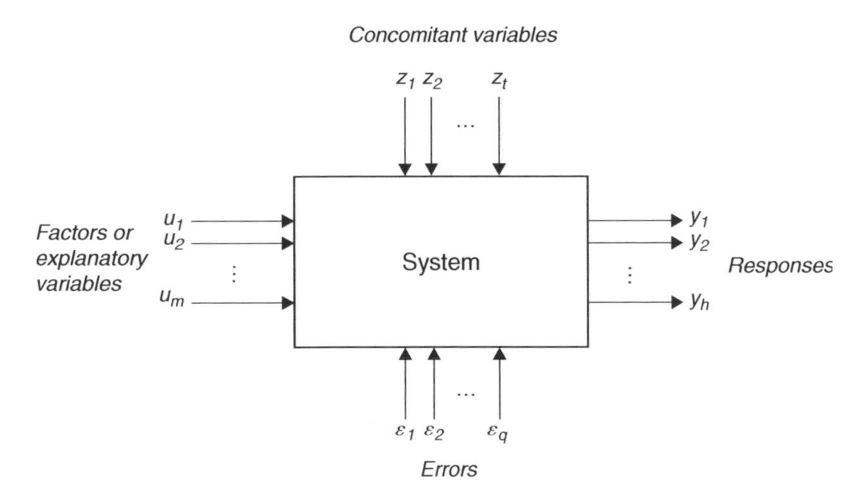

FIG. 2.1. Schematic representation of an experiment on a system. The relationship between the factors u and the measured responses y is obscured by the presence of errors ϵ. The response may also depend on the values of some concomitant variables z which cannot be controlled by the experimenter. The values of u are to be chosen by the experimenter who observes y but not ϵ.

weight, a concomitant variable which can be measured. In manufacturing processes the response may depend on the shift of operators, a variable which is difficult to measure and thus whose values are often unknown. In addition, the relationship between y and u, which is to be determined by the experiment, is obscured by the presence of unobservable random noise $\epsilon_1, \ldots, \epsilon_h$, often called errors.

Quantitative factors, also called predictors or explanatory variables, take values in a specified interval.

$$u_{i,\min} \le u_i \le u_{i,\max} \quad (i = 1, \ldots, m). \tag{2.1}$$

For instance, the K/C ratio in Example 1.1 is a quantitative factor taking values between $u_{i,\min} = 0$ and $u_{i,\max} = 2.5$. In this case, $u_{i,\min}$ is a physical limit, whereas $u_{i,\max}$ defines the region of interest to the experimenter. In general, the values of the upper and lower limits $u_{i,\max}$ and $u_{i,\min}$ depend upon the physical limitations of the system and upon the range of the factors thought by the experimenter to be interesting. For example, if pressure is one of the factors, the experimental range will be bounded by the maximum safe working pressure of the apparatus. However, $u_{i,\max}$ may be less than this value if such high pressures are not of interest. In clinical trials, the upper level for the dose of a drug will depend on the avoidance of toxicity and other side effects.

In order to apply general principles of design and also to aid interpretability of experimental results, it is convenient for most applications to scale the quantitative variables. The unscaled variables $u_1, \ldots, u_m$ are replaced by standardized, or coded, variables which are often, but not invariably, scaled to lie between -1 and 1. For such a range the coded variables are defined by

$$x_i = \frac{u_i - u_{i0}}{\Delta_i} \quad (i = 1, \ldots, m), \tag{2.2}$$

where

$$u_{i0} = (u_{i,\min} + u_{i,\max})/2$$

and

$$\Delta_i = u_{i,\max} - u_{i0} = u_{i0} - u_{i,\min}.$$

Designs will mostly be described in terms of the coded variables. An exception is the variable time: if it is a quantitative factor in the experiment, then it is often left uncoded. Even so, it is sometimes desirable to return to the original values of the factors, particularly for further use of

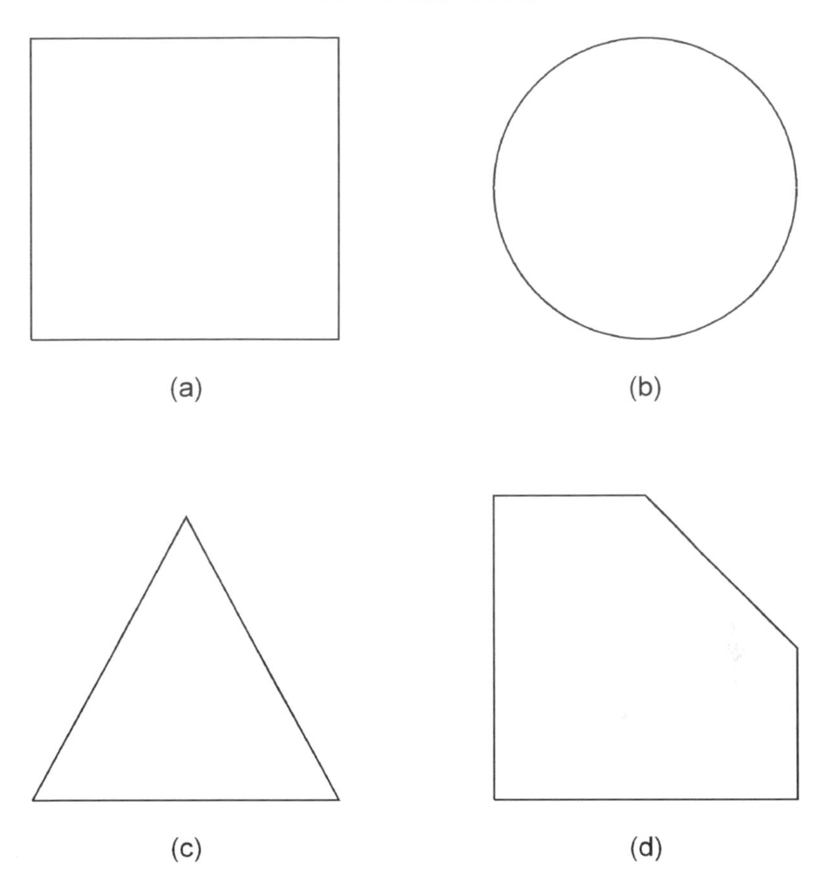

FIG. 2.2. Some design regions: (a) square (cubic or cuboidal for $m \geq 2$); (b) circular (spherical); (c) simplex for mixture experiments; (d) restricted to avoid simultaneous high values of the two factors.

the experimental results in calculations. The reverse transformation to (2.2) yields

$$u_i = u_{i0} + x_i \Delta_i \quad (i = 1, \ldots, m).$$

2.2 Design Regions

If the limits (2.1) apply independently to each of the m factors, the experimental region in terms of the scaled factors x_i will be a cube in m dimensions. For $m = 2$ this is the square shown in Figure 2.2.

The cubic design region is that most frequently encountered for quantitative variables. However, the nature of the experiment may sometimes cause more complicated specification of the factor intervals and of the design region. For example, the region will be spherical if it is defined by the relationship

$$\sum_{i=1}^{m} x_i^2 \leq r^2,$$

where the radius of the sphere is r. This circular design region for $m = 2$ is shown in Figure 2.2. Such a region suggests equal interest in departures in any direction from the centre of the sphere, which might be current experimental or operational conditions.

Different constraints on the experimental region arise in Chapter 16 when we consider mixture experiments in which the response depends only on the proportions of the components of a mixture and not at all on the total amount. One example is the octane rating of a petrol (gasoline) blend. An important feature of such experiments is that a change in the level of one of the factors necessarily leads to a change in the proportions of other factors. The constraints

$$\sum_{i=1}^{m} x_i = 1 \quad x_i \geq 0$$

imposed on the m mixture components make the design region a simplex in $(m - 1)$ dimensions. Figure 2.2(c) shows a design region for a three-component mixture.

In addition to quantitative factors, we shall also consider experiments with qualitative factors, such as the type of filler in the elastomer Example 1.2, which can take only a specified number of levels. Other examples are the gender of a patient in a clinical trial and the type of reactor used in a chemical experiment. Qualitative factors are often represented in designs by indicator or dummy variables. The model for one instance is presented in Example 5.3.

Many experiments involve both qualitative and quantitative factors, as did Examples 1.2, on elastomers, and Example 1.3 on the breaking strength of cotton fibres, the latter after some reinterpretation. Such experiments are the subject of Chapter 14. In addition, some of the quantitative factors might be mixture variables. The design regions may also be more complicated than those shown in Figure 2.2, often because of the imposition of extra constraints. An example, with which many readers may be familiar from their school-days, is the ability of reactions in organic chemistry to produce tar, rather than the desired product, when run at a high temperature for a long time by an unskilled operative. Such areas of the experimental region are best avoided leaving a region in the quantitative variables of the shape shown in Figure 2.2(d). A design for such a restricted region is given in Figure 12.3. Whatever the shape of the design region, which we call $\mathcal{X}$, the principles of experimental design remain the same. The algorithms of optimum design lead to a search over $\mathcal{X}$ for a design minimizing a function that often depends on the variances of the parameter estimates. The structure of $\mathcal{X}$ partly determines whether a standard design can be used, such as those of Chapter 7, or whether, and what kind of, a search is needed for an optimum design.

2.3 Random Error

The observations y_i obtained at the n design points in the region $\mathcal{X}$ are subject to error. This may be of two kinds.

1. **Systematic Errors.** Systematic errors or biases, due, for example, to an incorrectly calibrated apparatus. Such biases must, of course, be avoided. In part they come under the heading of the non-statistical aspects of the experiment discussed in §1.2. But randomization should be used to guard against the further sources of bias mentioned in §1.2. Such biases arise when treatments for comparisons are applied to units that differ in a systematic way (represented in Figure 2.1 as systematically different values of the concomitant variables z).

2. **Random Errors.** These are represented by the vector ϵ in Figure 2.1. In the simplest case, we suppose we have repeat measurements for which the model is

$$y_i = \mu + \epsilon_i \quad (i = 1, \ldots, N). \tag{2.3}$$

   Usually we assume additive and independent errors of constant variance, when least squares is the appropriate method of estimation of parameters such as μ (see Chapter 5).

To be specific, suppose $N = 5$ and let the readings of the yield of a chemical process be

$$y_1 = 0.91, \;\; y_2 = 0.92, \;\; y_3 = 0.90, \;\; y_4 = 0.93, \;\; y_5 = 0.92.$$

The results, obtained under supposedly identical conditions, show random fluctuation. To estimate μ the sample mean $\bar{y}$ is used where

$$\bar{y} = \frac{1}{N}\sum_{i=1}^{N} y_i.$$

The variance σ^2 of the readings is estimated by

$$s^2 = \frac{1}{N-1}\sum_{i=1}^{N}(y_i - \bar{y})^2.$$

For the five readings given above, $\bar{y} = 0.916$, $s^2 = 1.3 \times 10^{-4}$, and $s = 0.0114$. The estimates will be different for different samples.

Often the main aim of the experiment will be to find an approximating function, or model, relating y_i to the m factors u_i, with the estimation of the variance σ^2 being of secondary importance. Even so, knowledge of σ^2 is important to provide a measure of random variability against which to assess observed effects. Formal methods of assessment include confidence intervals for parameter estimates and predictions of the response, and significance tests such as those of the analysis of variance. These provide a mechanism for determining which terms should be included in a model.

Model (2.3) contains one unknown parameter μ. In general, there will be p parameters. In order to estimate these p parameters at least $N = p$ trials will be required of a univariate response at distinct points in the design region. For multivariate responses, that is $h > 1$, fewer trials may be required if the models for different responses share some parameters, as they do for the non-linear models that are the subject of §17.8. However, if σ^2 is not known, but has to be estimated from the data, more than the minimum number of trials will be required. The larger the residual degrees of freedom $\nu = N - p$, the better the estimate of σ^2, provided that the postulated model holds. If lack of fit of the model is of potential interest, σ^2 is better estimated from replicate observations. Analysis of the data then provides estimates both of σ^2 and of that part of the residual sum of squares due to lack of fit. The analysis in Chapter 8 of the data from Table 1.1 on the desorption of carbon monoxide provides an example.

Although by far the greatest part of this book follows (2.3) in assuming additive errors of constant variance, we do discuss design for other error structures. Often data need a transformation, perhaps most often of all by taking logarithms, before (2.3) is appropriate. The effects of transformation on design, particularly for non-linear models, are considered in Chapter 23. The binomial responses of Example 1.6 on the survival of beetles also require a different approach to design, that for generalized linear models which is the subject of Chapter 22.

2.4 Unbiasedness, Validity, and Efficiency

Critical features of the success of an experiment are the *unbiasedness*, *validity*, and *efficiency* of the results. The way the experiment is designed defines all these features. Customarily, the probability of bias in the results due to the unobservable and possibly unknown concomitant variables z of Figure 2.1 is reduced by randomization. For much of the book we ignore specific modelling of the concomitant variables. An exception is in the sequential design of clinical trials, the subject of §25.3, where each patient is assumed to arrive with a vector of concomitant variables or prognostic factors, over which some balance is required in the trial.

Randomization also contributes to obtaining results with required validity. For instance, in a clinical trial, enrolling a random selection of subjects from a population of interest would ensure that the results could be extended to that population. It might be administratively simpler only to enroll, for example, healthy young males. However, extension of results from such a sample to the whole population would be highly speculative and almost certainly misleading. Similarly, in Example 1.4, the choice of engines in which to insert particular valves will decide whether the results will apply to one, a few, or a larger class of car engines.

The concomitant variables formalize the need for blocking experiments, introduced in connection with Example 1.3 on the breaking strength of cotton fibres and Example 1.4 on the wear of engine valves. Blocking is usually beneficial in experimental situations where it is possible to identify groups, or blocks, of experimental units or conditions that need to be used in the experiment, such that within blocks the experimental units have similar values of z, while these values are different for different blocks. In agricultural experiments, blocks are typically composed of nearby plots in the same field. In industrial examples, they allow adjustment for various potentially important factors such as the behaviour of particular shifts of operators or the batch of raw material. In most applications the block effects are considered to be nuisance parameters and the accuracy of their estimation is not important. However, the variation between the blocks is accounted for by block effects included in the statistical model that is fitted to the data. Thus, the unexplained difference between the observations and the predictions obtained from the estimated statistical model may be reduced and the precision of estimation of model parameters of interest increased by correct blocking.

As discussed in Example 1.4, depending on the way the blocks are formed, their effects can be regarded as random or fixed. In some practical applications the experimenter may deal with both fixed and random blocking variables. The nature of the blocking variables has an important impact on the analysis of the data and the validity of the results. Chapter 15 describes the use of optimum design in the blocking of multi-factor experiments, when the existence of the z is explicitly included in the model.

Some of the history of the development of the concept of blocking, particularly associated with Fisher, is in Atkinson and Bailey (2001, §3). A useful discussion, in a historical context, of the various reasons for randomization is in §4 of the same paper. Bailey (1991) and its discussion covers randomization in designed experiments, in part at an advanced level. An expository account is Bailey (2006) which explains and develops the principles of randomization and blocking in the context of comparative experiments.

3

EXPERIMENTAL STRATEGIES

3.1 Objectives of the Experiment

In this book we are mainly concerned with experiments where the purpose is to elucidate the behaviour of a system by fitting an approximating function or model. The distinction is with experiments where the prime interest is in estimating differences, or other contrasts, in yield between units receiving separate treatments. Often the approximating function will be a low-order polynomial. But, as in Chapter 17, the models may sometimes be non-linear functions representing knowledge of the mechanism of the system under study. There are several advantages to summarizing and interpreting the results of an experiment through a fitted model.

1. A prediction can be given of the responses under investigation at any point within the design region. Confidence intervals can be used to express the uncertainty in these predictions by providing a range of plausible values.

2. We can find the values of the factors for which the optimum value (minimum or maximum or a specified target) of each response occurs. Depending upon the model, the values are found by either numerical or analytical optimization. The set of optimum conditions for each response is then a point in factor space, though not necessarily one at which the response was measured during the experiment. Optimization of the fitted model may sometimes lead to estimated optimum conditions outside the experimental region. Such extrapolations are liable to be unreliable and further experiments are needed to check whether the model still holds in this new untried region.

3. When there are several responses, it may be desired to find a set of factor levels that ensures optimum, or near optimum, values of all responses. If, as is often the case, the optima do not coincide, a compromise needs to be found. One technique is to weight the responses to reflect their relative importance and then to optimize the weighted combination of the responses.

TABLE 3.1. Example 3.1: the purification of nickel sulphate. The five factors and their coded and uncoded values

	Factor	Uncoded values u_i		Coded values x_i	
		Min	Max	Min	Max
1	Time of treatment (min.)	60	120	−1	1
2	Temperature (°C)	65	85	−1	1
3	Consumption of $CaCO_3$ (%)	100	200	−1	1
4	Concentration of zinc (g/dm^3)	0.1	0.4	−1	1
5	Mole ratio Fe/Cu	0.91	1.39	−1	1

4. A final advantage of the fitted model is that it allows graphical representation of the relationships being investigated. However, the conclusions of any analysis depend strongly on the quality of the fitted models and on the data. Hence they depend on the way in which the experiment is designed and performed.

These general ideas are described in a specific context in the next example, which also illustrates the use of the scaled variables introduced in §2.1.

Example 3.1 The Purification of Nickel Sulphate The purpose of the experiment was to optimize the purification of nickel sulphate solution, the impurities being iron, copper, and zinc, all in the bivalent state. Petkova *et al.* (1987) investigate the effect of five factors on six responses. Table 3.1 gives the maximum and minimum values of the unscaled factors u_i and the corresponding coded values x_i.

Since iron, copper, and zinc are impurities, high deposition of these three elements was required. These are given as the first three responses y_1, y_2, and y_3 in Table 3.2. Low loss of nickel was also important, and is denoted by y_4. Two further responses are y_5, the ratio of the final concentration of nickel to zinc, and y_6, the pH of the final solution. Target values were specified for all six responses.

From previous experience it was anticipated that second-order polynomials would adequately describe the response. The experimental design, given in Table 3.2, consists of the 16 trials of a 2^{5-1} fractional factorial plus star points, a form of composite design discussed in §7.6 for investigating second-order models. Table 3.2 also gives the observed values of the six responses. The 26 trials of the experiment were run in a random order, not in the standard order of the table, in which the first factor x_1 varies fastest and x_4 most slowly.

TABLE 3.2. Example 3.1: the purification of nickel sulphate. Experimental design and results

Factors					Responses					
x_1	x_2	x_3	x_4	x_5	y_1	y_2	y_3	y_4	y_5	y_6
1	1	1	1	1	94.62	99.98	99.83	9.19	104889	5.07
−1	1	1	1	−1	100.00	99.97	95.12	7.57	3462	4.94
1	−1	1	1	−1	100.00	99.99	93.81	7.68	2730	5.39
−1	−1	1	1	1	77.01	99.99	91.39	6.69	2084	5.05
1	1	−1	1	−1	89.96	82.63	24.58	1.38	239	2.62
−1	1	−1	1	1	81.89	86.97	99.78	3.27	85988	2.90
1	−1	−1	1	1	79.64	93.82	78.13	1.95	862	3.70
−1	−1	−1	1	−1	88.79	85.53	7.04	1.06	195	3.10
1	1	1	−1	−1	100.00	100.00	99.33	7.26	51098	4.92
−1	1	1	−1	1	93.23	99.93	97.99	7.38	22178	5.07
1	−1	1	−1	1	89.61	99.99	98.36	5.76	27666	5.29
−1	−1	1	−1	−1	99.95	99.92	91.35	5.01	4070	5.02
1	1	−1	−1	1	95.80	86.81	30.25	5.12	681	2.59
−1	1	−1	−1	−1	86.59	83.99	38.96	1.30	599	2.66
1	−1	−1	−1	−1	88.46	85.49	42.48	2.03	630	3.16
−1	−1	−1	−1	1	70.86	91.30	28.45	1.25	663	3.20
−1	0	0	0	0	97.68	99.97	95.02	5.68	4394	4.80
1	0	0	0	0	99.92	99.99	97.57	6.68	10130	5.18
0	−1	0	0	0	99.33	99.98	97.06	6.21	8413	5.08
0	1	0	0	0	99.38	99.90	96.83	7.05	7748	4.90
0	0	−1	0	0	80.10	87.74	19.55	1.10	324	3.08
0	0	1	0	0	98.57	99.98	99.31	6.40	35655	5.28
0	0	0	−1	0	98.64	99.95	97.09	6.00	28239	4.80
0	0	0	1	0	99.07	100.00	93.94	6.54	3169	4.81
0	0	0	0	−1	99.96	99.95	82.55	5.52	1910	5.04
0	0	0	0	1	97.68	100.00	89.06	6.30	3097	5.13

The five responses are: $y_1 - y_3$, deposition of iron, copper and zinc; y_4, loss of nickel; y_5, ratio of final concentration of nickel to final concentration of iron and y_6, pH of the final solution. The experiment is given in standard order.

The use of SAS in the analysis of experiments such as this is the subject of Chapter 8. Here the analysis consisted of fitting a separate second-order model to each response. Contour plots of the fitted responses against pairs of important explanatory variables indicated appropriate experimental conditions for each response, from which an area in the design region was found in which all the responses seemed to satisfy the experimental requirements. These conditions were $u_1 = 90$ minutes, $u_2 = 80^{\circ}$C and $u_4 = 0.175$ g/dm^3, with u_3 free to vary between 160% and 185% and the ratio u_5 lying between 0.99 and 1.24. Further experimentation under these conditions confirmed that the values of all six responses were satisfactory. ■

3.2 Stages in Experimental Research

The experiment described in §3.1 is one in which a great deal was known *a priori*. This information included:

1. the five factors known to affect the responses;
2. suitable experimental ranges for each factor;
3. an approximate model, in this case a second-order polynomial, for each response.

An appreciable part of this book is concerned with designs for second-order models, which are appropriate in the region of a maximum or minimum of a response. However, it may be that the results of the experiment allow a simpler representation with few, or no, second-order terms. An example of a statistical analysis leading to model simplification is presented in §8.3.

Such refinements of models usually occur far along the experimental path of iteration between experimentation and model building. In this section we discuss some of the earlier stage of experimental programmes. A typical sequence of experiments leading to a design such as that of Table 3.2 is discussed in the next section.

1. **Background to the Experiment.** The successful design of an experiment requires the evaluation and use of all prior information, even if only in an informal manner. What is known should be summarized and questions to be answered must be clearly formulated. Factors that may affect the response should be listed. The responses should both contain important information and be measurable.

Although such strictures may seem to be platitudes, the discipline involved in thinking through the purpose of the experimental programme is most valuable. If the programme involves collaboration, time used in clarifying knowledge and objectives is always well spent and often, in itself, highly informative.

2. **The Choice of Factors.** At the beginning of an experimental programme there will be many factors that, separately or jointly, may have an effect on the response. Some initial effort is often spent in screening out those factors that matter from those that do not. First-order designs, such as those of §7.2 and the 2^{6-3} fractional factorial of Table 7.3 are suitable at this stage. Second-order designs, such as the composite design of Table 3.2, are used at a later stage when trials are made near to the minimum or maximum of a response.

It is assumed that quantitative factors can be set exactly to any value in the design region, independently of one another, although there may be combinations of values that are undesirable or unattainable. It is also assumed that factors that are not specifically varied by design remain unchanged throughout the experiment. An exception is in the design of quantitative mixture experiments, such as those of Chapter 16, where changing the proportion of one component must change the proportion of at least one other component.

The intervals over which quantitative factors are varied during the experiment need to be chosen with care. If they are too small, the effect of the factor on the response may be swamped by experimental error. On the other hand, if the intervals are too wide, the underlying relationship between the factors and the response can become too complicated to be represented using a reasonably simple model. In addition, regions may be explored in which the values of the response are outside any range of interest.

3. **The Reduction of Error.** If there is appreciable variability between experimental units, these can sometimes be grouped together into blocks of more similar units, as in Examples 1.3 and 1.4, with an appreciable gain in accuracy. Some examples of the division of 2^m factorials into blocks are given in §7.3. More general blocking of response surface designs is described in Chapter 15.

The choice of an appropriate blocking structure is often of great importance, especially in agricultural experiments, and can demand much skill from the experimenter. In technological experiments, batches of raw material are frequently an important source of variability, the effect of which can be reduced by taking batches as a blocking factor. In clinical trials, trial centres such as clinics or hospitals, may need to be treated as blocking factors. An alternative approach is to use numerical values of nuisance variables at the design stage. If the purity of a raw material is its important characteristic, the experiment can be designed using values measured on the various batches. Similarly an experiment can be designed to behave well in the presence of a quadratic trend in time by including the terms of the quadratic in the model, or by seeking to find a design orthogonal to the trend (Atkinson and Donev 1996).

4. **The Choice of a Model.** Optimum experimental designs depend upon the model relating the response to the factors. The model needs to be sufficiently complicated to approximate the main features of the data, without being so complicated that unnecessary effort is involved in estimating a multitude of parameters. Some widely applicable models are described in the next chapter.

5. **Design Criterion and Size of the Design.** The size of the design is constrained by resources, often cost and time. The precision of parameter estimates increases with the number of trials, but also depends on the location of the design points. Several design criteria are described in Chapter 10. These lead to designs maximizing information about specific aspects of the model. Compound design criteria, the subject of Chapter 21, allow maximization of a balance between the various aspects.

6. **Choice of an Experimental Design.** In many cases the required experimental design can be found in the literature, either printed or electronic. Chapter 7 describes some standard designs and §7.7 describes how to access the vast number of designs available in SAS. If a standard design is used, it is important that it takes into account all the features of the experiment, such as structure of the experimental region and the division of the units into blocks. If a standard design is not available, the methods of optimum experimental design will provide an appropriate design. In either case, randomization in the application of treatments to units will be important.

7. **Conduct of the Experiment.** The values of the responses should be measured for all trials. The measurements of individual responses are usually statistically independent although the components of multivariate responses such as those in Example 3.1 may be correlated. However, if the same model is fitted to all responses, the correlation can be ignored and the responses treated as independent. On the other hand, if several observations are made over time on a single unit, the assumption of independence may be violated and the time series element in the data should be allowed for in design, as in Chapter 24, and in the analysis.

If any values of the factors are set incorrectly, the actual values should be recorded and used in the analysis. If obtaining correct settings of the factors is likely to present difficulties, experimental designs should be used with only a few settings of each factor. For multifactor polynomial models, optimum experimental designs with each factor at only three or five levels, for example, can be found by searching over a grid of candidate points. The designs are often only slightly less efficient than those found by searching over a continuous region. Table 12.1 exemplifies this for various numbers of observations when the model is a second-order polynomial in two factors.

8. **Analysis of Data.** The results of the experiment can be summarized in tables such as those of Chapter 1. Very occasionally, no further analysis is

required. However, almost invariably, a preliminary graphical investigation will be informative, to be followed by a more formal statistical analysis yielding parameter estimates and associated confidence intervals. Examples are given in Chapter 8.

Experimentation is iterative. The preceding list of points suggests a direct path from problem formulation to solution. However, at each stage, the experimenter may have to reconsider decisions made at earlier stages of the investigation. Problems arising at intermediate stages may add to the eventual understanding of the problem.

If the model fitted at stage 8 appears inadequate, this may be because the model is too simple, or there may be errors and outliers in the data owing, for example, to failures in measurement and recording devices, or in data transmission and entry. In any case, the design may need to be augmented and stages 6–8 above repeated until a satisfactory model is achieved.

3.3 The Optimization of Yield

Experiments for finding the conditions of maximum yield are often sequential and nicely illustrate the stages of an experimental programme. In the simplest case, described here, all factors are quantitative, with the response being a smooth function of the settings of the factors.

1. **Screening Experiments.** First-order designs are often used to determine which of the many potential factors are important. The 2^{6-3} design of Table 7.3 has already been mentioned. Other screening designs include the Plackett–Burman designs introduced in §7.5.

2. **Initial First-Order Design.** As the result of the screening stage, a few factors will emerge as being most important. In general there will be m such factors. For illustration we take $m = 2$. The path of a typical experiment is represented in Figure 3.1. The initial design consists of a 2^m factorial, or maybe a fraction if $m \geq 5$, with perhaps three centre points. If the average response at the centre is much the same as the average of the responses at the factorial points and there are no interactions, the results can be represented by a first-order surface, that is a plane.

3. **Steepest Ascent.** The experiments indicated by triangles in Figure 3.1 form a path of steepest ascent and are performed sequentially in a direction perpendicular to the contours of the plane of values

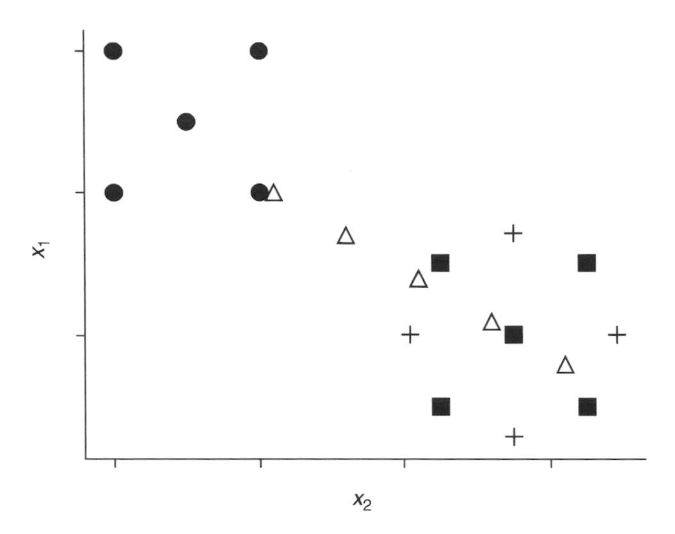

FIG. 3.1. The optimization of yield: • initial first-order design with centre points; Δ path of steepest ascent; ■ second first-order design; + star points providing a second-order design.

of the expected response from the fitted first-order model in stage 2. Progress in the direction of steepest ascent continues until the response starts to decrease, when the maximum in this direction will have been passed. In both this stage and the previous one, assessments of differences between observed or fitted responses have to be made relative to the random error of the measurements.

4. **Second First-Order Design.** The squares in the figure represent a second first-order design, centred at or near the conditions of highest yield found so far. We suppose that, in this example, comparison of the average responses at the centre of the design and at the factorial points indicates curvature of the response surface. Therefore the experiment has taken place near a maximum, perhaps of the whole surface, but certainly in the direction of steepest ascent.

5. **Second-Order Design.** The addition of trials at other levels of the factors, the star points in Figure 3.1, makes possible the estimation of second-order terms. These may indicate a maximum in or near the experimental region. Or it may be necessary to follow a second path of steepest ascent along a ridge in the response surface orthogonal to the path of stage 3.

The formal description of the use of steepest ascent and second-order designs for the experimental attainment of optimum conditions was introduced by Box and Wilson (1952). Their work came from the chemical industry, where it is natural to think of the response as a yield to be maximized. In other situations the response might be a percentage of unacceptable product; for example, etched wafers in the production of chips in the electronics industry. Here the response should be minimized. Another response to be minimized is the proportion of patients in drug development who react unfavourably to a proposed medication. In that case the methods of design for generalized linear models (Chapter 22) would have to be combined with the methods of this section.

3.4 Further Reading

Many statistical books on experimental design, especially Cox (1958) and Box, Hunter, and Hunter (2005), contain material on the purposes and strategy of experimentation. Wu and Hamada (2000) in addition has much material on screening and first-order designs. Experiments in which observations are made over time on single units are often called repeated measures. For the analysis of such experiments see, amongst others, Lindsey (1999). The analysis of multiple response experiments and the determination of experimental conditions providing satisfactory values of all responses is addressed in §6.6 of Myers and Montgomery (2002). Chapter 6 and succeeding chapters of Box and Draper (1987) give a full treatment of the material on the experimental attainment of optimum conditions outlined in §3.3. We describe a method for the optimum designs of experiments with multivariate responses in §10.10. Chapter 20 is concerned with model checking, with a discussion of the usefulness, or otherwise, of centre points for model checking in §20.3.1.

4

THE CHOICE OF A MODEL

4.1 Linear Models for One Factor

Optimum experimental designs depend upon the model or models to be fitted to the data, although not usually, for linear models, on the values of the parameters of the models. This chapter is intended to give some advice on the choice of an appropriate form for a model. Whether or not the choice is correct can, of course, only be determined by analysis of the experimental results. But, as we shall see, designs that are optimum for one model are often almost optimum for a wide class of models.

The true underlying relationship between the observed response y and the factors x is usually unknown. Therefore we choose as a model an approximating function that is likely to accommodate the main features of the response over the region of interest—for example, the rough locations of the minimum and maximum values. Our concern will be mostly with polynomial models that are linear in the parameters. However, this section does end with comments on non-linear models that can be linearized by transformation. Non-linear models are the subject of §4.2. Often polynomial models may be thought of as Taylor series approximations to a true relationship which may well be non-linear. We begin this section with the simplest linear model, the first-order model for a single factor.

Figure 4.1 shows the relationship between the expected response $\eta(x)$ and x for the four first-order models

$$\eta(x) = 16 + 7.5x \tag{4.1}$$

$$\eta(x) = 18 - 4x \tag{4.2}$$

$$\eta(x) = 12 + 5x \tag{4.3}$$

$$\eta(x) = 10 + 0.1x. \tag{4.4}$$

These models describe monotonically increasing or decreasing functions of x in which the rate of increase of $\eta(x)$ does not depend on the value of x. The values of the two parameters determine the slope and intercept of each line. The use of least squares for estimating the parameters of such models once data have been collected is described in §5.1. Unless the error variance

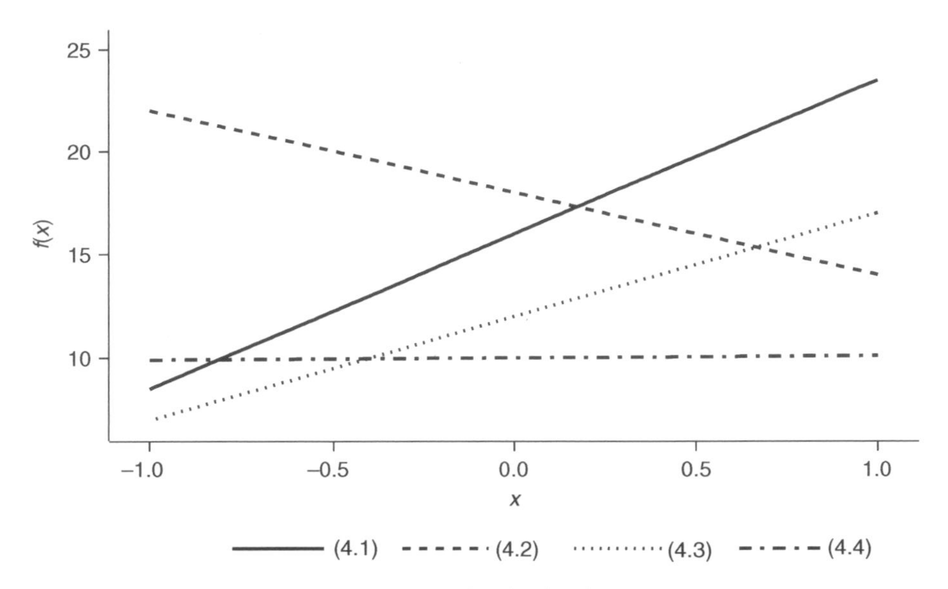

FIG. 4.1. Four first-order models (4.1)–(4.4). In the presence of observational error the slope of 0.1 for model (4.4) might be mistaken for zero.

is very small or the number of observations is very large, it will be difficult to detect the relationship between $\eta(x)$ and x in Model (4.4).

Increasing x in Models (4.1), (4.3), or (4.4) causes $\eta(x)$ to increase without bound. Often, however, responses either increase to an asymptote with x, or even pass through a maximum and then decrease. Some suitable curves with a single maximum or minimum are shown in Figure 4.2. The three models are

$$\eta(x) = 25 - 14x + 6x^2 \tag{4.5}$$

$$\eta(x) = 20 - 10x + 40x^2 \tag{4.6}$$

$$\eta(x) = 50 + 5x - 35x^2. \tag{4.7}$$

A second-order, or quadratic, model is symmetrical about its extreme, be it a maximum or a minimum. For Model (4.5), with a positive coefficient of x^2, the minimum is at $x = 7/6$, outside the range of plotted values. For Model (4.6) the extreme is again a minimum, as is indicated by the positive coefficient of 40 for x^2. For Model (4.7) the extremum is at 1/14 and, since the coefficient of x^2 is negative, it is a maximum. The larger the absolute value of the quadratic coefficient, the more sharply the single maximum or minimum is defined.

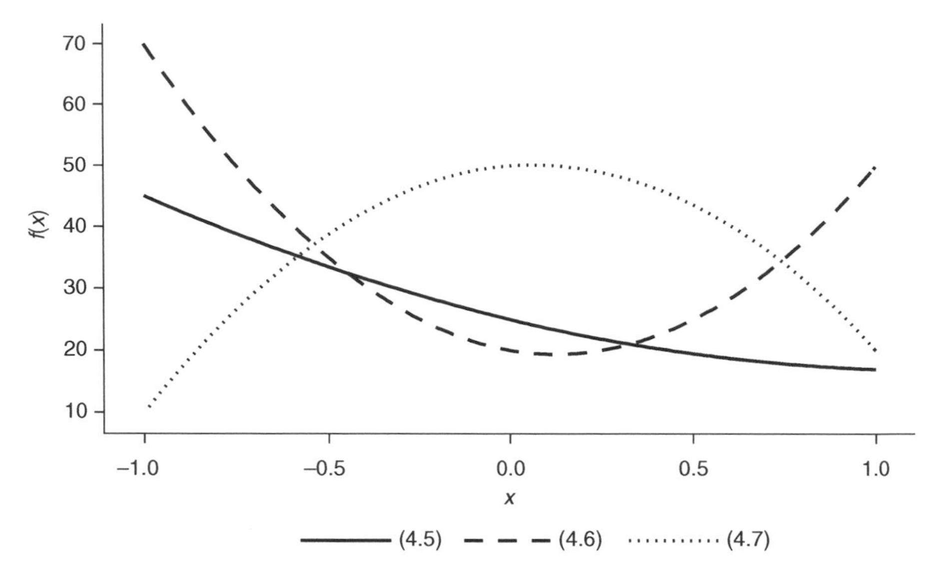

FIG. 4.2. Three second-order models (4.5)–(4.7), two showing typical symmetrical extrema.

More complicated forms of response-factor relationship can be described by third-order polynomials. Examples are shown in Figure 4.3 for the models

$$\eta(x) = 90 - 85x + 16x^2 + 145x^3 \tag{4.8}$$

$$\eta(x) = 125 + 6x + 10x^2 - 80x^3 \tag{4.9}$$

$$\eta(x) = 62 - 25x - 70x^2 - 54x^3. \tag{4.10}$$

These three figures show that the more complicated the response relationship to be described, the higher the order of polynomial required. Although high-order polynomials can be used to describe quite simple relationships, the extra parameters will usually not be justified when experimental error is present in the observations to which the model is to be fitted. In general, the inclusion of unnecessary terms inflates the variance of predictions from the fitted model. Increasing the number of parameters in the model may also increase the size of the experiment, so providing an additional incentive for the use of simple models.

Experience indicates that in very many experiments the response can be described by polynomial models of order no greater than two; curves with multiple points of inflection, like those of Figure 4.3, are rare. A more frequent form of departure from the models of Figure 4.2 is caused by asymmetry around the single extreme point. This is often more parsimoniously

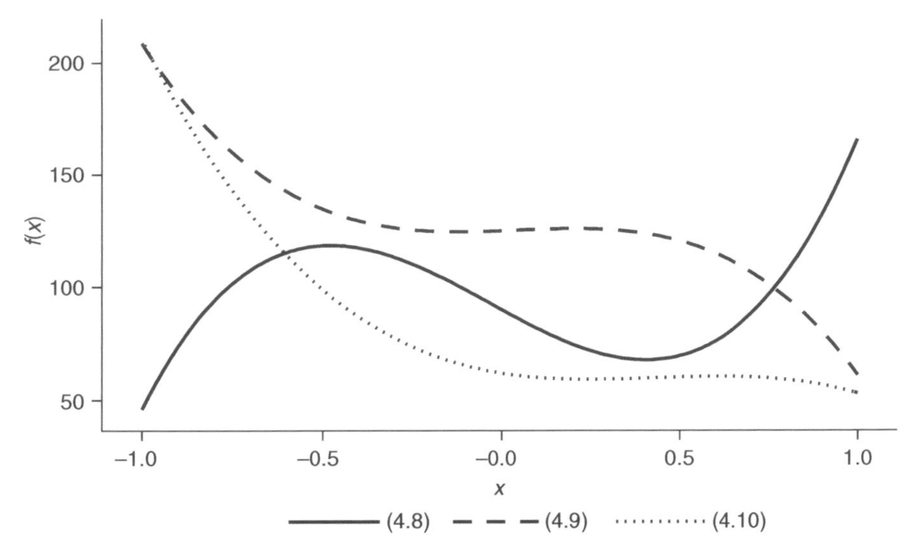

FIG. 4.3. Three third-order models (4.8)–(4.10). Such cubic models are rarely necessary.

modelled by a transformation of the factor, for example to $x^{1/2}$ or log x, than by the addition of a term in x^3. In this book we are mostly concerned with polynomial models of order no higher than two.

In addition to polynomial models, there is an important class of models in which the parameters appear non-linearly that, after transformation, become linear. For example, the model

$$\eta(x) = \beta_0 x_1^{\beta_1} x_2^{\beta_2} \dots x_m^{\beta_m} \tag{4.11}$$

is non-linear in the parameters $\beta_1, \dots, \beta_m$. Such models are used in the Cobb–Douglas relationship economics, in chemistry to describe the kinetics of chemical reactions and in technology and engineering to provide relationships between dimensionless quantities. Taking logarithms of both sides of (4.11) yields

$$\log \eta(x) = \log \beta_0 + \beta_1 \log x_1 + \cdots + \beta_m \log x_m,$$

which may also be written in the form

$$\tilde{\eta}(x) = \tilde{\beta}_0 + \beta_1 \tilde{x}_1 + \cdots + \beta_m \tilde{x}_m \tag{4.12}$$

where

$$\tilde{\eta}(x) = \log \eta(x) \quad \text{and} \quad \tilde{x}_j = \log x_j \ \ (j = 1, \dots, m).$$

Thus (4.11) is a first-order polynomial in the transformed variables $\tilde{x}$. However, the equivalence of (4.11) and (4.12) in the presence of experimental error also requires a relationship between the errors in the two equations. If those in (4.11) are multiplicative they will become additive in the transformed model (4.12).

4.2 Non-linear Models

There are often situations when models in which the parameters occur non-linearly are to be preferred to attempts at linearization such as (4.12). This is particularly so if the errors of data from the non-linear model are homoscedastic; they will then be rendered hetereoscedastic by the linearizing transformation. Where the non-linear model arises from theory, estimation of the parameters will be of direct interest. In other cases, it may be that the response surface can only be described succinctly by a non-linear model. As an example, the model

$$\eta(x, \beta) = \beta_0\{1 - \exp(-\beta_1 x)\} \qquad (4.13)$$

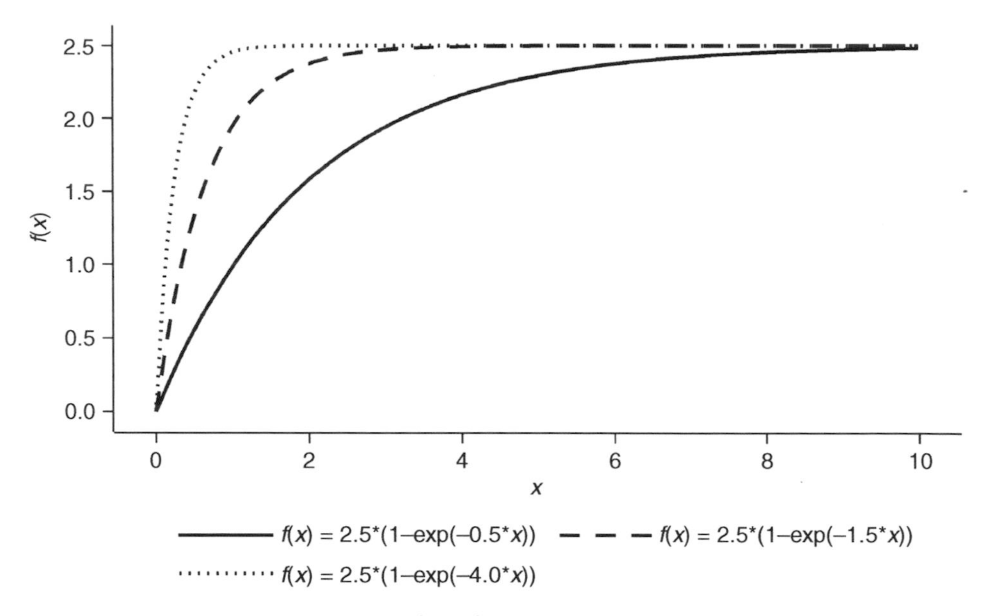

FIG. 4.4. A non-linear model (4.13) with an asymptote reached at greater speed as β_1 increases.

is plotted in Figure 4.4 for $\beta_0 = 2.5$ and three values of β_1, 0.5, 1.5, and 4. For all three sets of parameter values the asymptote has the same value of 2.5, but this value is approached more quickly as β_1 increases. Simple polynomial models of the type described in the previous section are not appropriate for models such as this that include an asymptote; as $x \to \infty$, the response from the polynomial model will itself go to $\pm\infty$, however many polynomial terms are included.

Indeed, one advantage of non-linear models is that they often contain few parameters when compared with polynomial models for fitting the same data. A second advantage is that, if a non-linear model is firmly based in theory, extrapolation to values of x outside the region where data have been collected is unlikely to produce seriously misleading predictions. Unfortunately the same is not usually true for polynomial models. A disadvantage of non-linear models is that optimum designs for the estimation of parameters depend upon the unknown values of the parameters. Designs for non-linear models are the subject of Chapter 17.

4.3 Interaction

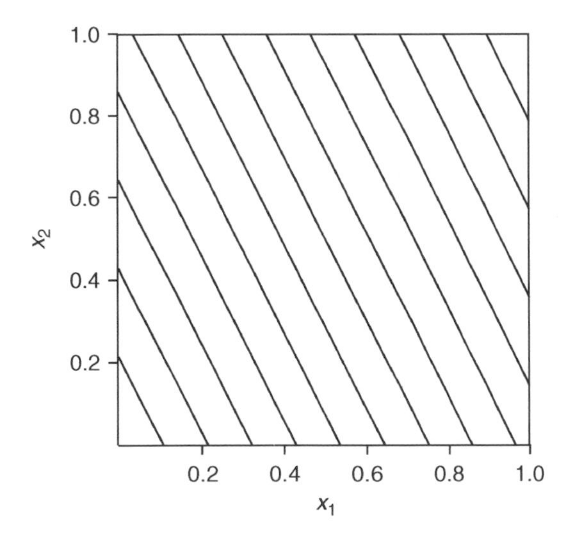

FIG. 4.5. Contours of the first-order two-factor model (4.14.) There is no interaction between x_1 and x_2.

The simplest extension of the polynomials of §4.1 is to the first-order model in m factors. Figure 4.5 gives the equispaced contours for the two-factor model

$$\eta(x) = 1 + 2x_1 + x_2. \tag{4.14}$$

The effects of the two factors are additive. Whatever the value of x_2, a unit increase in x_1 will cause an increase of two in $\eta(x)$.

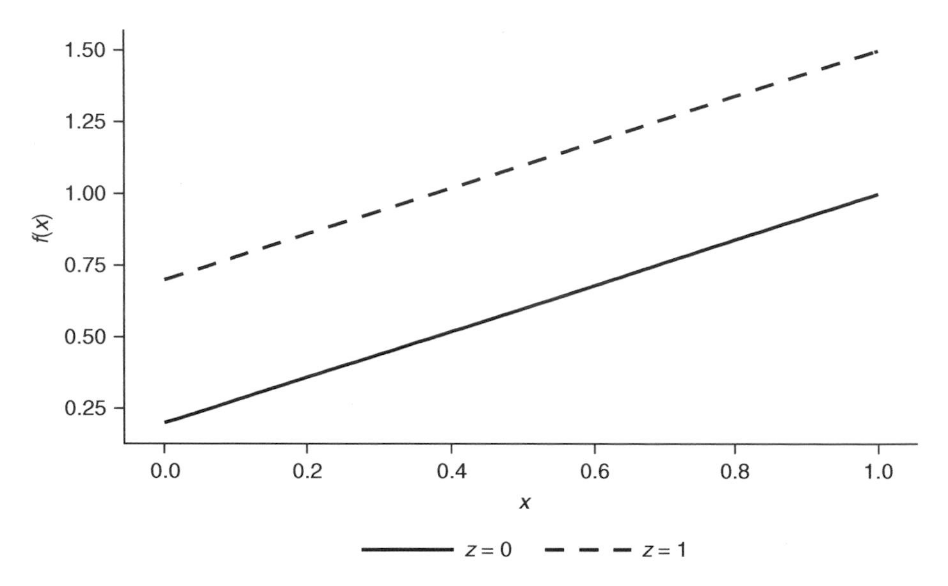

FIG. 4.6. Model (4.15) for one qualitative and one quantitative factor. There is no interaction between the factors.

Often, however, factors interact, so that the effect of one factor depends on the level of the other. Suppose that there are again two factors, one quantitative and the other, z, a qualitative factor at two levels, 0 and 1. Figure 4.6 shows a plot of $\eta(x)$ against x for the model

$$\eta(x) = 0.2 + 0.8x + 0.5z \quad (z = 0, 1). \tag{4.15}$$

For this model without interaction the effect of moving from the low to the high level of z is to increase $\eta(x)$ by 0.5. However, at either level, the rate of increase of $\eta(x)$ with x is the same. This is in contrast to Figure 4.7, a plot for the model

$$\eta(x) = 0.2 + 0.8x + 0.3z + 0.7xz \quad (z = 0, 1). \tag{4.16}$$

The effect of the interaction term xz is that the rate of increase of $\eta(x)$ with x depends upon the value of z. Instead of the two parallel lines of Figure 4.6,

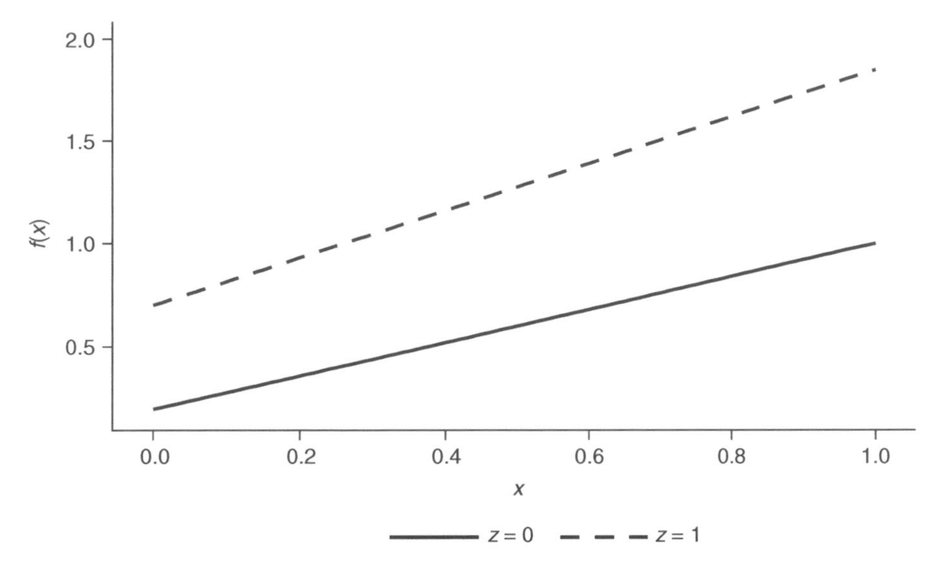

FIG. 4.7. Model (4.16) for one qualitative and one quantitative factor. As a result of interaction between the factors, the two lines are not parallel.

Figure 4.7 shows two straight lines that are not parallel. In the case of a strong interaction between x and z, the sign of the effect of x might even reverse between the two levels of z.

Interaction for two quantitative factors is illustrated in Figure 4.8, where the model is

$$\eta(x) = 1 + 2x_1 + x_2 + 2x_1x_2. \tag{4.17}$$

The effect of the interaction term x_1x_2 is to replace the straight line contours of Figure 4.5 by hyperbolae. For any fixed value of x_1, the effect of increasing x_2 is constant, as can be seen by the equispaced contours. Similarly the effect of x_1 is constant for fixed x_2. However, the effect of each variable depends on the level of the other.

Interactions frequently occur in the analysis of experimental data. An advantage of designed experiments in which all factors are varied to give a systematic exploration of factor levels is that interactions can be estimated. Designs in which one factor at a time is varied, all others being held constant, do not yield information about interactions. They will therefore be inefficient, if not useless, when interactions are present.

Interactions of third, or higher, order are possible, with the three-factor interaction involving terms like $x_1x_2x_3$. It is usually found that there are fewer two-factor interactions than main effects, and that higher-order interactions are proportionately less common. Pure interactions, that is

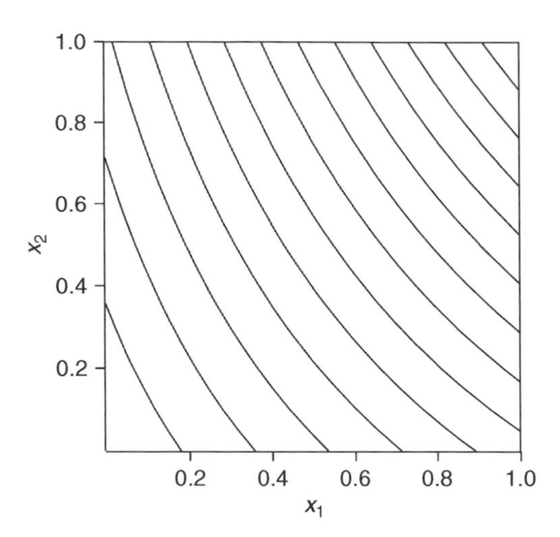

FIG. 4.8. Contours of the first-order two-factor model (4.17) with interaction between the quantitative factors.

interactions between factors, the main effects of which are absent, are rare. One reason is that, if the pure interaction is present in terms of the scaled factors x, rewriting the interaction as a function of the unscaled factors u automatically introduces main effects of the factors that are present in the interaction.

4.4 Response Surface Models

Experiments in which all factors are quantitative frequently take place at or near the maximum or minimum of the response, that is in the neighbourhood of conditions that are optimum for some purpose. In order to model the curvature present, a full second-order model is required. Figure 4.9 shows contours of the response surface

$$\begin{aligned}\eta(x) &= 1.91 - 2.91x_1 - 1.6x_2 + 2x_1^2 + x_2^2 + x_1x_2 \\ &= 2(x_1 - 0.6)^2 + (x_2 - 0.5)^2 + (x_1 - 0.6)(x_2 - 0.5) + 0.64. \qquad (4.18)\end{aligned}$$

The positive coefficients of x_1^2 and x_2^2 indicate that these elliptical contours are modelling a minimum. In the absence of the interaction term x_1x_2 the axes of the ellipse would lie along the co-ordinate axes. If, in addition, the coefficients of x_1^2 and x_2^2 are equal, the contours are circular. An advantage of writing the model in the form (4.18) is that it is clear that the ellipses are centred on $x_1 = 0.6$ and $x_2 = 0.5$.

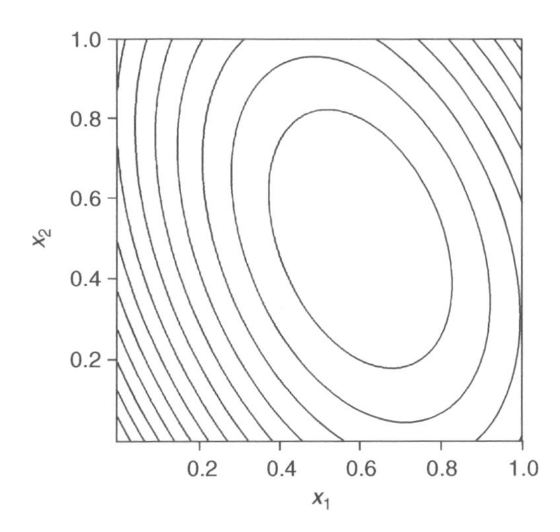

FIG. 4.9. A second-order response surface: contours of the polynomial (4.18) which includes quadratic terms.

The second-order model (4.18) is of the same order in both factors. The requirement of being able to estimate all the coefficients in the model dictates the use of a second-order design such as the 3^2 factorial or the final first-order design with added star points of Figure 3.1. Often, however, the fitted model will not need all the terms and will be of different order in the various factors.

Example 4.1 Freeze Drying Table 4.1 gives the results of a 3^2 factorial experiment on the conditions of freeze drying taken from Savova *et al.* (1989). The factors are the amount of glycerine x_1 per cent and the effect of the speed of freeze drying x_2 °C/min. The response y was the percentage of surviving treated biological material. The model that best describes the data is

$$\hat{y} = 90.66 - 0.5x_1 - 9x_2 - 1.5x_1x_2 - 3.5x_1^2, \tag{4.19}$$

which is second-order in x_1, first order in x_2, and also includes an interaction term. ■

Since the response in Example 4.1 has a maximum value of 100, it might be preferable to fit a model that takes account of this bound, perhaps by using the transformation methods derived from those of §23.2.2 or by fitting a nonlinear model related to that portrayed in Figure 4.4. Transformation methods may also lead to models of varying order in the factors. For example, the model with log y as a response fitted by Atkinson (1985, p. 131) to Brownlee's stack loss data (Brownlee 1965, p. 454) is exactly of the form

TABLE 4.1. Example 4.1: freeze drying. Percentage of surviving treated biological material

x_1 (amount of glycerine)	x_2 (speed of freeze drying) °C/min		
%	10	20	30
10	96	85	82
20	100	92	80
30	96	88	76

(4.19). Atkinson and Riani (2000, p. 113) prefer a first-order model, two outliers and the square root transformation. Sometimes generalized linear models, Chapter 22, provide an alternative to models based on normality. In particular, in §22.5 we find designs for gamma response surface models.

There is a distinct literature on the design and analysis of experiments for response surface models like (4.18). In addition to Box and Draper (1987), booklength treatments include Khuri and Cornell (1996) and Myers and Montgomery (2002). Recent developments and further references are in Khuri (2006). Myers, Montgomery, and Vining (2001) illustrate the use of generalized linear models in the analysis of response surface data.

5

MODELS AND LEAST SQUARES

5.1 Simple Regression

The plots in Chapter 4 illustrate some of the forms of relationship between an m-dimensional factor x and the true response $\eta(x)$, which will often be written as $\eta(x, \beta)$ to stress dependence on the vector of p unknown parameters β. For a first-order model $p = m + 1$. Measurements of $\eta(x)$ are subject to error, giving observations y. Often the experimental error is additive and the model for the observations is

$$y_i = \eta(x_i, \beta) + \epsilon_i \quad (i = 1, \ldots, N). \tag{5.1}$$

If the error term is not additive, it is frequently possible to make it so by transformation of the response. For example, taking the logarithm of y makes multiplicative errors additive. An instance of the analysis of data which is improved by transformation is in §8.3, when we fit models to Example 1.2, Derringer's elastomer data, with both the original and a logged response.

The absence of systematic errors implies that $\mathrm{E}(\epsilon_i) = 0$, where E stands for expectation. The customary second-order assumptions are:

$$\begin{array}{ll} \text{(i)} & \mathrm{E}(\epsilon_i \epsilon_j) = \mathrm{cov}(\epsilon_i, \epsilon_j) = 0 \quad (i \neq j) \quad \text{and} \\ \text{(ii)} & \mathrm{var}(\epsilon_i) = \sigma^2. \end{array} \tag{5.2}$$

This assumption of independent errors of constant variance will invariably need to be checked. Violations of independence are most likely when the data form a series in time or space. Designs for correlated observations are the subject of Chapter 24.

The second-order assumptions justify use of the method of least squares to estimate the vector parameter β. The least squares estimator $\hat{\beta}$ minimizes the sum over all N observations of the squared deviations

$$S(\beta) = \sum_{i=1}^{N} \{y_i - \eta(x_i, \beta)\}^2, \tag{5.3}$$

so that

$$S(\hat{\beta}) = \min_{\beta} S(\beta). \tag{5.4}$$

The formulation in (5.3) and (5.4) does not imply any specific structure for $\eta(x, \beta)$. If the model is non-linear in some or all of the parameters β, the numerical value of $\hat{\beta}$ has to be found iteratively (see §17.10). However, if the model is linear in the parameters, explicit expressions can be found for $\hat{\beta}$.

The plot in Figure 1.1 of the 22 readings on the desorption of carbon monoxide from Table 1.1 suggests that, over the experimental region, there is a straight line relationship between y and x of the form

$$\eta(x_i, \beta) = \beta_0 + \beta_1 x_i \quad (i = 1, \ldots, N). \tag{5.5}$$

For this simple linear regression model

$$S(\beta) = \sum_{i=1}^{N} (y_i - \beta_0 - \beta_1 x_i)^2.$$

The minimum is found by differentiation, giving the pair of derivatives

$$\frac{\partial S}{\partial \beta_0} = -2 \sum_{i=1}^{N} (y_i - \beta_0 - \beta_1 x_i) \tag{5.6}$$

$$\frac{\partial S}{\partial \beta_1} = -2 \sum_{i=1}^{N} (y_i - \beta_0 - \beta_1 x_i) x_i. \tag{5.7}$$

At the minimum both derivatives are zero. Solution of the resulting simultaneous equations yields the least squares estimators

$$\hat{\beta}_1 = \frac{\sum y_i (x_i - \bar{x})}{\sum (x_i - \bar{x})^2}$$

and

$$\hat{\beta}_0 = \bar{y} - \hat{\beta}_1 \bar{x}, \tag{5.8}$$

where the sample averages are $\bar{x} = \sum x_i / N$ and $\bar{y} = \sum y_i / N$. Therefore the fitted least squares line passes through $(\bar{x}, \bar{y})$ and has slope $\hat{\beta}_1$.

The distribution of $\hat{\beta}$ depends upon the distribution of the errors ϵ_i. Augmentation of the second-order assumptions (5.2) by the condition

$$\text{(iii)} \qquad \text{the errors } \epsilon_i \sim N(0, \sigma^2) \tag{5.9}$$

yields the normal-theory assumptions. The estimators $\hat{\beta}$ which, from (5.8) are linear combinations of normally distributed observations, are then themselves normally distributed. Even if the errors are not normally distributed, the central limit theorem assures that, for large samples, the distribution of the least squares estimators is approximately normal.

The least squares estimators are unbiased, that is $\mathrm{E}(\hat{\beta}) = \beta$, provided that the correct model has been fitted. The variance of $\hat{\beta}_1$ is

$$\mathrm{var}(\hat{\beta}_1) = \frac{\sigma^2}{\sum(x_i - \bar{x})^2}. \tag{5.10}$$

Usually σ^2 will have to be estimated, often from the residual sum of squares, giving the residual mean square estimator

$$s^2 = \frac{S(\hat{\beta})}{N-2}$$

on $N - 2$ degrees of freedom. The estimate will be too large if the model is incorrect. This effect of model inadequacy can be avoided by estimating σ^2 solely from replicated observations. An attractive feature of the design of Table 1.1 is that it provides a replication mean square estimate of σ^2 on 16 degrees of freedom. Another possibility is to use an external estimate of σ^2, based on experience or derived from previous experiments. Experience shows that such estimates are usually unrealistically small.

Whatever its source, let the estimate of σ^2 be s^2 on ν degrees of freedom. Then to test the hypothesis that β_1 has the value β_{10},

$$\frac{\hat{\beta}_1 - \beta_{10}}{\{s^2/\sum(x_i - \bar{x})^2\}^{1/2}} \tag{5.11}$$

is compared with the t distribution on ν degrees of freedom. The $100(1-\alpha)\%$ confidence limits for β_1 are

$$\hat{\beta}_1 \pm \frac{t_{\nu,\alpha} s}{\{\sum(x_i - \bar{x})^2\}^{1/2}}.$$

The prediction at the point x, not necessarily included in the data from which the parameters were estimated, is

$$\hat{y}(x) = \hat{\beta}_0 + \hat{\beta}_1 x = \bar{y} + \hat{\beta}_1(x - \bar{x}) \tag{5.12}$$

with variance

$$\mathrm{var}\{\hat{y}(x)\} = \sigma^2 \left\{ \frac{1}{N} + \frac{(x - \bar{x})^2}{\sum(x_i - \bar{x})^2} \right\}.$$

The N least squares residuals are defined to be

$$e_i = y_i - \hat{y}_i = y_i - \bar{y} - \hat{\beta}_1(x_i - \bar{x}).$$

The use of residuals in checking the model assumed for the data is exemplified in Chapter 8.

TABLE 5.1. Analysis of variance for simple regression

Source	Degrees of freedom	Sum of squares	Abbreviation	Mean square	F
Regression	1	$\sum(\hat{y}_i - \bar{y})^2$	SSR	SSR	SSR/s^2
Residual (error)	$N-2$	$\sum(y_i - \hat{y}_i)^2$	SSE	SSE $/(N-2) = s^2$	
Total (corrected)	$N-1$	$\sum(y_i - \bar{y})^2$	SST		

It is often convenient, particularly for more complicated models, to summarize the results of an analysis, including hypothesis tests such as (5.11), in an analysis of variance table. The decomposition

$$\sum_{i=1}^{N}(y_i - \bar{y})^2 = \sum_{i=1}^{N}(\hat{y}_i - \bar{y})^2 + \sum_{i=1}^{N}(y_i - \hat{y}_i)^2,$$

leads to Table 5.1. The entries in the column headed 'Mean square' are sums of squares divided by degrees of freedom. The F test for regression is, in this case, the square of the t test of (5.11). A numerical example of such a table is given in §8.2 as part of the analysis of the data on the desorption of carbon monoxide introduced as Example 1.1.

5.2 Matrices and Experimental Design

To extend the results of the previous section to linear models with $p > 2$ parameters, it is convenient to use matrix algebra. The basic notation is established in this section.

The linear model will be written

$$\mathrm{E}(y) = F\beta \tag{5.13}$$

where, in general, y is the $N \times 1$ vector of responses, β is a vector of p unknown parameters and F is the $N \times p$ extended design matrix. The ith row of F is $f^{\mathrm{T}}(x_i)$, a known function of the m explanatory variables.

Example 5.1 Simple Regression For $N = 3$, the simple linear regression model (5.5) is

$$\mathrm{E}\begin{bmatrix} y_1 \\ y_2 \\ y_3 \end{bmatrix} = \begin{bmatrix} 1 & x_1 \\ 1 & x_2 \\ 1 & x_3 \end{bmatrix} \begin{bmatrix} \beta_0 \\ \beta_1 \end{bmatrix}.$$

Here $m = 1$ and $f^{\mathrm{T}}(x_i) = (1 \quad x_i)$.

TABLE 5.2. Designs for a single quantitative factor

Design	Design points x				Number of trials N
5.1	−1	0	1		3
5.2	−1	1	1		3
5.3	−1	−1/3	1/3	1	4

In order to design the experiment it is necessary to specify the design matrix

$$D = \begin{bmatrix} x_1 \\ x_2 \\ x_3 \end{bmatrix}.$$

The entries of F are then determined by D and by the model. Suppose that the factor x is quantitative with $-1 \leq x \leq 1$. The design region is then written $\mathcal{X} = [-1, 1]$. A typical design problem is to choose N points in $\mathcal{X}$ so that the linear relationship between y and x given by (5.5) can be estimated as precisely as possible. One possible design for this purpose is Design 5.1 in Table 5.2 which consists of trials at three equally spaced values of x with design matrix

$$D_1 = \begin{bmatrix} -1 \\ 0 \\ 1 \end{bmatrix}.$$

Another possibility is Design 5.2, which has two trials at one end of the design region and one at the other. The design matrix is then

$$D_2 = \begin{bmatrix} -1 \\ 1 \\ 1 \end{bmatrix}.$$

■

Example 5.2 Quadratic Regression If the model is

$$\mathrm{E}\,(y_i) = \beta_0 + \beta_1 x_i + \beta_2 x_i^2, \tag{5.14}$$

allowing for curvature in the dependence of y on x, trials will be needed for at least three different values of x in order to estimate the three parameters. The equally spaced four-trial Design 5.3 with design matrix

$$D_3 = \begin{bmatrix} -1 \\ -1/3 \\ 1/3 \\ 1 \end{bmatrix}$$

TABLE 5.3. Models for a single quantitative factor

Model	Algebraic form
First order, Example 5.1	$\mathrm{E}(y) = \beta_0 + \beta_1 x$
Quadratic, Example 5.2	$\mathrm{E}(y) = \beta_0 + \beta_1 x + \beta_2 x^2$
Quadratic, with one qualitative factor, Example 5.3	$\mathrm{E}(y) = \alpha_j + \beta_1 x + \beta_2 x^2$ $(j = 1, \ldots, l)$

would allow detection of departures from the quadratic model. For D_3 the extended design matrix for the quadratic model is

$$F = \begin{bmatrix} 1 & -1 & 1 \\ 1 & -1/3 & 1/9 \\ 1 & 1/3 & 1/9 \\ 1 & 1 & 1 \end{bmatrix},$$

where the final column gives the values of x_i^2 (Table 5.3). ■

Example 5.3 Quadratic Regression with a Single Qualitative Factor The simple quadratic model (5.14) can be extended by assuming that the response depends not only on the quantitative variable x, but also on a qualitative factor z at l unordered levels. These might, for example, be l patients or l different chemical reactor designs to be compared over a range of x values. Designs for such models are the subject of Chapter 14.

Suppose that $l = 2$. If the effect of z is purely additive so that the response curve is moved up or down, as in Figure 5.1, the model is

$$\mathrm{E}(y_i) = \alpha_j + \beta_1 x_i + \beta_2 x_i^2 \quad (i = 1, \ldots, N; j = 1, 2) \tag{5.15}$$

or, in matrix form

$$\mathrm{E}(y) = X\gamma = Z\alpha + F\beta. \tag{5.16}$$

In general the matrix Z in (5.16), of dimension $N \times l$, consists of indicator variables for the levels of z.

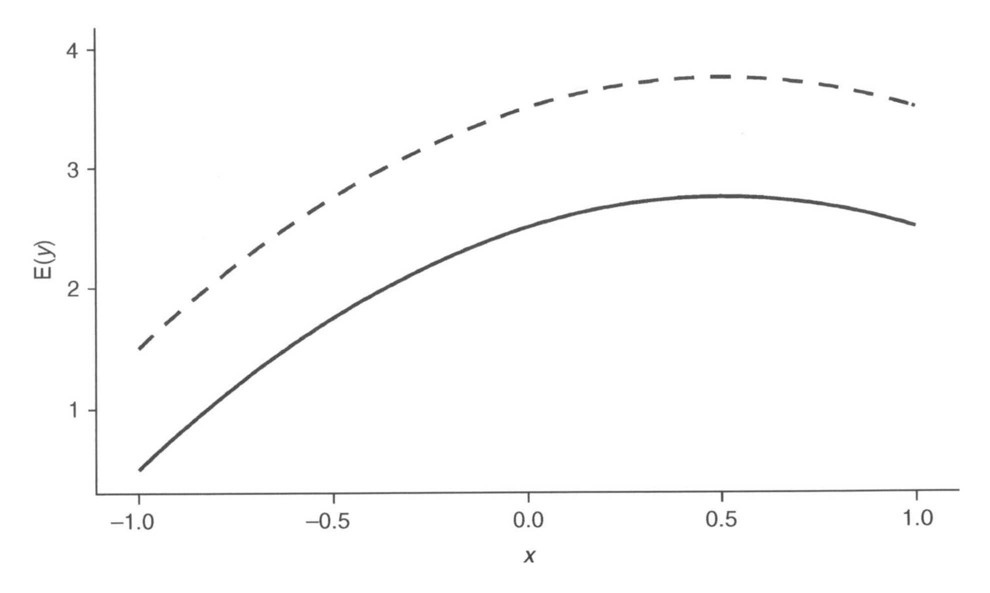

FIG. 5.1. Example 5.3: quadratic regression with a single qualitative factor. Response E(y) when the qualitative factor has two levels.

Suppose that the three-level Design 5.1 is repeated once at each level of z. Then

$$X = \begin{array}{c} \quad Z \;\; : \;\; F \\ \left[\begin{array}{cccc} 1 & 0 & -1 & 1 \\ 1 & 0 & 0 & 0 \\ 1 & 0 & 1 & 1 \\ 0 & 1 & -1 & 1 \\ 0 & 1 & 0 & 0 \\ 0 & 1 & 1 & 1 \end{array}\right] \end{array}. \tag{5.17}$$

The ith row of X is $x_i^{\mathrm{T}} = (z_i^{\mathrm{T}}, f_i^{\mathrm{T}})$.

A more complicated model with a similar structure might be appropriate for the analysis of Derringer's data on the viscosity of elastomer blends (Example 1.2). A natural approach to designing experiments with some qualitative factors is to find a good design for the x alone and then to repeat it at each level of z. Derringer's data has this structure. However, constraints on how many trials can be made at a given level of z often prohibit such equal replication.

The parameterization of the effect of the qualitative variable in (5.17) is not the only possible one. For example, the parameterization

$$X = \overset{\begin{matrix} Z & : & F \end{matrix}}{\begin{bmatrix} 1 & -1 & -1 & 1 \\ 1 & -1 & 0 & 0 \\ 1 & -1 & 1 & 1 \\ 1 & 1 & -1 & 1 \\ 1 & 1 & 0 & 0 \\ 1 & 1 & 1 & 1 \end{bmatrix}}$$

leads to an equivalent linear model with a different interpretation of the parameters α in (5.16). The form used in (5.17) is particularly convenient for generalizing to $l > 2$ levels as in §14.1. ■

5.3 Least Squares

This section gives the extension of the results of §5.1 to the linear model with p parameters (5.13). For this model the sum of squares to be minimized is

$$S(\beta) = (y - F\beta)^{\mathrm{T}}(y - F\beta). \tag{5.18}$$

The least squares estimator of β, found by differentiation of (5.18) satisfies the p least squares, or normal, equations

$$F^{\mathrm{T}}F\hat{\beta} = F^{\mathrm{T}}y. \tag{5.19}$$

The $p \times p$ matrix $F^{\mathrm{T}}F$ is the information matrix for β. The larger $F^{\mathrm{T}}F$, the greater is the information in the experiment. Experimental design criteria for maximizing aspects of information matrices are discussed in Chapter 10.

By solving (5.19) the least squares estimator of the parameters is found to be

$$\hat{\beta} = (F^{\mathrm{T}}F)^{-1}F^{\mathrm{T}}y. \tag{5.20}$$

If the model is not of full rank, $F^{\mathrm{T}}F$ cannot be uniquely inverted, and only a set of linear combinations of the parameters can be estimated, perhaps a subset of β. In the majority of examples in this book inversion of $F^{\mathrm{T}}F$ is not an issue. The covariance matrix of the least squares estimator is

$$\mathrm{var}\,\hat{\beta} = \sigma^2(F^{\mathrm{T}}F)^{-1}. \tag{5.21}$$

The variance of $\hat{\beta}_j$ is proportional to the jth diagonal element of $(F^{\mathrm{T}}F)^{-1}$ with the covariance of $\hat{\beta}_j$ and $\hat{\beta}_k$ proportional to the (j,k)th off-diagonal element. If interest is in the comparison of experimental designs, the value of σ^2 is not relevant, since the value is the same for all proposed designs for a particular experiment.

Tests of hypotheses about the individual parameters β_j can use the t test (5.11), but with the variance of $\hat{\beta}_j$ from (5.21) in the denominator. For tests about several parameters, the F test is used. The related $100(1-\alpha)\%$ confidence region for all p elements of β is of the form

$$(\beta - \hat{\beta})^{\mathrm{T}} F^{\mathrm{T}} F (\beta - \hat{\beta}) \leq p s^2 F_{p,\nu,\alpha}, \tag{5.22}$$

where s^2 is an estimate of σ^2 on ν degrees of freedom and $F_{p,\nu,\alpha}$ is the $\alpha\%$ point of the F distribution on p and ν degrees of freedom.

In the p-dimensional space of the parameters, (5.22) defines an ellipsoid, the boundary of which is the contour of constant residual sum of squares

$$S(\beta) - S(\hat{\beta}) = p s^2 F_{p,\nu,\alpha}.$$

The volume of the ellipsoid is inversely proportional to the square root of the determinant $|F^{\mathrm{T}}F|$. For the single slope parameter in simple regression, the variance, given by (5.10), is minimized if $\sum(x_i - \bar{x})^2$ is large. From (5.21), $|(F^{\mathrm{T}}F)^{-1}| = 1/|F^{\mathrm{T}}F|$ is called the generalized variance of $\hat{\beta}$. Designs which maximize $|F^{\mathrm{T}}F|$ are called D-optimum (for **D**eterminant). They are discussed in Chapter 10 and are the subject of Chapter 11.

The shape, as well as the volume, of the confidence region depends on $F^{\mathrm{T}}F$. The implication of various shapes of confidence region, and their dependence on the experimental design are described in Chapter 6. Several criteria for optimum experimental design and their relationship to confidence regions are discussed in Chapter 10.

The predicted value of the response, given by (5.12) for simple regression, becomes

$$\hat{y}(x) = \hat{\beta}^{\mathrm{T}} f(x) \tag{5.23}$$

when β is a vector, with variance

$$\mathrm{var}\{\hat{y}(x)\} = \sigma^2 f^{\mathrm{T}}(x)(F^{\mathrm{T}}F)^{-1} f(x). \tag{5.24}$$

These formulae are exemplified for the designs and models of §5.2. But first we derive a few further results about the properties of the least squares fit of the general model (5.13) that are useful in the analyses of Chapter 8.

The vector of N least squares residuals e can be written in several revealing forms including

$$e = y - \hat{y} = y - F\hat{\beta} = y - F(F^{\mathrm{T}}F)^{-1}F^{\mathrm{T}}y = (I - H)y, \tag{5.25}$$

where

$$H = F(F^{\mathrm{T}}F)^{-1}F^{\mathrm{T}} \tag{5.26}$$

is a projection matrix, often called the *hat matrix* since $\hat{y} = Hy$. If σ^2 is estimated by s^2 based on the residuals e,

$$(N-p)s^2 = e^{\mathrm{T}}e = \sum_{i=1}^{N} e_i^2 = y^{\mathrm{T}}(I-H)y = S(\hat{\beta}). \qquad (5.27)$$

The estimator s^2 is unbiased, provided the model (5.13) holds. Regression variables that should have been included in F, but were omitted, and outliers are both potential causes of estimates of σ^2 that will be too large.

Example 5.1 Simple Regression continued For simple regression (5.5), the information matrix is

$$F^{\mathrm{T}}F = \begin{bmatrix} \sum 1 & \sum x_i \\ \sum x_i & \sum x_i^2 \end{bmatrix} = \begin{bmatrix} N & \sum x_i \\ \sum x_i & \sum x_i^2 \end{bmatrix},$$

where again all summations are over $i = 1, \ldots, N$. The determinant of the information matrix is

$$\begin{aligned} |F^{\mathrm{T}}F| &= \begin{vmatrix} N & \sum x_i \\ \sum x_i & \sum x_i^2 \end{vmatrix} \\ &= N\Sigma x_i^2 - (\Sigma x_i)^2 \\ &= N\Sigma(x_i - \bar{x})^2. \end{aligned} \qquad (5.28)$$

Thus the covariance matrix of the least squares estimates $\hat{\beta}_0$ and $\hat{\beta}_1$ is

$$\sigma^2(F^{\mathrm{T}}F)^{-1} = \frac{\sigma^2}{|F^{\mathrm{T}}F|}\begin{bmatrix} \sum x_i^2 & -\sum x_i \\ -\sum x_i & N \end{bmatrix}, \qquad (5.29)$$

where each element is to be multiplied by $\sigma^2/|F^{\mathrm{T}}F|$. In particular, the variance of $\hat{\beta}_1$, which is element (2, 2) of (5.29), reduces to (5.10).

For the three-point Design 5.1,

$$F^{\mathrm{T}}F = \begin{bmatrix} 3 & 0 \\ 0 & 2 \end{bmatrix} \quad |F^{\mathrm{T}}F| = 6 \quad \text{and}$$

$$(F^{\mathrm{T}}F)^{-1} = \begin{bmatrix} 1/3 & 0 \\ 0 & 1/2 \end{bmatrix}. \qquad (5.30)$$

For this symmetric design the estimates of the parameters are uncorrelated, whereas, for Design 5.2, again with $N = 3$ but with only two support points,

$$F^{\mathrm{T}}F = \begin{bmatrix} 3 & 1 \\ 1 & 3 \end{bmatrix} \quad |F^{\mathrm{T}}F| = 8 \quad \text{and}$$

$$(F^{\mathrm{T}}F)^{-1} = \begin{bmatrix} 3/8 & -1/8 \\ -1/8 & 3/8 \end{bmatrix}. \qquad (5.31)$$

Thus the two estimates are negatively correlated.

TABLE 5.4. Determinants and variances for designs for a single quantitative factor

	Number of trials N	$\lvert F^{\mathrm{T}}F \rvert$	max $d(x,\xi)$ over $\mathcal{X}$
First-order model			
Design 5.1: (-1 0 1)	3	6	2.5
Design 5.2: (-1 1 1)	3	8	3
Quadratic model			
Design 5.1: (-1 0 1)	3	4	3
Design 5.3: (-1 $-1/3$ $1/3$ 1)	4	7.023	3.814

From (5.24) the variance of the predicted response from Design 5.1 is

$$\frac{\mathrm{var}\{\hat{y}(x)\}}{\sigma^2} = (1 \quad x)\left[\begin{array}{cc} 1/3 & 0 \\ 0 & 1/2 \end{array}\right]\left(\begin{array}{c} 1 \\ x \end{array}\right) = \frac{1}{3} + \frac{x^2}{2}.$$

In comparing designs it is often helpful to scale the variance for σ^2 and the number of trials and to consider the standardized variance

$$d(x,\xi) = N\frac{\mathrm{var}\{\hat{y}(x)\}}{\sigma^2} = 1 + \frac{3x^2}{2}. \tag{5.32}$$

This quadratic has a maximum over the design region $\mathcal{X}$ of 2.5 at $x = \pm 1$. In contrast, the standardized variance for the non-symmetric Design 5.2 is

$$d(x,\xi) = \frac{3}{8}(3 - 2x + 3x^2), \tag{5.33}$$

a non-symmetric function that has a maximum over $\mathcal{X}$ of 3 at $x = -1$.

These numerical results are summarized in the top two lines of Table 5.4. If $|F^{\mathrm{T}}F|$ is to be used to select a design for the first-order model, Design 5.2 is preferable. If, however, the criterion is to minimize the maximum of the standardized variance $d(x,\xi)$ over $\mathcal{X}$, a criterion known as G-optimality, Design 5.1 would be selected. This example shows that a design that is optimum for one purpose may not be so for another. The General Equivalence Theorem of §9.2 establishes a relationship between G- and D-optimality and provides conditions under which the relationship holds. ■

Example 5.2 Quadratic Regression continued For the quadratic regression model in one variable (5.14)

$$F^{\mathrm{T}}F = \begin{bmatrix} N & \sum x_i & \sum x_i^2 \\ \sum x_i & \sum x_i^2 & \sum x_i^3 \\ \sum x_i^2 & \sum x_i^3 & \sum x_i^4 \end{bmatrix}.$$

The symmetric three-point Design 5.1 yields

$$F^{\mathrm{T}}F = \begin{bmatrix} 3 & 0 & 2 \\ 0 & 2 & 0 \\ 2 & 0 & 2 \end{bmatrix} \tag{5.34}$$

with $|F^{\mathrm{T}}F| = 4$ and

$$(F^{\mathrm{T}}F)^{-1} = \begin{bmatrix} 1 & 0 & -1 \\ 0 & 1/2 & 0 \\ -1 & 0 & 3/2 \end{bmatrix}. \tag{5.35}$$

Now $f^{\mathrm{T}}(x) = (1 \quad x \quad x^2)$ and, from (5.35), the standardized variance

$$d(x, \xi) = 3 - 9x^2/2 + 9x^4/2. \tag{5.36}$$

This symmetric quartic has a maximum over $\mathcal{X}$ of 3 at $x = -1, 0$ or 1, which are the three design points. Further, this maximum value is equal to the number of parameters p.

Design 5.3 is again symmetric, but $N = 4$ and

$$F^{\mathrm{T}}F = \begin{bmatrix} 4 & 0 & 20/9 \\ 0 & 20/9 & 0 \\ 20/9 & 0 & 164/81 \end{bmatrix}$$

with $|F^{\mathrm{T}}F| = 7.023$. The standardized variance is

$$d(x, \xi) = 2.562 - 3.811x^2/2 + 5.062x^4,$$

again a symmetric quartic, but now the maximum value over $\mathcal{X}$ is 3.814 when $x = \pm 1$.

These results for the second-order model, summarized in the bottom two lines of Table 5.4, again seem to indicate that the two designs are better for different criteria. However, Design 5.1 is for three trials whereas Design 5.3 is for four. The variances $d(x, \xi)$ are scaled to allow for this difference in N. To scale the value $|F^{\mathrm{T}}F|$ from Design 5.3 for comparison with three-trial designs, we multiply by $(3/4)^3$, obtaining 2.963. Thus, on a per trial basis, Design 5.1 is preferable to Design 5.3 for the quadratic model. The implications of comparison of designs per trial are explored in Chapter 9 when we consider exact and continuous designs. ■

Example 5.3 Quadratic Regression with a Single Qualitative Factor continued The least squares results of this section extend straightforwardly to the model for qualitative and quantitative factors (5.15). Replication of Design 5.1 at the two levels of z yields the information matrix

$$X^{\mathrm{T}}X = \begin{bmatrix} 3 & 0 & 0 & 2 \\ 0 & 3 & 0 & 2 \\ 0 & 0 & 4 & 0 \\ 2 & 2 & 0 & 4 \end{bmatrix}$$

which is related to the structure of (5.34). In general the upper $l \times l$ matrix is diagonal for designs with a single qualitative factor. The 2×2 lower submatrix results from the two replications of the design for the quantitative factors, here x and x^2. This structure is important for the designs with both qualitative and quantitative factors that are the subject of Chapter 14. ■

5.4 Further Reading

Least squares and regression are described in many books at a variety of levels. Weisberg (2005) is firmly rooted in applications. Similar material, at a greater length, can be found in Draper and Smith (1998). The more mathematical aspects of the subject are well covered by Seber (1977). Particular aspects of data analysis motivate some modern treatments: robust methods for Ryan (1997), graphics for Cook and Weisberg (1999), and the forward search for Atkinson and Riani (2000). An appropriate SAS book is Littell, Stroup, and Freund (2002).

6

CRITERIA FOR A GOOD EXPERIMENT

6.1 Aims of a Good Experiment

The results of Chapter 5 illustrate that the variances of the estimated parameters in a linear model depend upon the experimental design, as does the variance of the predicted response. An ideal design would provide small values of both variances. However, as the results of Table 5.4 show, a design which is good for one property may be less good for another. Usually one or a few important properties are chosen and designs found that are optimum for these properties. In this chapter we first list some desirable properties of an experimental design. We then illustrate the dependence of the ellipsoidal confidence regions (5.22) and of the variance of the predicted response (5.24) on the design. Finally, the criteria of D-, G-, and V-optimality are described, and examples of optimum designs given for both simple and quadratic regression.

Box and Draper, (1975, 1987 Chapter 14), list 14 aims in the choice of an experimental design. Although their emphasis is on response-surface designs, any, all, or some of these properties of a design may be important.

1. Generate a satisfactory distribution of information throughout the region of interest, which may not coincide with the design region $\mathcal{X}$.
2. Ensure that the fitted value, $\hat{y}(x)$ at x, be as close as possible to the true value $\eta(x)$ at x.
3. Make it possible to detect lack of fit.
4. Allow estimation of transformations of both the response and of the quantitative experimental factors.
5. Allow experiments to be performed in blocks.
6. Allow experiments of increasing size to be built up sequentially. Often, as in Figure 3.1, a second-order design will follow one of first-order.
7. Provide an internal estimate of error from replication.
8. Be insensitive to wild observations and to violations of the usual normal theory assumptions.

9. Require a minimum number of experimental runs.
10. Provide simple data patterns that allow ready visual appreciation.
11. Ensure simplicity of calculation.
12. Behave well when errors occur in the settings of the experimental variables.
13. Not require an impractically large number of levels of the experimental factors.
14. Provide a check on the 'constancy of variance' assumption.

Different aims will, of course, be of different relative importance as circumstances change. Thus Point 11, requiring simplicity of calculation, will not much matter if good software is available for the analysis of the experimental results. But, in this context, 'good' implies the ability to check that the results have been correctly entered into the computer. The restriction on the number of levels of individual variables (Point 13) is likely to be of particular importance when experiments are carried out by unskilled personnel, for example on a production process.

This list of aims will apply for most experiments. Two further aims which may be important with experiments for quantitative factors are:

15. Orthogonality: the designs have a diagonal information matrix, leading to uncorrelated estimates of the parameters.
16. Rotatability: the variance of $\hat{y}(x)$ depends only on the distance from the centre of the experimental region.

Orthogonality is too restrictive a requirement to be attainable in many of the examples considered in this book, such as the important designs for second-order models in §11.5. However, orthogonality is a property of many commonly used designs such as the 2^m factorials and designs for qualitative factors. Rotatabilty was much used by Box and Draper (1963) in the construction of designs for second- and third-order response surface models. We discuss an example of a rotatable design in §6.4 where we introduce variance–dispersion graphs for the comparison of designs.

6.2 Confidence Regions and the Variance of Prediction

For the moment, of the many objectives of an experiment, we concentrate on the relationship between the experimental design, the confidence ellipsoid for the parameters given by (5.22), and the variance of the predicted

TABLE 6.1. Some designs for first- and second-order models when the number of factors $m = 1$

Design	Number of trials N	Values of x							
6.1	3	−1	0	1					
6.2	6	−1	−1	0	0	1	1		
6.3	8	−1	−1	−1	−1	−1	−1	1	1
6.4	5	−1	−0.5	0	0.5	1			
6.5	7	−1	−1	−0.9	−0.85	−0.8	−0.75	1	
6.6	2	−1	1						
6.7	4	−1	−1	0	1				
6.8	4	−1	0	0	1				

response (5.24). Several of the aims listed in §6.1 will be used in later chapters to assess and compare designs.

Table 6.1 gives eight single-factor designs for varying N; some of the designs are symmetrical and some are not, just as some are more concentrated on a few values of x than are others. We compare the first six for simple regression, that is for the first-order model in one quantitative factor. Suppose that the fitted model is

$$\hat{y}(x) = 16 + 7.5x. \tag{6.1}$$

Contour plots for the parameter values β for which

$$(\beta - \hat{\beta})^{\mathrm{T}} F^{\mathrm{T}} F(\beta - \hat{\beta}) = \delta^2 = 1$$

are given in Figures 6.1 and 6.2 for the first six designs of the table. From (5.22) $\delta^2 = ps^2 F_{p,\nu,\alpha}$, so that the size of the confidence regions will increase with s^2. However, the relative shapes of the regions for different designs will remain the same.

These sets of elliptical contours are centred at $\hat{\beta} = (16, 7.5)^{\mathrm{T}}$. Comparison of the ellipses for Designs 6.1 and 6.2 in Figure 6.1 shows how increasing the number of trials N decreases the size of the confidence region; Design 6.2 is two replicates of Design 6.1. Both these designs are orthogonal, so that $F^{\mathrm{T}}F$ is diagonal and the axes of the ellipses are parallel to the co-ordinate axes. On the other hand, the two designs yielding Figure 6.2 both have several trials at or near the lower end of the design region. As a result, the designs are not orthogonal; $F^{\mathrm{T}}F$ has non-zero off diagonal elements and the axes

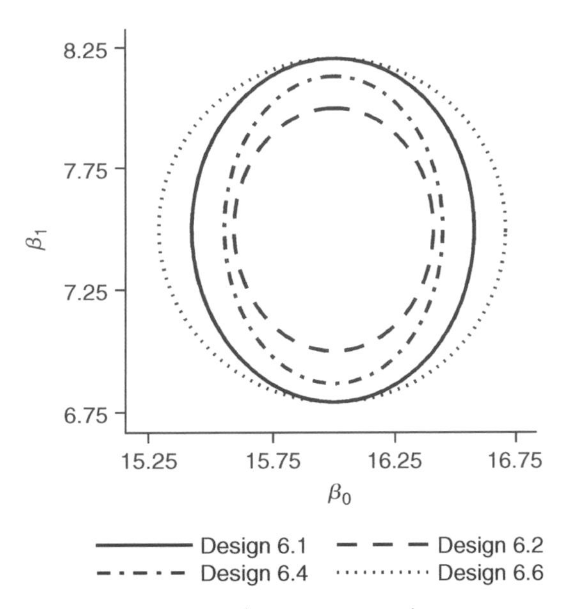

FIG. 6.1. Confidence ellipses $(\beta-\hat{\beta})^{\mathrm{T}}F^{\mathrm{T}}F(\beta-\hat{\beta}) = 1$ for first-order models (simple regression) fitted to the symmetric designs of Table 6.1.

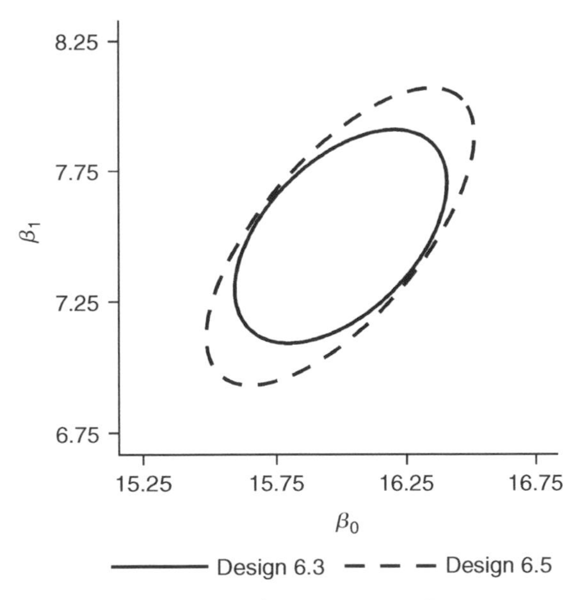

FIG. 6.2. Confidence ellipses $(\beta-\hat{\beta})^{\mathrm{T}}F^{\mathrm{T}}F(\beta-\hat{\beta}) = 1$ for first-order models (simple regression) fitted to the asymmetric designs of Table 6.1.

of the ellipses do not lie along the co-ordinate axes. These axes are also of markedly different lengths. The effect is that some linear combinations of the parameters are estimated with small variance and others are estimated much less precisely. However, all four designs in Figure 6.1 are orthogonal. The two trial Design 6.6 yields the largest region of all; it is the design with fewest trials, even though they do span the experimental region. Shrunk versions of Design 6.6 with trials at a and $-a, 0 < a < 1$, give ellipses larger than that in Figure 6.1, but with the same proportions.

For designs when more than two parameters are of importance, the ellipses of Figures 6.1 and 6.2 are replaced by ellipsoids. Graphical methods of assessment then need to be augmented or replaced by analytical methods. In particular, the lengths of the axes of the ellipsoids are proportional to the square roots of the eigenvalues of $(F^{\mathrm{T}}F)^{-1}$, which are the reciprocals of the eigenvalues of $F^{\mathrm{T}}F$. A design in which the eigenvalues differ appreciably will typically produce long, thin ellipsoids. The squares of the volumes of the confidence ellipsoids are proportional to the product of the eigenvalues of $(F^{\mathrm{T}}F)^{-1}$, which is equal to the inverse of the determinant of $F^{\mathrm{T}}F$. Hence, in terms of these eigenvalues, a good design should have a large determinant, giving a confidence region of small content, with the values all reasonably equal. These ideas are formalized in Chapter 10.

We now consider the standardized variance of the predicted response $d(x, \xi)$ introduced in (5.32). Figures 6.3 and 6.4 show these functions, plotted

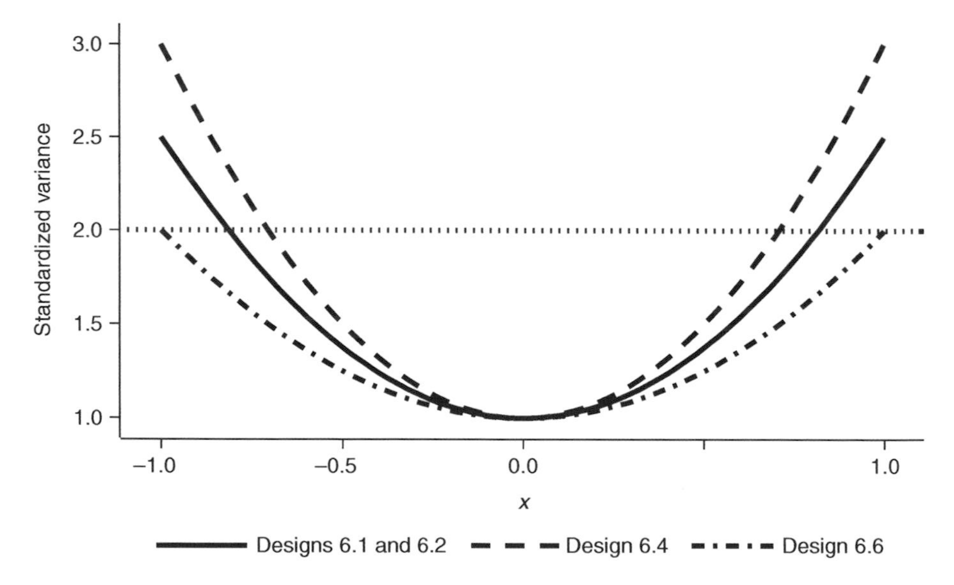

FIG. 6.3. Standardized variances $d(x, \xi)$ for four symmetrical designs of Table 6.1; simple regression.

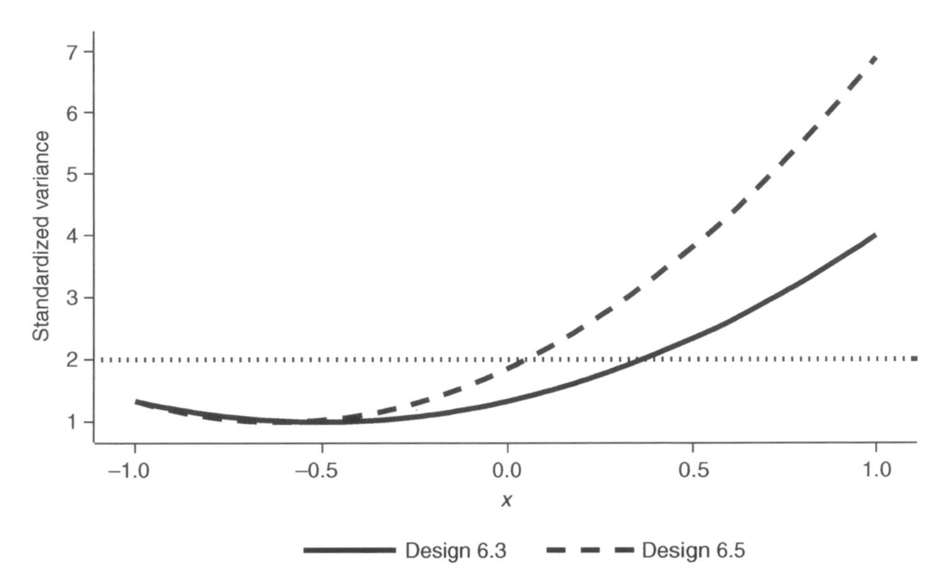

FIG. 6.4. Standardized variances $d(x, \xi)$ for two asymmetrical designs of Table 6.1; simple regression.

over $\mathcal{X}$, for the six designs in Table 6.1 which gave rise to the ellipses of Figures 6.1 and 6.2. The symmetrical designs 6.1, 6.2, 6.4, and 6.6 all give symmetrical plots over $\mathcal{X}$ as shown in Figure 6.3. Because the variances are standardized by the number of trials, the values are the same for both Design 6.1 and Design 6.2. Of these symmetrical designs, Design 6.4 shows the highest variance of the predicted response over all of $\mathcal{X}$, except for the value of $d(0, \xi)$ which is unity for all four symmetrical designs. The plots in Figure 6.4 illustrate how increasing the number of trials in one area of $\mathcal{X}$, in this case near $x = -1$, reduces the variance in that area but leads to a larger variance elsewhere.

Of the symmetrical designs, Design 6.4 has its five trials spread uniformly over $\mathcal{X}$, but as we have seen in Figure 6.3, that does not lead to the estimate of $\hat{y}(x)$ with the smallest variance over all of $\mathcal{X}$. On the contrary, the figure shows that Design 6.6, which has equal numbers of trials at each end of the interval and none at the centre, leads to the design with smallest $d(x, \xi)$ over the whole of $\mathcal{X}$. Such concentration on a few design points is characteristic of many optimum designs. We show in §6.5 that no design can do better than this for the first-order model.

When the model is second order in one factor, the standardized variance $d(x, \xi)$ becomes a quartic. Figures 6.5 and 6.6 give plots of this variance for some of the designs of Table 6.1, including some of those used for the plots

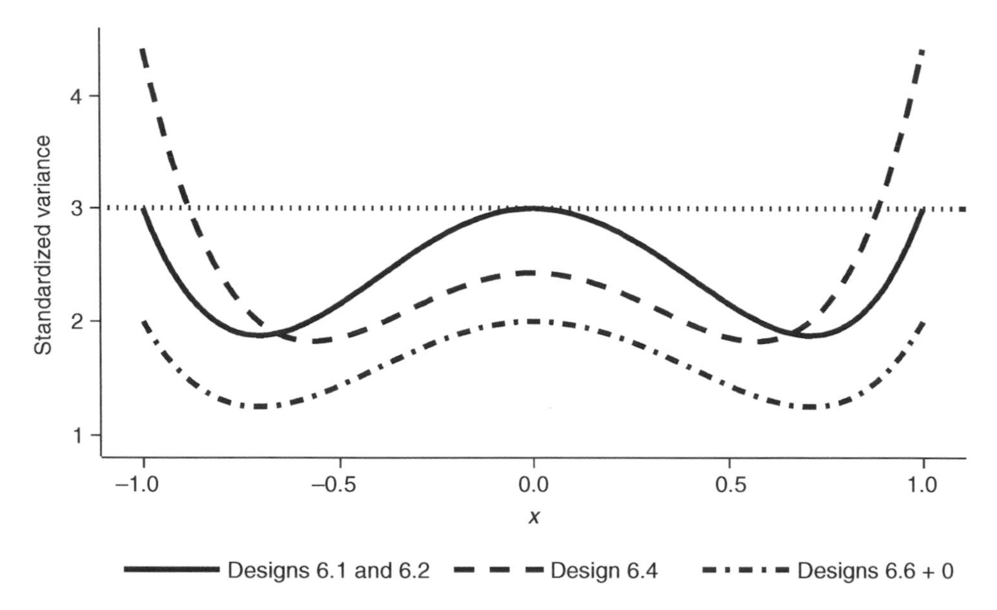

FIG. 6.5. Standardized variances $d(x, \xi)$ for three symmetrical designs of Table 6.1; quadratic regression.

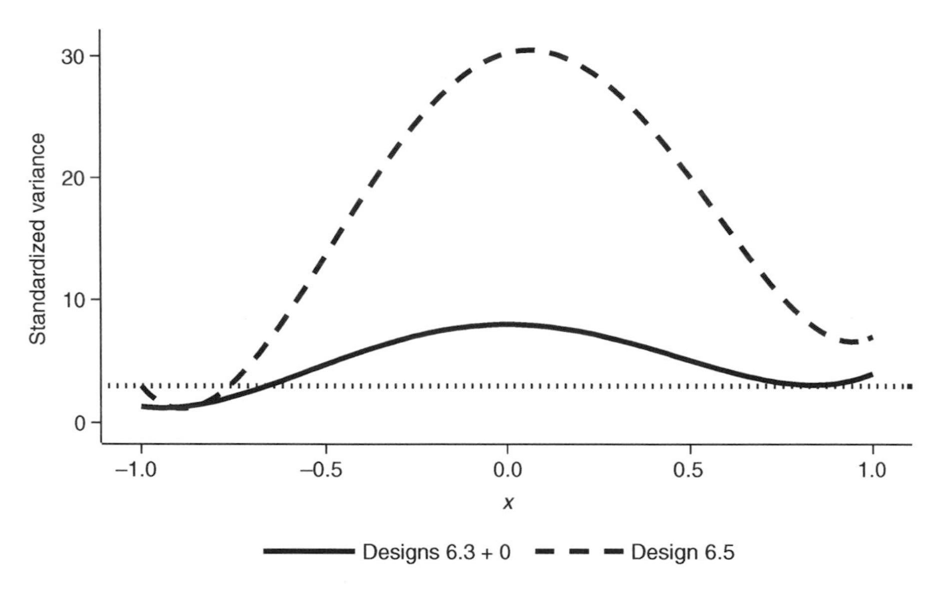

FIG. 6.6. Standardized variances $d(x, \xi)$ for two asymmetrical designs of Table 6.1; quadratic regression.

for the first-order model. As can be seen in Figure 6.5, Designs 6.1 and 6.2 ensure that the maximum of $d(x, \xi)$ over the design region is equal to three, which is the number of parameters in the model. Kiefer and Wolfowitz (1960) show that this is the smallest possible value for the maximum. Design 6.4, like Design 6.1, is symmetrical, but distributes five trials uniformly over the design region. Figure 6.5 shows that the maximum value of $d(x, \xi)$ for this design is 4.428, greater than the value of three for Designs 6.1 and 6.2, and that the variance is relatively reduced in the centre of $\mathcal{X}$. Locating the trials mainly at one end of the interval causes the variance to be large elsewhere, as in the plot for Design 6.5 in Figure 6.6.

Finally, we look at the properties of designs in which one trial is added to the three-point Design 6.1 which has support points $-1, 0$, and 1. In the first case the added trial is at $x = -1$, a design we call 6.7, whereas in the other it is at $x = 0$ (Design 6.8). The two resulting plots for the variances are quite different. In the first case, as Figure 6.6 shows, the variance is reduced near to the lower boundary of $\mathcal{X}$, while it increases elsewhere. On the other hand the symmetric Design 6.8, Figure 6.5, ensures low variance in the centre of $\mathcal{X}$, that is near the replicated design point. However, $d(x, \xi)$ increases sharply towards the ends of the region.

6.3 Contour Plots of Variances for Two-Factor Designs

When there are two factors instead of one, the graphs of the previous section become contour plots of $d(x, \xi)$ over $\mathcal{X}$. Figure 6.7 gives such a plot for a first-order model in two factors when the design is a 2^2 factorial. The circular contours show that the design is rotatable, point 16 of §6.1, the value of $d(x, \xi)$ depending only on the distance from the centre of the design region. Sections of this plot through the origin have the quadratic shape shown in Figure 6.3 for first-order models in one factor. The maximum value of $d(x, \xi)$ is three, the number of parameters in the model. These maxima occur at the points of the 2^2 factorial.

Contour plots for second-order models are more complicated. Figure 6.8 is the contour plot for the six-parameter second-order model including interaction when the design is a 3^2 factorial. There are now nine local maxima of $d(x, \xi)$ which occur at the nine support points of the design. In this case, sections of this plot through the origin have the quartic shape shown in Figure 6.5 for symmetric designs for second-order models in one factor. The values of the maxima of $d(x, \xi)$ vary with the design points. At the support of the 2^2 factorial they are 7.25 and 5 at the other five design points. The minimum values are 3.21 around $(\pm 0.65, \pm 0.65)$. The implications of the values at the local maxima for the construction of optimum designs in

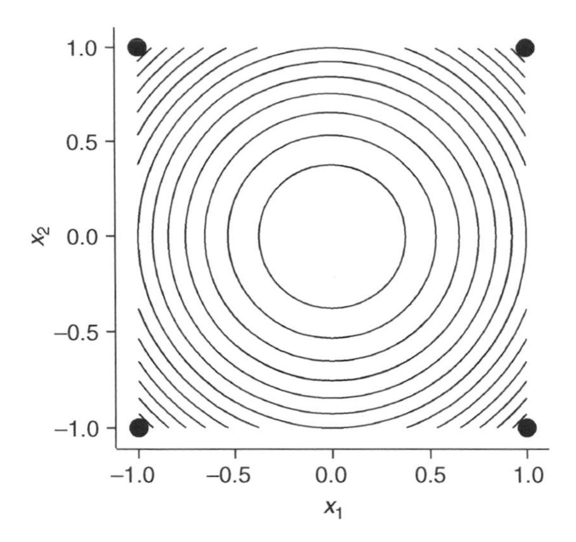

FIG. 6.7. Standardized variance $d(x, \xi)$ for 2^2 factorial; first-order model. A rotatable design; $d(x, \xi) = 3$ at the points of the design which are marked by circles.

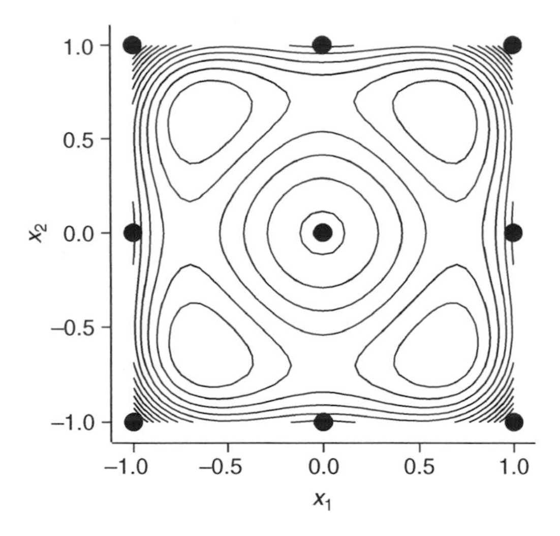

FIG. 6.8. Standardized variance $d(x, \xi)$ for 3^2 factorial; second-order model. The maxima of the variance are at the points of the design which are marked by circles. There are local minima near $(\pm 0.65, \pm 0.65)$.

one factor are illustrated in §11.2. The implications for designs for this two-factor model are explored in Example 12.1 in §12.2; the augmentation of the 3^2 factorial with a further trial at each of the points of the 2^2 factorial yields a design with virtually constant variance over the design points.

6.4 Variance–Dispersion Graphs

The contour plot of Figure 6.8 is not particularly easy to interpret and comparison of several such graphs for competing designs can be difficult. So, for $m > 2$, and even for $m = 2$, it is easier to look at summaries of the contour plots. One such summary is the variance–dispersion graph, a graph of the behaviour of $d(x, \xi)$ as we move away from the centre of the experimental region. For a series of spherical shells expanding from the centre of $\mathcal{X}$ we typically look at the minimum, average, and maximum of $d(x, \xi)$, over the shell and plot the three quantities against the radius of the shell. We thus look at the dispersion, or range, of the variance as a function of distance.

Figure 6.9 shows the variance–dispersion plot derived from the contour plot of Figure 6.7 for the two-factor first-order model with the 2^2 factorial design. Because the design is rotatable, the variance is constant over each sphere, so that the minimum, average, and maximum of $d(x, \xi)$ are the same for a given radius and we obtain the single curve shown in the figure. This

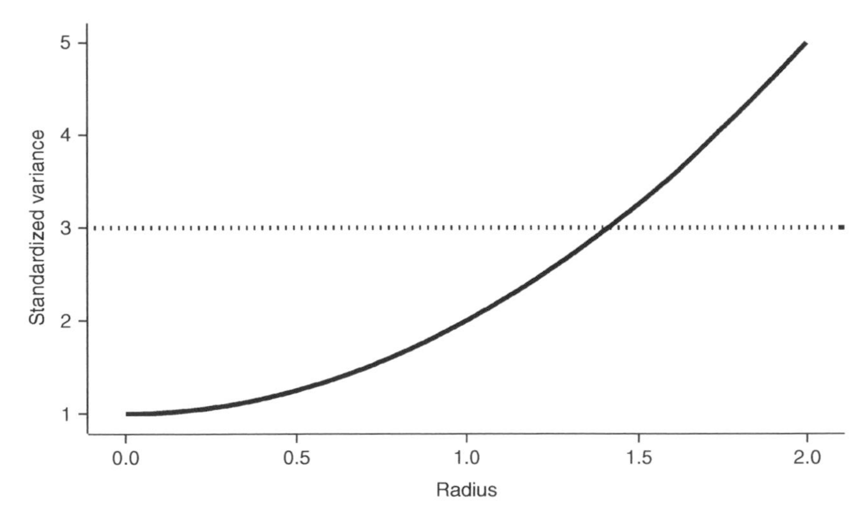

FIG. 6.9. Variance–dispersion graph for 2^2 factorial and a first-order model. Since the design is rotatable the curves for the minimum, mean, and maximum of $d(x, \xi)$ coincide.

starts low and rises to a value of 3 at $x = \sqrt{2}$ in a quadratic manner similar to the right-hand half of the plots of $d(x, \xi)$ in Figure 6.5. We have extended the plot to $x = 2$ as a reminder that variances of prediction continue to increase as we move away from the centre of $\mathcal{X}$.

The variance–dispersion graph derived from the contour plot of Figure 6.8 for the 3^2 factorial and a second-order model is in Figure 6.10. Now there are separate curves for the minimum, mean, and maximum of $d(x, \xi)$ all following, approximately, the quartic shape of the right-hand half of Figure 6.3 for the single-factor quadratic model. The three curves are, as they must be, coincident at the origin. Again we have extended the radius to 2. For distances greater than one the calculation includes points outside the design region. For some of these the variance, for this non-rotatable design, is appreciably higher than it is for points within $\mathcal{X}$. An alternative is to exclude from the figure values from those parts of the spherical shells that lie outside $\mathcal{X}$, although such plots are not customary.

In principle, calculating the minimum, maximum, and average standardized variance on spherical shells is a non-trivial numerical problem. Given the moments of the sphere, $d_{\text{avg}}(\xi)$ is easy to calculate (see Myers and Montgomery, 2002), but calculating $d_{\text{min}}(\xi)$ and $d_{\text{max}}(\xi)$ necessarily requires constrained non-linear optimization. An alternative is to calculate $d(x, \xi)$ for discrete points uniformly spread around the sphere and then to calculate $d_{\text{avg}}(\xi)$, $d_{\text{min}}(\xi)$, and $d_{\text{max}}(\xi)$ as the average, minimum, and maximum of these discrete values. This is how Figure 6.10 was computed,

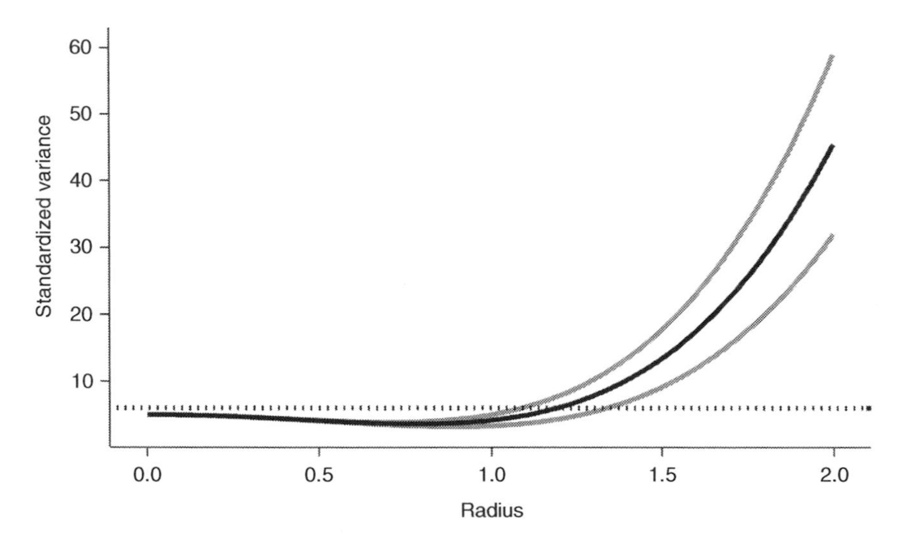

FIG. 6.10. Variance–dispersion graph. Minimum, mean, and maximum of $d(x, \xi)$ for 3^2 factorial; second-order model.

using 1000 points uniformly distributed on the circle. It is also how Figure 6.9 was calculated although, with such a rotatable design, it is enough to calculate the variance along a single ray from the centre of the design. For higher dimensional spheres a uniform distribution of points is more difficult to achieve. The ADX Interface in SAS takes many points uniformly distributed in the m-dimensional cube and projects them to the surface of the sphere. While the resulting set of points is not truly uniform on the sphere, it is typically quite dense. The characteristics of $d(x, \xi)$ over these points should be quite similar to the characteristics over the sphere.

6.5 Some Criteria for Optimum Experimental Designs

The examples in this chapter and in Chapter 5 show how different designs can be in the values they yield of $|F^{\mathrm{T}}F|$, in the plot of $d(x, \xi)$ over the design region $\mathcal{X}$, in the variance–dispersion graph derived from that plot and in the maximum value of that variance over $\mathcal{X}$. An ideal design for these models would simultaneously minimize the generalized variance of the parameter estimates and minimize $d(x, \xi)$ over $\mathcal{X}$. Usually a choice has to be made between these desiderata. Three possible design criteria that relate to these properties follow:

- **D-optimality:** a design is D-optimum if it maximizes the value of $|F^{\mathrm{T}}F|$, that is the generalized variance of the parameter estimates is minimized (Chapter 11).
- **G-optimality:** a G-optimum design minimizes the maximum over the design region $\mathcal{X}$ of the standardized variance $d(x, \xi)$. For some designs this maximum value equals p (§10.7).
- **V-optimality:** an alternative to G-optimality is V-optimality in which the average of $d(x, \xi)$ over $\mathcal{X}$ is minimized (§10.6).

The criteria of G- and V-optimality thus find designs to minimize one aspect of $d(x, \xi)$ displayed in the variance dispersion graphs of §6.4. A use of these graphs is to compare the properties of various designs, including those that optimize the different criteria.

The mathematical construction and evaluation of designs according to the criteria of D- and G-optimality, and the close relationship between these criteria, are the subjects of Chapter 9. More general criteria are discussed in Chapter 10. We conclude the present chapter by assembling the numerical results obtained so far for first- and second-order polynomials in one variable.

Example 6.1 Simple Regression: Example 5.1 continued The fitted simple regression model was written in (5.12) as

$$\hat{y}(x) = \bar{y} + \hat{\beta}_1(x - \bar{x}).$$

This suggests rewriting model (5.5) with centred x as

$$\text{E}(y) = \alpha + \beta_1(x - \bar{x}), \tag{6.2}$$

which again gives (5.12) as the fitted model. When the model is written in this orthogonal form (6.2), the diagonal information matrix has the revealing structure

$$F^{\text{T}}F = \begin{bmatrix} N & 0 \\ 0 & \sum(x_i - \bar{x})^2 \end{bmatrix},$$

so that the D-optimum design is that which minimizes the variance of $\hat{\beta}_1$ (5.10) when the value of N is fixed. If $\mathcal{X} = [-1, 1]$, $\sum(x_i - \bar{x})^2$ is maximized by putting half the trials at $x = +1$ and the other half at $x = -1$. When $N = 2$ this is Design 6.6 of Table 6.1. The plot of $d(x, \xi)$ for this design in Figure 6.3 has a maximum value over $\mathcal{X}$ of two, the smallest possible maximum. Therefore this design is also G-optimum.

Provided that N is even, replications of this design are D- and G-optimum. If N is odd, the situation is more complicated. The results of §5.3 for $N = 3$, summarized in Table 5.4, show that Design 5.2, in which one of the points $x = 1$ or $x = -1$ is replicated, is to be preferred for D-optimality. However the three-point Design 5.1, or equivalently 6.1, is preferable for G-optimality, yielding the symmetric curve for Design 6.1 of $d(x, \xi)$ in Figure 6.3 with a maximum value of 2.5.

A design in which the distribution of trials over $\mathcal{X}$ is specified by a measure ξ, regardless of N, is called continuous or approximate. The equivalence of D- and G-optimum designs holds, in general, only for continuous designs. Designs for a specified number of trials N are called exact. Section 9.1 gives a more detailed discussion of the differing properties of exact and continuous designs.

In the current example the exact D-optimum design for even N puts $N/2$ trials at $x = -1$ and $N/2$ at $x = +1$. If N is not even, for example 3 or 5, this equal division is, of course, not possible and the exact optimum design will depend on the value of N. Often, as here, a good approximation to the D-optimum design is found by an integer approximation to the D-optimum continuous design. For $N = 5$, two trials could be put at $x = -1$ and three at $x = +1$, or vice versa. But, in more complicated cases, especially when there are several factors and N is barely greater than the number of parameters in the model, the optimum exact design has to be found by numerical optimization for each N. Algorithms for the numerical construction of exact designs are the subject of Chapter 12. The numerical results for Example 12.1 in §12.2 illustrate the dependence of the D-optimum exact design for a two-factor second-order response surface on the value of N. ■

Example 6.2 Quadratic Regression: Example 5.2 continued For the one-factor quadratic model (5.14), Figure 6.5 is a plot of $d(x, \xi)$ for the three-point symmetric Designs 5.1 or 6.1. The maximum value of $d(x, \xi)$ is three and this design is D- and G-optimum. Again, replication of this design will provide the optimum designs provided N is divisible by three. For $N = 4$, the D- and G-optimum designs are not the same, an example discussed fully in §9.3. ■

As well as showing the distinction between exact and continuous designs, these examples serve to display some of the properties of D-optimum designs. One is that the design depends on the model. A second is that the number of distinct values of x is often equal to the number of parameters p in the model, particularly if there is only one factor. The designs therefore provide no method of checking goodness of fit. The introduction of trials at extra values of x in order to provide such checks reduces the efficiency of the design for the primary purposes of response and parameter estimation. Designs achieving a balance between parameter estimation and model checking are described in Chapters 20 and 21.

7

STANDARD DESIGNS

7.1 Introduction

The examples in Chapter 6 show that it is not always possible to find a design that is simultaneously optimum for several criteria. When there are only one or two factors, plots of the variance function, similar to those in Chapter 6, can be used to assess different aspects of a design's behaviour. But, when there are several factors and the model is more complicated, such plots become more difficult to interpret. The variance–dispersion graphs of §6.4 provide another graphical means of design comparison. But, in general, designs satisfying one or several criteria are best found by numerical optimization of a design criterion, a function of the design points and of the number of trials. Algorithms for the construction of continuous designs, which do not depend on the total number of trials, are described in Chapter 9. The more complicated algorithms for exact designs are the subject of Chapter 12. This chapter presents a short survey of some standard designs which do not require the explicit use of these algorithms. The construction of these, and other, standard designs in SAS, is described in §7.7. If particular properties of the designs are important, the equivalence theorems of Chapter 10 can be used to assess and compare these designs.

7.2 2^m Factorial Designs

This design consists of all combinations of points at which the factors take coded values of ± 1, that is all combinations of the high and low levels of each factor. The most complicated model that can be fitted to the experimental results contains first-order terms in all factors and two-factor and higher-order interactions up to that of order m. For example, if $m = 3$, the model is

$$\mathrm{E}(y) = \beta_0 + \beta_1 x_1 + \beta_2 x_2 + \beta_3 x_3 + \beta_{12} x_1 x_2 + \beta_{13} x_1 x_3 + \beta_{23} x_2 x_3 + \beta_{123} x_1 x_2 x_3. \tag{7.1}$$

The design is given in 'standard order' in Table 7.1. That is, the level of x_1 changes most rapidly and that of x_m, here x_3, changes only once.

TABLE 7.1. 2^m factorial design in three factors

	Factors				
Trial number	A x_1	B x_2	C x_3	Treatment combination	Response
1	-1	-1	-1	(1)	$y_{(1)}$
2	$+1$	-1	-1	a	y_a
3	-1	$+1$	-1	b	y_b
4	$+1$	$+1$	-1	ab	y_{ab}
5	-1	-1	$+1$	c	y_c
6	$+1$	-1	$+1$	ac	y_{ac}
7	-1	$+1$	$+1$	bc	y_{bc}
8	$+1$	$+1$	$+1$	abc	y_{abc}

The order of these treatment combinations should be randomized when the design is used. Two notations for the design are used in the table. In the second the factors are represented by capital letters; the interaction between the first two factors can then be written as AB or as x_1x_2. The presence of the corresponding lower-case letter in the treatment combination indicates that the factor is at its high level. This notation is useful when blocking 2^m factorial designs and for the construction of fractional factorials.

The 2^m designs are powerful experimental tools that are both easy to construct and easy to analyse. The designs are orthogonal, so that the information matrix $F^{\mathrm{T}}F$ is diagonal. Each diagonal element has the value $N = 2^m$. Because of the orthogonality of the design, each treatment effect can be estimated independently of any other. Further, the estimators have a simple form; since each column of the extended design matrix F consists of $N/2$ elements equal to $+1$ and the same number equal to -1, the elements of $F^{\mathrm{T}}y$ consist of differences of sums of specific halves of the observations. For example, in the notation of Table 7.1,

$$\hat{\beta}_1 = \{(y_a + y_{ab} + y_{ac} + y_{abc}) - (y_{(1)} + y_b + y_c + y_{bc})\}/8. \qquad (7.2)$$

Thus the estimate is half the difference in the response at the high level of A (treatment combinations including the symbol a) and that at the low level (treatments without the symbol a). This structure extends to estimation of any of the coefficients in the model (7.1). Estimation of $\hat{\beta}_{12}$ requires the vector x_1x_2, found by multiplying together the columns for x_1 and x_2 in Table 7.1. In order, the elements are

$$(+1 \quad -1 \quad -1 \quad +1 \quad +1 \quad -1 \quad -1 \quad +1),$$

so that

$$\hat{\beta}_{12} = \{(y_{(1)} + y_{ab} + y_c + y_{abc}) - (y_a + y_b + y_{ac} + y_{bc})\}/8. \tag{7.3}$$

The first group of four units in (7.3) are those with an even parity of treatment letters with ab (either 0 or 2) whereas the second group have an odd parity, in this case one letter in common with ab.

This structure extends straightforwardly to higher values of m. So does the D-optimality of the design. The maximum value of the variance of the predicted response is at the corners of the experimental region, where all x_j^2 are equal to one. Since the variance of each parameter estimate is σ^2/N, the maximum value of the standardized variance $d(x, \xi)$ is p at each design point.

If the full model, with all interaction terms, is fitted, $p = 2^m$. But, once the model has been fitted, the parameter estimates can be tested for significance. The omission of non-significant terms then leads to simplification of the model; both the value of p and the maximum variance of prediction are reduced. Because of the orthogonality of the design, the parameters β do not have to be re-estimated when terms are deleted. However the residual sum of squares, and so the estimate of σ^2 will change. An example of such an analysis, in the absence of an independent estimate of error, is given in §8.4.

A second assumption is that the full factorial model is adequate. However, some curvature may be present in the quantitative factors that would require the addition of quadratic terms to the model. In order to check whether such terms are needed, three or four 'centre points' are often added to the design. These experiments at $x_j = 0, (j = 1, \ldots, m)$ also provide a few degrees of freedom for the estimation of σ^2 from replicate observations. A systematic approach to generating designs for detecting lack of fit is described in §20.2; §19.3 describes methods for finding optimum designs for the augmentation of models. Both procedures provide alternatives to the replication of centre points.

The concept of a centre point is usually not meaningful for qualitative factors. If qualitative factors are present, one strategy is to include centre points in the quantitative factors at each level of the qualitative ones.

Factors at Four Levels Factorial designs with factors at more than two levels are much used, particularly for qualitative factors. If some of the factors are at four levels, the device of 'pseudo-factors' can be used to preserve the convenience for fractionation and blocking of the 2^m series. For example, suppose that a qualitative factor has four levels $T_1, \ldots, T_4$. These can be represented by two pseudo-factors, each at two levels:

Level of qualitative factor	T_1	T_2	T_3	T_4
Pseudo-factor level	(1)	a	b	ab

In interpreting the analysis of such experiments, it needs to be remembered that AB is not an interaction.

7.3 Blocking 2^m Factorial Designs

The family of 2^m factorials can readily be divided into 2^f blocks of size 2^{m-f}, customarily with the sacrifice of information on high-level interactions.

Consider again the 2^3 experiment of the preceding section. The estimate of the three-factor interaction is

$$\hat{\beta}_{123} = \{(y_a + y_b + y_c + y_{abc}) - (y_{(1)} + y_{ab} + y_{ac} + y_{bc})\}/8. \tag{7.4}$$

If the experimental units are divided into two groups, as shown in Table 7.2(a), with treatments a, b, c and abc in one block and the remaining four treatments in the other block, then any systematic difference between the two blocks will be estimated by (7.4). The three-factor interaction is said to be *confounded* with blocks; they are both estimated by the same linear combination of the observations. Because of the orthogonality of the design, the blocking does not affect the estimates of the other effects; all are free of the effect of blocking .

It is customary to use high-order interactions for blocking. Table 7.2(b) gives a division of the 16 trials of a 2^4 experiment into two blocks of eight such that the four-factor interaction is confounded with blocks. The *defining contrast* of this design is $I = ABCD$. This means that the two blocks consist of those treatment combinations which respectively have an odd and an even number of characters in common with $ABCD$.

As a last example we divide the 32 trials of a 2^5 design into four blocks of eight. This is achieved by dividing the trials according to two defining contrasts, when their product is also a defining contrast. We choose the two three-factor interactions ABC and CDE. The third defining contrast is given by

$$I = ABC = CDE = ABC^2DE = ABDE, \tag{7.5}$$

the product of any character with itself being the identity. Table 7.2(c) gives the four blocks, which each contain all treatments with a specific combination of an odd and even number of characters in common with ABC and CDE. Given the first block, the other blocks are found by multiplication by a treatment combination that changes the parity in the first two columns. Thus multiplication of the treatments in the first block by a gives an even

TABLE 7.2. Blocking 2^m factorial experiments

(a) 2^3 in two blocks $I = ABC$

Number of symbols in common with ABC	Block number	Treatment combinations			
Odd	1	a	b	c	abc
Even	2	(1)	ab	ac	bc

(b) 2^4 in two blocks $I = ABCD$

Number of symbols in common with $ABCD$	Block number	Treatment combinations							
Odd	1	a	b	c	abc	d	abd	acd	bcd
Even	2	(1)	ab	ac	bc	ad	bd	cd	$abcd$

(c) 2^5 in four blocks $I = ABC = CDE = ABDE$

Number of symbols in common with		Block	Treatment combinations							
ABC	CDE	number								
Odd	Odd	1	c	abc	ad	bd	ae	be	cde	$abcde$
Even	Odd	2	ac	bc	d	abd	e	abe	$acde$	$bcde$
Odd	Even	3	a	b	cd	$abcd$	ce	$abce$	ade	bde
Odd	Odd	4	(1)	ab	acd	bcd	ace	bce	de	$abde$

number of symbols in common with ABC and treatment combinations ac, bc, and so on. As in (7.5), the product of any symbol with itself is the identity. In practice, the choice of which interactions to include in the defining contrast depends upon which interactions can be taken as negligible or not of interest.

7.4 2^{m-f} Fractional Factorial Designs

A disadvantage of the 2^m factorial designs is that the number of trials increases rapidly with m. As a result, very precise estimates are obtained of all parameters, including high-order interactions. If these interactions are known to be negligible, information on the main effects and lower-order interactions can be obtained more economically by running only a fraction of the complete $N = 2^m$ trials.

A half-fraction of the 2^4 design can be obtained by running one of the two blocks of Table 7.2(b). Each effect of the full 2^4 design will now be *aliased* with another effect in that they are estimated by the same linear combination of the observations. The defining contrast for this factorial in two blocks was $I = ABCD$. The two 2^{4-1} fractional factorials are generated by the relationship $I = -ABCD$ for the design given by the first block in Table 7.2(b) and $I = ABCD$ for the second. The alias structure is found by multiplication into the generator. In this case $I = ABCD$ gives the alias structure

$$\begin{array}{llll} A = BCD & B = ACD & C = ABD & D = ABC \\ AB = CD & AC = BD & AD = BC. & \end{array} \tag{7.6}$$

If the three-factor interactions are negligible, unbiased estimates of the main effects are obtained. However, (7.6) shows that the two-factor interactions are confounded in pairs. Interpretation of the results of fractional factorial experiments is often helped by the experience that interactions between important factors are more likely than interactions between factors that are not individually significant. If interpretation of the estimated coefficients remains ambiguous, further experiments may have to be undertaken. In this example the other half of the 2^{4-1}design might be performed, or perhaps one half fraction of that design.

Running one of the blocks of the 2^5 design in Table 7.2(c) gives a 2^{5-2} factorial, again with eight trials. For this quarter replicate, each effect is confounded with three others, given by multiplication into the generators of the design. Multiplication into (7.5) gives the alias structure for the main effects in the fourth fraction of Table 7.2(c) as

$$\begin{array}{ccccccc} A & = & BC & = & ACDE & = & BDE \\ B & = & AC & = & BCDE & = & ADE \\ C & = & AB & = & DE & = & ABCDE \\ D & = & ABCD & = & CE & = & ABE \\ E & = & ABCE & = & CD & = & ABD. \end{array}$$

For the 2^{5-2} design consisting of the first block of Table 7.2(c), the alias structure follows from the generators

$$I = -ABC = -CDE = ABDE,$$

giving, for example,

$$A = -BC = -ACDE = BDE,$$

which is the same structure as before, but with some signs changed.

TABLE 7.3. 2^{m-f} factorial design in six factors and $f = 3$; 'first-order design'

N	x_1	x_2	x_3	x_4 $(= x_1x_2x_3)$	x_5 $(= x_1x_2)$	x_6 $(= x_2x_3)$
1	−1	−1	−1	−1	+1	+1
2	+1	−1	−1	+1	−1	+1
3	−1	+1	−1	+1	−1	−1
4	+1	+1	−1	−1	+1	−1
5	−1	−1	+1	+1	+1	−1
6	+1	−1	+1	−1	−1	−1
7	−1	+1	+1	−1	−1	+1
8	+1	+1	+1	+1	+1	+1

For this design the shortest word amongst the generators has length three. The design is then said to be of *'resolution 3'*. In a resolution 3 design at least some main effects are aliased with two-factor interactions, but no main effects are aliased with each other. In a resolution 4 design some two-factor interactions are aliased with each other but, at worst, main effects are aliased with three-factor interactions. With several factors the choice of alias structure may, for example for resolution 3 designs, influence the number of main effects that are aliased with two-factor interactions.

An alternative method of generating a 2^{m-f} fractional factorial is to start with a factorial in $m - f$ factors and to add f additional factors. For example, the first block of the 2^{4-1} design of Table 7.2(b), for which $D = -ABC$, could be generated by imposing this relationship on a 2^3 factorial. In the alternative notation that is more convenient for most of this book, this corresponds to putting $x_4 = -x_1x_2x_3$. The levels of x_4 are then determined by the level of the three-factor interaction between the factors of the original 2^3 experiment. The second 2^{4-1} fractional design is found by putting $x_4 = x_1x_2x_3$.

A design for $m = 6$ and $f = 3$ is shown in Table 7.3. This has generators $x_4 = x_1x_2x_3, x_5 = x_1x_2$, and $x_6 = x_2x_3$. The alias structure is found by multiplying these generators together to give the full set of eight aliases, perhaps more clearly written in letters as

$$I = ABCD = ABE = BCF = CDE = ADF = ACEF = BDEF.$$

Then, for example, multiplication by the symbol A shows that the alias structure for A is

$$A = BCD = BE = ABCF = ACDE = DF = CEF = ABDEF.$$

The design is of resolution 3. In the absence of any two-factor and higher-order interactions, the estimates of the main effects are unbiased. Such designs, called main-effect plans or designs, are often used in the screening stage of an experimental programme, mentioned in §3.2. We now describe the Plackett–Burman designs that provide an extension of main-effect designs to more values of N.

7.5 Plackett–Burman Designs

A disadvantage of the method of construction of the main-effect design in Table 7.3 is that it only works when N is a power of 2. Plackett and Burman (1946) provide orthogonal designs for factors at two levels for values of N that are multiples of 4 up to $N = 100$, with the exception of the design for $N = 92$, for which see Baumert, Golomb, and Hall (1962).

The Plackett–Burman designs are mostly formed by the cyclical shifting of a generator which forms the first row of the design. The first exception given by Plackett and Burman (1946) is when $N = 28$. For $N = 12$ the generator is

$$+ + - + + + - - - + - \tag{7.7}$$

which specifies the levels of up to 11 factors. The second row of the design is, as shown in Table 7.4, found by moving (7.7) one position to the right. Continuation of the process generates 11 rows. The 12th row consists of all factors at their lowest levels. Equivalent designs are found by reversing all + and − signs and by permuting rows and columns. As usual, the design should be randomized before use, by permuting both the rows and the order in which factors are allocated to columns.

The Plackett–Burman generators include values of N that are powers of 2, such as 16 and 32, and so provide an alternative to generation of first-order designs from 2^m factorials when $N = 2^m$. The generators for $N = 16$ and 20 are

$$\begin{array}{l} + + + + - + - + + - - + - - - \\ + + + + + - + - + + - - + - + - - - -. \end{array} \tag{7.8}$$

The design for $N = 16$ generated from the key in (7.8) is obtained from the 2^4 factorial by reversing the + and − signs and permuting the rows and columns. Since the design matrices for all these designs are orthogonal,

TABLE 7.4. First-order Plackett and Burman design for up to 11 factors in 12 trials

	Factors										
Trial	1	2	3	4	5	6	7	8	9	10	11
1	+	+	−	+	+	+	−	−	−	+	−
2	−	+	+	−	+	+	+	−	−	−	+
3	+	−	+	+	−	+	+	+	−	−	−
4	−	+	−	+	+	−	+	+	+	−	−
5	−	−	+	−	+	+	−	+	+	+	−
6	−	−	−	+	−	+	+	−	+	+	+
7	+	−	−	−	+	−	+	+	−	+	+
8	+	+	−	−	−	+	−	+	+	−	+
9	+	+	+	−	−	−	+	−	+	+	−
10	−	+	+	+	−	−	−	+	−	+	+
11	+	−	+	+	+	−	−	−	+	−	+
12	−	−	−	−	−	−	−	−	−	−	−

the effect of each factor is estimated as if it were the only one in the experiment. It therefore follows that the designs are D-optimum.

The Plackett–Burman designs are limited by the requirement that N be a multiple of 4. This is not usually an important practical constraint. However, extensions of main effect plans to general N, as well as the Plackett–Burman designs, are available in SAS. See §7.7.

7.6 Composite Designs

If a second-order polynomial in m factors is to be fitted, observations have to be taken at more than two levels of each factor, as was the case in the design of Table 3.2. One possibility is to replace the two-level factorial designs of §7.2 with 3^k factorials that consist of all combinations of each factor at the levels $-1, 0,$ and 1. As m increases, such designs rapidly require an excessive number of trials. The composite designs provide another family of designs requiring appreciably fewer trials.

Composite designs consist of the points of a 2^{m-f} fractional factorial for $f \geq 0$, and $2m$ 'star' points. These star points have $m-1$ zero co-ordinates and one equal to either α or $-\alpha$. When the design region is cubic (taken to include both the square and the hypercube) $\alpha = 1$. When the design region is spherical $\alpha = m^{1/2}$. If $m = 1$ a centre point must be included. However three or four centre points are often included in the design, whatever the value of m, giving 'central composite designs'. These centre points provide, as they do when used to augment 2^m factorials, an estimate of the error variance σ^2 based solely on replication. They also provide a test for lack of fit. If there is evidence of lack of fit, one possibility is to consider fitting a third-order polynomial to the results of further observations. However, it is often preferable to investigate transformation of the response, which frequently leads to readily interpretable models with a reduced number of parameters. An example is given in §8.3. Transformation of the explanatory variables, for example from dose to logdose, sometimes also leads to simpler models.

Central composite designs are widely used in the exploration of response surfaces around the neighbourhood of maxima and minima. The exact value of α and the number of centre points depend upon the design criterion. Box and Draper (1987, p. 512) give a table of values of those design characteristics that yield rotatable central composite designs. The resulting values of α are close to $m^{1/2}$ over the range $2 \leq m \leq 8$. The number of centre points in the absence of blocking is in the range 2–4. The effect of the centre points is to decrease the efficiency of the designs as measured by D-optimality. A further distinction with D-optimum designs is that the Box and Draper designs are shrunk away from the edges of the experimental region in order to reduce the effect of bias from higher-order terms omitted from the model. This protection is bought at the cost of reduced efficiency as measured by D- or G-optimality; D-optimum designs for linear models span the experimental region.

To conclude this section we give, in Table 7.5, an example of a five-factor central composite design for a cubic region. The design includes the points of a 2^{5-1} fractional factorial, $2m$ star points at the centres of the faces of the cube that would be formed by the points of the full 2^5 factorial and four centre points. The total number of trials is $N = 2^{5-1} + 2 \times 5 + 4 = 30$. This number is great enough to allow estimation of the 21 parameters of the second-order model

$$\mathrm{E}(y) = \beta_0 + \sum_{i=1}^{5} \beta_i x_i + \sum_{i=1}^{4} \sum_{j=i+1}^{5} \beta_{ij} x_i x_j + \sum_{i=1}^{5} \beta_{ii} x_i^2. \tag{7.9}$$

The generator for the fractional factorial part of the design was taken as $x_5 = x_1 x_2 x_3 x_4$. In practice, the 30 trials of Table 7.5 should be run in random order.

Central composite designs such as that of Table 7.5 have a simple geometric structure. However, they frequently have high D-efficiency only for

TABLE 7.5. Central composite design based on a 2^{5-1} fractional factorial ($m = 5, f = 1$): cubic region, four centre points

Trial number	x_1	x_2	x_3	x_4	x_5
1	−1	−1	−1	−1	−1
2	+1	−1	−1	−1	+1
3	−1	+1	−1	−1	+1
4	+1	+1	−1	−1	−1
5	−1	−1	+1	−1	+1
6	+1	−1	+1	−1	−1
7	−1	+1	+1	−1	−1
8	+1	+1	+1	−1	+1
9	−1	−1	−1	+1	+1
10	+1	−1	−1	+1	−1
11	−1	+1	−1	+1	−1
12	+1	+1	−1	+1	+1
13	−1	−1	+1	+1	−1
14	+1	−1	+1	+1	+1
15	−1	+1	+1	+1	+1
16	+1	+1	+1	+1	−1
17	−1	0	0	0	0
18	+1	0	0	0	0
19	0	−1	0	0	0
20	0	+1	0	0	0
21	0	0	−1	0	0
22	0	0	+1	0	0
23	0	0	0	−1	0
24	0	0	0	+1	0
25	0	0	0	0	−1
26	0	0	0	0	+1
27	0	0	0	0	0
28	0	0	0	0	0
29	0	0	0	0	0
30	0	0	0	0	0

models with a regular structure such as (7.9). The designs also exist for only a few, relatively large values of N. Smaller alternatives have been proposed. The *small composite designs* of Hartley (1959) and Draper and Lin (1990) have the same general structure as central composite designs, but replace the factorial portion with a smaller 2^k fraction in which only the two-factor interactions may be estimated. The *hybrid designs* of Roquemore (1976) combine a central composite design in $k-1$ factors with values of the kth factor chosen to make the design rotatable. Both small composite designs and hybrid designs are available in SAS.

If designs are required for models such as (7.9) but for a different value of N, they need to be generated using the algorithms described in Chapter 12, the SAS implementations of which are described in Chapter 13. Algorithmic methods will also be needed if the model does not have the regular structure of (7.9), but contains terms of differing orders in the various factors, as in (4.19). Algorithmic methods are also required when the design region is of an irregular shape, as in Example 12.2.

7.7 Standard Designs in SAS

The primary tool for constructing standard designs in SAS is the ADX Interface. ADX employs various underlying SAS tools to create designs, one of the most important being the FACTEX procedure for regular q^k fractional designs. Finally, the MkTex macro described by Kuhfeld and Tobias (2005) can construct a very large variety of *orthogonal arrays*, which are useful as designs themselves or as candidate sets for optimum designs.

7.7.1 Creating Standard Designs in the ADX Interface

The ADX Interface can construct all of the standard designs discussed in §§7.2 through 7.6. Designs are categorized under broad headings that correspond to the goals of the types of experiments for which they are appropriate. These headings and the standard designs available under each one are listed below.

- Two-level designs
  - Regular 2^{m-f} full and fractional designs, with and without blocks, as discussed in §7.2–§7.4.
  - Plackett–Burman designs based on Hadamard matrices, as discussed in §7.5.

- ◦ Regular split-plot full and fractional designs; see Bingham and Sitter (2001) and Huang, Chen, and Voelkel (1998). These designs are appropriate when there are restrictions on how the factors can change from plot to plot, as often happens, for example, when the experimental material is the product of an industrial process.

- Response surface designs
  - ◦ Central composite, both orthogonal and uniform precision; small composite designs; and hybrid designs, as discussed in §7.6.
  - ◦ Box–Behnken designs (Box and Behnken 1960) - fractions of 3^k designs based on balanced incomplete block designs. These designs are favoured for the relative ease with which they may be implemented and for their symmetry, but are not particularly efficient for either estimation or prediction.

- Mixture designs
  - ◦ Simplex centroid and simplex lattice designs, as discussed in Chapter 16.

- Mixed level factorial designs
  - ◦ Full factorials.
  - ◦ Orthogonal arrays providing fractional designs for mixed level factors, as discussed in §7.7.3.

In order to construct a design of a specific type with the ADX Interface, under the 'File' menu select the option to create a design of the appropriate general class. The resulting screen shows an (initially empty) outline of the tasks involved in designing and analysis an experiment (Figure 7.1). Click on 'Select Design . . .', set the number of factors (or any other design parameter) to the desired values, and then choose from the list of designs displayed (Figure 7.2). When you leave the design selection screen, the constructed design (unrandomized) will be displayed in the left-hand part of the design screen.

SAS Task 7.1. Use the ADX Interface to create the following designs:

1. a 2^3 full factorial;
2. a 2^5 full factorial in four blocks;
3. a 2^{5-2} fractional factorial first-order design;
4. a 2^{5-1} fractional factorial second-order design;
5. the central composite design of Table 7.5.

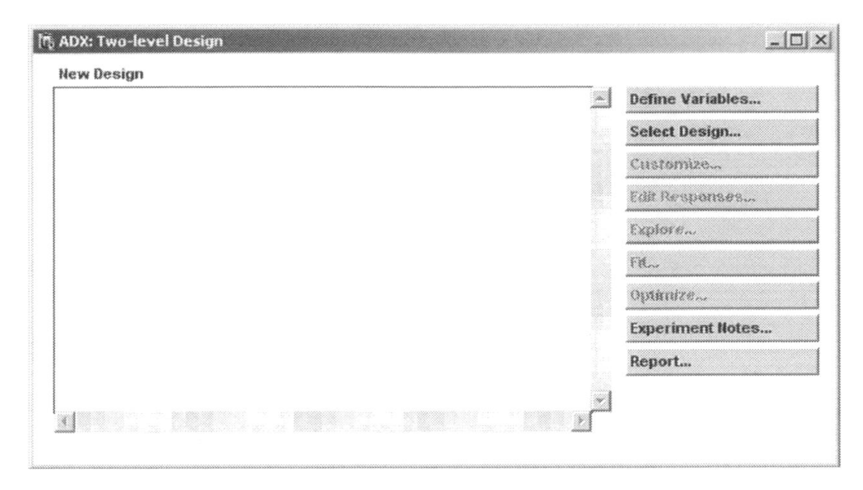

FIG. 7.1. ADX Interface: new design screen.

ADX: Two-Level Design Specifications

Design List Options

Number of factors: 7

Design Details...

Show designs of type

☑ Fractional factorial designs

☑ Plackett-Burman designs

☑ Show blocked designs

Factors	Runs	Type	Resolution: Estimable Effects	Blocks	Block Size
7	12	Plackett-Burman	3:Main Effects Only	2	6
7	24	Plackett-Burman	4:Some 2FI	4	6
7	16	1/8 Fraction	4:Some 2FI	2	8
7	16	1/8 Fraction	4:Some 2FI	4	4
7	16	1/8 Fraction	4:Some 2FI	8	2
7	32	1/4 Fraction	4:Some 2FI	2	16
7	32	1/4 Fraction	4:Some 2FI	4	8
7	32	1/4 Fraction	4:Some 2FI	8	4
7	32	1/4 Fraction	4:Some 2FI	16	2
7	64	1/2 Fraction	6:All 2FI	2	32

FIG. 7.2. ADX Interface: design selection screen.

Note that the central composite design contains more than four centre points by default, and the axial value is set to 2. In order to reproduce the design of Table 7.5, you will need to change the axial scaling on the design selection screen, and the number of centre points on the design customization screen.

7.7.2 The FACTEX Procedure

The algebra for constructing blocked and fractional 2^m designs discussed in §§7.2 and 7.3 can be extended to designs with factors all at q levels, where q is the power of a prime number $(2, 3, 4, 5, 7, \ldots)$. The mathematical characterization of any such design is that it is a linear subspace of the space of all m-dimensional vectors with elements from the finite field of size q. Their statistical feature is that any two effects are either orthogonal to one another or are completely confounded. The FACTEX procedure constructs general q^m designs according to given specifications for characteristics of the final design—namely, which effects should be estimable in the final design, and which other effects they should not be confounded with. The key computational component of FACTEX is an implementation of an algorithm similar to that of Franklin and Bailey (1985) for searching for the design's generating relations. The ADX Interface uses FACTEX to create fractional and blocked two-level designs and the two-level components of central composite designs.

While the designs constructed by FACTEX are G-optimum—and hence D- and A-optimum—for the estimable effects, they are motivated by other concerns than information optimality. Therefore, PROC FACTEX will not often be an important tool for the designs we discuss in this book. An exception is in constructing optimum designs for large numbers of factors. As we shall see, a fundamental task in practical optimum design is searching a finite candidate set of potential design points for an optimum selection of points. When the number of factors m and their potential levels q is not too large, we can simply use the set of all q^m points as a candidate set. But as m increases, a search over all q^m candidates becomes infeasible. One technique in such cases is to use an appropriate q^{m-n} fraction as the candidate set.

7.7.3 The MkTex Macro

Regular q^m designs as discussed in the last section are *orthogonal arrays*, meaning that every subset of λ columns constitutes a full factorial array, for some number λ, called the *strength* of the array. There are many other types of orthogonal arrays than just regular q^m designs, not necessarily involving factors with a prime power number of levels, nor even factors that all have the same number of levels. Orthogonal arrays are useful standard designs, although not typically for the kinds of experiments discussed in this book. The MkTex macro, described in Kuhfeld and Tobias (2005), collects a very large variety of construction methods for orthogonal arrays. MkTex incorporates nearly 600 different recipes for orthogonal arrays, resulting in about 700 different 'parent' designs. From these parents, it constructs over 100,000 different orthogonal arrays. The FACTEX procedure described in the previous section is involved in many of these constructions. Moreover, MkTeX also uses D-optimality to construct *near-orthogonal* arrays when no truly orthogonal array exists for given factors and design size. Although MkTeX's results are not typically the sort of design discussed in this book, the underlying techniques for constructing them are the same as those discussed in Chapter 12.

7.8 Further Reading

The study of fractional factorial designs started with Finney (1945). A detailed treatment of the 2^{m-f} fractional factorial designs of §7.4 is given by Box and Hunter (1961*a*, *b*). The use of these and other fractional designs in screening raises questions of the properties of the designs under various assumptions about the number of active factors and their relationship to the aliasing structure. See, for example, Tsai, Gilmour, and Mead (2000), Cheng and Tang (2005) and, for these and other aspects of factorial designs, Mukerjee and Wu (2006). Some of the papers in Dean and Lewis (2006) cover more general aspects of screening. Wu and Hamada (2000) includes examples of the use of Plackett and Burman designs.

8

THE ANALYSIS OF EXPERIMENTS

8.1 Introduction

This book is primarily concerned with the design of experiments rather than with their analysis. We have typically assumed a linear model for uncorrelated observations of constant variance

$$\mathrm{E}(y) = F\beta, \quad \mathrm{var}(y) = \sigma^2 I \tag{8.1}$$

and have been concerned with designs that optimize aspects of a least squares fit for this model. However, in this chapter we use several sets of data in order to illustrate other components of a complete analysis for a typical experiment.

The examples are as follows:

- The data on carbon monoxide production (Example 1.1) is used to discuss outlier detection, testing for lack of fit, and testing for specific values of linear coefficients.
- Derringer's elastomer data (Example 1.2) illustrates the importance of transformations of the response.
- A saturated 2^{5-1} fractional factorial experiment allows us to demonstrate how significant effects are detected when there are few or no degrees of freedom for estimating the nominal level of error.

In this chapter we are concerned with least squares regression analysis, which is the analytic methodology assumed for most of the optimum designs that we develop. Least squares makes some very particular and possibly very stringent assumptions about the hypothetical random mechanism that generates the data. We will discuss diagnostics for how well the data conform to these assumptions, as well as some possible remedies for assumption violations that still allow a simple least squares analysis to be performed. It should be noted, however, that often the best remedy for violations of the standard assumptions of simple least squares is to fit a more complicated model, either a *generalized linear model* when the response is intrinsically non-normal or a fully fledged *non-linear model*. These approaches are discussed in Chapters 22 and 17 respectively.

8.2 Example 1.1 Revisited: The Desorption of Carbon Monoxide

The standard linear model is a highly idealized description of the structure of any set of data. It is almost always an approximation at best, though it may be a useful one. In view of this, the goal of checking assumptions is not to prove or disprove conclusively whether the model is correct, but to detect departures from the model which may limit its utility. It should be noted, however, that discovering departures from standard assumptions—an unexpected interaction, for example—may prove to be the most valuable results of an experiment.

Often the best way to detect violations of assumptions is simply to look at the data, that is, to compute statistics that measure anomalous features and to display them graphically. A simple measure of each observation's deviation from the model is the least squares residual for that observation, e_i (5.25). Just plotting residuals against the run order can indicate individual outliers, whereas plotting them against factors or predicted responses can reveal trends that indicate inadequacy in the model.

Consider the data on carbon monoxide production given in Table 1.1, but with a slight change—namely, adding 1 to the response for the seventh run, making it 1.95 instead of 0.95. The scatter plot in Figure 1.1 seems to show a clear linear relationship between CO desorption and K/C ratio. The simple linear regression model (5.5) is thus a sensible starting point for modelling these data. Figure 8.1 shows the residuals plotted against the fitted values $\hat{y}_i$ for a simple linear model in K/C ratio with these perturbed CO data. Note that the residual for the run that we changed is a good deal larger than the rest. Having detected this possible problem, the experimenter studies it in detail and decides whether to keep the observation, or delete it, or try to correct it.

Since least squares residuals do not all have the same variance, comparisons between them may be misleading. Although extreme differences between these variances are rare in designed experiments, residuals are often standardized to have the same variance. If we rewrite (5.24) as

$$\mathrm{var}\{\hat{y}(x_i)\} = \sigma^2 h_i,$$

where h_i is the ith diagonal element of the hat matrix H (5.26) then

$$\mathrm{var}(e_i) = \sigma^2(1 - h_i)$$

and the studentized residuals

$$t_i = \frac{e_i}{s\sqrt{1 - h_i}}$$

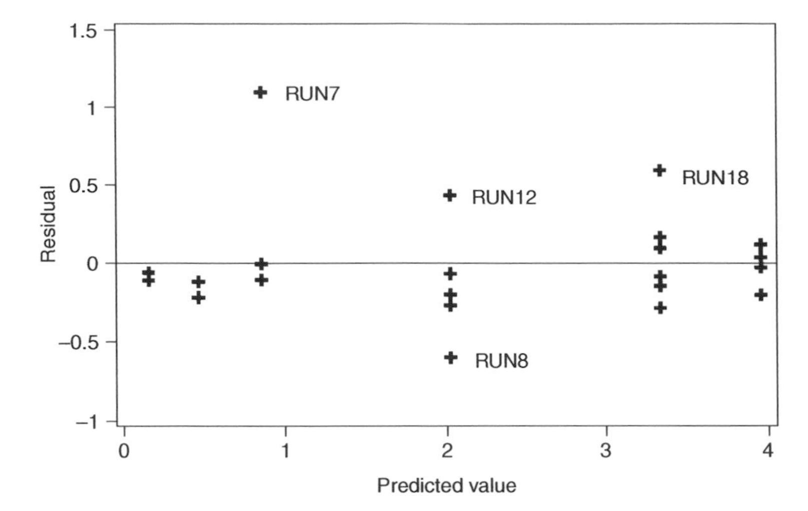

FIG. 8.1. The desorption of carbon monoxide: least squares residuals against $\hat{y}_i$, perturbed data.

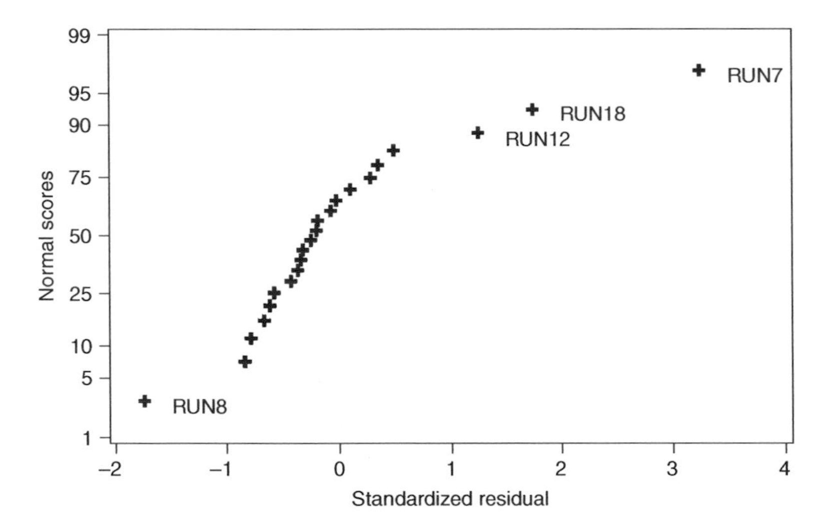

FIG. 8.2. The desorption of carbon monoxide: normal plot of studentized residuals, perturbed data.

do all have the same variance. Figure 8.2 is a normal plot of studentized residuals for the perturbed CO data. The points on this plot should all lie roughly on a straight line if the residuals represent a random normal

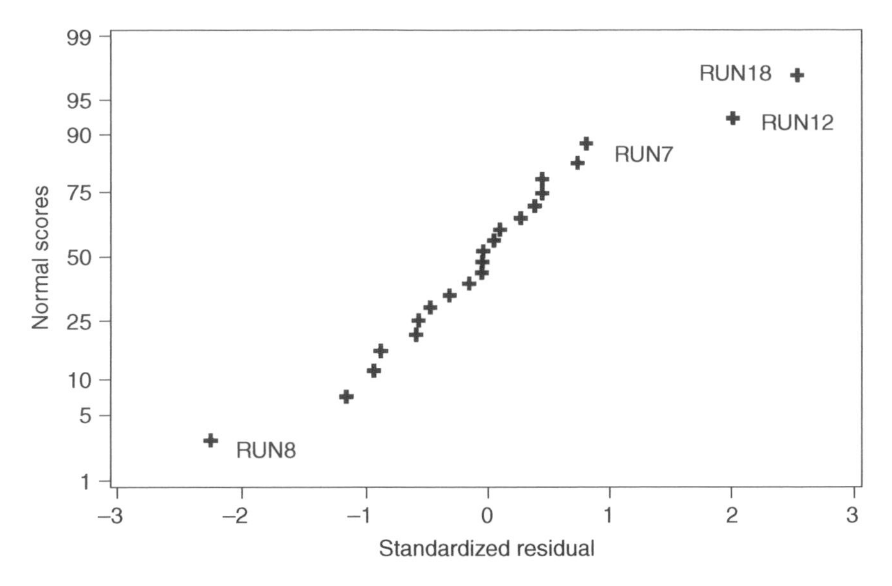

FIG. 8.3. The desorption of carbon monoxide: normal plot of studentized residuals, corrected data.

sample. Clearly they do not in this case: most of the points do seem to line up, but several do not lie on this line, with the run we changed being again the most deviant. Figure 8.3 presents the same normal plot, but with the response value for the seventh run changed back to the original 0.95. Now the seventh run follows the trend of the rest of the data and there seems to be no appreciable departure from a line. Runs 8, 12, and 18 are extreme, but it is unclear whether they are uncharacteristic of the rest of the observations. This is a problem with interpreting normal plots: it can be difficult to decide whether a particular plot is sufficiently 'unstraight' to indicate departures from the model. If this is important, simulations of normal data can be used to build up an envelope providing bounds within which the plot should lie. Numerous examples are given by Atkinson (1985). In the present example it is not crucial to establish whether runs 8, 12, and 18 are outliers. The overall relationship between response and explanatory variable is unambiguously established. The major effect of deleting these three observations is to reduce the estimate of the nominal level of noise σ^2 and thus to give slightly smaller confidence intervals for the parameter estimates. However, for these data, too much attention should not be given to the fine structure of the residuals; the numerical values of Table 1.1 were extracted manually from a plot in the original paper and so are subject to non-random patterns in the last digit.

TABLE 8.1. The desorption of carbon monoxide: analysis of variance

Source	DF	Sum of squares	Mean square	F	Pr > F
Model	1	42.6921	42.6921	697.6139	<.0001
Error	20	1.2239	0.0612		
Lack of fit	4	0.0998	0.0250	0.3552	0.8366
Pure error	16	1.1241	0.0703		
Total	21	43.9160			

Continuing with the standard analysis for the original CO data (with the outlying response corrected), the first row of Table 8.1 gives the standard ANOVA F test for the K/C ratio coefficient. The p-value of <0.0001 for the term indicates overwhelming evidence of a relationship between carbon monoxide production and the initial K/C ratio. Is a linear relationship sufficient? When there is replication in the data, this question can be answered by comparing the model error to the 'pure error'. In this case, there are 22 observations at only six distinct values of K/C ratio, giving a pure error estimate of σ^2 with $22-6=16$ degrees of freedom. Fitting the linear model absorbs two degrees of freedom, so that four degrees of freedom are available for testing the goodness of fit of the linear model. The resulting analysis of variance is also given in Table 8.1. The results—in particular, the p-value for lack of fit of 0.8367—indicate no evidence of systematic lack of fit of the model.

While the lack of fit test provides a useful confirmation in this case, in many response surface experiments graphical examination of residuals is actually better for judging lack of fit than the standard replication-based F test. There are several reasons for this. For one thing, residuals are available even if there is no replication in the design. Also, the shape of the plot may indicate why the model fails to fit well, and what can be done to improve it. Finally, a traditional F test for lack of fit may not even address the question of interest. Any linear model is realistically only an approximation to the true mechanism generating the data. The important question is, is this approximation close enough to serve the experimenter's purposes, whether that be to identify important factors or to find the region of optimal response? On the one hand, with enough data, an F test will flag significant lack of fit for the approximation, regardless of its practical adequacy. On the other hand, the F test will often fail to detect sizeable lack of fit only because of insufficient data. It is much better to make a habit of examining graphical displays of the characteristics of a fit rather than always relying on simple summaries to detect problems.

TABLE 8.2. The desorption of carbon monoxide: estimated coefficients

Parameter	Estimate	Standard error	t Value	Pr > $\|t\|$
Intercept	-0.0380	0.1004	-0.38	0.7086
K/C Ratio	1.6031	0.0607	26.41	<0.0001

Table 8.2 gives the coefficients for the estimated line, both the intercept and the slope relative to the K/C ratio. If all assumptions of the linear model seem valid, the least squares analysis typically ends here. These fitted coefficients are taken to be an adequate description of this relationship. Further analysis would proceed by studying this fitted model itself to determine, for example, an optimal level of the K/C ratio. However, there are some special features of this chemical system which make further least squares analysis interesting.

Finally, it might be expected that carbon monoxide will not be desorbed when the K/C ratio is zero, so that the intercept β_0 in (5.5) should be zero. The t-test of Table 8.2 does indeed indicate that this is so. The model then becomes

$$\eta(x, \beta) = \beta_1 x, \tag{8.2}$$

regression through the origin. However, there is a further simplification. The postulated mechanism for the reaction suggests that $\beta_1 = 1.5$. For the fitted model (8.2), $\hat{\beta}_1 = 1.584$ with an estimated standard error of 0.0312. Comparing $(\hat{\beta}_1 - 1.5)/0.0312$ to a t distribution with 21 degrees of freedom yields a p-value of 0.014. This constitutes statistically significant evidence that the true value is not 1.5.

The concluding analysis of the data would involve repeating the residual plot of Figure 8.3 for this reduced model. No new points arise. But there are a few general comments on the design of this experiment. If the purpose is solely to estimate the slope of the line, assuming an intercept of zero, then the design minimizing the variance of $\hat{\beta}_1$ puts all trials at the maximum value of x. If it is required to check the straight line model against the quadratic alternative

$$\eta(x, \beta) = \beta_1 x + \beta_2 x^2,$$

then two values of x are required. This design is discussed as Example 9.1 in Chapter 9.

TABLE 8.3. Viscosity of elastomer blends: significance of second-order terms

Contrast	DF	Sum of squares	Mean square	F	Pr > F
Second-order model	3	3620	1207	102.25	<.0001

8.3 Example 1.2 Revisited: The Viscosity of Elastomer Blends

The analysis of a slightly more complicated data set, the data on the viscosity of elastomer blends from Table 1.2, illustrates other diagnostics for the underlying assumptions of a linear model. The data set gives the viscosity of elastomer-filler blends for three kinds of fillers, as a function of two quantitative factors, the amount of naphthenic oil and the amount of filler. For illustration, this analysis concentrates on the 23 results for filler B, which constitute a 4×6 factorial with one missing observation. The interesting feature of the analysis is that a more powerful analysis of the data is obtained by working with the logarithm of the response rather with the viscosity values as originally measured.

We begin with a first-order model

$$\eta(x) = \beta_0 + \beta_1 x_1 + \beta_2 x_2, \tag{8.3}$$

for the measured viscosity values, where x_1 is the filler level and x_2 is the level of naphthenic oil. Plots of the data and of the residuals, as discussed in §8.2, indicate fairly severe departures from the assumptions of the linear model. In particular, Figure 8.4 shows that while there is a clear relationship between viscosity and filler level, the relationship appears to be slightly curved, with the variance increasing as both the filler level and the average viscosity increase. The plot of least squares residuals against fitted values for this model (Figure 8.5) shows that something more is needed than a first-order model with constant error variance. Finally, the normal plot of the standardized residuals does not lie on a single straight line (Figure 8.6). Fitting a full second-order model

$$\eta(x) = \beta_0 + \beta_1 x_1 + \beta_2 x_2 + \beta_{11} x_1^2 + \beta_{22} x_2^2 + \beta_{12} x_1 x_2, \tag{8.4}$$

alleviates these problems to some degree; the model fits the data appreciably better, reducing the residual sum of squares to 201, as opposed to 3,821 for the first-order model. The significance of this reduction is calculated in Table 8.3, giving a highly significant F value of 102. However, the points in a normal plot for the standardized residuals still do not approximate a simple straight line (Figure 8.7), indicating that the least squares assumption of identically distributed normal errors may be violated.

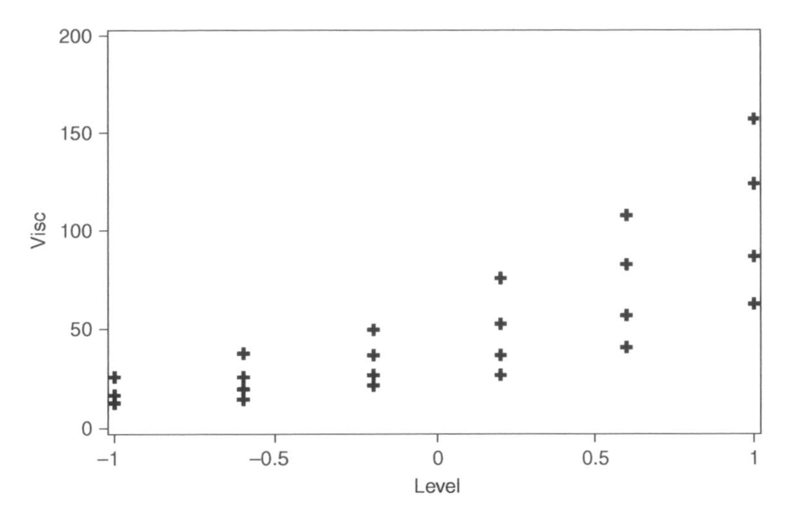

FIG. 8.4. Viscosity of elastomer blends: viscosity against amount of filler (scaled units) for filler B.

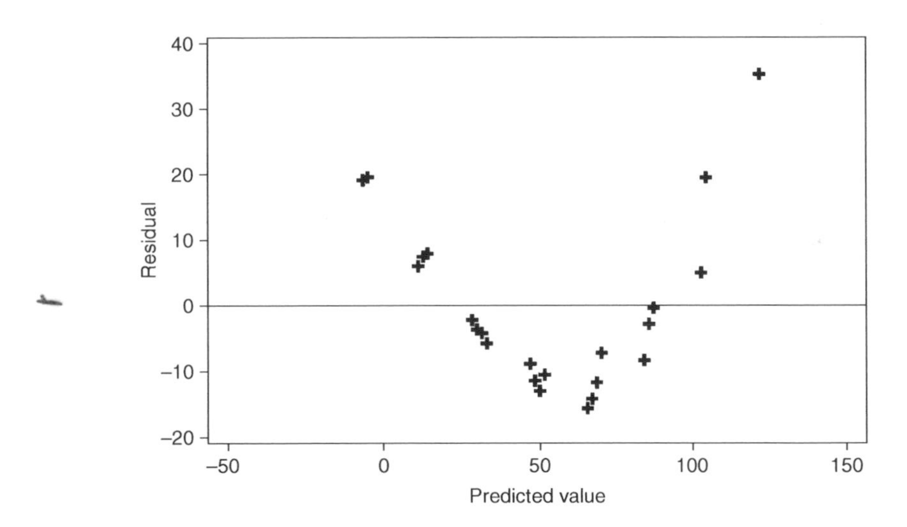

FIG. 8.5. Viscosity of elastomer blends: least squares residuals against fitted values for the first-order model (8.3).

Notice that the observed values of viscosity are necessarily non-negative and have a ratio of just over 10:1 between the smallest and largest values. For data such as these, it is a good idea to explore whether a transformation

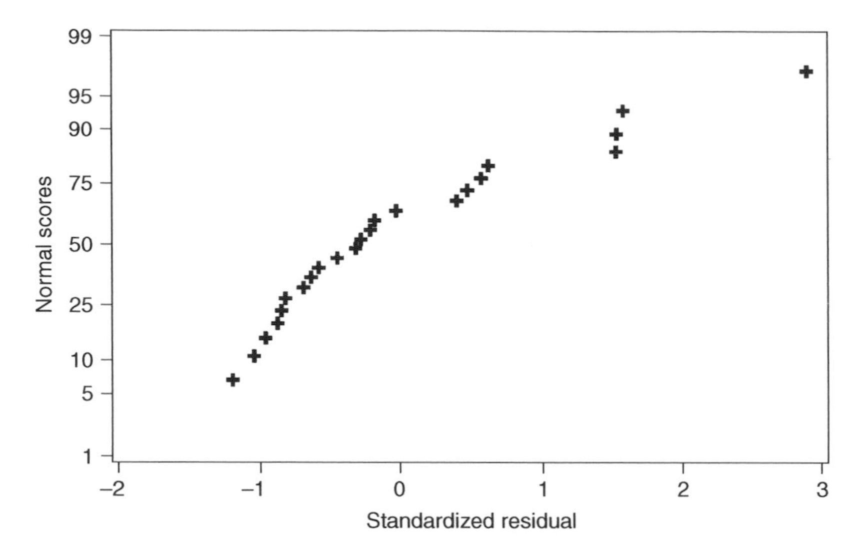

FIG. 8.6. Viscosity of elastomer blends: normal plot of residuals from first-order model (8.3).

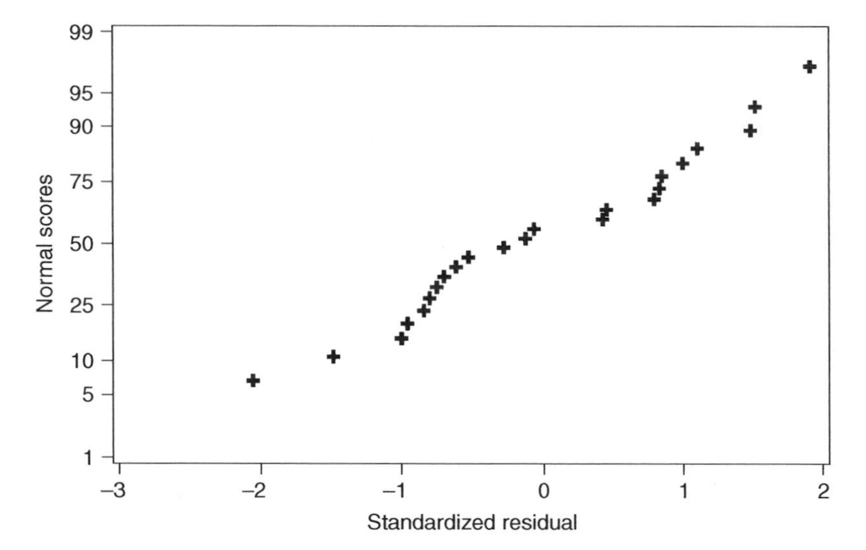

FIG. 8.7. Viscosity of elastomer blends: normal plot of residuals from second-order model (8.4). Despite the improved fit, the plot is still not straight.

of the response might conform better to the assumptions of a least squares analysis than the original values. Box and Cox (1964) defined a family of transformations that is sufficiently rich for most data and yet is parameterized by a single variable λ. The Box–Cox transformation of y for given λ is defined as

$$y(\lambda) = \begin{cases} (y^\lambda - 1)/\lambda & \lambda \neq 0 \\ \log y & \lambda = 0, \end{cases} \tag{8.5}$$

The model is that some Box–Cox transformation of the original response follows the usual assumptions of a linear model. Formally this is a non-linear model, but the non-linearity has only one parameter (λ); the usual approach is to determine a likely value for λ and then to proceed with a least squares analysis of this transformed value. Likely values of λ can be found using a maximum likelihood analysis, which takes a simple and intuitive form. The marginal log-likelihood for λ is proportional to the logarithm of the residual sum of squares (RSS) for least squares regression with the normalized form of the Box–Cox transformation

$$z(\lambda) = \begin{cases} \frac{y^\lambda - 1}{\lambda \dot{y}^{\lambda - 1}} & (\lambda \neq 0) \\ \dot{y} \log y & (\lambda = 0) \end{cases},$$

where $\dot{y}$ is the geometric mean of the original y values. The value of λ which corresponds to the minimum RSS for $z(\lambda)$ is the maximum likelihood estimate. Moreover, the values of λ for which the log-likelihood $-\frac{1}{2} N \log \hat{\sigma}^2(\lambda)$ is within $\chi^2_1(0.95) = 1.92$ of this optimum constitute an approximate 95% confidence interval for λ. Figure 8.8 shows the log-likelihood for λ between -2 and 2 for the viscosity data with the first-order model (8.3), along with a confidence interval for the true value. The maximum likelihood estimate is $\lambda = 0.05$, but the confidence interval contains the much more intuitively appealing value of $\lambda = 0$, corresponding to a logarithmic transformation of viscosity. Indeed, viscosity measurements are frequently log transformed for the purposes of analysis. Further discussion of Box–Cox transformation is in §23.2.2.

Moreover, if the second-order model (8.4) is fitted to log(Viscosity), it is found that the second-order terms are actually not needed. Table 8.4 gives t values for all coefficients of the second-order model for both Viscosity and log(Viscosity). The difference is amazing. By taking logarithms, a simple additive model has been obtained which, in addition, satisfies the constraint that the response must be non-negative. Further evidence for the desirability of the transformation comes from plots of residuals against fitted values, similar to Figure 8.5, which are not shown here. That for log(Viscosity) is structure free, confirming that all information about the data is in the fitted linear model and the random scatter about that model.

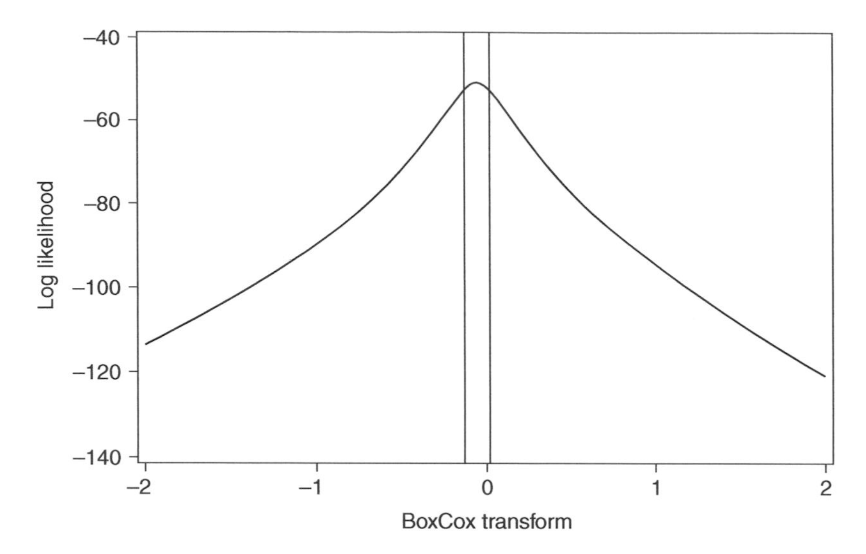

FIG. 8.8. Viscosity of elastomer blends: Box–Cox analysis for first-order model (8.3).

TABLE 8.4. Viscosity of elastomer blends: t values for coefficients of full second-order model

Parameter	Viscosity	log Viscosity
Intercept	25.52	171.05
x_1	38.26	55.98
x_2	−22.05	−31.42
$x_1{*}x_1$	12.64	1.43
$x_2{*}x_2$	3.78	1.04
$x_1{*}x_2$	−14.07	−0.72

This analysis concludes our discussion of the data of Table 1.2. We have only looked at a part of the example, that to do with the second filler. A more complete analysis of the entire data set would investigate analyses for the other two fillers, preferably leading to a single model with the same transformation for all three fillers. Any inferences, such as confidence intervals, should be made on the transformed data, although they may be then back-transformed to the original scale for ease of interpretation.

The design of experiments when it is desired to estimate the transformation parameter λ is investigated in §23.3.

8.4 Selecting Effects in Saturated Fractional Factorial Designs

This section illustrates the analysis of a 2^{5-1} fractional factorial design. In the analysis of the data on the production of carbon monoxide in §8.2 replicate observations provided an estimate of the error variance σ^2. In the analysis of the elastomer data of §8.3, the residual sum of squares from the model served the same purpose. However, fractional factorial designs often have very few runs relative to the number of potential main effects and interactions. As a result, after estimating these effects, there are no degrees of freedom available for estimating the error, and thus no way to discern whether an estimated effect is larger than the nominal level of noise. Unless there is an external estimate of error, progress is only possible if some of the effects are assumed to be negligible, allowing the sum of their squares to be pooled and used as an estimate of error. In this section we illustrate the use of various tools for assessing such negligibility.

The example is taken from Box, Hunter, and Hunter (2005) and is used by them to introduce the idea of a fractional factorial design. Data from a full 2^5 design are available. However, to make the point that little is lost by using only a half-replicate of the full factorial, Box *et al.* analyse the half of the data comprising the 2^{5-1} factorial given in Table 8.5.

The formulae for estimating the parameters of linear models fitted to 2^m factorials and their fractions are given in §7.2. It is often assumed that third- and higher-order interactions are negligible, and for sufficiently large factorials the sums of squares corresponding to these degrees of freedom can be pooled to provide an estimate of error. However, the 2^{5-1} design is *saturated* for the second-order model—that is, all 16 degrees of freedom are used up by estimating the over-all mean, the 5 main effects, and the 10 two-factor interactions. We briefly survey four different methods for determining which effects are estimating error in saturated designs.

All of the methods discussed in the remainder of this section depend on the fact that the estimates of the negligible effects should look like a random sample from a normal distribution, and the real effects like outliers from this distribution. In §8.2 we discussed the use of a normal plot of the residuals for detecting outliers. A normal plot of the effects can be used in the same way. Any appreciably non-zero parameter should have an estimate which does not follow the null distribution and which will be revealed by the plot as an outlier. The remaining estimates which follow the null distribution can be dropped from the model, adding their effect to the estimate of residual variance. Figure 8.9 shows the normal plot of the estimates for the reactor data. The plot shows a relatively straight section, corresponding to effects which can be assumed to be negligible. The slope of the line through these points corresponds to the estimate of error based

TABLE 8.5. Reactor data: a 2^{5-1} fractional factorial

x_1	x_2	x_3	x_4	x_5	Response (% reacted) y
−	−	−	−	+	56
+	−	−	−	−	53
−	+	−	−	−	63
+	+	−	−	+	65
−	−	+	−	−	53
+	−	+	−	+	55
−	+	+	−	+	67
+	+	+	−	−	61
−	−	−	+	−	69
+	−	−	+	+	45
−	+	−	+	+	78
+	+	−	+	−	93
−	−	+	+	+	49
+	−	+	+	−	60
−	+	+	+	−	95
+	+	+	+	+	82

Factor	−	+
x_1 feed rate (1/min)	10	15
x_2 catalyst (%)	1	2
x_3 agitation rate (rev/min)	100	200
x_4 temperature (°C)	140	180
x_5 concentration (%)	3	6

on the corresponding effects. There are also five effects that do not fall near this line. In order of magnitude these are the main effects of factors 2 and 4, their interaction, the interaction of factors 4 and 5, and the main effect of factor 5. Thus the normal plot reveals three reaction factors, together with two of their interactions, which seem to be important for explaining the reaction rate in this experiment, over and above background noise.

Normal plots are interpreted by deciding which points fit a straight line. In Figure 8.9 there is some ambiguity, when drawing a line through most of the points, whether the main effect of factor 5 and its interaction with factor 4 should be included; there is some ambiguity as

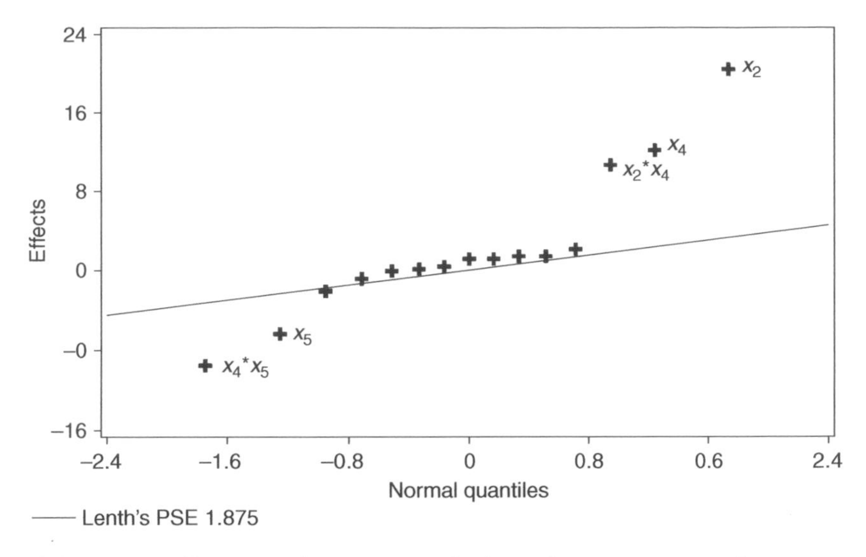

FIG. 8.9. Reactor data: normal plot of parameter estimates.

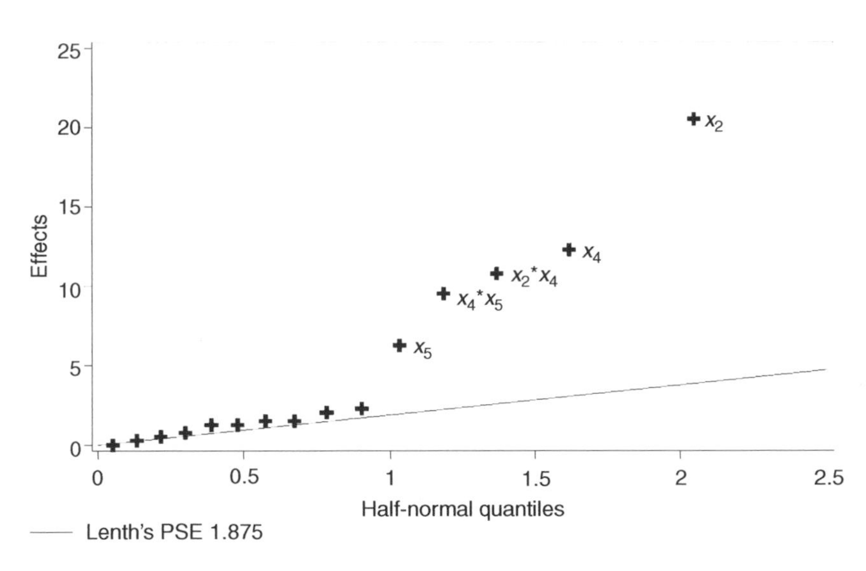

FIG. 8.10. Reactor data: half-normal plot of parameter estimates; as Figure 8.9 but absolute values.

to whether these effects are real. For small designs, a clearer idea of which part of the plot is straight is sometimes obtained from a half-normal plot such as Figure 8.10. A potential disadvantage of plotting the absolute values of the parameter estimates is that information on the signs is lost. The half-normal plot seems to show even more clearly that

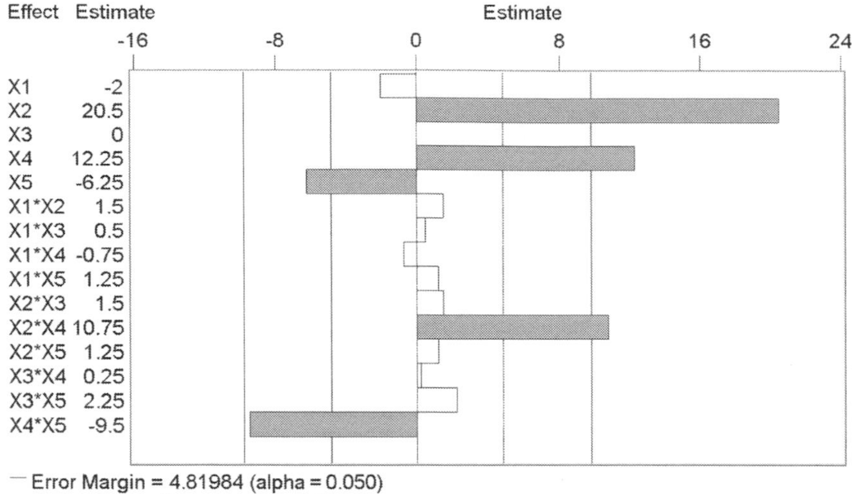

FIG. 8.11. Reactor data: Lenth plot of parameter estimates.

the effects 5 and 45 do not follow the same distribution as the smaller effects.

In the absence of an estimate of error, using normal plots to select effects relies on a subjective evaluation of the line that fits most effects, which corresponds to their variance. The two methods that we discuss next replace this subjectivity with objective measures, one with a traditional significance test flavour and the other with a Bayesian interpretation.

The method proposed by Lenth (1989) applies robust methods to estimate the variance for the majority of the effects, the so-called *pseudo standard error* (PSE). First, let M_β be the the median of the absolute values of all the estimated effects. Lenth suggests using $1.5 \times M_\beta$ as a preliminary estimate of the standard error. In the second stage only the effects within 2.5 times the preliminary estimate are retained. The PSE is now similarly computed as 1.5 times the median of this trimmed sample of the absolute effects. The PSE is used to compute bounds for deciding which effects are significant. Figure 8.11 shows a plot of the effects for the reactor experiment, including Lenth's bounds, with the largest effects highlighted. Lenth's bounds come in two varieties: one pair at $|\beta_i| = 4.8$ calibrated to control Type 1 inference errors in an individual sense; and another, more conservative pair at $|\beta_i| = 9.8$ which control for errors in an over-all, experiment-wise sense. Notice that the ambiguous effects, 5 and 45, are inferred to be significant in an individual sense, but their values are not larger than can be expected

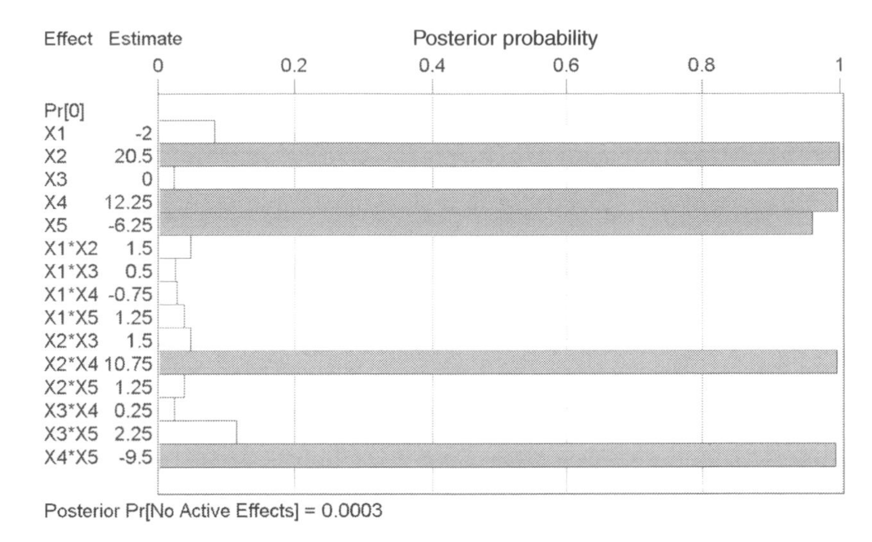

FIG. 8.12. Reactor data: Bayes effect plot of parameter estimates.

from the maximum of 15 such effects. However, they are not so extreme, being the fourth and fifth largest values of absolute effects.

Finally, Box and Meyer (1986) propose an approach to analysing saturated designs that takes a prior probability for the proportion of non-null, or 'active' effects which are viewed as a sample from a distribution with a higher variance than that for the null effects. A Bayesian interpretation is used to produce posterior probabilities for whether each effect is in the distribution with the larger variance. Figure 8.12 shows a plot of such posterior probabilities as horizontal bars for each effect in the reactor experiment, again with the largest effects highlighted. In this case, the prior is that effects have an 80% probability of coming from a distribution with variance σ^2 and a 20% probability of coming from a distribution with variance $10\sigma^2$. Consistent with the subjective interpretation of the normal and half-normal plots, the Box–Meyer analysis indicates that the same effects noted before have a high posterior probability ($> 90\%$) of having the larger variance; of the other effects, the 35 interaction has the highest posterior probability, between 10% and 20%.

This concludes our discussion of alternative ways to select which model effects affect the response. For the reactor data we found just 5 of the 15 potential effects to be important, leaving 10 degrees of freedom for estimating error in a traditional analysis of variance. It is however not clear what extra is to be gained from such an analysis based on summary statistics when compared with our analyses which look at individual effects. A difficulty with this ANOVA is that only small effects are selected to estimate error; the test statistics and their p-values may therefore be biased. Nevertheless, a plot of the residuals from the selected model may still be useful for assessing model adequacy, and such a plot is given in Box, Hunter, and Hunter (2005).

8.5 Robust Design

The design of experiments for quality has aroused appreciable interest in manufacturing industry, often under the name 'Taguchi methods'. The fundamental issue is that variability in the performance of a product is often caused by variability in factors which are difficult or impossible to control in practice—either fluctuations in the manufacturing process or variability in the environment in which the product is used. For example, a good photographic film will provide a sharp non-grainy image with true colours when used and developed under the conditions specified by the manufacturer. However, a quality product is taken to be one which is *robust* to a variety of conditions. Thus, according to this definition, a quality film will provide a good image even if the exposure is incorrect and the developing solutions are at the wrong concentration and temperature.

It is convenient to divide the experimental factors into two groups. The *design factors* are the variables which have been the subject of most of the designs in earlier chapters. They are assumed to be controllable in practice, so that it makes sense to estimate and optimize how the response depends on them, as with the response surface designs of §3.3. But now to these are to be added *environmental factors* (referring to either the manufacturing environment or the operational environment of the product) which are not controllable in practice, although they can be manipulated for the purposes of experimentation.

Experiments for so-called 'robust design' systematically vary the environmental factors along with the design factors. The analysis then examines how the design factors affect both the variance and the mean of the response over the levels of the environmental factors, seeking to minimize the variance while bringing the mean to target. When the focus is on manufacturing variability, then such experimentation is sometimes called 'off-line quality control'. The distinction is with 'on-line control', where production is monitored to remove substandard product. Off-line control removes the wastage caused by rejection of substandard product and by the effort in detecting it.

A common form of robust design experiment (e.g. Taguchi 1987) is the product of simple factorials, or their fractions, in the two sets of variables. Often the factorial for the design factors is called an inner array. This generates a series of trial products, each of which is then assessed for performance

TABLE 8.6. Cake mix example: design and average taste index

	Design factors								
	Amounts of								
	Flour	Sugar	Egg	Environmental factors					
				x_4 Oven temperature	0	−	+	−	+
	x_1	x_2	x_3	x_5 Baking time	0	−	−	+	+
1	0	0	0		6.7	3.4	5.4	4.1	3.8
2	−	−	−		3.1	1.1	5.7	6.4	1.3
3	+	−	−		3.2	3.8	4.9	4.3	2.1
4	−	+	−		5.3	3.7	5.1	6.7	2.9
5	+	+	−		4.1	4.5	6.4	5.8	5.2
6	−	−	+		5.9	4.2	6.8	6.5	3.5
7	+	−	+		6.9	5.0	6.0	5.9	5.7
8	−	+	+		3.0	3.1	6.3	6.4	3.0
9	+	+	+		4.5	3.9	5.5	5.0	5.4

under the conditions of the design in the environmental variables, which form an outer array.

Although the idea of such designs is simple, interesting points arise both in design and analysis. These are illustrated with an example.

This example, taken from Box *et al.* (1989), concerns the development of a new cake mix for the consumer market. The product needs to be robust to incorrect cooking conditions, represented by the environmental factors oven temperature x_4 and baking time x_5. The three design factors, which are under the control of the manufacturer, are the amounts of flour, sugar, and egg powder, denoted x_1, x_2, and x_3 respectively. The experimental design, given in Table 8.6, consists of a 2^3 factorial with centre point in the design factors crossed with a 2^2 factorial plus centre point in the environmental factors. The factor levels 0 in the table correspond to the intended composition of the mix and the suggested cooking conditions. The response is a taste index with large values desired. In passing we note that the variables in the three design factors would more properly be investigated using a mixture experiment.

Inspection of the results of Table 8.6 suggests that trials 7 and 9 produce mixtures which are least susceptible to variation in the environmental factors x_4 and x_5, but that trial 7 has the higher average and so is the best mix for the market.

TABLE 8.7. Cake mix example—summary statistics for the effects of environmental factors

x_1	x_2	x_3	$\bar{y}$	s_y^2	log s_y^2
0	0	0	4.68	1.84	0.608
−	−	−	3.52	6.00	1.792
+	−	−	3.66	1.15	0.142
−	+	−	4.74	2.19	0.783
+	+	−	5.20	0.87	−0.134
−	−	+	5.38	2.12	0.750
+	−	+	5.90	0.46	−0.766
−	+	+	4.36	3.30	1.195
+	+	+	4.86	0.44	−0.814

This informal analysis of the results of this experiment suffices to extract the relevant information. However, in more complicated experiments of this type, a more sophisticated analysis may be required. As an example of the considerations involved, a more formal analysis is now given of the results of Table 8.6.

The canonical analysis for robust design experiments proceeds by computing summary characteristics of the response over the levels of the environmental factors for each combination of levels of the design factors. Table 8.7 gives, for each of the nine combinations of the design factors, the mean response $\bar{y}$ and the estimated variance s_y^2, together with log s_y^2. These results confirm the informal impression of the robustness of mixes 7 and 9 to the environmental factors. In contrast, mix 6 has a high mean score, but also a large variance, so that it might produce unsatisfactory cakes if the cooking instructions were not followed sufficiently scrupulously.

The results of Table 8.7 come from a 2^3 factorial plus centre point. Table 8.8 gives the parameter estimates from the analysis of this design for each of the three responses $\bar{y}$, s_y^2, and log s_y^2. Because there is only one degree of freedom for error, confounded with the lack-of-fit degree of freedom, we use half-normal plots of effects, described in §8.3, to indicate which effects are important.

The plot for the means (Figure 8.13) shows that the important effects, in order, are −23, 3, and 1. This suggests that x_3 should be at the high level, x_2 at the low level, and x_1 at the high level in order to obtain the maximum average response, which correctly identifies mix 7. The normal plot for the estimated variances (Figure 8.14) suggests that x_1 is the important factor for

TABLE 8.8. Cake mix example—estimated effects of design factors on the summary statistics of Table 8.7

Effect	$\bar{y}$	s_y^2	log s_y^2
Mean	4.700	2.043	0.395
1	0.203	−1.334	−0.761
2	0.088	−0.366	−0.111
3	0.423	−0.486	−0.277
12	0.038	0.291	0.030
13	0.053	0.206	−0.120
23	−0.603	0.657	0.210
123	−0.043	−0.593	−0.153

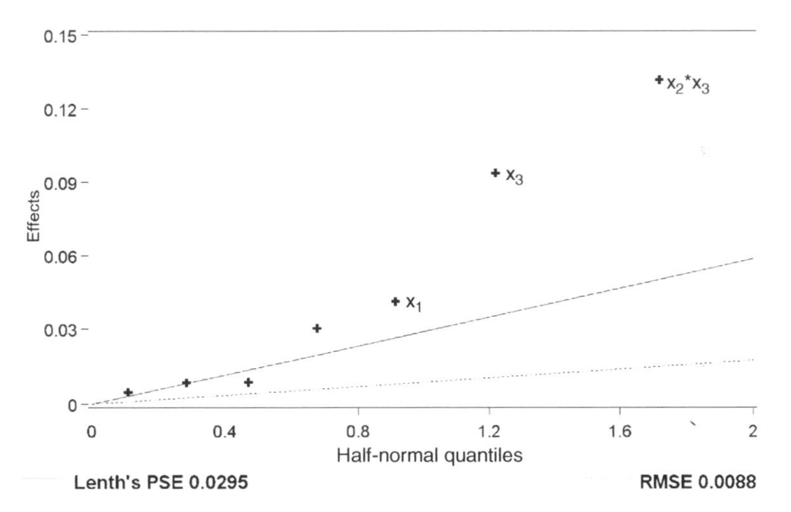

FIG. 8.13. Cake experiment: half-normal plot of parameter estimates for $\bar{y}$.

variance, and that it should be at its higher level. However, the plot seems not to go through the origin. This is hardly surprising since the estimates of variance are distributed as scaled ξ_4^2 random variables. This plot confirms that x_1 is the overriding important factor in determining the variance. The remaining effects appear to fall on a straight line, even if it does not go through the origin. A much closer approximation to normality of the effects can be obtained from the analysis of log s_y^2. The corresponding half-normal plot for this summary statistic (Figure 8.15) is quite similar to the one for s_y^2, except that x_1 seems to stand out even more distinctly as the important factor for variance.

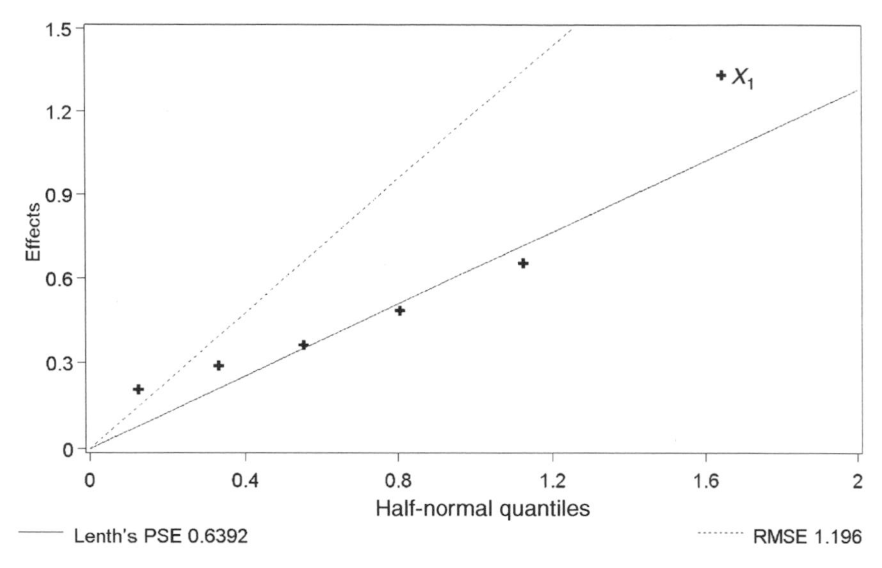

FIG. 8.14. Cake experiment: half-normal plot of parameter estimates for s_y^2.

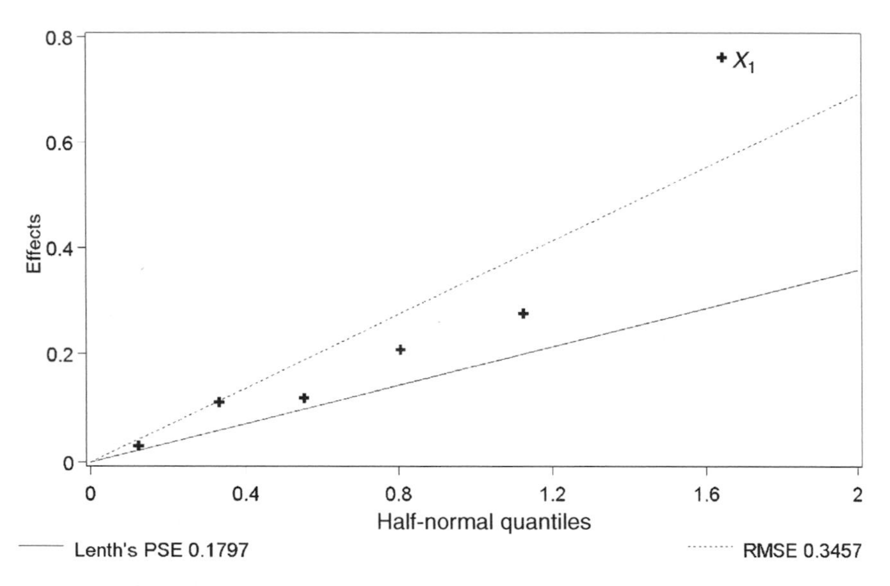

FIG. 8.15. Cake experiment: half-normal plot of parameter estimates for $\log s_y^2$.

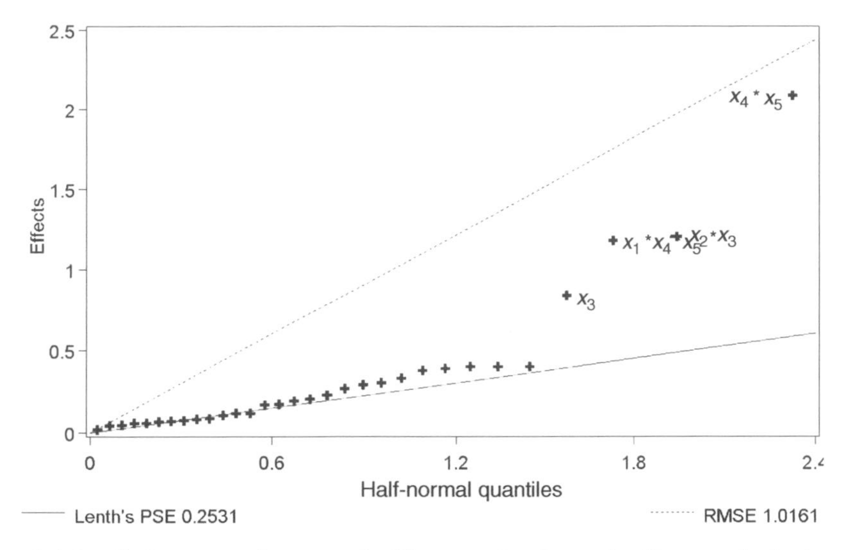

FIG. 8.16. Cake experiment: half-normal plot of effects of design and environmental factors from analysis as a full 2^5 factorial.

These analyses have maintained the distinction between the design factors and the environmental factors. An alternative approach, in line with the implicit analyses for the designs in the rest of this book, is to treat the design as a 2^5 factorial with a variety of centre points.

Figure 8.16 gives a half-normal plot of the 31 resulting effects. This plot picks out the interaction of the environmental factors x_4 and x_5 as being of greatest importance, followed by the x_2x_3 interaction and two others. However, this combined analysis is of little use in suggesting a product which is robust to the environmental factors, whilst having a good taste. Although it might suggest different recommended values for the cooking conditions x_4 and x_5, the separation of the factors into two groups and the responses into mean and variance due to environmental factors are crucial to the understanding of this kind of experiment.

The analysis of the cake mix experiment is straightforward because there is no conflict between the design conditions which increase average taste and those which decrease the variance due to the environmental factors. The high level of x_1 is needed to keep the variance down. To increase the mean, x_3 should be at the high level, x_2 at the low level, and x_1 again at the high level. Sometimes, however, there will be a conflict between design conditions which satisfy a target value for the mean response and those which are robust to environmental factors. In such cases much use has been made of so-called signal-to-noise ratios of the form $\log(\bar{y}^2/s_y^2)$. If the effects of the design factors are as well behaved as they are for the cake mix example, maximization of the signal-to-noise ratio leads to simultaneous maximization of y and minimization of s_y^2. But, in general, maximization of the ratio will lead to maximization of the arbitrary linear combination $2\log\bar{y} - \log s_y^2$. It will usually be better, when possible, to associate costs with both differences of y from the target value and with the variability s_y^2, and then to treat the index as a new response to be maximized. Another possibility is that of transforming the data. We saw in the analysis of the viscosity data, particularly from Figure 8.4, that one of the indications of the desirability of a transformation of y was a relationship between mean and variance, which could be broken by a suitable transformation. Here too, although the circumstances are different, the relationship between mean and variance may sometimes be broken by transformation of the original response before calculation of means and variances. Box *et al.* (1989, p. 370) give a discussion, as do Logothetis and Wynn (1989, p. 253).

The cake mix example, and the introductory example on photographic film, are close to the standard technological applications of experimental design in the remainder of this book. A systematic exposition of the design of experiments for quality in engineering design is given by Logothetis and Wynn (1989). A more recent treatment of related material is Vuchkov and Boyadjieva (2001). Three books of selected papers describing the methods outlined in this section and the accumulation of understanding and misunderstanding associated with the phrase "Taguchi methods" are Barnard *et al.* (1989), Bendell *et al.* (1989), and Dehnad (1989). The treatment of Wu and Hamada (2000) likewise emphasizes industrial applications. However, similar ideas can also be used in biological experiments. For example, for a bioassay to be successful, it is important that the assay window relative to the assay variability is large. Then finding appropriate levels for operational factors like cell numbers, sample times, and sample speed can be done in a similar way.

In the analysis of the cake mix example we have used the half-normal plots of effects that were introduced in §8.4 for the analysis of saturated factorial designs. Behind this analysis lay the customary assumption that the observations were homoscedastic. In the cake mix example, the analysis of the means $\bar{y}$ summarized in Figure 8.13 therefore comes from an unweighted analysis of the $\bar{y}$. We then used a separate analysis to see how the error standard deviation varied with the factors. This showed that the observations do not have constant variance, so that the correct analysis of the means must involve weights. The effect of this weighting on the parameter estimates is likely to be slight, but the variances of parameter estimates and the distribution of test statistics will be more strongly affected. Severe problems of interpretation can arise if there is a combination of effects in the means and in dispersion in the absence of replication. Then these two sets of effects may well be confounded with each other. Several methods of analysis in this situation for full factorials are compared by Brenneman and Nair (2001). McGrath and Lin (2001) consider dispersion effects in unreplicated fractional factorials.

Finally, we note that, from the experimental design perspective of this book much remains to be done to improve the seemingly rather crude designs, such as that of Table 8.6, found by crossing factorials. The use of fractional factorials would obviously lead to a reduction in experimental effort. But if the purpose of the experiment is to explore regions of maximum average yield and low variance, second-order designs will be important in identifying better products and improved recommended conditions of use.

8.6 Analysing Data with SAS

There are many tools in SAS that can fit the linear models involved in the analysis of the majority of experimental designs. The most familiar tools for statistical analysis in SAS are the components of the SAS programming language known as *procedures* or just *PROCs*. A PROC takes a data set as input and produces tables and graphics as well as data sets as output. For example, SAS syntax to create a dataset for Example 1.1 on CO desorption and to fit a simple linear model is

```
data CO;
   input KCRatio n @@;
   do i = 1 to n; input Desorbed @@; output; end;
datalines;
0.05 2 0.05 0.10
0.25 2 0.25 0.35
0.50 3 0.75 0.85 0.95
1.25 5 1.42 1.75 1.82 1.95 2.45
2.10 6 3.05 3.19 3.25 3.43 3.50 3.93
2.50 4 3.75 3.93 3.99 4.07
;

proc reg data=CO;
   model Desorbed = KCRatio;
run;
```

Note that it is not necessary to be explicit about the intercept term in this model specification; it is assumed to be in the model by default and requires an extra option to remove when it is *not* appropriate. Thus, in order to fit the no-intercept model (8.2) for these data, the appropriate PROC REG code adds the NOINT option to the MODEL statement, as follows:

```
proc reg data=CO;
   model Desorbed = KCRatio / noint;
run;
```

Graphical techniques for diagnosing outliers and influential observations are also available in selected SAS procedures, including PROC REG. These graphics can be relatively expensive to compute, so they are not produced by default. They are requested by flanking the PROC code with appropriate ODS statements, as in the following:

```
ods html;
ods graphics on;

proc reg data=CO;
   model Desorbed = KCRatio;
run;

ods graphics off;
ods html close;
```

The resulting graphics are shown in Figures 8.17, 8.18, and 8.19.

> **SAS Task 8.1.** Write SAS code to make a data set to hold the results from the analysis of filler B of the elastomer data as in §8.3. Then use PROC REG to fit a model for the logarithm of viscosity that is linear in the two factors. Use graphics to confirm that the residuals are structure free.

Procedures tend to offer a somewhat generic and syntax-heavy approach to statistical analysis, leaving it up to the user to specifically request optional features of an analysis. As an alternative, in addition to procedures, SAS includes a variety of point-and-click interfaces for doing statistics. Besides easing the user input, these interfaces encourage a more interactive style of statistical analysis, making it easy to explore the validity of assumptions and to re-run the analysis. In particular, the ADX Interface is especially designed for designed experiments, and although ADX exploits SAS procedures 'under the hood (or bonnet)', as it were, the user is not required to know SAS syntax in order to select and construct a design, check assumptions, fit a model, select important effects, and optimize the response. All of the graphics in §§8.2 through 8.4 were produced by the ADX Interface, and in the remainder of this section we will discuss how to use ADX to perform these analyses.

The ADX is invoked by selecting 'Design of Experiments' under the 'Solutions' menu in the SAS display manager. At the top level, ADX displays a

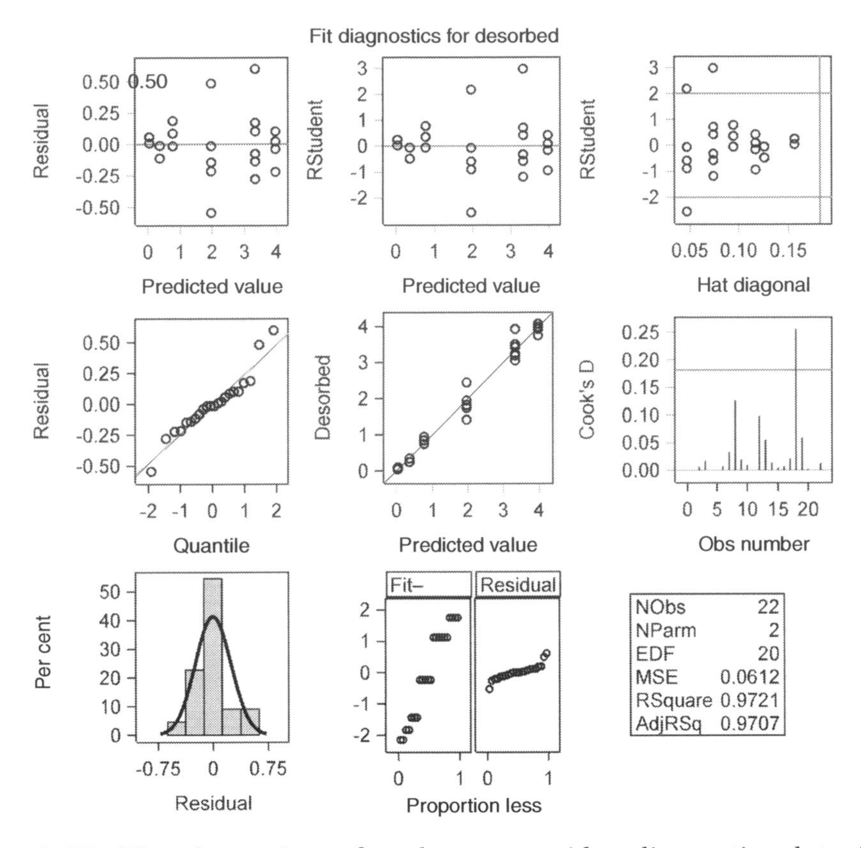

FIG. 8.17. The desorption of carbon monoxide: diagnostic plots from PROC REG.

folder of designs that have already been constructed. If you populate this folder with the example designs from the ADX documentation, your top level folder may look like Figure 8.20—a window containing icons for experiments already created. These icons correspond to experiments with their own designs and data; double-clicking on one will open up ADX's interface for constructing, fitting, analysing, and optimizing this experiment. You can start a new design by clicking on the appropriate small icon for the desired type of design, above the ADX window. You can also import existing data to be analysed as a designed experiment by selecting 'Import Design. . . ' under the 'File' menu. Importing is the easiest way to handle the CO desorption dataset.

SAS Task 8.2. Use the 'Import Design' facility of ADX to make a design out of the CO data set defined earlier. Note that the ADX import facility requires a design to have two or more factors, so select both

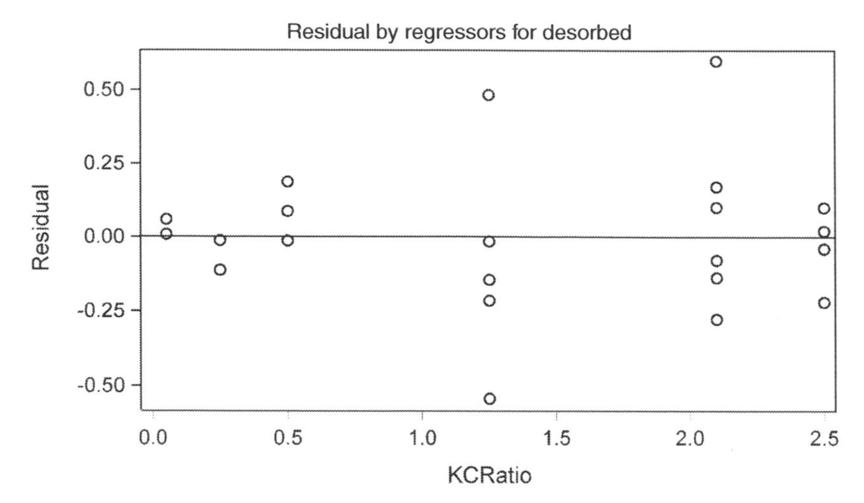

FIG. 8.18. The desorption of carbon monoxide: diagnostic plots from PROC REG.

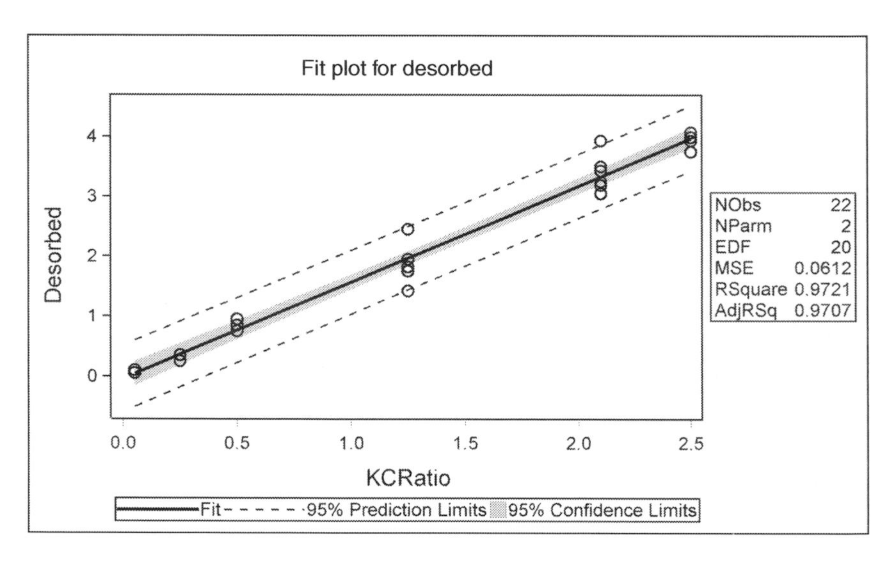

FIG. 8.19. The desorption of carbon monoxide: diagnostic plots from PROC REG.

KCRatio and N as factors, but define a master model that contains only the main effect of KCRatio.

In ADX, assumptions are checked automatically during the process of fitting the default model. So when you open the CO design and select

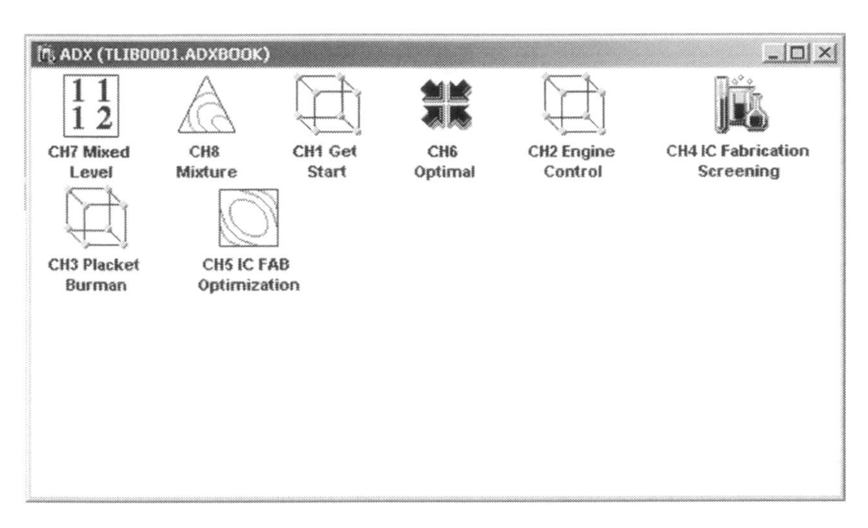

FIG. 8.20. ADX Interface: design folder.

'Fit', ADX performs the following analyses in addition to fitting the linear model:

- Compute residuals and check for outliers.
- Compute measures of influence for each observation and check for overly influential points.
- Perform a Box–Cox analysis to check whether a transformation is required.

If any of these checks reveals problems, ADX will bring up a screen to display the issue and allow the user to address it. Observations are automatically flagged as possible outliers by default if the studentized residual exceeds a critical value based on the t-distribution but corresponding to a percentile that uses a Bonferroni-type adjustment to achieve an over-all Type 1 error probability of no more than 5%. Even if the assumption-checking screens do not come up by default, they are available by selecting 'Check Assumptions' under the Model tab for the Fit screen.

SAS Task 8.3. Using the imported CO data set in ADX, select 'Edit Responses...' to change the response value of the seventh observation from 0.95 to 1.95. Then select 'Fit...' to check the assumptions of this model, discovering that the modified observation is an outlier. Correct the response value and fit again.

If there are determined to be no problems with assumptions, ADX summarizes the fit with an effect selection screen, allowing the user to distinguish appreciable effects from negligible ones. This is particularly useful when models have many potential terms, as in the reactor experiment. A variety of effect selection tools is available, including all those discussed in §8.4.

SAS Task 8.4. Use the ADX Interface to analyse the entire design for Example 1.2, the experiment on the viscosity of elastomer-filler blends, including data for all three fillers. Determine a parsimonious model, including any necessary transformation of the response, and find settings of the factors which give a predicted viscosity of 50.

PART II

THEORY AND APPLICATIONS

9

OPTIMUM DESIGN THEORY

9.1 Continuous and Exact Designs

In previous chapters we have presented ideas, mostly informal, about good experimental design, leading up to a discussion in Chapters 5 and 6 of some optimum designs for linear and quadratic regression in one factor. In this chapter we begin the second part of the book with a more formal discussion of the General Equivalence Theorem, which is the central result on which the theory of the optimum design of experiments depends. The theory applies to a wide variety of design criteria, many of which are called by a letter of the alphabet, so that the subject is sometimes called 'alphabetic optimality'. These criteria are the subject of the next chapter where we also give references to the history of optimum design in §10.11. In order to avoid repetition, in this chapter we work with a general criterion Ψ which is to be minimized. A special case is D-optimality and its relationship to G-optimality. Because of the importance of D-optimality, Chapter 11 is devoted to this criterion, with examples of designs and algorithms for constructing them. We begin this chapter with a more formal statement of the important distinction between exact and continuous designs, and give the notation for them. On their first exposure to this material, readers may prefer to go straight to Chapter 11 after the end of this section.

In §6.5 we showed that the D-optimum design for the first-order model put half the trials at $x = -1$ and the other half at $x = +1$ when $\mathcal{X} = [-1, 1]$, provided that the number of trials N was even. For odd N the number is divided as equally as possible. Thus for increasing odd N there will be a sequence of exact designs starting with a 2:1 division of the trials when $N = 3$, which approaches the equal division for even N. The mathematical problem of finding the optimum design is simplified by considering only this asymptotic or continuous design, thus ignoring the constraint that the number of trials at any design point must be an integer.

Continuous designs are represented by the measure ξ over $\mathcal{X}$. If the design has trials at n distinct points in $\mathcal{X}$, we write

$$\xi = \left\{ \begin{array}{ccc} x_1 & x_2 \ldots x_n \\ w_1 & w_2 \ldots w_n \end{array} \right\}, \tag{9.1}$$

where the first line gives the values of the factors at the design points with the w_i the associated design weights. Since ξ is a measure, $\int_{\mathcal{X}} \xi(dx) = 1$ and $0 \leq w_i \leq 1$, with $\sum w_i = 1$ for all i. It is implicit in (9.1) that the measure has finite support, that is that $n < \infty$, a condition satisfied by optimum designs that are functions of a finite number of information matrices.

The D-optimum design for the first-order model found in §6.5 is written in the form given by (9.1) as

$$\xi = \left\{ \begin{array}{cc} -1 & 1 \\ 1/2 & 1/2 \end{array} \right\}, \tag{9.2}$$

whereas the design for three trials, two of which are at the upper value of x, is, as a measure and so independently of $N = 3$,

$$\xi = \left\{ \begin{array}{cc} -1 & 1 \\ 1/3 & 2/3 \end{array} \right\}.$$

If we wish to stress that a measure refers to an exact design, realizable in integers for a specific N, the measure is written

$$\xi_N = \left\{ \begin{array}{cc} x_1 & x_2 \ldots x_n \\ r_1/N & r_2/N \ldots r_n/N \end{array} \right\}, \tag{9.3}$$

where r_i is the integer number of trials at x_i and $\sum r_i = N$.

In practice, all designs are exact. For moderate N good exact designs can frequently be found by integer approximation to the optimum continuous measure ξ^*. The details of approximation rules are considered by Pukelsheim and Rieder (1992). Often, for simple one-factor models with p parameters, there will be p design points with equal weights $1/p$, so that the exact design with $N = p$ trials is optimum. However, if the design weights are not rational, it will not be possible to find an exact design for any finite N that is identical to the continuous optimum design.

Example 9.1 Quadratic Regression Through the Origin The model is

$$\eta(x) = \beta_1 x + \beta_2 x^2 \tag{9.4}$$

over the region $\mathcal{X} = [0, 1]$. The purpose is to estimate β_2 with minimum variance, which corresponds to checking whether curvature is present in data such as that for Example 1.1 (the desorption of carbon monoxide). As we show in §9.5, the optimum continuous measure is

$$\xi^* = \left\{ \begin{array}{cc} \sqrt{2} - 1 & 1 \\ \sqrt{2}/2 & (2 - \sqrt{2})/2 \end{array} \right\}. \tag{9.5}$$

Although good approximations to this design can be found for various values of N, there is no exact version of ξ^* for finite N. ■

Difficulties in finding exact designs usually arise when N is close to the number of support points of the optimum continuous design, leading to a poor approximation to ξ^*. Chapter 12 describes special algorithms for finding optimum exact designs and gives a variety of examples. In comparing exact designs it should be borne in mind that it is the value of the design criterion Ψ which is of importance, rather than the closeness of ξ_N to ξ^*.

For an N-trial design the information matrix for β in the model $\mathrm{E}(y)=F\beta$ was defined in §5.3 as $F^{\mathrm{T}}F$, where

$$F^{\mathrm{T}}F = \sum_{i=1}^{N} f(x_i)f^{\mathrm{T}}(x_i) \tag{9.6}$$

and $f^{\mathrm{T}}(x_i)$ is the ith row of F. For the continuous design ξ the information matrix is

$$\begin{aligned} M(\xi) &= \int_{\mathcal{X}} f(x)f^{\mathrm{T}}(x)\xi(dx) = \sum_{i=1}^{n} w_i M(\bar{\xi}_i) \\ &= \sum_{i=1}^{n} w_i f(x_i)f^{\mathrm{T}}(x_i), \end{aligned} \tag{9.7}$$

where the measure $\bar{\xi}_i$ puts unit mass at x_i. Because of the presence of the weights w_i, the last form in (9.7), summed over the n design points, becomes the normalized version of (9.6) for the exact design ξ_N. That is

$$M(\xi_N) = \frac{F^{\mathrm{T}}F}{N}.$$

The variance of the predicted response for an N-trial design was given in (5.24) as

$$\mathrm{var}\{\hat{y}(x)\} = \sigma^2 f^{\mathrm{T}}(x)(F^{\mathrm{T}}F)^{-1}f(x).$$

For continuous designs the standardized variance of the predicted response is

$$d(x,\xi) = f^{\mathrm{T}}(x)M^{-1}(\xi)f(x), \tag{9.8}$$

a function of both the design ξ and the point at which the prediction is made. If the design is exact,

$$d(x,\xi_N) = f^{\mathrm{T}}(x)M^{-1}(\xi_N)f(x) = \frac{N\mathrm{var}\{\hat{y}(x)\}}{\sigma^2},$$

the standardized variance introduced in (5.32).

9.2 The General Equivalence Theorem

In the theory for continuous designs we consider minimization of the general measure of imprecision $\Psi\{M(\xi)\}$. Under very mild assumptions, the most important of which are the compactness of $\mathcal{X}$ and the convexity and differentiability of Ψ, designs that minimize Ψ also satisfy a second criterion.

One example is D-optimality, in which

$$\Psi\{M(\xi)\} = \log|M^{-1}(\xi)| = -\log|M(\xi)|,$$

so that the determinant of the information $M(\xi)$ is maximized. Taking the logarithm of the determinant leads to minimization of a convex function, so that any minimum found will be global rather than local. Continuous designs that are D-optimum are also G-optimum, that is they minimize the maximum over $\mathcal{X}$ of the variance $d(x, \xi)$ (9.8).

The General Equivalence Theorem can be viewed as a consequence of the result that the derivatives are zero at a minimum of a smooth function over an unconstrained region. However, the function to be minimized depends on the measure ξ through the information matrix $M(\xi)$. Let the measure $\bar{\xi}$ put unit mass at the point x and let the measure ξ' be given by

$$\xi' = (1 - \alpha)\xi + \alpha\bar{\xi}. \tag{9.9}$$

Then, from (9.7),

$$M(\xi') = (1 - \alpha)M(\xi) + \alpha M(\bar{\xi}). \tag{9.10}$$

Accordingly, the derivative of Ψ in the direction $\bar{\xi}$ is

$$\phi(x, \xi) = \lim_{\alpha \to 0^+} \frac{1}{\alpha}[\Psi\{(1 - \alpha)M(\xi) + \alpha M(\bar{\xi})\} - \Psi\{M(\xi)\}]. \tag{9.11}$$

The General Equivalence Theorem then states the equivalence of the following three conditions on ξ^*: (9.12)

1. The design ξ^* minimizes $\Psi\{M(\xi)\}$.
2. The design ξ^* maximizes the minimum over $\mathcal{X}$ of $\phi(x, \xi)$.
3. The minimum over $\mathcal{X}$ of $\phi(x, \xi^*) = 0$, this minimum occurring at the points of support of the design.

As a consequence of 3, we obtain the further condition:

4. For any non-optimum design ξ the minimum over $\mathcal{X}$ of $\phi(x, \xi) < 0$.

This theorem provides methods for the construction and checking of designs. However, it says nothing about n, the number of support points of the design. A bound on this number can be obtained from the nature of $M(\xi)$, which is a symmetric $p \times p$ matrix. Because of the additive nature of information matrices (9.7), the information matrix of a design can be represented as a weighted sum of, at most, $p(p+1)/2$ rank-one information matrices $M(\bar{\xi}_i)$.

Optimum designs are not necessarily unique. In Table 22.2 we show two D-optimum designs, each with four points of support. These two designs have the same information matrix and so will any convex linear combination of the design matrices. These designs, which in this case have six points of support, are therefore also D-optimum. In this example $p = 3$, so that $n = p(p+1)/2$. But, even if an optimum design is found that contains more than this number of points, a design with the same information matrix, and so the same value of $\Psi\{M(\xi)\}$ can be found that has support at no more than $p(p+1)/2$ points. Usually optimum designs contain fewer points. For many D-optimum designs, especially for non-linear models in one factor, the designs contain p points, each with weight $1/p$.

The bound on the number of design points depends on the linear structure of $M(\xi)$ and on the generation of a rank-one matrix $M(\bar{\xi}_i)$ by each design point. However, experiments in which several responses are measured generate matrices of higher rank so that the number of support points of the design may be reduced if the models for the different responses contain parameters in common. Examples of this kind, for non-linear models, are described in §17.9. For single-response models the bound of $p(p+1)/2$ points of support holds for criteria that are a function of a single information matrix, as are nearly all the criteria discussed in this book. However, several criteria for the Bayesian designs of Chapter 18 are non-linear functions of several information matrices. As examples show, the number of design points is not bounded by $p(p+1)/2$. However, the optimum designs do satisfy a General Equivalence Theorem of the form given by (9.12).

Example 9.2 Quadratic Regression—Example 5.2 continued The D-optimum continuous design for the quadratic model in one variable with $\mathcal{X} = [-1, 1]$ is

$$\xi^* = \left\{ \begin{array}{ccc} -1 & 0 & 1 \\ 1/3 & 1/3 & 1/3 \end{array} \right\}. \tag{9.13}$$

The claim that this design is D-optimum can be checked using the General Equivalence Theorem.

For D-optimality, when $\Psi\{M(\xi)\} = -\log|M(\xi)|$, the derivative function is

$$\phi(x, \xi) = p - d(x, \xi), \tag{9.14}$$

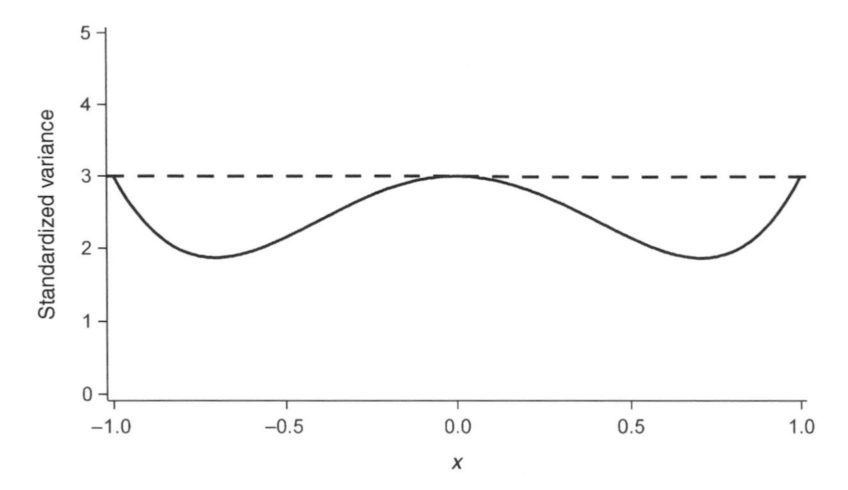

FIG. 9.1. Example 9.2: quadratic regression. Variance function $d(x, \xi^*)$ for the D-optimum continuous design. The maximum value of 3 occurs at the design points.

where $d(x, \xi)$ is the standardized variance function (9.8). From condition (3) of (9.12) that $\phi(x, \xi^*) \geq 0$, it follows that

$$d(x, \xi^*) \leq p. \tag{9.15}$$

For the continuous design (9.13), $d(x, \xi)$ is the quartic given by (5.36). The plot in Figure 9.1 shows that the maximum value of $d(x, \xi^*)$ is indeed p, in this case 3, so that the design is D-optimum as claimed. Further, Condition 3 of the theorem is satisfied, since these maxima occur at the points of support of the design. ■

Plots like Figure 9.1 of the derivative functions of the criteria to be described in Chapter 10 can be used in a similar manner to explore the properties of designs and to check the optimality of continuous designs, particularly when $\mathcal{X}$ is of low dimension. However, it is essential that $\mathcal{X}$ is searched carefully, as is shown by the following example, again for D-optimality.

Example 9.3 Simple Regression—Example 5.1 continued For simple regression consider the design

$$\xi = \left\{ \begin{array}{cc} -a & a \\ 1/2 & 1/2 \end{array} \right\}.$$

with $0 < a < 1$ and $\mathcal{X} = [-1, 1]$. To check the D-optimality of this design calculate

$$d(x, \xi) = 1 + x^2/a^2. \tag{9.16}$$

Then at the points of the design $x = a$ and $d(x, \xi) = 2$, the number of parameters in the model. However the design, as we know from §6.5, is not D-optimum. The maximum of (9.16) occurs at $x = \pm 1$ and is $1 + 1/a^2$ which is greater than 2. Condition 4 of the General Equivalence Theorem (9.12) confirms that this design is not D-optimum. This example illustrates that it is necessary to check the value of the derivative function not only at the design points but over the whole of $\mathcal{X}$. ■

9.3 Exact Designs and the General Equivalence Theorem

The General Equivalence Theorem holds for continuous designs represented by the measure ξ. It does not, in general, hold for exact designs. For D-optimality the implication is that there will be some values of N for which one design will be D-optimum and another G-optimum. Examples were given in §5.3. We now explore one of these in greater detail and calculate a G-optimum design.

Example 9.4 Quadratic Regression, $N = 4$ When $\mathcal{X} = [-1, 1]$ the D-optimum continuous design (9.13) puts weight 1/3 at the three points $x = -1, 0$, and 1. This division provides exact D-optimum designs when N is a multiple of 3, but not otherwise. When $N = 4$ the exact D-optimum design consists of replicating any one of the three design points to give a design for which $|M(\xi_4^*)| = 1/8$. The General Equivalence Theorem cannot be used to prove that this design is D-optimum. Suppose that the centre point is replicated, giving the design

$$\xi_4^* = \left\{ \begin{array}{ccc} -1 & 0 & 1 \\ 1/4 & 1/2 & 1/4 \end{array} \right\}. \tag{9.17}$$

Then, as the black curve in Figure 9.2 shows, the local maxima of the variance function are $d(-1, \xi_4^*) = d(1, \xi_4^*) = 4$, with $d(0, \xi_4^*) = 2$, so that the maximum value, which occurs at the unreplicated points, is four which is greater than $p = 3$. In fact, there is no easy demonstration that (9.17) is the exact D-optimum design. In most examples, computer searches of the kind described in Chapter 12 have to suffice.

However, it is possible, in this example, to find the G-optimum exact design for $N = 4$. The black curve in Figure 9.2 shows that the design (9.17) concentrates too much design weight at the centre of the design region for

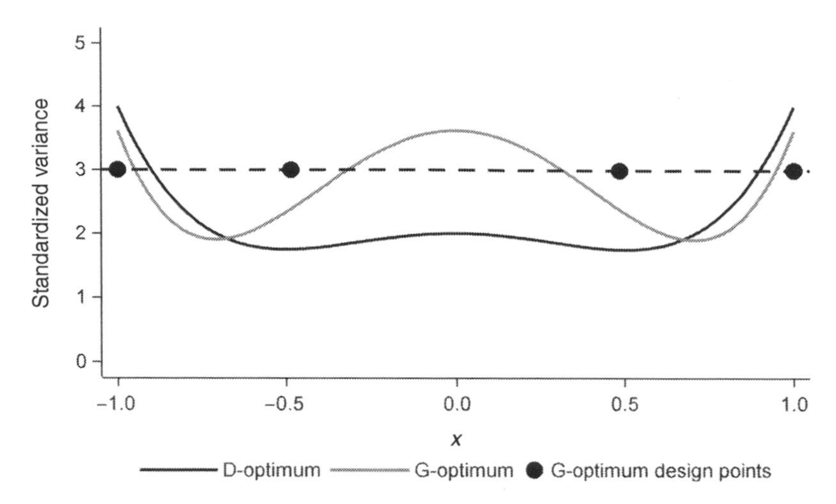

FIG. 9.2. Example 9.2: quadratic regression, $N=4$. Variance function $d(x, \xi^*)$ for the D-optimum and G-optimum exact designs, with the points of the G-optimum design.

the design to be G-optimum. The symmetrical four-point design

$$\xi_4 = \left\{ \begin{array}{cccc} -1 & -a & a & 1 \\ 1/4 & 1/4 & 1/4 & 1/4 \end{array} \right\} \tag{9.18}$$

can have the value of a chosen so that $d(x, \xi)$ has the same value at $x = 0$ and $x = 1$. This requirement yields $a^2 = \sqrt{5} - 2$, when $d(0, \xi) = d(1, \xi) = 3.618$.

The grey curve in Figure 9.2 shows that this is the maximum value of $d(x, \xi)$ over $\mathcal{X}$. Although this value is greater than three, this is the exact G-optimum design for $N = 4$. A feature that complicates the search for G-optimum exact designs is that, as here, the maxima of $d(x, \xi_N)$ do not necessarily occur at the design points. A last property of this design is that $|M(\xi_4)|^{1/3} = 0.4485 < 0.5$, the value for the D-optimum designs such as (9.17). ■

In most examples computer searches are needed to find exact designs for specific N. Because ξ_N is discrete the search for the design minimizing $\Psi\{M(\xi_N)\}$ may find local minima corresponding to different designs with similar values of $\Psi\{M(\xi_N)\}$. For exact G-optimum designs the difficulty of the optimization is increased by the need to search over $\mathcal{X}$ to identify the maximum of $d(x, \xi_N)$ for each trial design measure ξ_N. The search for this second maximum is replaced by evaluation of $d(x, \xi_N)$ over a grid in $\mathcal{X}$. This can be relatively coarse if an exact design is sought which is only required to be close to G-optimality. Evaluation of $d(x, \xi_N)$ over a grid is also used for the comparison of designs that behave almost equally well according to some other criterion, for example exact D-optimality.

9.4 Algorithms for Continuous Designs and the General Equivalence Theorem

We return to consideration of algorithms for finding continuous measures ξ that minimize $\Psi\{M(\xi)\}$. From the General Equivalence Theorem (9.12) it follows that at the optimum design all gradients $\phi(x, \xi^*)$ are non-negative. Away from the optimum there will be some directions in which $\phi(x, \xi) < 0$. If, as in §9.2, we let the measure $\bar{\xi}_k$ put unit mass at the point x_k, chosen so that $\phi(x_k, \xi_k) < 0$, setting

$$\xi_{k+1} = (1 - \alpha_k)\xi_k + \alpha_k \bar{\xi}_k \tag{9.19}$$

will lead to a decrease in Ψ, provided that the step length α_k for the kth step is chosen to be sufficiently small.

The algorithms following from (9.19) are a family of descent algorithms. Usually $\bar{\xi}_k$ is chosen so that x_k is the point for which $\phi(x, \xi_k)$ is a minimum, when the algorithm becomes one of steepest descent. Although it is possible to search in this direction and to find α_k in (9.19) for which $\Psi\{M(\xi_{k+1})\}$ is minimized, a fixed sequence of step lengths is often used. In the original algorithms for the construction of D-optimum designs $\alpha_k = (k+1)^{-1}$. Let

$$\bar{d}(\xi) = \max_{x \in \mathcal{X}} d(x, \xi). \tag{9.20}$$

Then, from (9.14), the steepest descent algorithm for D-optimality will successively add mass to the design measure corresponding to the point where $\bar{d}(\xi)$ is attained. The choice of weights $\alpha_k = (k+1)^{-1}$ yields an algorithm in which the optimum design is constructed by the sequential addition of trials at the points where $d(x, \xi_k)$ is a maximum. This corresponds to the forward sequential algorithm discussed in §12.4 and implemented with `METHOD=SEQUENTIAL` in PROC OPTEX in SAS.

Algorithms based on (9.11) only use information from the first derivative of the objective function Ψ. The convergence of these first-order algorithms can be slow, compared with algorithms that also use information from second derivatives. However, ultimate convergence is usually not required. Starting from an arbitrary design, the algorithm typically adds mass at a restricted number of conditions found by search over a grid of values in $\mathcal{X}$. These conditions suggest the support of the optimum design. Once this pattern becomes clear, the arbitrary starting design can be rejected and the algorithm restarted. Alternatively a second-order algorithm can be employed to minimize $\Psi\{M(\xi)\}$. The advantage of the first-order algorithm is that it reduces the search for an optimum design to a sequence of optimizations in m dimensions.

9.5 Function Optimization and Continuous Design

It is sometimes possible to find an optimum continuous design by algebraic minimization. We begin this section with such an example, before describing some transformations that are helpful in the construction of optimum designs by numerical optimization.

Example 9.1 continued. Quadratic Regression Through the Origin The design for estimating β_2 with minimum variance in the model (9.4) is given in (9.5). To find this design analytically consider the measure

$$\xi = \left\{ \begin{array}{cc} a & 1 \\ w & 1-w \end{array} \right\}. \tag{9.21}$$

It is clear that one point of support of the design must be at $x = 1$, since any design for which this is not the case can be scaled up, with a consequent increase in all elements of the information matrix.

For quadratic regression through the origin it follows from (9.7) that the information matrix is

$$M(\xi) = \left[\begin{array}{cc} \int x^2\xi(dx) & \int x^3\xi(dx) \\ \int x^3\xi(dx) & \int x^4\xi(dx) \end{array} \right].$$

For the measure (9.21) this becomes

$$M(\xi) = \left[\begin{array}{cc} 1-w+wa^2 & 1-w+wa^3 \\ 1-w+wa^3 & 1-w+wa^4 \end{array} \right]. \tag{9.22}$$

Then, from (9.22),

$$\begin{aligned} |M(\xi)| &= (1-w+wa^2)(1-w+wa^4) - (1-w+wa^3)^2 \\ &= w(1-w)a^2(1-a)^2. \end{aligned}$$

The combination of this result with (9.22) shows that element $(2,2)$ of $M^{-1}(\xi)$ is

$$\frac{1-w+wa^2}{w(1-w)a^2(1-a)^2}. \tag{9.23}$$

The design for which var $\hat{\beta}_2$ is minimized will therefore minimize (9.23). Differentiation of this expression followed by setting the derivatives equal to zero leads to the design (9.5).

The final stage in finding this optimum design is to check that the design is indeed optimum amongst all designs, not just among two-point designs of the form (9.21). The calculations for the appropriate equivalence theorem are given in Example 10.2 of §10.3. They show that this two-point design is indeed optimum among all designs with $\mathcal{X} = [0,1]$. ■

The design for which the variance of the estimate of a single parameter is minimized is D_1-optimum, a special case of D_S-optimality, in which the generalized variance of a subset of s of the p parameters is minimized. The corresponding equivalence theorem is given in §10.3.

If it were not possible to minimize (9.23) analytically, a second-order numerical method, such as a quasi-Newton algorithm, could be used to find the minimum of this function of a and w. However, the minimization would be subject to the constraints $0 \leq w \leq 1$ and $0 \leq a \leq 1$. Although general methods for constrained optimization could be used, it is often possible, for straightforward design problems, to use transformations that lead to an unconstrained search in a suitably defined space.

Suppose that x is constrained to be non-negative, that is $0 \leq x < \infty$. For example, x might be a quantitative factor such as time or dose of a drug. Then the transformation

$$x = e^z \quad (-\infty < z < \infty) \tag{9.24}$$

ensures that the constraint on x is satisfied for any value of z. A disadvantage of (9.24) is that $x = 0$ corresponds to $z = -\infty$. A preferable transformation is to put

$$x = z^2 \quad (-\infty < z < \infty) \tag{9.25}$$

when $x = 0$ is in the centre of the range of z. If (9.25) is used with a numerical search algorithm, the steps in z should not be so large that the algorithm oscillates between positive and negative values of z.

For the cubic design region with $-1 \leq x_i \leq 1$ $(i = 1, 2, \ldots, m)$ an appropriate transformation is to put

$$x_i = \sin z_i \quad (i = 1, 2, \ldots, m) \tag{9.26}$$

or equivalently $x_i = \cos z_i$. Searching in the unconstrained space of the z_i leads to a constrained search in x space. Again, care is needed in the choice of step length for the search.

If $\mathcal{X}$ is the m-dimensional unit sphere centred at the origin, the transformation is more complicated. One possibility, using polar co-ordinates, is

$$
\begin{aligned}
x_1 &= \sin z_1 \\
x_2 &= \sin z_2 \cos z_1 \\
&\vdots \\
x_i &= \sin z_i \prod_{j=1}^{i-1} \cos z_j \quad (i = 2, \ldots, m).
\end{aligned} \tag{9.27}
$$

As a result of (9.27) the x_i satisfy the constraints

$$
-1 \leq x_i \leq 1 \;\; (i = 1, 2, \ldots, m) \quad \text{and} \quad \sum_{i=1}^{m} x_i^2 \leq 1 \tag{9.28}
$$

for all z_i.

The search for an optimum design measure involves constraints not only on the values of x imposed by $\mathcal{X}$ but also on the design weights w_i. In place of (9.28) these constraints are

$$
0 \leq w_i \leq 1 \;\text{ and }\; \sum_{i=1}^{n} w_i = 1. \tag{9.29}
$$

A particularly simple and efficient transformation begins with the square transformation to non-negativity, and then normalizes by the sum of the squares to achieve the sum-to-one constraint

$$
w_j = \frac{z_j^2}{\sum_{i=1}^{n} z_i^2}. \tag{9.30}
$$

A characteristic of this transformation is that as a function of the vector z, the information matrix is invariant to scale changes: $M\{w(az)\} = M\{w(z)\}$ for $a \neq 0$. This singularity in the dependence of the information on the parameters of the design measure is not usually a problem, but if it is, an alternative transformation is

$$
\begin{aligned}
w_1 &= \sin^2 z_1 \\
w_2 &= \sin^2 z_2 \cos^2 z_1 \\
&\vdots \\
w_i &= \sin^2 z_i \prod_{j=1}^{i-1} \cos^2 z_j \quad (i = 2, \ldots, n-1) \\
&\vdots \\
w_n &= \prod_{j=1}^{n-1} \cos^2 z_j. \qquad (9.31)
\end{aligned}
$$

There are only $n-1$ values of z in (9.31) as the weights sum to unity. Either transformation (9.30) or (9.31) can also be used for searching over the design region for mixture experiments.

It is our experience that these transformations, combined with a quasi-Newton method using numerical derivatives, provide a powerful method for finding optimum continuous designs. If the structure of the design is approximately known, convergence of the algorithm can be made faster by choosing narrower intervals for the x_i than those given by (9.26). If, for x_i, these are centred on the value x_{i0} the transformation is

$$x_i = x_{i0} + d_i \sin z_i.$$

Certainly d_i must be less than one and, preferably, much less. The value should also be such that $x_i \in \mathcal{X}$ for all z_i.

9.6 Finding Continuous Optimum Designs Using SAS/IML Software

In this section we will discuss some SAS techniques for finding continuous optimal designs, using the numerical optimization approach mentioned in §9.5. SAS has various general optimization tools, but we will focus on the subroutines for this purpose included in SAS/IML software. IML is a matrix programming language, and thus provides good facilities for defining the relevant optimality criteria.

A complete introduction to programming in IML is outside the scope of this book. But the language is fairly intelligible on its own to anyone familiar with matrix notation and perhaps another matrix programming language such as MatLab, so a detailed code example should serve to motivate the approach. For more complete information on programming in IML, see SAS Institute Inc. (2007*b*).

Suppose we want to find the D-optimum design measure for a quadratic model in one variable on $[-1, 1]$. Formally, a continuous design measure is an infinite-dimensional function, but for computational purposes we operate on a discrete grid over the design region. The following statements use a DATA step to create such a grid, with increment $1/10$, read it into IML, and then create the points Z for a quadratic model over the grid.

```
data Grid;
   do x = -1 to 1 by .1;
      output;
      end;
run;

proc iml;
   use Grid;
   read all var ("x") into xg;
   Ng = nrow(xg);
   Zg = j(Ng,1) || xg || xg##2;
```

The READ statement above reads all 21 grid points into the column vector `xg`. The last line uses the all-ones vector `j()`, the horizontal concatenation operator `||`, and element-wise squaring `##2` to create the matrix `Zg` of the quadratic model over the grid.

The IML procedure is interactive, so after submitting the statements above, the IML environment is still active. The following additional statements define the function to be optimized.

```
   start DCritW(w) global(Zg);
      return(det(Zg`*diag(w)*Zg));
   finish;
```

The START statement names the function and its parameters, which in this case is just the vector of weights, and also pulls in the matrix `Zg` of the quadratic model over the grid created above. The return statement computes the determinant of the (weighted) information matrix and returns it as the function value, and the FINISH statement flags that the function definition is complete.

As defined, optimizing the function defined earlier will require constrained weights—namely, weights that are non-negative and sum to 1. The IML non-linear optimization routines can accommodate constraints, but they can make the optimization problem ill-behaved—slow to converge and particularly subject to getting trapped in local optima. As discussed in §9.5, an alternative approach is to transform the weights to satisfy the constraints automatically. The following code first defines a function that uses (9.30) to transform arbitrary values `z` into non-negative weights that sum to one, and then uses that function to define another version of the D-optimality criterion.

```
start WTrans(z);
   z2 = z##2;
   return(z2/sum(z2));
finish;

start DCritW(z) global(Zg);
   w = WTrans(z);
   return(log(det(Zg`*diag(w)*Zg)));
finish;
```

We can optimize this function without specifying constraints. All that is required is a starting value for the optimization. A convenient one is the design that puts equal weights on all candidate points. In the following statements this is provided to the NLPQN subroutine, which performs a quasi-Newton non-linear optimization. This method calls for the derivatives of the objective function, but if they are not provided then it computes finite difference approximations for them.

```
zInit = j(nrow(Zg),1) / nrow(Zg);
call nlpqn(rc,          /* Return code                 */
           zOpt,        /* Returned optimal parameters */
           "DCritW",    /* Function to optimize        */
           zInit,       /* Initial value of parameters */
           1);          /* Specify a maximization      */
wOpt = WTrans(zOpt)`;
print xg wOpt;
```

The optimization runs very quickly in this case, and prints the results shown in Table 9.1. Of course, the many tiny weights are effectively zero, so that the resulting design measure is numerically equivalent to the optimal one (9.13). Discrepancies of this order between theoretical and numerical results are common and to be expected.

In order to optimize over a different grid, the only change required to the code above is in how the data set `Grid` is defined. Note however that as the size of the grid increases, the number of parameters in the optimization also increases, impacting both the quality of the optimum and the computing time required to find it. If the goal is to compute the theoretically optimum *value* of the D-criterion, as opposed to close examination of the continuous optimal design *points*, then a relatively coarse grid usually suffices.

TABLE 9.1. Continuous optimum one-factor quadratic design: raw weights

Support points x	Weights w
-1	0.33333
-0.9	2.9×10^{-11}
-0.8	4.0×10^{-12}
-0.7	4.9×10^{-10}
-0.6	2.8×10^{-9}
-0.5	3.6×10^{-11}
-0.4	9.6×10^{-14}
-0.3	6.0×10^{-11}
-0.2	3.4×10^{-13}
-0.1	7.1×10^{-8}
-1.4×10^{-6}	0.33334
0.1	7.1×10^{-8}
0.2	3.6×10^{-13}
0.3	6.0×10^{-11}
0.4	9.4×10^{-14}
0.5	3.6×10^{-11}
0.6	2.8×10^{-9}
0.7	4.9×10^{-10}
0.8	4.0×10^{-12}
0.9	2.9×10^{-11}
1	0.33333

SAS Task 9.1. Use PROC IML to find D-optimum one-factor design measures over $[-1, 1]$ for cubic and quartic models. Note that if the points of the true optimum measure are not contained in the grid, the numerical optimum may approximate it by putting partial weight on neighbouring grid points.

10

CRITERIA OF OPTIMALITY

10.1 A-, D-, and E-optimality

In this chapter we describe the more important special cases of the design criterion $\Psi\{M(\xi)\}$ of Chapter 9 and give derivative functions $\phi(x, \xi)$ for the particular General Equivalence Theorems analogous to (9.12).

The most important design criterion in applications is that of D-optimality, in which the generalized variance of the parameter estimates, or its logarithm $-\log |M(\xi)|$, is minimized. The relationship between G-optimality and this criterion was extensively discussed in Chapter 9. Two other criteria that have a statistical interpretation in terms of the information matrix $M(\xi)$ are A- and E-optimality. In A-optimality $\mathrm{tr}\, M^{-1}(\xi)$, the total variance of the parameter estimates, is minimized, equivalent to minimizing the average variance. In E-optimality the variance of the least well estimated linear combination $a^{\mathrm{T}}\hat{\beta}$ is minimized subject to the constraint that $a^{\mathrm{T}}a = 1$. Thus the E in the name stands for 'extreme'.

D-optimality was motivated in Chapter 6 by reference to the ellipsoidal confidence regions for the parameters of the linear model. A D-optimum design minimizes the content of this confidence region and so minimizes the volume of the ellipsoid. However, other properties of the confidence region may be of interest. A long thin ellipsoid orientated along, or close to, the parameter axes will result in comparatively poor estimation of one or more parameters, the variances of which will then be unduly large. If the ellipsoid is orientated at an appreciable angle to the axes, the variance of each individual parameter estimate may be satisfactorily small; however, owing to the correlations between the estimates, there will be linear combinations of the parameters, corresponding to the directions of the long axes of the ellipsoid, which will be imprecisely estimated. It is these situations that are respectively addressed by A- and E-optimality.

The above ideas can be expressed more formally by considering the eigenvalues $\lambda_1, \ldots, \lambda_p$ of $M(\xi)$. The eigenvalues of $M^{-1}(\xi)$ are then $1/\lambda_1, \ldots, 1/\lambda_p$ and are proportional to the squares of the lengths of the axes of the confidence ellipsoid. In terms of these eigenvalues the three criteria are as follows:

A. Minimize the sum, or average, of the variances of the parameter estimates

$$\min \sum_{i=1}^{p} \frac{1}{\lambda_i};$$

D. Minimize the generalized variance of the parameter estimates

$$\min \prod_{i=1}^{p} \frac{1}{\lambda_i};$$

E. Minimize the variance of the least well-estimated linear combination $a^{\mathrm{T}}\hat{\beta}$ with $a^{\mathrm{T}}a = 1$

$$\min \max_i \frac{1}{\lambda_i}.$$

All three criteria can be regarded as special cases of the more general criterion of choosing designs to minimize

$$\Psi_k(\xi) = \left(p^{-1} \sum_{i=1}^{p} \lambda_i^k\right)^{1/k} \qquad (0 \leq k < \infty). \tag{10.1}$$

For A-, D-, and E-optimality the values of k are 1, 0, and ∞ respectively when the limiting operations are properly defined. Kiefer (1975) uses this family to study the variation in structure of the optimum design as the optimality criterion changes in a smooth way.

In order to state the equivalence theorems for these and other criteria it is convenient to rewrite the derivative (9.11) as

$$\phi(x, \xi) = \Delta(\xi) - \psi(x, \xi),$$

where

$$\begin{aligned} \Delta(\xi) &= -\mathrm{tr} M \tfrac{\partial \Psi}{\partial M} \qquad \psi(x, \xi) = -f^{\mathrm{T}}(x) \tfrac{\partial \Psi}{\partial M} f(x) \\ \Psi &= \Psi\{M(\xi)\} \qquad\qquad M = M(\xi). \end{aligned} \tag{10.2}$$

The functions for A- D- and E-optimality are given in Table 10.1, together with the functions for some other criteria. In Table 10.1, A-optimality is the special case of linear optimality with $A = I$, the identity matrix.

An advantage of D-optimality is that the optimum designs for quantitative factors do not depend on the scale of the variables, even though the value of $M(\xi^*)$ does. Affine invariant linear transformations of $f(x)$ leave the optimum design unchanged, which is not, in general, the case for A- and E-optimum designs. In principle this is a serious drawback to the other

TABLE 10.1. Functions in the General Equivalence Theorem (10.2) for several optimality criteria

Criterion	Ψ	$\frac{\partial \Psi}{\partial M}$	Δ
A	$\mathrm{tr}\ M^{-1}$	M^{-2}	$\mathrm{tr}\ M^{-1}$
D	$\log \lvert M^{-1} \rvert$	M^{-1}	p
Linear	$\mathrm{tr}\ AM^{-1}$	$M^{-1}AM^{-1}$	$\mathrm{tr}\ AM^{-1}$
E	$\min \lambda_i(M) = \lambda_{\min}$	$rr^{T\,*}$	$\lambda_{\min}$
Generalized D ($\mathrm{D_A}$)	$\log \lvert A^T M^{-1} A \rvert$	$M^{-1}A(A^T M^{-1} A)^{-1} A^T M^{-1}$	$s = \mathrm{rank}\ A$
Generalized G	$\max_{x \in \mathcal{Z}} w(x) d(x, \xi) = C(\xi)$	$M^{-1} \int_{\mathcal{Z}} w(x) f(x) f^T(x) dx\, M^{-1}$	$C(\xi)$
q	$q^{-1} \mathrm{tr}\ M^{-q}$	M^{-q-1}	$\mathrm{tr}\ M^{-q}$

* r is the eigenvector for $\lambda_{\min}$.

two criteria; it seems undesirable that an optimum design should depend on whether a factor is measured in inches or centimetres. However, as we have seen, it is customary to work with scaled variables x_j rather than with the unscaled variables u_j of §2.1. As a consequence, the units of measurement become irrelevant. For designs with all factors qualitative, such as block designs, the problem of scale does not arise and A- and E-optimum designs are frequently employed. One reason is the relative ease of proof of the optimality of such designs in non-standard situations. D-optimum designs are more readily constructed for experiments with quantitative factors. We now consider several useful extensions to D-optimality.

10.2 $\mathrm{D_A}$-optimality (Generalized D-optimality)

Sometimes interest is not in all p parameters of the model, but only in s linear combinations of β which are the elements of $A^{\mathrm{T}}\beta$, where A is $p \times s$ with rank $s < p$. The covariance matrix for these s linear combinations is $M_A(\xi) = A^{\mathrm{T}} M^{-1}(\xi) A$. If $s = 1$, designs minimizing the three criteria of §10.1 for $M_A(\xi)$ all reduce to the c-optimum designs of §10.4 in which the variance of the estimated linear combination of the parameters is minimized. When $s > 1$, the A-optimum design minimizing the trace of $M_A(\xi)$ is an example of the linear designs of §10.5. In this section we define Generalized D-optimum designs minimizing

$$\Psi\{M(\xi)\} = \log |A^{\mathrm{T}} M^{-1}(\xi) A|. \qquad (10.3)$$

To emphasize the dependence of the design on the matrix of coefficients A, this criterion is called $\mathrm{D_A}$-optimality (Sibson 1974). The analogue of the variance function $d(x, \xi)$ (9.8) is

$$d_A(x, \xi) = f^{\mathrm{T}}(x) M^{-1} A (A^{\mathrm{T}} M^{-1} A)^{-1} A^{\mathrm{T}} M^{-1} f(x). \tag{10.4}$$

If we let

$$\bar{d}_A(\xi) = \sup_{x \in \mathcal{X}} d_A(x, \xi),$$

then $\bar{d}_A(\xi^*_{DA}) = s$, where ξ^*_{DA} is the continuous $\mathrm{D_A}$-optimum design. When the design is optimum, the maxima of $d_A(x, \xi^*_{DA})$ occur at the points of support of the design and the other aspects of the General Equivalence Theorem (9.12) also hold for this new criterion. One application of $\mathrm{D_A}$-optimality, described in §25.3, is the allocation of treatments in clinical trails when differing importance is given to estimation of the effects of the treatments and of the effects of the prognostic factors. We now consider an important special case of $\mathrm{D_A}$-optimality.

10.3 $\mathrm{D_S}$-optimality

$\mathrm{D_S}$-optimum designs are appropriate when interest is in estimating a subset of s of the parameters as precisely as possible. Let the terms of the model be divided into two groups

$$\mathrm{E}(Y) = f^{\mathrm{T}}(x)\beta = f_1^{\mathrm{T}}(x)\beta_1 + f^{\mathrm{T}}(x)_2\beta_2, \tag{10.5}$$

where the β_1 are the parameters of interest. The $p - s$ parameters β_2 are then treated as nuisance parameters. One example is when β_1 corresponds to the experimental factors and β_2 corresponds to the parameters for the blocking variables. Examples of $\mathrm{D_S}$-optimum designs for blocking are given in Chapter 15. A second example is when experiments are designed to check the goodness of fit of a model. The tentative model, with terms $f_2(x)$ is embedded in a more general model by the addition of terms $f_1(x)$. In order to test whether the simpler model is adequate, precise estimation of β_1 is required. A fuller description of this procedure is in §21.5.

To obtain expressions for the design criterion and related variance function, we partition the information matrix as

$$M(\xi) = \begin{bmatrix} M_{11}(\xi) & M_{12}(\xi) \\ M_{12}^{\mathrm{T}}(\xi) & M_{22}(\xi) \end{bmatrix}. \tag{10.6}$$

The covariance matrix for the least squares estimate of β_1 is $M^{11}(\xi)$, the $s \times s$ upper left submatrix of $M^{-1}(\xi)$. It is easy to verify, from results on the inverse of a partitioned matrix (e.g., Fedorov 1972, p. 24), that

$$M^{11}(\xi) = \{M_{11}(\xi) - M_{12}(\xi)M_{22}^{-1}(\xi)M_{12}^{\mathrm{T}}(\xi)\}^{-1}.$$

The $\mathrm{D_S}$-optimum design for β_1 accordingly maximizes the determinant

$$|M_{11}(\xi) - M_{12}(\xi)M_{22}^{-1}(\xi)M_{12}^{\mathrm{T}}(\xi)| = \frac{|M(\xi)|}{|M_{22}(\xi)|}. \tag{10.7}$$

The right-hand side of (10.7) leads to the expression for the variance

$$d_s(x,\xi) = f^{\mathrm{T}}(x)M^{-1}(\xi)f(x) - f_2^{\mathrm{T}}(x)M_{22}^{-1}(\xi)f_2(x). \tag{10.8}$$

For the $\mathrm{D_S}$-optimum design ξ_{Ds}^*

$$d_s(x, \xi_{Ds}^*) \leq s, \tag{10.9}$$

with equality at the points of support of the design. These results follow from those for $\mathrm{D_A}$-optimality by taking $A = (I_s \quad 0)^{\mathrm{T}}$, where I_s is the $s \times s$ identity matrix.

A mathematical difficulty that arises with $\mathrm{D_S}$-optimum designs and with some other designs, such as the c-optimum designs of §10.4, is that $M(\xi^*)$ may be singular. As a result, only certain linear combinations or subsets of the parameters may be estimable. The consequent difficulties in the proofs of equivalence theorems are discussed, for example, by Silvey (1980, p. 25) and Pázman (1986, p. 122). In the numerical construction of optimum designs the problem is avoided by regularization of the information matrix through the addition of a small multiple of the identity matrix. That is, we let

$$M_\epsilon(\xi) = M(\xi) + \epsilon I \tag{10.10}$$

for ϵ small, but large enough to permit inversion of $M_\epsilon(\xi)$ (Vuchkov 1977). Then, for example, the first-order algorithm (9.19) can be used for the numerical construction of optimum designs. An example is given in §17.5, where designs are found for various properties of a non-linear model, for which the information matrix is singular.

We conclude this section with two examples of $\mathrm{D_S}$-optimum designs.

Example 10.1 Quadratic Regression: Example 5.2 continued The D-optimum continuous design for quadratic regression in one variable with $\mathcal{X} = [-1, 1]$ is given in (9.13). It consists of weights 1/3 at $x = -1$, 0, and 1. It is appropriate for estimating all three parameters of the model with minimum generalized variance. The $\mathrm{D_S}$-optimum design for β_2 would be used if it were known that there was a relationship between y and x and interest was in whether the relationship could be adequately represented by a straight line or whether some curvature was present. The design leads to estimation of β_2 with minimum variance and so to the most powerful test of $\beta_2 = 0$. This $\mathrm{D_S}$-optimum design for β_2 is

$$\xi^*_{Ds} = \begin{bmatrix} -1 & 0 & 1 \\ 1/4 & 1/2 & 1/4 \end{bmatrix}. \tag{10.11}$$

In this example the points of support of the D- and $\mathrm{D_S}$-optimum designs are thus the same, but the weights are different.

To verify that (10.11) is $\mathrm{D_S}$ optimum we use the condition on the variance $d_s(x, \xi^*_{Ds})$ given by (10.9). Since interest is in the quadratic term, the coefficients of the constant and linear terms are nuisance parameters. For the design (10.11) the information matrix thus partitions as

$$M(\xi^*_{Ds}) = \begin{bmatrix} A & B \\ B^{\mathrm{T}} & D \end{bmatrix} = \left[\begin{array}{c|cc} \sum x^4 & \sum x^2 & 0 \\ \hline \sum x^2 & 1 & 0 \\ 0 & 0 & \sum x^2 \end{array}\right] = \left[\begin{array}{c|cc} 1/2 & 1/2 & 0 \\ \hline 1/2 & 1 & 0 \\ 0 & 0 & 1/2 \end{array}\right], \tag{10.12}$$

where the vertical and horizontal lines in the matrices show the division of terms according to (10.6). Then

$$M^{-1}(\xi^*_{Ds}) = \begin{bmatrix} 4 & -2 & 0 \\ -2 & 2 & 0 \\ 0 & 0 & 2 \end{bmatrix}$$

and

$$M^{-1}_{22}(\xi^*_{Ds}) = \begin{bmatrix} 1 & 0 \\ 0 & 2 \end{bmatrix}.$$

So, from (10.8),

$$\begin{aligned} d_s(x, \xi^*_{Ds}) &= 4x^4 - 2x^2 + 2 - (2x^2 + 1) \\ &= 4x^4 - 4x^2 + 1. \end{aligned} \tag{10.13}$$

This quartic equals unity at $x = -1$, 0, or 1. These are the three maxima over $\mathcal{X}$ since, for $-1 < x < 1$, $x^4 \leq x^2$, with equality only at $x = 0$. Thus (10.11) is the $\mathrm{D_S}$-optimum design for the quadratic term. ■

Example 10.2 Quadratic Regression through the Origin: Example 9.1 continued The $\mathrm{D_S}$-optimum design of the previous example provides an unbiased estimator of the quadratic term with minimum variance. The design is appropriate for testing whether the quadratic term should be added to a first-order model. Similarly, the design given by (9.5) is appropriate

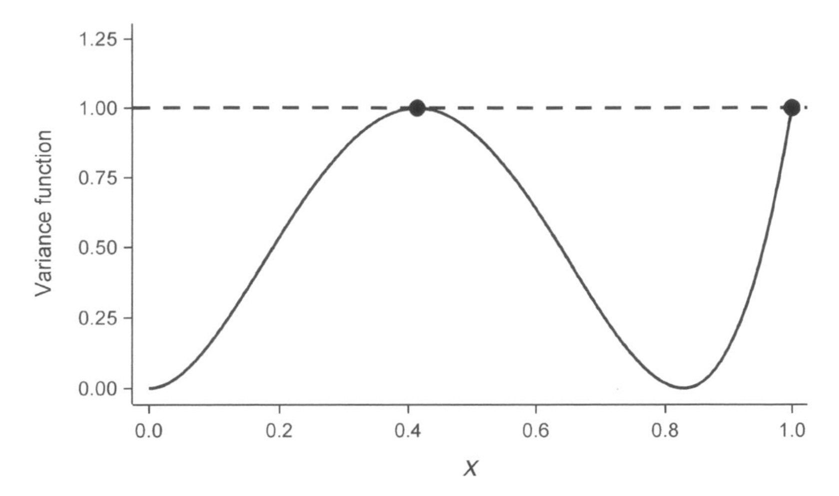

FIG. 10.1. Example 10.2 Quadratic regression through the origin. Variance function $d_s(x, \xi_s^*)$ for the $\mathrm{D_S}$-optimum design for β_2; $s = 1$.

for checking curvature in the two-parameter model when the quadratic regression is forced to go through the origin.

The design (9.5) was calculated in §9.5 by direct minimization of $\mathrm{var}\,\hat{\beta}_2$. To show that this design is $\mathrm{D_S}$ optimum for β_2 we again calculate $d_s(x, \xi_{Ds}^*)$. Substitution of (9.5) in (10.8) yields the numerical expression

$$\begin{aligned} d_s(x, \xi_{Ds}^*) &= 25.73x^2 - 56.28x^3 + 33.97x^4 - 2.42x^2 \\ &= 23.31x^2 - 56.26x^3 + 33.97x^4. \end{aligned}$$

Figure 10.1 is a plot of this curve, which does indeed have a maximum value of unity at the design points, which are marked by black dots. The difference between Figure 10.1 and the variance curves of Chapter 6, such as those of Figure 6.3, is instructive. Here, because the model passes through the origin, the variance of prediction must be zero when $x = 0$. ■

These two examples illustrate some of the properties of $\mathrm{D_S}$-optimum designs. However, in both cases, $s = 1$. The designs are therefore optimum according to several criteria discussed in this chapter, for example c-optimality, discussed in §10.4, in which the variance of a linear combination of the parameter estimates is minimized. However, for $s \geq 2$, $\mathrm{D_S}$-optimum designs will, in general, be different from those satisfying other criteria.

10.4 c-optimality

In c-optimality interest is in estimating the linear combination of the parameters $c^{\mathrm{T}}\beta$ with minimum variance, where c is a known vector of constants. The design criterion to be minimized is thus

$$\operatorname{var} c^{\mathrm{T}}\hat{\beta} \propto c^{\mathrm{T}}M^{-1}(\xi)c, \tag{10.14}$$

where c is $p \times 1$. The equivalence theorem states that, for the optimum design,

$$\{f^{\mathrm{T}}(x)M^{-1}(\xi^*)c\}^2 \leq c^{\mathrm{T}}M^{-1}(\xi^*)c, \tag{10.15}$$

for $x \in \mathcal{X}$.

Examples of c-optimum designs for a non-linear model are given in §17.5. A disadvantage of c-optimum designs is that they are often singular. For example, if c is taken to be $f(x_0)$ for a specific $x_0 \in \mathcal{X}$, the criterion becomes minimization of the variance of the predicted response at x_0. One way to achieve this is to perform all trials at x_0, a singular optimum design which is non-informative about any other aspects of the model and data. In §21.9 we use compound optimality to provide designs informative, in an adjustable way, about both the parameters and the particular features of the model that are of interest.

The linear optimality criterion of the next section is an extension of c-optimality to designs for two or more linear combinations.

10.5 Linear Optimality: C- and L-optimality

Let L be a $p \times q$ matrix of coefficients. Then minimization of the criterion function

$$\operatorname{tr}\{M^{-1}(\xi)L\} \tag{10.16}$$

leads to a linear, or L-optimum, design. The linearity in the name of the criterion is thus in the elements of the covariance matrix $\{M^{-1}(\xi)\}$. We now consider the relationship of this criterion to some of the other criteria of this chapter.

If L is of rank $s \leq q$ it can be expressed in the form $L = AA^{\mathrm{T}}$ where A is a $p \times s$ matrix of rank s. Then

$$\operatorname{tr}\{M^{-1}(\xi)L\} = \operatorname{tr}\{M^{-1}(\xi)AA^{\mathrm{T}}\} = \operatorname{tr}\{A^{\mathrm{T}}M^{-1}(\xi)A\}. \tag{10.17}$$

This form suggests a relationship with the $\mathrm{D_A}$-optimum designs of §10.2, where the determinant, rather than the trace, of $A^{\mathrm{T}}M^{-1}(\xi)A$ was minimized. An alternative, but unappealing, name for this design criterion could therefore be $\mathrm{A_A}$-optimality, with A-optimality recovered when $L = I$, the identity matrix.

Another special case of (10.17) arises when $s = 1$, so that A becomes the c of the previous section. If several linear combinations of the parameters are of interest, these can be written as the rows of the $s \times p$ matrix C^{T}, when the criterion function to be minimized is $\mathrm{tr}\, C^{\mathrm{T}} M^{-1}(\xi) C$, whence the name C-optimality.

In the notation of (10.17), the equivalence theorem states that, for the optimum design,

$$f^{\mathrm{T}}(x) M^{-1}(\xi^*) A A^{\mathrm{T}} M^{-1}(\xi^*) f(x) \leq \mathrm{tr}\,\{A^{\mathrm{T}} M^{-1}(\xi^*) A\}, \qquad (10.18)$$

the generalization of the condition for c-optimality given in (10.15).

10.6 V-optimality: Average Variance

A special case of c-optimality mentioned earlier was minimization of the quantity $f^{\mathrm{T}}(x_0)\ M^{-1}(\xi) f(x_0)$, the variance of the predicted response at x_0. Suppose now that interest is in the average variance over a region $\mathcal{R}$. Suppose further that averaging is with respect to a probability distribution μ on $\mathcal{R}$. Then the design should minimize

$$\int_{\mathcal{R}} f^{\mathrm{T}}(x) M^{-1}(\xi) f(x) \mu(dx) = \int_{\mathcal{R}} d(x, \xi) \mu(dx). \qquad (10.19)$$

If we let

$$L = \int_{\mathcal{R}} f^{\mathrm{T}}(x) f(x) \mu(dx),$$

which is a non-negative definite matrix, it follows that (10.19) is another example of the linear optimality criterion (10.17). A design which minimizes (10.19) is called I-optimum (for 'Integrated') or V-optimum (for 'Variance').

The idea of V-optimality was mentioned briefly in §6.5. In practice the importance of the criterion is often as a means of comparing designs found by other criteria. The numerical value of the design criterion (10.19) is usually approximated by averaging the variance over a grid in $\mathcal{R}$.

10.7 G-optimality

G-optimality was introduced in §6.5 and used in §9.4 in the iterative construction of a D-optimum design. The definition is repeated here for completeness.

Let

$$\bar{d}(\xi) = \max_{x \in \mathcal{X}} d(x, \xi).$$

Then the design that minimizes $\bar{d}(\xi)$ is G-optimum. For continuous designs this optimum design measure ξ_G^* will also be D-optimum and $\bar{d}(\xi_G^*) = p$. However, Example 9.2 and Figure 9.1 show that this equivalence may not hold for exact designs. For an exact G-optimum design $\xi_{G,N}^*$ we may have $\bar{d}(\xi_{G,N}^*) > p$. As Figure 9.2 showed, the maxima may not occur at the design points and so search over a grid may again be necessary to determine the value of $\bar{d}(\xi)$.

10.8 Compound Design Criteria

The criteria considered so far in this chapter are examples of the convex design criterion $\Psi\{M(\xi)\}$ introduced in §9.2. We now extend those results to convex linear combinations of these design criteria.

Let $\Psi_i\{M_i(\xi)\}, (i = 1, \ldots, h)$ be a set of h convex design criteria defined on a common experimental region $\mathcal{X}$ and let $a_i, i = 1, \ldots, h$ be a set of non-negative weights. Then the linear combination

$$\Psi(\xi) = \sum_{i=1}^{h} a_i \Psi_i\{M_i(\xi)\} \tag{10.20}$$

is itself a convex design criterion, to which the General Equivalence Theorem (9.12) can be applied. This leads to a

General Equivalence Theorem for Compound Criteria. The following three conditions on ξ^* are equivalent:

1. The design ξ^* minimizes $\Psi(\xi)$ given in (10.20).
2. Let $\phi_i(x, \xi)$ be the derivative of $\Psi_i\{M_i(\xi)\}$ defined by (9.11). Then with

   $$\phi(x, \xi) = \sum_{i=1}^{h} a_i \phi_i(x, \xi),$$

   the design ξ^* maximizes the minimum over $\mathcal{X}$ of $\phi(x, \xi)$.
3. The minimum over $\mathcal{X}$ of $\phi(x, \xi^*) = 0$, this minimum occurring at the points of support of the design.

Condition 4 of (9.12) also holds, that is:

4. For any non-optimum design ξ the minimum over $\mathcal{X}$ of $\phi(x, \xi) < 0$.

(10.21)

In the definition of $\Psi(\xi)$ the criteria $\Psi_i\{\cdot\}$ can be the same or different, as can the information matrices $M_i(\cdot)$. An example in which all $\Psi_i\{\cdot\}$ are the same is given in the next section when we consider generalized D-optimality. In the designs for parameter estimation and model discrimination of §21.8 the individual criteria are however different.

10.9 Compound D$_A$-optimality

As an example of a compound optimality criterion we extend the D$_A$-optimality of §10.2. In (10.21) we take

$$\Psi(\xi) = \sum_{i=1}^{h} a_i \log |A_i^{\mathrm{T}} M_i^{-1}(\xi) A_i|. \tag{10.22}$$

This criterion was called S-optimality by Läuter (1974). The equivalence theorem for the optimum design ξ^* states that, for all $x \in \mathcal{X}$

$$\begin{aligned} &\textstyle\sum_{i=1}^{h} a_i f_i^{\mathrm{T}}(x) M_i^{-1}(\xi^*) A_i \{A_i^{\mathrm{T}} M_i^{-1}(\xi^*) A_i\}^{-1} A_i^{\mathrm{T}} M_i^{-1}(\xi^*) f_i(x) \\ &\quad \leq \textstyle\sum_{i=1}^{h} a_i s_i, \end{aligned} \tag{10.23}$$

where s_i is the rank of A_i. This criterion was used by Atkinson and Cox (1974) to design experiments for the simultaneous estimation of parameters in a variety of models, whilst estimating parameter subsets for model discrimination. In the quadratic examples of §10.3 the a_i could be used to reflect the balance between estimation of all the parameters in the model and the precise estimation of β_2, the parameters of the model that is being checked. This topic is discussed in more detail in §21.5.

10.10 D-optimum Designs for Multivariate Responses

The preceding design criteria assume that observations are made on only one response. We conclude this chapter by considering D-optimality when instead measurements are on a vector of h responses.

For models that are linear in the parameters the model for univariate responses (5.1) is

$$y_i = f^{\mathrm{T}}(x_i)\beta + \epsilon_i \quad (i = 1, \ldots, N), \tag{10.24}$$

with $\mathrm{E}(\epsilon_i) = 0$. Since the errors are independent, with constant variance,

$$\mathrm{E}(\epsilon_i \epsilon_l) = 0 \quad (i \neq l) \text{ and } \mathrm{E}(\epsilon_i^2) = \sigma^2.$$

The multivariate generalization of (10.24) is that the h responses for observation i are correlated, but that observations i and l are independent, $i \neq l$. Thus the observations follow the model

$$y_{iu} = f_u^{\mathrm{T}}(x_i)\beta + \epsilon_{iu}, \tag{10.25}$$

with

$$\mathrm{E}(\epsilon_{iu}) = 0, \quad \mathrm{E}(\epsilon_{iu}\epsilon_{iv}) = \sigma_{uv},$$

and, since the observations at different design points are independent,

$$\mathrm{E}(\epsilon_{iu}\epsilon_{lu}) = \mathrm{E}(\epsilon_{iu}\epsilon_{lv}) = 0,$$

$u, v = 1, \ldots, h$. The variance–covariance matrix of the responses is

$$\Sigma = \{\sigma_{uv}\}_{u,v=1,\ldots,h}. \tag{10.26}$$

Estimation of the parameter vector β is by generalized least squares with weights Σ^{-1}.

The contribution to the information matrix of responses u and v is, from (9.7),

$$M_{uv}(\xi) = \int_{\mathcal{X}} f_u(x) f_v^{\mathrm{T}}(x) \xi(dx)$$

and (Draper and Hunter 1966) the information matrix for all h responses is

$$M(\xi) = \sum_{u=1}^{h} \sum_{v=1}^{h} \sigma^{uv} M_{uv}(\xi), \tag{10.27}$$

where $\Sigma^{-1} = \{\sigma^{uv}\}_{u,v=1,\ldots,h}$.

The results of Fedorov (1972, p. 212) show that a form of the usual equivalence theorem applies for the D-optimality of designs maximizing $|M(\xi)|$ (10.27). If the definition of the standardized variance of prediction $d(x, \xi)$ in (9.8) is extended to

$$d_{uv}(x, \xi) = f_u^{\mathrm{T}}(x) M^{-1}(\xi) f_v(x), \tag{10.28}$$

with $M(\xi)$ given by (10.27), the equivalence theorem (9.12) applies to

$$d(x, \xi) = \sum_{u=1}^{h} \sum_{v=1}^{h} \sigma^{uv} d_{uv}(x, \xi).$$

From (9.15) the D-optimum design ξ^* maximizing $|M(\xi)|$ is such that $d(x, \xi^*) \leq p$ for $x \in \mathcal{X}$.

Although experiments frequently have multivariate responses, the correlations between responses often have no effect on the experimental design. This arises because, if all model functions $f_u(x)$ are the same, weighted least squares reduces to ordinary least squares (Rao 1973, p. 545), even though Σ is not the identity matrix, and the univariate design is optimum. It is only if it is assumed, at the design stage, that different models will be needed for the various responses, that the covariance matrix Σ plays a role in determining the optimum design. If different models are fitted to the different responses, even if there are no parameters in common, generalized least squares with known Σ is optimum. In econometrics, this form of least squares analysis is known as seemingly unrelated regression (Zellner 1962).

We return to D-optimum designs for multivariate responses in §17.9 where the non-linear models for the various responses are distinct functions of a few parameters. As we have stated, when one response is measured, optimum designs have at least p points of support. For non-linear models in a single factor it is often the case that $n = p$. However, (10.27) shows that each response contributes a rank one matrix, weighted by σ^{ii}, to the information matrix at x_i. Provided that $f_u(x_i)$ and $f_v(x_i)$ do not lie in the same subspace, the h responses at each point of support therefore contribute a rank h matrix. For some non-linear models the value of n may then be less than p.

10.11 Further Reading and Other Criteria

The history of optimum experimental design starts with Smith (1918) who, in an amazing paper which was 30 years before its time, calculated G-optimum designs for one-factor polynomials up to order six and showed that the designs were optimum. See §11.4 for a description of these designs. Kirstine Smith, a Dane, worked with Karl Pearson. Biographical details and a description of the historical importance of her non-design paper Smith (1916) can be found in Hald (1998, p. 712), Atkinson and Bailey (2001, §3) and at `http://www.webdoe.cc/publications/kirstine.php`.

Smith's paper seems not to have had an immediate effect. Wald (1943) compared designs using the determinant of the information matrix and so was led to D-optimality. The mathematical theory of weighing designs (Hotelling 1944; Mood 1946) was developed at much the same time as the more general results of Plackett and Burman (1946) described in §7.5. Elfving (1952) investigated c- and A-optimality for a two-variable regression model without intercept. His results were generalized by Chernoff (1953) to locally D-optimum designs for non-linear models; locally optimum because the design depends on the unknown values of the parameters of the non-linear model.

Guest (1958) generalized Smith's results on G-optimum designs for one-factor polynomials, showing that the designs are supported at the roots of Legendre polynomials (see §11.4). Hoel (1958) who, like Guest, mentions Smith's work, considered D-optimum designs for the same polynomials and found that he obtained the same designs as Guest. The equivalence of G- and D-optimality was shown by Kiefer and Wolfowitz (1960).

The alphabetical nomenclature for design criteria used here was introduced by Kiefer (1959). That paper, together with the publication of the equivalence theorem, ushered in a decade of rapid development of optimum design making use of results from the theory of convex optimization. Whittle (1973*a*) provides a succinct proof of a very general version of the theorem. Silvey (1980, Chapter 3) gives a careful discussion and proof of the theorem of §9.2. The argument for the bound on the number of support points of the design depends upon the application of Carathéodory's Theorem to the representation of an arbitrary design matrix as a convex combination of unitary design matrices (Silvey 1980, Appendix 2). Wynn (1985) reviews Kiefer's work on optimum experimental design as an introduction to a volume that reprints Kiefer's papers on the subject. The section 'Comments and References', in effect the unnumbered sixteenth chapter, of Pukelsheim (1993) provides an extensive, fully referenced discussion of the development of optimum design. Further references are given in the survey paper Atkinson (1988) and in the introduction to Berger and Wong (2005).

Box (1952) in effect finds D-optimum designs for a multifactor first-order model and discuses rotation of the design to avoid biases from omitted higher-order terms. Bias is important in the development of response surface designs by Box and Draper (1959, 1963), where the mean squared error of prediction, J, over a specified region is divided into a variance component V and a bias component B. Designs are found which give differing balance between B and V, although the properties of the pure variance designs are not studied in detail. Designs minimizing V were introduced by Studden (1977) in the context of optimum design theory and were called I-optimum for integrated variance. The term V-optimality is used interchangeably. Box and Lucas (1959) provide locally D-optimum designs for non-linear models with examples from chemical kinetics. We give examples of such designs in Chapter 17.

Algorithms, in particular the first-order algorithm of §9.4, have been important in the numerical construction of designs. The algorithm for D-optimality was introduced by Fedorov (1972) and by Wynn (1970). Wu and Wynn (1978) prove convergence of the algorithm for more general design criteria.

A geometrical interpretation of c-optimum designs was given by Elfving (1952) and developed by Silvey and Titterington (1973) and Titterington (1975). For D-optimality the support points of the design lie on the ellipsoid of least volume that contains the design region. For $\mathrm{D_S}$-optimality the ellipsoid is replaced by a cylinder.

This chapter covers the majority of criteria to be met with in the rest of this book, one exception being the T-optimum model discrimination designs of Chapter 20. The criteria of this chapter are all functions of the single information matrix $M(\xi)$. If ξ_1 and ξ_2 are two design measures such that $M(\xi_1) - M(\xi_2)$ is positive definite, then ξ_1 will be a better design that ξ_2 for any criterion function Ψ. If a ξ_1 can be found for which the difference is, at least, non-negative definite for all ξ_2 and positive definite for some ξ_2, then ξ_1 is a globally optimum design. In most situations in this book this is too strong a requirement to be realized, although it holds for some designs for qualitative factors, such as Latin squares. An introduction is in Wynn (1985) with more recent results in Giovagnoli and Wynn (1985, 1996) and in Pukelsheim (1993, p. 426).

Finally we return to Elfving (1952) who introduces the cost of experimentation in a simple way. The information matrix in (9.7) was standardized by the number of observations. To be explicit we can write

$$M_N(\xi) = \sum_{i=1}^{n} w_i M(\bar{\xi}_i), \tag{10.29}$$

where, for an exact design, $w_i = n_i/N$. Suppose that an observation at x_i incurs a cost $c(x_i)$ and that there is a restriction on the total cost

$$\sum_{i=1}^{n} n_i c(x_i) \leq C. \tag{10.30}$$

We now normalize the information matrix by the total cost C, rather than by N, and let

$$M_C(\xi) = \frac{N M_N(\xi)}{C} = \sum_{i=1}^{n} w_i M_C(\bar{\xi}_i), \tag{10.31}$$

where

$$w_i = \frac{n_i c(x_i)}{C} \quad \text{and} \quad M_C(\bar{\xi}) = \frac{M(\bar{\xi})}{c(x)}.$$

Therefore standard methods of optimum design can be used once the costs have been defined. To obtain exact designs with frequencies n_i requires rounding of the values $w_i C/c(x_i)$ to the nearest integer subject to the constraint (10.30). Examples for non-linear models are given by Fedorov, Gagnon, and Leonov (2002) and by Fedorov and Leonov (2005).

The only constraint on the number of trials in (10.31) is on the total cost. Imposition of the second constraint $\sum_{i=1}^{n} n_i \leq N$ leads to a more complicated design problem (Cook and Fedorov 1995).

11

D-OPTIMUM DESIGNS

11.1 Properties of D-optimum Designs

In this section we list a variety of results for D- and G-optimum designs. In §11.2 an illustration is given of the use of the variance function in the iterative construction of a D-optimum design. The following section returns to the example of the desorption of carbon monoxide with which the book opened. A comparison is made of the design generating the data of Table 1.1 with several of the D- and D_S-optimum designs derived in succeeding chapters. The last two sections of the chapter discuss D-optimum designs which might be useful in practice, particularly for second-order models. But, to begin, we consider some definitions and general properties of D-optimum designs.

1. The D-optimum design ξ^* maximizes $|M(\xi)|$ or, equivalently, minimizes $|M^{-1}(\xi)|$. It is sometimes more convenient to consider the convex optimization problems of maximizing $\log|M(\xi)|$ or minimizing $-\log|M(\xi)|$.

2. The D-efficiency of an arbitrary design ξ is defined as

$$D_{\text{eff}} = \left\{ \frac{|M(\xi)|}{|M(\xi^*)|} \right\}^{1/p}. \tag{11.1}$$

   The comparison of information matrices for designs that are measures removes the effect of the number of observations. Taking the pth root of the ratio of the determinants in (11.1) results in an efficiency measure which has the dimensions of a variance, irrespective of the dimension of the model. So two replicates of a design measure for which $D_{\text{eff}} = 0.5$ would be as efficient as one replicate of the optimum measure. In order to compare design ξ_1 with design ξ_2, the relative D-efficiency

$$D_{\text{rel-eff}} = \left\{ \frac{|M(\xi_1)|}{|M(\xi_2)|} \right\}^{1/p} \tag{11.2}$$

can be used. Unlike the D-efficiency, the relative D-efficiency (11.2) can take values greater than one, in which case design ξ_1 is better than design ξ_2 with respect to the determinant criterion.

3. A generalized G-optimum design over the region $\mathcal{R}$ is one for which

$$\max_{x \in \mathcal{R}} w(x)d(x, \xi^*) = \min_{\xi} \max_{x \in \mathcal{R}} w(x)d(x, \xi).$$

Usually $\mathcal{R}$ is taken as the design region $\mathcal{X}$ and $w(x) = 1$, when the General Equivalence Theorem (9.2) holds. Then, with

$$\bar{d}(\xi) = \max_{x \in \mathcal{X}} d(x, \xi),$$

the G-efficiency of a design ξ is given by

$$G_{\text{eff}} = \frac{\bar{d}(\xi^*)}{\bar{d}(\xi)} = \frac{p}{\bar{d}(\xi)}. \tag{11.3}$$

4. The D-optimum design need not be unique. If ξ_1^* and ξ_2^* are D-optimum designs, the design

$$\xi^* = c\xi_1^* + (1 - c)\xi_2^* \quad (0 \leq c \leq 1)$$

is also D-optimum. All three designs will have the same information matrix. An example is in Table 22.2.

5. The D-optimality criterion is model dependent. However, the D-efficiency of a design (and hence its D-optimality) is invariant to non-degenerate linear transformations of the model. Thus a design D-optimum for the model $\eta = \beta^{\mathrm{T}} f(x)$ is also D-optimum for the model $\eta = \gamma^{\mathrm{T}} g(x)$, if $g(x) = Af(x)$ and $|A| \neq 0$. Here β and γ are both $p \times 1$ vectors of unknown parameters.

6. Let n denote the number of support points of the design. We have already discussed the result that there exists a D-optimum ξ^* with $p \leq n \leq p(p + 1)/2$, although, from point 4 above, there may be optimum designs with the same information matrix but with n greater than this limit.

7. The D-efficiency of the D-optimum N-trial exact design ξ_N^* satisfies

$$\frac{\{N!/(N - p)!\}^{1/p}}{N} \leq D_{\text{eff}}(\xi_N^*) \leq 1.$$

8. If the design ξ^* is D-optimum with the number of support points $n = p$, then $\xi_i^* = 1/p, \quad (i = 1, \ldots, n)$. The design will clearly be a D-optimum

exact design for $N = p$. For these designs

$$\text{cov}(\hat{y}_i, \hat{y}_j) \propto d(x_i, x_j) = f^{\mathrm{T}}(x_i)M^{-1}(\xi^*)f(x_j) = 0 \qquad (11.4)$$
$$(i, j = 1, 2, \ldots, n; i \neq j).$$

This result, which also holds for non-optimum ξ with $\xi_i = 1/p$ and $n = p$, is of particular use in the construction of mixture designs with blocks (§16.5).

Other results on D-optimum designs can be found in the references cited at the end of this chapter. It is important to note, from a practical point of view, that D-optimum designs often perform well according to other criteria. The comparisons made by Donev and Atkinson (1988) for response surface designs are one example.

11.2 The Sequential Construction of Optimum Designs

In this section we give an example of the sequential construction of a D-optimum continuous design. We use the special case of the first-order algorithm of §9.4 which sequentially adds a trial at the point where $d(x, \xi_N)$ is a maximum. In this way a near-optimum design is constructed. However, there is no intention that the experiment should actually be conducted in this manner, one trial at a time. The purpose is to find the optimum design measure ξ^*.

The algorithm can be described in a way which is helpful for the algorithms for exact designs of the next chapter. Let the determinant of the information matrix after N trials be

$$|M(N)| = |F^{\mathrm{T}}F|.$$

Then addition of one further trial at x yields the determinant

$$|M(N+1)| = |F^{\mathrm{T}}F + f(x)f^{\mathrm{T}}(x)|.$$

This can be rewritten (see, for example, Rao (1973), p. 32) as a multiplicative update of $|M(N)|$,

$$\begin{aligned} |M(N+1)| &= |F^{\mathrm{T}}F|\{1 + f(x)(F^{\mathrm{T}}F)^{-1}f^{\mathrm{T}}(x)\} \qquad (11.5) \\ &= |M(N)|\left\{1 + \frac{d(x, \xi_N)}{N}\right\}. \end{aligned}$$

Thus, the addition to the design of a trial where $d(x, \xi_N)$ is a maximum will result in the largest possible increase in $|M(N)|$.

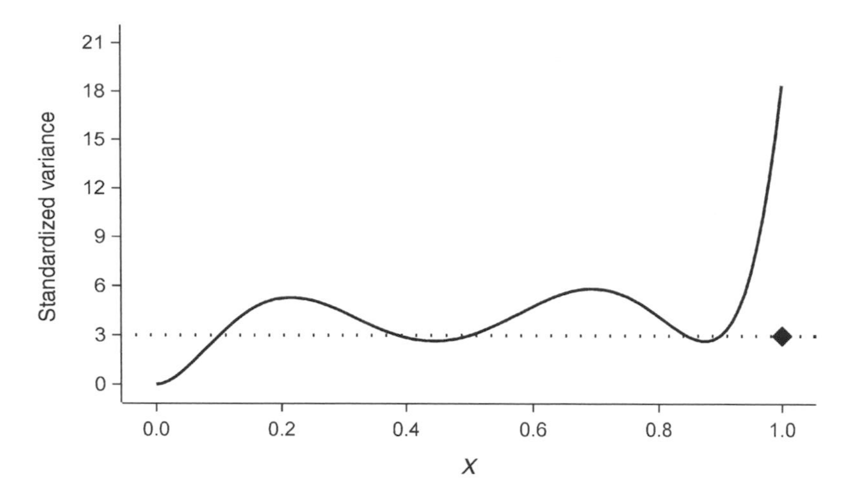

FIG. 11.1. Example 11.1: cubic regression through the origin. Sequential construction of the D-optimum design; $d(x, \xi_3)$. The diamond marks the maximum value.

Example 11.1 Cubic Regression through the Origin As an example of the use of (11.5) in constructing a design we take

$$\eta(x) = \beta_1 x + \beta_2 x^2 + \beta_3 x^3,$$

with $\mathcal{X} = [0, 1]$. The model is chosen because it provides a simple illustration of the procedure for constructing designs in a situation for which we have not yet discussed D-optimality. Such a third-order polynomial without an intercept term is unlikely to be required in practice; transformations of response or explanatory variable are likely to be preferable.

The starting point for the algorithm is not crucial. We take the symmetrical three-point design

$$\xi_3 = \left\{ \begin{array}{ccc} 0.1 & 0.5 & 0.9 \\ 1/3 & 1/3 & 1/3 \end{array} \right\}. \tag{11.6}$$

Figure 11.1 shows the resulting plot of $d(x, \xi_3)$. As with Figure 10.1 for quadratic regression through the origin, the variance is zero at the origin. The maximum value of $d(x, \xi_3)$ is 18.45 at $x = 1$, reflecting in part the fact that the design does not span the design region.

When a trial at $x = 1$ is added to the initial design (11.6), the plot of $d(x, \xi_4)$ is as shown in Figure 11.2. Comparison with Figure 11.1 shows that the variance at $x = 1$ has been appreciably reduced by addition of the extra trial. The two local maxima in the curve are now of about equal importance.

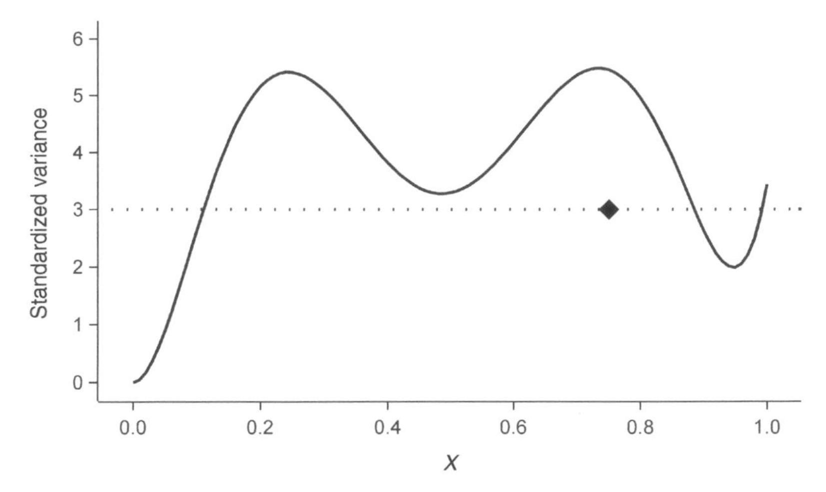

FIG. 11.2. Example 11.1: cubic regression through the origin. Sequential construction of the D-optimum design; $d(x, \xi_4)$. The diamond marks the maximum value.

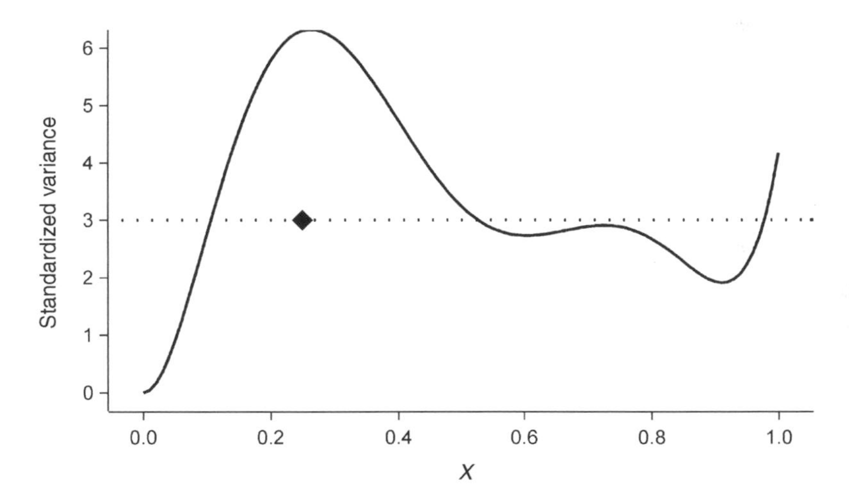

FIG. 11.3. Example 11.1: cubic regression through the origin. Sequential construction of the D-optimum design; $d(x, \xi_5)$. The diamond marks the maximum value.

Rounding x to the nearest 0.05, the maximum value of 5.45 is at $x = 0.75$. If this point is added to the design, the resulting five-point design gives rise to the variance curve of Figure 11.3. The maximum variance is now at

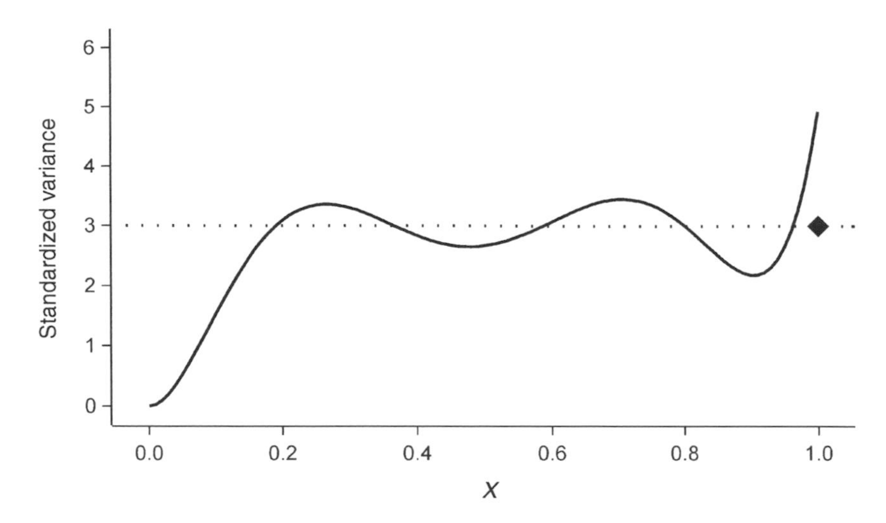

FIG. 11.4. Example 11.1: cubic regression through the origin. Sequential construction of the D-optimum design; $d(x, \xi_6)$. The diamond marks the maximum value.

$x = 0.25$. The six-point design including this trial gives the plot of $d(x, \xi_6)$ of Figure 11.4. As with Figure 11.1, the maximum value is at $x = 1$, which would be the next trial to be added to the design.

The process can be continued. Table 11.1 shows the construction of the design for up to 12 trials. The search over $\mathcal{X}$ is in steps of size 0.05. The algorithm quickly settles down to the addition, in turn, of trials at or near $1, 0.7$, and 0.25. The value of $\bar{d}(\xi_N)$ decreases steadily, but not monotonically, towards 3 and is, for example, 3.102 for $N = 50$.

Two general features are of interest: one is the structure of the design and the other is its efficiency. The structure can be seen in Figure 11.5, a histogram of the values of x obtained up to $N = 50$. The design is evolving towards equal numbers of trials at three values around $0.25, 0.7$, and 1. The starting values for the design, marked by black bars in the histogram, are clearly poor. There are several possibilities for finding the D-optimum design more precisely, which are discussed in §§9.4 and 9.5.

1. Delete the poor starting design, and either start again with a better approximation to the optimum design or continue from the remainder of the design of Figure 11.5. The deletion, as well as the addition, of trials is important in some algorithms for exact designs, such as DETMAX, described in §12.5.

TABLE 11.1. Example 11.1. Sequential construction of a D-optimum design for a cubic model through the origin

N	x_{N+1}	$\bar{d}(\xi_N)$	G_{eff}	D_{eff}
3	1	18.45	0.163	0.470
4	0.75	5.45	0.550	0.679
5	0.25	6.30	0.476	0.723
6	1	4.93	0.609	0.791
7	0.7	3.94	0.761	0.828
8	0.25	4.43	0.677	0.841
9	1	4.01	0.748	0.866
10	0.7	3.58	0.837	0.881
11	0.25	3.92	0.766	0.887
12	1	3.68	0.814	0.900

The initial design has trials at 0.1, 0.5, and 0.9.

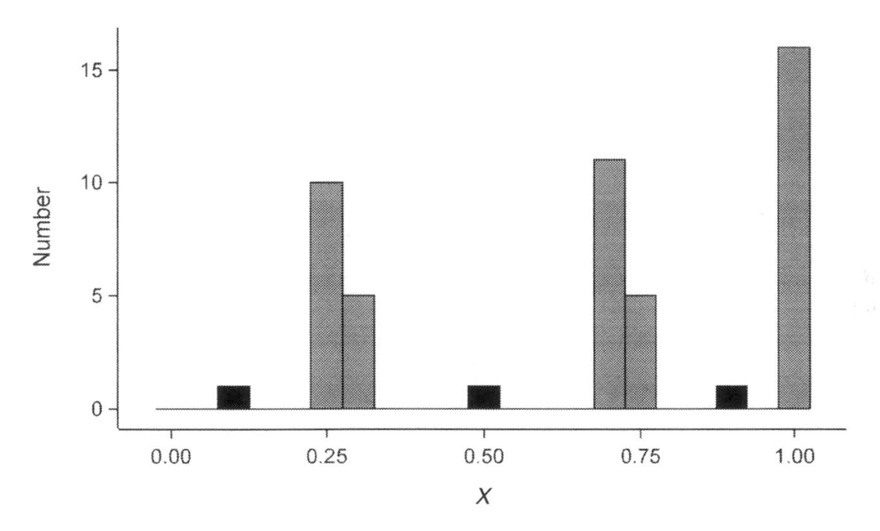

FIG. 11.5. Example 11.1: cubic regression through the origin. Histogram of design points generated by sequential construction up to $N = 50$; black bars, the initial three-trial design.

2. Use a numerical method to find an optimum continuous design with a starting point for the algorithm suggested by Figure 11.5.

3. Analytical optimization.

In this case we explore the third method. It is clear from the results of Figure 11.5 that the optimum continuous design will consist of equal weight at three values of x, one of which will be unity. The structure is the same as that for the D-optimum designs described in earlier chapters for other polynomials in one variable. Here $|M(\xi)|$ is a function of only two variables, and techniques similar to those of Example 9.1 of §9.5 can be used to find the optimum design. Elementary, but lengthy, algebra yields the design

$$\xi^* = \left\{ \begin{array}{ccc} (5-\sqrt{5})/10 & (5+\sqrt{5})/10 & 1 \\ 1/3 & 1/3 & 1/3 \end{array} \right\}, \tag{11.7}$$

i.e. equal weight at $x = 0.2764, 0.7236$ and 1. A plot of $d(x, \xi^*)$ shows that this is the D-optimum design with $\bar{d}(\xi^*) = 3$ at the design points. If the assumption that the design was of this form were incorrect, the plot would have revealed this through the existence of a value of $d(x, \xi) > 3$.

The D-efficiency of the sequentially constructed design, as defined in (11.1), is plotted in Figure 11.6. Although the efficiency of the initial design (11.6) is only 0.470, the efficiency rises rapidly towards unity. An interesting feature is that the progress towards the optimum may not be monotonic. This feature is evident in the plot of G-efficiency (Figure 11.7). As the plots of variance in Figures 11.1 to 11.4 indicate, these efficiency values are lower than those for D-efficiency. They also exhibit an interesting pattern of groups of three increasing efficiency values. The highest of each group corresponds to the balanced design with nearly equal weight at the three support points. Optimum additions of one further trial causes the design to be slightly unbalanced in this respect, and leads to a decrease in G-efficiency. As the weight of the added trial is $1/N$, the resulting non-monotonic effect decreases as N increases. A last comment on design efficiency is that the neighbourhood of the D-optimum design is usually fairly flat when considered as a function of ξ, so that designs that seem rather different may have similar D-efficiencies. ■

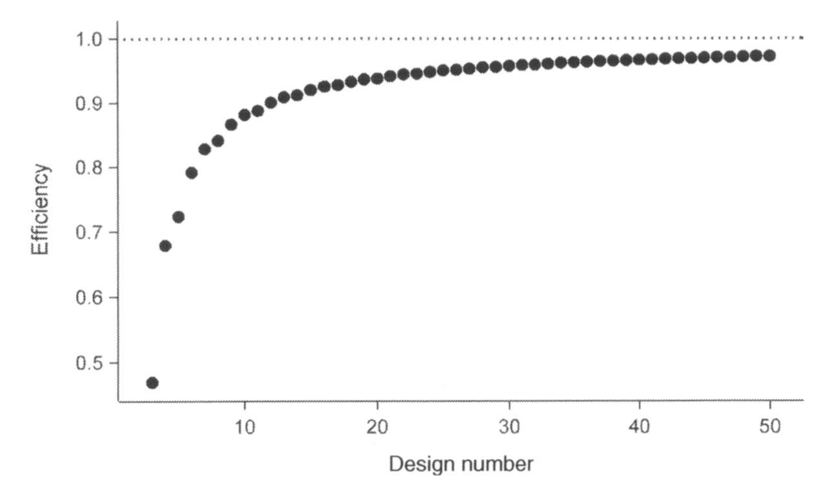

FIG. 11.6. Example 11.1: cubic regression through the origin. D-efficiency of the sequentially constructed designs.

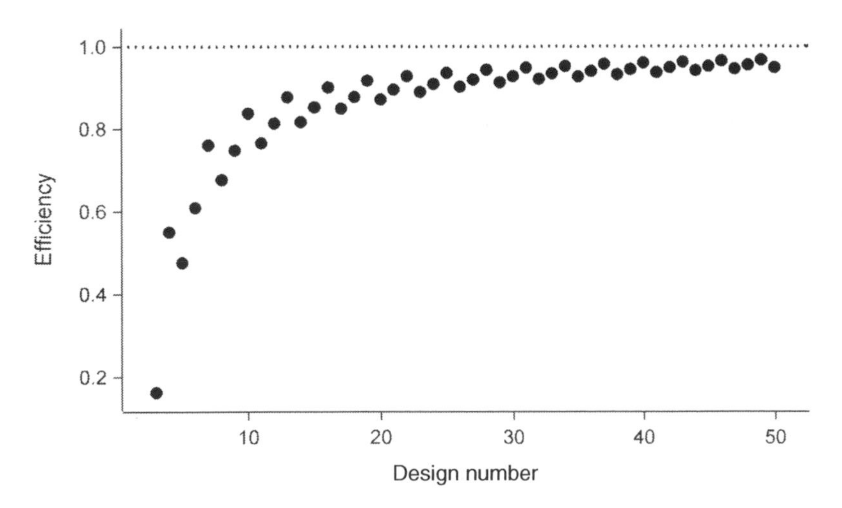

FIG. 11.7. Example 11.1: cubic regression through the origin. G-efficiency of the sequentially constructed designs.

TABLE 11.2. Example 1.1. Efficiency for a variety of purposes of the design* of Table 1.1 for measuring the desorption of carbon monoxide

Model	Optimality	Weight at design points				Efficiency
$\eta(x)$	criterion	0	$\surd(2)-1$	0.5	1	(%)
$\beta_0+\beta_1 x$	D	1/2	—	—	1/2	69.5
$\beta_0+\beta_1 x+\beta_2 x^2$	D	1/3	—	1/3	1/3	81.7
$\beta_0+\beta_1 x+\beta_2 x^2$	$\mathrm{D_S}$ for β_2	1/4	—	1/2	1/4	47.4
$\beta_1 x$	D	—	—	—	1	43.7
$\beta_1 x+\beta_2 x^2$	D	—	—	1/2	1/2	62.4
$\beta_1 x+\beta_2 x^2$	$\mathrm{D_S}$ for β_2	—	$\surd 2/2$	—	$1-\surd 2/2$	47.2

* The design region is scaled to be $\mathcal{X}=[0,1]$. The design of Table 1.1 is then

$$\xi_{22}=\left\{\begin{array}{cccccc} 0.02 & 0.1 & 0.2 & 0.5 & 0.84 & 1.0 \\ 2/22 & 2/22 & 3/22 & 5/22 & 6/22 & 4/22 \end{array}\right\}.$$

11.3 An Example of Design Efficiency: The Desorption of Carbon Monoxide. Example 1.1 Continued

The design of Table 1.1 for studying the desorption of carbon monoxide is typical of many in the scientific and technical literature: the levels of the factors and the number of replicates at each design point seem to have been chosen with no very clear objective in mind. As the examples of optimum designs in this book show, precisely defined objectives lead to precisely defined designs, the particular design depending upon the particular objectives. In this section the efficiency of the design in Table 1.1 is calculated for a number of plausible objectives.

The models considered for these data in Chapter 8 included first- and second-order polynomials, either through the origin or with allowance for a non-zero intercept. The D-optimum designs for all of these models require at most three design points: $0, 0.5$, and 1 as does the $\mathrm{D_S}$-optimum design for checking the necessity of the second-order model with a non-zero intercept. The $\mathrm{D_S}$-optimum design for checking the second-order model through the origin introduces one extra design point, $\surd 2-1=0.414$; both the $\mathrm{D_S}$-optimum designs have unequal weighting on the design points. Even so, the design of Table 1.1, with six design points, can be expected to be inefficient for these purposes. That this is so is shown by the results of Table 11.2. Three of the D-efficiencies are below 50% and only one is much above.

The design is more efficient for models which are allowed to have a non-zero intercept. However, since carbon monoxide is not desorbed in the absence of the catalyst, it would be safe to design the experiment on the assumption that $\beta_0 = 0$. Then, the bottom half of the table shows that the optimum design concentrates trials at unity and $\sqrt{2} - 1$ or 0.5. With its greater spread of design points, the actual design achieves efficiencies of 40–60% for these models, indicating that about half the experimental effort is wasted. As we have seen, a first-order model fits these data very well. It is therefore unnecessary to design for any more complicated models than those listed in the table. However, there remains the question as to whether a design can be found that is efficient for all, or several, of these criteria. In particular, can designs be found which are efficient both for estimating the parameters of a first-order model and for checking the fit of that model? We discuss such topics further in Chapter 23 on compound design criteria.

11.4 Polynomial Regression in One Variable

In the remaining two sections of this chapter we present designs for polynomial models. In this section the model is a dth-order polynomial in one factor. In the next section it is a second-order polynomial in m factors.

The model is

$$\mathrm{E}(y) = \beta_0 + \sum_{j=1}^{d} \beta_j x^j, \tag{11.8}$$

with $\mathcal{X} = [-1, 1]$. We begin with D-optimum designs and then consider $\mathrm{D_S}$-optimum designs for the highest-order terms in (11.8).

The D-optimum continuous design for $d = 1$ and $d = 2$ have appeared several times. For $d = 1$ half the trials are at $x = 1$ and the other half are at $x = -1$. For $d = 2$, a quadratic model, a third of the trials are at $x = -1, 0$, and 1. In general, $p = d + 1$ and the design puts mass $1/p$ at p distinct design points. Guest (1958) shows that the location of these points depends upon the derivative of the Legendre polynomial $P_d(x)$. This set of orthogonal polynomials is defined by the recurrence

$$(d + 1)P_{d+1}(x) = (2d + 1)xP_d(x) - dP_{d-1}(x) \tag{11.9}$$

with $P_0(x) = 1$ and $P_1(x) = x$ (see, for example, Abramowitz and Stegun 1965, p. 342). From (11.9)

$$P_2(x) = \frac{3x^2 - 1}{2}$$

and

$$P_3(x) = \frac{5x^3 - 3x}{2}.$$

Guest (1958) shows that the points of support of the D-optimum design for the dth order polynomial are at ± 1 and the roots of the equation

$$P'_d(x) = 0.$$

Equivalently (Fedorov 1972, p. 89), the design points are roots of the equation

$$(1 - x^2)P'_d(x) = 0.$$

For example, when $d = 3$, the design points are at ± 1 and those values for which

$$P'_3(x) = \frac{15x^2 - 3}{2} = 0.$$

That is, $x = \pm 1/\sqrt{5}$. Table 11.3 gives analytical and numerical expressions for the optimum x values up to the sixth-order polynomial. The designs up to this order were first found by Smith (1918) in a remarkable paper set in its historical context in §10.11; the design criterion was what is now called G-optimality. Her description of the criterion is as follows: 'in other words the curve of standard deviation with the lowest possible maximum value within the working range of observations is what we shall attempt to find'. In a paper of 85 pages she found designs not only for constant error standard deviation, but also for deviations of the asymmetrical form $\sigma(1 + ax), (0 \leq a < 1)$ and of the symmetrical form $\sigma(1 + ax^2), (a > -1)$. D-optimum designs for general non-constant variance are described in §22.2.

Table 11.3 exhibits the obvious feature that the optimum design depends upon the order of the polynomial model. In §21.4 designs are found, using a compound design criterion, which are simultaneously as efficient as possible for all models up to the sixth order. The results are summarized in Table 21.1. We conclude the present section with the special case of D_S-optimum designs for the highest-order term in the model (11.8).

For the quadratic polynomial, that is, $d = 2$, the D_S-optimum design for β_2 when $\mathcal{X} = [-1, 1]$ puts half of the trials at $x = 0$, with a quarter at $x = -1$ and $x = +1$. The extension to precise estimation of β_d in the dth order polynomial (Kiefer and Wolfowitz 1959) depends on Chebyshev polynomials. The design again has $d + 1$ points of support, but now with

$$x_j = -\cos\left(\frac{j\pi}{d}\right) \quad (0 \leq j \leq d).$$

The D_S-optimum design weight w^* is spread equally over the $d - 1$ points in the interior of the region with the same weight divided equally between

TABLE 11.3. Polynomial regression in one variable: points of support of D-optimum designs for dth-order polynomials

d	x_1	x_2	x_3	x_4	x_5	x_6	x_7
2	-1			0			1
3	-1	$-a_3$				a_3	1
4	-1	$-a_4$		0		a_4	1
5	-1	$-a_5$	$-b_5$		b_5	a_5	1
6	-1	$-a_6$	$-b_6$	0	b_6	a_6	1

$a_3 = 1/\sqrt{5} = 0.4472$
$a_4 = \sqrt{(3/7)} = 0.6547$
$a_5 = \sqrt{\{(7+2\sqrt{7})/21\}} = 0.7651, \quad b_5 = \sqrt{\{(7-2\sqrt{7})/21\}} = 0.2852$
$a_6 = \sqrt{\{(15+2\sqrt{15})/33\}} = 0.8302, \quad b_6 = \sqrt{\{(15-2\sqrt{15})/33\}} = 0.4688$

$x = \pm 1$, that is,

$$w^*(-1) = w^*(1) = \frac{1}{2d}$$

$$w^*\left\{\cos\left(\frac{j\pi}{d}\right)\right\} = \frac{1}{d} \quad (1 \leq j \leq d-1),$$

which generalizes the 1/4, 1/2, 1/4 weighting when $d = 2$. If exact designs are required, for N not a multiple of $2d$, the numerical methods of Chapter 12 are required.

11.5 Second-order Models in Several Variables

The second-order polynomial in m factors is

$$\mathrm{E}(y) = \beta_0 + \sum_{j=1}^{m} \beta_j x_j + \sum_{j=1}^{m-1} \sum_{k=j+1}^{m} \beta_{jk} x_j x_k + \sum_{j=1}^{m} \beta_{jj} x_j^2.$$

Continuous D-optimum designs for this model over the sphere, cube, and simplex are given by Farrell, Kiefer, and Walbran (1968), who also give designs for higher-order polynomials. In this section we first consider designs when $\mathcal{X}$ is a sphere and then when it is cube. In both cases the description of the optimum continuous design is augmented by a table of small exact designs. Designs over the simplex are the subject of Chapter 16 on mixture experiments.

D-optimum continuous designs over the sphere have a very simple structure. A measure $2/\{(m+1)(m+2)\}$ is put on the origin, that is, the centre

TABLE 11.4. Second-order polynomial in m factors: continuous D-optimum designs for spherical experimental region

m	p	$\|M(\xi^*)\|$	d_{ave}	$d_{\max}$
2	6	2.616×10^{-2}	4.40	6
3	10	2.519×10^{-7}	7.14	10
4	15	7.504×10^{-15}	10.71	15
5	21	4.440×10^{-25}	15.17	21

point of the design. The rest of the design weight is uniformly spread over the sphere of radius $\sqrt{m}$ which forms the boundary of $\mathcal{X}$. Table 11.4 gives the values of $|M(\xi^*)|$ for these optimum designs for small m, together with the values of $\bar{d}(\xi^*)$, which equal p, and the values of $d_{\text{ave}}(x, \xi^*)$ found, for computational convenience, by averaging over the points of the 5^m factorial with vertices ± 1. Although this averaging excludes part of $\mathcal{X}$, it does provide an informed basis for the comparison of designs. Greater detail about the behaviour of the variance function can be found from the variance–dispersion plots of §6.4.

Exact designs approximating the continuous designs are found by the addition of centre points and star points, with axial co-ordinate $\sqrt{m}$, to the points of the 2^m factorial. Table 11.5 gives nine designs, for several of which the D-efficiency is 98% or better. The addition of several centre points to the designs, which is often recommended to provide an estimate of σ^2, causes a decrease in D-efficiency. For example, for $m = 3$ the optimum weight at the centre is $2/\{(m+1)(m+2) = 1/10\}$, so that the addition of one or two centre points to the 2^3 factorial with star points provides a good exact design. However, increasing the number of centre points does initially have the desirable effects of reducing the average and maximum values of $d(x, \xi)$. In interpreting the results of Tables 11.4 and 11.5 it needs to be kept in mind that the values of d_{ave} and the maximum variance $d_{\max}$ are calculated only at the points of the 5^m factorial. In particular, the values of $d_{\max}$ for the number of centre points $N_0 = 1$ are an underestimate of $\bar{d}(\xi)$, which equals p only for the optimum continuous design.

The situation for cubic design regions is slightly more complicated. Farrell, Kiefer, and Walbran (1968) show that the optimum continuous design is supported on subsets of the points of the 3^m factorial, with the members of each subset having the same number of non-zero co-ordinates. Define $[k]$ to be the set of points of the 3^m factorial with k non-zero co-ordinates. For example, $[0]$ contains a single point, the centre point, and $[m]$ is the set of points of the 2^m factorial, that is, with all m co-ordinates

TABLE 11.5. Second-order polynomial in m factors: central composite exact designs for spherical experimental region

m	p	N	N_0	D-efficiency (%)	d_{ave}	$d_{\max}$
2	6	9	1	98.6	5.52	9.00
		11	3	96.9	4.35	6.88
		13	5	89.3	4.57	8.13
3	10	15	1	99.2	8.16	15.00
		17	3	97.7	6.54	10.52
		19	5	91.9	6.71	11.76
4	15	25	1	99.2	12.37	25.00
		27	3	98.9	10.63	15.75
		29	5	95.2	10.83	16.92

The design consists of a 2^m factorial with $2m$ star points and N_0 centre points. All values of d_{ave}, as well as $d_{\max}$ are calculated over the points of the 5^m factorial.

equal to ± 1, which are the corner points of the 3^m factorial. Only three subsets are required, over each of which a specified design weight is uniformly distributed. The D-optimum continuous designs then have support on three subsets $[k]$ to be chosen with

$$\begin{array}{llll} 0 \le k \le m-2 & k = m-1 & k = m & (2 \le m \le 5) \\ 0 \le k \le m-3 & k = \left\{ \begin{array}{c} m-2 \\ \text{or} \\ m-1 \end{array} \right. & k = m & (m \ge 6) \end{array} . \tag{11.10}$$

Of the designs satisfying (11.10), those with support ([0], $[m-1]$, $[m]$) require fewest support points. These are the centre point, the midpoints of edges, and the corner points of the 3^m factorial, respectively. This family was studied by Kôno (1962). The weights for the D-optimum continuous designs for $m \le 5$ are given in Table 11.6. It is interesting to note that the central composite designs, which belong to the family ([0], [1], $[m]$), cannot provide the support for a D-optimum continuous design when $m > 2$.

Since the D-optimum continuous designs have support on the points of the 3^m factorial, it is reasonable to expect that good exact designs for small N can be found by searching over the points of the 3^m factorial. Properties of designs for second-order models for $m \le 5$ found using SAS (see §13.2 for

TABLE 11.6. Second-order polynomial in m factors: cubic experimental region. Weights and number of support points n for continuous D-optimum designs supported on points of the 3^m factorial with [0], $[m-1]$, and $[m]$ non-zero co-ordinates

Number of	Design weights			
factors m	w_0	w_{m-1}	w_m	n
2	0.096	0.321	0.583	9
3	0.066	0.424	0.510	21
4	0.047	0.502	0.451	49
5	0.036	0.562	0.402	113

computational matters) are given in Table 11.7. The D- and G-efficiencies relative to the continuous designs in Table 11.4 are also given. For fixed m, and therefore fixed number of parameters p, the D-efficiency is smallest for $N = p$. The addition of one or a few trials causes an appreciable increase in the efficiency of the design, in addition to the reduced variance of parameter estimates coming from a design with more support points. This effect decreases as m increases. The G-efficiencies and the values of d_{ave} were calculated over a 5^m grid, as in Tables 11.4 and 11.5. The general behaviour of G-efficiency is similar to that of D-efficiency; for instance, moving from $N = p$ to $N = p+1$ can produce a large increase. However, as comparison of Figures 11.6 and 11.7 showed, the behaviour of G-efficiency is more volatile than that of D-efficiency, although the trend to increasing efficiency with N is evident. Small values of the average variance d_{ave} are desirable and these behave much like the reciprocal of G-efficiency, yielding better values as N increases for fixed m.

The designs in Tables 11.5, and 11.7 should meet most practical situations where a second-order polynomial model is fitted and N is restricted to be not much greater than p. Methods of dividing the designs into blocks are given in Chapter 15.

The calculation of the designs in these tables requires the use of numerical algorithms for optimum design, in our case those in SAS. Historically, before these algorithms were widely implemented, the emphasis was on evaluating the properties of existing designs, particularly the ([0], [1], $[m]$) family of central composite designs. For example, Table 21.5 of Atkinson and Donev (1992) gives the D-efficiencies of designs with this support as the number of observations at the centre, star, and factorial points is changed. The table also lists the D_S-efficiencies of the designs for second-order and quadratic

TABLE 11.7. Second-order polynomial in m factors: cubic experimental region. Properties of exact D-optimum designs

m	p	N	D_{eff}	G_{eff}	d_{ave}
2	6	6	0.8849	0.3636	7.69
		7	0.9454	0.6122	5.73
		8	0.9572	0.6000	5.38
3	10	10	0.8631	0.2903	11.22
		14	0.9759	0.8929	7.82
		15	0.9684	0.7754	8.85
		16	0.9660	0.7417	8.81
		18	0.9703	0.6815	9.22
		20	0.9779	0.8257	8.93
4	15	15	0.8700	0.4522	19.28
		18	0.9311	0.5115	14.50
		24	0.9669	0.6639	14.16
		25	0.9773	0.6891	14.07
		27	0.9815	0.6984	13.89
5	21	21	0.9055	0.3001	23.50
		26	0.9519	0.6705	19.86
		27	0.9539	0.6493	19.71

terms. If these efficiencies are indeed important, we would recommend generating the design with a compound criterion that includes a suitable weighting between D- and D_S-optimality. Table 21.3 gives designs generated by such a procedure for a second-order model with $p = 4$ and three values of N.

11.6 Further Reading

The study of D-optimality has been central to work on optimum experimental designs since the beginning (e.g. Kiefer 1959). An appreciable part of the material in the books by Fedorov (1972), Silvey (1980), and Pázman (1986) likewise stresses D-optimality. Farrell, Kiefer, and Walbran (1968), in addition to the results quoted in this chapter, give a summary of earlier work on D-optimality. This includes Kiefer and Wolfowitz (1959) and Kiefer

(1961) which likewise concentrate on results for regression models, including extensions to D_S-optimality.

Silvey (1980) compares first-order algorithms of the sort exemplified in §11.2. If, perhaps as a result of such sequential construction, the support of the optimum design is clear, the weights of the continuous design can be found by numerical optimization; see Chapters 12 and 13. Alternatively, a special algorithm can be used such as that of Silvey, Titterington, and Torsney (1978). References to further developments of this algorithm are given by Torsney and Mandal (2004).

12

ALGORITHMS FOR THE CONSTRUCTION OF EXACT D-OPTIMUM DESIGNS

12.1 Introduction

The construction of a design that is optimum with respect to a chosen criterion is an optimization problem which can be attacked in several ways. Most of the available algorithms borrow ideas from generic methods for function optimization where the objective function is defined by the stipulated criterion of optimality. For example, the Fedorov algorithm (§12.5) and its modifications described in §12.6 are steepest ascent algorithms. They were used to calculate many of the examples in this book.

The implementation of algorithms for the construction of experimental designs takes into account the specific nature of the design problem. Some methods for continuous designs were described in §§9.4 and 9.5. In this chapter we describe algorithms for the construction of exact D-optimum designs. The problem is introduced in the next section. The search is usually carried out over a grid of candidate points. The sequential algorithms described in §12.4 produce designs so that a design with $N+1$ or $N-1$ trials is obtained from a design with N trials by adding or deleting an observation. However, a design that is optimum for a particular optimality criterion for N trials usually cannot be obtained from the optimum design with $N+1$ or $N-1$ trials in such a simple manner. Therefore the search for an optimum design is usually carried out for a specified design size. Most of these algorithms, which are the subject of §12.5, consist of three phases. In the first phase a starting design of N_0 trials is generated. This is then augmented to N trials and, in the third phase, subjected to iterative improvement. The basic formulae common to the three phases are presented in §12.3. These phases are detailed for the KL and the BLKL exchange algorithms in §12.6. Because the design criteria for exact designs do not lead to convex optimization problems, the algorithms of these sections may converge to local optima. The probability of finding the globally optimum design is increased by running the search many times from different starting points, possibly chosen at random. Other algorithms for the construction of exact designs are summarized in §12.8.

12.2 The Exact Design Problem

The exact D-optimum design measure ξ_N^* maximizes

$$|M(N)| = |F^{\mathrm{T}}F|, \tag{12.1}$$

where F is the $N \times p$ extended design matrix. As before, F is a function of m factors which may be quantitative factors continuously variable over a region, qualitative factors, or mixture variables. Because the design is exact, the quantities Nw_i are integer at all design points. Since the design may include replication the number of distinct design points may be less than N.

The optimum design is found by searching over the design region $\mathcal{X}$. For simple problems an analytical solution is sometimes possible as it was for cubic regression through the origin (9.5). But the complexity of the problem is usually such that numerical methods have to be used. A problem that has to be tackled is that the objective function (12.1) has many local optima.

Example 12.1 Second-order Response Surface in Two Factors Box and Draper (1971) use function maximization of the kind described in §9.5 to find exact D-optimum designs for second-order models in $m = 2$ and $m = 3$ factors with $\mathcal{X}$ a square or cube. When $m = 2$ the second-order model has $p = 6$ parameters. The exact optimum designs given by them for $N = 6, \ldots, 9$ are as follows:

- $N = 6: (-1,-1), (1,-1), (-1,1), (-\alpha,-\alpha), (1,3\alpha), (3\alpha,1)$ where $\alpha = \{4 - \sqrt{13}\}/3 = 0.1315$. Equally optimum designs are obtained by rotation of this design through $\pi/2, \pi$, or $3\pi/2$;
- $N = 7: (\pm 1, \pm 1), (-0.092, 0.092), (1, -0.067), (0.067, -1)$;
- $N = 8: (\pm 1, \pm 1), (1, 0), (0.082, 1), (0.082, -1), (-0.215, 0)$;
- $N = 9$: the 3^2 factorial with levels -1, 0, and 1.

The continuous version of this design problem was the subject of §9.4, where it was stated that, for general m, the D-optimum continuous design was supported on subsets of the points of the 3^m factorial; these numerical results show that the general result does not hold for exact designs. The exact designs are illustrated in Figure 12.1, with some designs rotated to highlight the common structure as N increases. Apart from the design for $N = 6$, the designs are very close to fractions of the 3^2 factorial. However, even the design for $N = 6$ contains three such points.

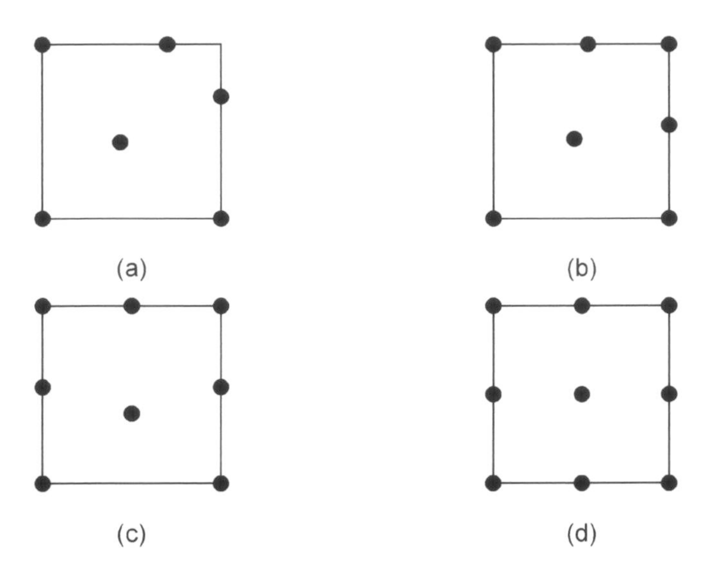

FIG. 12.1. Example 12.1: second-order response surface in two factors. Exact D-optimum designs found by optimization over the continuous region $\mathcal{X}$: (a) $N = 6$; (b) $N = 7$; (c) $N = 8$; (d) $N = 9$.

The design weights for the continuous D-optimum design for this problem are

4 corner points $(\pm 1, \pm 1)$	0.1458	
4 centre points of sides $(0, \pm 1; \pm 1, 0)$	0.0802	(12.2)
1 centre point $(0, 0)$	0.0960	

The D-efficiencies of the designs of Figure 12.1 relative to this design are given in Table 12.1. A good exact design for $N = 13$ is found by putting two, rather than one, observation at each of the four factorial points. The D-efficiency of this design is also given in Table 12.1. ■

As the dimension of the problem and the number of factors increase, the time needed to search over the continuous region for the exact design rapidly becomes unacceptable. Following the indication of results such as those of Box and Draper, the search over the continuous region $\mathcal{X}$ is often replaced by a search over a list of candidate points. The list, usually a coarse grid in the experimental region, frequently includes the points of the D-optimum continuous design. Even if the dimension of the problem is not such as to make the continuous optimization prohibitively difficult, there are practical reasons for restricting consideration to certain discrete values of the factors. In the example above the list might well consist of only the nine points of the

3^2 factorial. The design problem of Example 12.1 is then the combinatorial one of choosing the N out of the available nine points which maximize $|M(N)|$. In general, the problem is that of selecting N points out of a list of N_C candidate points. Since replication is allowed, the selection is with replacement.

Example 12.1 Second-order Response Surface in Two Factors continued An alternative to the exact designs given by Box and Draper are the designs of Table 11.6 found by searching over the grid of the 3^2 factorial. The designs are plotted in Figure 12.2, again with some rotation, now to emphasize the relationship with the designs of Figure 12.1. Searching over the 25-point grid generated from the 5^2 factorial led to the same designs as searching the 3^m grid. The D-efficiencies of the designs from searching over the grid and over the continuous region are given in Table 12.1. As can be seen, even for $N = p = 6$, little is lost by using the 3^2 grid rather than the continuous design region. Comparison of Figures 12.1 and 12.2 shows that the two sets of designs have the same symmetries and that the Box–Draper designs involve only slight distortions of the fractions of the 3^m factorial. Use of the finer 5^2 factorial has no advantage over that of the 3^2 grid. ■

12.3 Basic Formulae for Exchange Algorithms

Numerical algorithms for the construction of exact D-optimum designs by searching over a list of N_C candidate points customarily involve the iterative improvement of an initial N-trial design. The initial design of size N can be constructed sequentially from a starting design of size N_0, either by the addition of points if $N_0 < N$, or by the deletion of points if $N_0 > N$. Improvement of the design in the third phase is made by an exchange in

TABLE 12.1. Example 12.1: second-order response surface in two factors. D-efficiencies of designs of Figures 12.1 and 12.2 found by searching over a continuous square design region and over the 3^2 factorial

N	Points of 3^2 factorial	Continuous square region
6	0.8849	0.8915
7	0.9454	0.9487
8	0.9572	0.9611
9	0.9740	0.9740
13	0.9977	0.9977

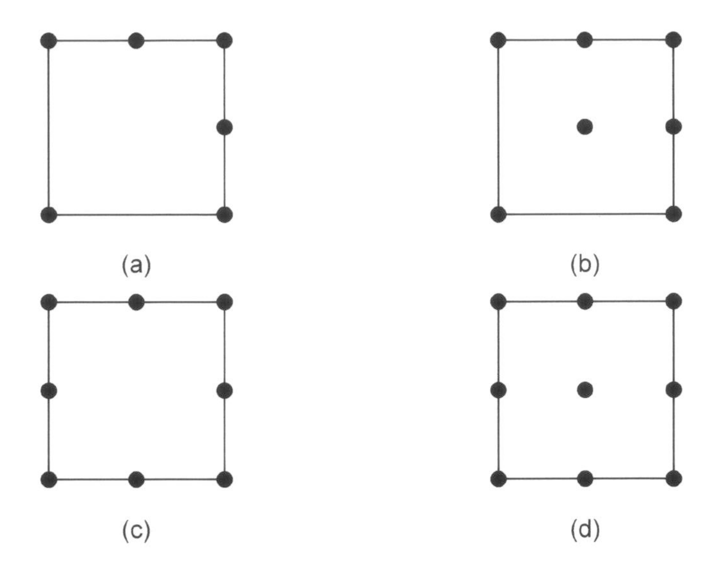

FIG. 12.2. Example 12.2: second-order response surface in two factors. Exact D-optimum designs found by searching the points of the 3^m and 5^m factorials: (a) $N = 6$; (b) $N = 7$; (c) $N = 8$; (d) $N = 9$.

which points in the design are replaced by those selected from the candidate list, with the number of points N remaining fixed. The common structure is that the algorithms add a point x_l, to the design, delete a point x_k from it, or replace a point x_k from the design with a point x_l, from the list of candidate points. For D-optimality the choice of the points x_k and x_l depends on the variance of the predicted response at these points, the determinant of the information matrix, and the values of elements of its inverse. The search strategy is determined by the algorithm, several of which are described in the next sections. In this section we give the formulae which provide updated information at each iteration.

Let i, $(i \geq 0)$, be the number of iterations already performed. To combine the sequential and interchange steps in a single formula we require constants c_k and c_l such that

1. $c_l = (N+1)^{-1}$, $c_k = 0$ if a point x_l is added to the design;
2. $c_k = (N+1)^{-1}$, $c_l = 0$ if the design point x_k is deleted;
3. $c_k = c_l = (N+1)^{-1}$ if the design point x_l is exchanged with the point x_k.

Let the vector of model terms which form the rows of F at these points be $f_k^{\mathrm{T}} = f^{\mathrm{T}}(x_k)$ and $f_l^{\mathrm{T}} = f^{\mathrm{T}}(x_l)$. The following formulae then give the relation between the information matrices, their determinants, and the inverses of the information matrices at iterations i and $i+1$:

$$M(\xi_{i+1}) = \frac{1-c_l}{1-c_k} M(\xi_i) + \frac{1}{1-c_k}(c_l f_l f_l^{\mathrm{T}} - c_k f_k f_k^{\mathrm{T}}) \tag{12.3}$$

$$\begin{aligned} |M(\xi_{i+1})| = \Bigg[& \left\{1 + \frac{c_l}{1-c_l} d(x_l, \xi_i)\right\}\left\{1 - \frac{c_k}{1-c_k} d(x_k, \xi_i)\right\} \\ & + \frac{c_k c_l}{(1-c_l)^2} d^2(x_k, x_l, \xi_i)\Bigg] \left(\frac{1-c_l}{1-c_k}\right)^p |M(\xi_i)| \end{aligned} \tag{12.4}$$

$$M^{-1}(\xi_{i+1}) = \frac{1-c_k}{1-c_l}\left\{M^{-1}(\xi_i) - \frac{M^{-1}(\xi_i) A M^{-1}(\xi_i)}{qz + c_k c_l d^2(x_l, x_k, \xi_i)}\right\} \tag{12.5}$$

where

$$\begin{aligned} d(x_l, x_k, \xi_i) &= f_l^{\mathrm{T}} M^{-1}(\xi_i) f_k \\ q &= 1 - c_l + c_l d(x_l, \xi_i) \\ z &= 1 - c_l - c_k d(x_k, \xi_i) \end{aligned}$$

and

$$A = c_l z f_l f_l^{\mathrm{T}} + c_k c_l d(x_l, x_k, \xi_i)(f_l f_k^{\mathrm{T}} + f_k f_l^{\mathrm{T}}) - c_k q f_k f_k^{\mathrm{T}}. \tag{12.6}$$

For example, if a point x_l is added to an N-point design with information matrix $M(\xi_i)$,

$$|M(\xi_{i+1})| = \{1 + d(x_l, \xi_i)\}\left(\frac{N}{N+1}\right)^p |M(\xi_i)|.$$

If a point x_k is removed from an $N+1$-point design with information matrix $M(\xi_i)$,

$$|M(\xi_{i+1})| = \{1 - d(x_k, \xi_i)\}\left(\frac{N+1}{N}\right)^p |M(\xi_i)|.$$

Finally, if a point x_l is added while a point x_k is removed from a design with information matrix $M(\xi_i)$, the determinant of the information matrix of the new design is

$$|M(\xi_{i+1})| = [\{1 - d(x_k, \xi_i)\}\{1 + d(x_l, \xi_i)\} + d^2(x_k, x_l, \xi_i)]|M(\xi_i)|.$$

The computer implementation of these formulae requires care: updating the design and the inverse of its information matrix, in addition to recalculation of the variances at the design points, can consume computer time and space. With the high and continually increasing speed of computers this is rarely a crucial issue.

12.4 Sequential Algorithms

An exact design for N trials can be found either by the sequential addition of trials to a smaller design, or by the sequential deletion of trials from a larger design. If necessary, the exact N-trial design can then be improved by the methods described in the next two sections. The formulae of §12.3 apply for either addition or deletion.

1. *Forward procedure.* Given a starting design with N_0 trials, the N-trial exact design ($N > N_0$) is found by application of the algorithm illustrated in §11.2, that is, each trial is added sequentially at the point where the variance of the predicted response is greatest:

$$d(x_l, \xi_i) = \max_{\mathcal{X}} d(x, \xi_i). \tag{12.7}$$

   As was shown in §9.4, continuation of this process leads, as $i \to \infty$, to the D-optimum continuous design ξ^*. The exact design yielded by the application of (12.7) can then be regarded as an approximation to ξ^* which improves as N increases.

2. *Backwards procedure.* This starts from a design with $N_0 >> p$ and proceeds by the sequential deletion of points with low variance. At each iteration we delete the design point x_k at which the variance of the predicted response is a minimum:

$$d(x_k, \xi_i) = \min_{\mathcal{X}} d(x, \xi_i). \tag{12.8}$$

   As (12.4) shows, the decrease in the value of the determinant of the information matrix at each iteration is the minimum possible. Often the list of candidate points is taken as the starting design for this procedure. However, if N is so large that the N-trial design might contain some replication, the starting design could contain two or more trials at each candidate point.

A common feature of both these one-step-ahead procedures is that they do not usually lead to the best exact N-trial design. The backwards procedure is clumsy in comparison with the forwards procedure because of the size of the starting design. The performance of the forwards procedure can be improved by using different starting designs. In order to generate a series of starting designs Galil and Kiefer (1980) suggest selecting j points ($0 < j < p$) at random from the candidate set. For this design $F^{\mathrm{T}}F$ will be singular. However, the $N \times N$ matrix FF^{T} can be used instead with the forwards procedure until the design is no longer singular. Thereafter (12.7) is used directly. Thus different runs of the algorithm will produce a variety of exact N-trial designs, the best of which will be selected. The KL exchange algorithm described in §12.6 uses another method to obtain a variety of starting designs. The method, which relies upon the regularization of singular $F^{\mathrm{T}}F$, can be adapted to allow division of the trials of the design into blocks of specified size.

In some practical situations, particularly when the number of trials is appreciably greater than the number of parameters, the forwards sequential procedure yields a satisfactory design. In others, the design will need to be improved by one of the non-sequential methods of the next two sections.

12.5 Non-sequential Algorithms

Non-sequential algorithms are intended for the improvement of an N-trial exact design. This is achieved by deleting, adding, or exchanging points, according to the rules of the particular algorithm, to obtain an improved N-trial design. Because the procedures are non-sequential, it is possible that the best design of N trials might be quite different from that obtained for $N-1$ or $N+1$ trials.

Van Schalkwyk (1971) proposed an algorithm which at each iteration deleted the point x_k from the design to cause minimum decrease in the determinant of the information matrix as in (12.8). The N-trial design is then recovered by adding the point x_l which gives a maximum increase of the determinant, thus satisfying (12.7) for $i = N - 1$. Progress ceases when the same point is deleted and then re-entered. Mitchell and Miller (1970) and Wynn (1972) suggest a similar algorithm in which the same actions are performed, but in the opposite order: first the point x_l, is added and then the point x_k is deleted, with the points for addition and deletion being decided by the same rule as in van Schalkwyk's algorithm.

The idea of alternate addition and deletion of points is extended in the DETMAX algorithm (Mitchell 1974) to 'excursions' of various sizes in which a chosen number of points is sequentially added and then deleted. The size of the excursions increases as the search proceeds, usually up to a maximum of six. Of course, a size of unity corresponds to the algorithm described at the end of the previous paragraph. Galil and Kiefer (1980) describe computational improvements to the algorithm and generate D-optimum exact designs for second-order models in three, four, and five factors.

DETMAX, like the other algorithms described so far, separates the searches for addition and deletion. The two operations are considered together in the exchange algorithm suggested by Fedorov (1972, p. 164): at each iteration the algorithm evaluates all possible exchanges of pairs of points x_k from the design and x_l, from the set of candidate points.

The exchange giving the greatest increase in the determinant of the information matrix, assessed by application of (12.4), is undertaken: the process continues as long as an interchange increases the determinant. As one way of speeding up the algorithm, Cook and Nachtsheim (1980) consider each design point in turn, perhaps in random order, carrying out any beneficial exchange as soon as it is discovered. They call the resulting procedure a modified Fedorov exchange. Johnson and Nachtsheim (1983) further reduce the number of points to be considered for exchange by searching over only the k ($k < N$) design points with lowest variance. The algorithm of the next section generalizes Johnson and Nachtsheim's modification of Fedorov's original exchange procedure.

12.6 The KL and BLKL Exchange Algorithms

As the exchange algorithms for exact designs are finding local optima of functions with many extrema, improvement can come from an increased number of searches with different starting designs as well as from more precise identification of local optima. Experience suggests that, for fixed computational cost, there are benefits from a proper balance between the number of tries and the thoroughness of the local search. For example, Fedorov's exchange algorithm is made slow by the large number of points to be considered at each iteration—a maximum of $N(N_C - 1)$ in the absence of replication in the design—and by the need to follow each successful exchange by updating the design, the covariance matrix $M^{-1}(\xi)$, and the variance of the predicted values at the design and candidate points. The thoroughness of the search contributes to the success of the algorithm. However, the search can be made faster by noting that the points most likely to be exchanged are design points with relatively low variance of the predicted response and candidate points for which the variance is relatively high. This idea underlies the KL exchange and its extension to blocking, which we have called BLKL.

The algorithm passes through three phases:

1. Generation of the starting design.
2. Sequential generation of the initial N-trial design.
3. Improvement of the N-trial design by exchange.

Sometimes there may be points which the experimenter wishes to include in the design. The purpose might be to check the model, or they might represent data already available. The first phase starts with $N^{(1)}$ ($N^{(1)} \geq 0$) such points. The random starts to the search for the optimum come from choosing $N^{(2)}$ points at random from the candidate set, where $N^{(2)}$ is itself a randomly chosen integer $0 \leq N^{(2)} \leq \min(N - N^{(1)}, [p/2])$ and [A] is the integer part of A.

The initial N-trial design is completed by sequential addition of those $N - (N^{(1)} + N^{(2)})$ points which give maximum increase to the determinant of the information matrix. For $N < p$ the design will be singular and is regularized, as in (10.10), by replacement of $F^{\mathrm{T}}F$ by $F^{\mathrm{T}}F + \epsilon I_p$, where ϵ is a small number, typically between 10^{-4} and 10^{-6}. If the design is to be laid out in blocks, the search for the next design point is confined to those parts of the candidate set corresponding to non-full blocks.

In the third phase the exchange of points x_k from the design and x_l from the candidate list is considered. As in other algorithms of the exchange type, that exchange is performed which leads to the greatest increase in the determinant of the information matrix. The algorithm terminates when there is no longer any exchange which would increase the determinant. The points x_k and x_l considered for exchange are determined by parameters K and L such that

$$1 \leq k \leq K \leq N - N^{(1)}$$

and

$$1 \leq l \leq L \leq N_C - 1.$$

The point x_k is that with the kth lowest variance of prediction among the $N - N^{(1)}$ design points, with the initial $N^{(1)}$ points not considered for exchange, while x_l has the lth highest variance among the N_C candidate points. If blocking is required, the orderings of points should theoretically be over each block with exchanges limited to pairs within the same block. However, we have not found this refinement to be necessary.

When $K = N$ and $L = N_C - 1$, the KL exchange coincides with Fedorov's procedure. However, by choosing $K < N$ and $L < N_C - 1$, the number of pairs of points to be considered at each iteration is decreased. Although this must diminish the probability of finding the best possible exact design at each try, the decrease can be made negligible if K and L are properly chosen. The advantage is the decrease in computational time.

There are two possible modifications to this algorithm:

1. Make all beneficial exchanges as soon as they are discovered, updating the design after each exchange.

2. Choose the K design points and L candidate points at random rather than according to their variances.

The first modification brings the algorithm close to the modified Fedorov procedure of Cook and Nachtsheim (1980) when $K = N$ and $L = 1$ and becomes the K exchange (Johnson and Nachtsheim, 1983) for $0 < K < N$

and $L = 1$. Our extension includes the choice of L and so provides extra flexibility.

The second modification, that of the random choice of points, could be used when further increase in the speed of the algorithm is required. In this case the variance of the predicted value is calculated only for the $K + L$ points, rather than for all points in the design and the candidate set. In the unmodified algorithm this larger calculation is followed by ordering of the points to identify the $K + L$ for exchange. This modification should yield a relatively efficient algorithm when the number of design or candidate points is large.

The best values of K and L depend, amongst other variables, on the number of factors, the degrees of freedom for error $\nu = N - p$, and the number of candidate points N_C. For example, when $\nu = 0$, the variance of the predicted response at all design points is the same: there is then no justification for taking $K < N$. However, as ν increases, the ratio K/N should decrease. The best value of L increases with the size of the problem, but never exceeded $N/2$ in the examples considered by Donev (1988). In most cases values of K and L much smaller than these limits are sufficient, particularly if the number of tries is high. In an example reported in detail by Donev the unmodified Fedorov algorithm was used for repeated tries on two test problems in order to find the values of K and L required so that the optimum exchange was always made. In none of the tries would taking K or L equal to unity have yielded the optimum exchange—usually much larger values were necessary.

When the observations have to be divided into blocks generated by a single qualitative variable, the BLKL algorithm searches for the D-optimum design with pre-specified block sizes. If there are loose or no restrictions on the sizes of the blocks, the optimum block sizes are also found. Examples of such designs and extensions of the scope of application are discussed in Chapters 15 and 16.

12.7 Example 12.2: Performance of an Internal Combustion Engine

One of the main advantages of using algorithms to construct experimental designs, rather than trying to use a standard design, is that designs can be tailor-made to satisfy the exact needs of the experimenter. In this example, which is a simplified version of a problem which arose in testing car engines, there are two factors, spark advance x_1 and torque x_2. Both are independently variable, so that a square design region for which standard designs exist is theoretically possible. However, certain factor combinations

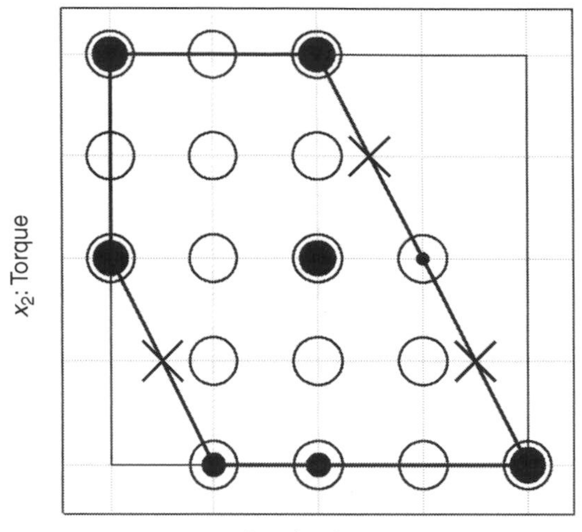

FIG. 12.3. Internal combustion engine performance: • 20-trial D-optimum design points, with marker size corresponding to 1, 2, or 3 observations; ∘ unused candidate points; × second-stage candidate points (also not used).

have to be excluded, either because the engine would not run under these conditions, or because it would be damaged or destroyed. The excluded combinations covered trials when the two factors were both at either high or low levels. The resulting pentagonal design region is shown in Figure 12.3. It is known from previous experiments that a second-order model in both factors is needed to model the response and that 20 trials give sufficient accuracy.

Selecting a list of candidate points in such situations is not easy because it is not clear which points will support the exact D-optimum design. We used an initial candidate set that contained the 17 points of the 5^2 factorial that satisfy the constraints. The candidate set is shown by circles in Figure 12.3. Dots within circles denote the points of the optimum design, with open circles denoting those members of the candidate set which were not used. The resulting D-optimum design has support at 8 of the total 17 points: five points support three trials each, two two and one one. Replication is far from equal. Because of the oblique nature of the constraints, some of the candidate points are distant from the constraints. The three extra candidate points marked by crosses were therefore added to give a denser candidate set.

Repeating the search over this enlarged set of 20 points yielded the same optimum design as before. The extra candidate points were not used.

There are two general points about this example. The first is that exclusion of candidate points on a grid, which lie outside the constraints of the design region, may result in too sparse a candidate set. In our example this was not the case. More important is the design of Figure 12.3. This has eight support points. Perhaps almost as efficient a design could have been found by inspired distortion of the nine-trial D-optimum design for the square region. However, guessing the design weights, given those of (12.2), would have required supernatural levels of inspiration. Our experience suggests that such attempts to adapt standard designs are the initial reaction of experimenters faced with constrained design problems yielding irregular regions. But, even if such inspiration did descend on the experimenter, and problems with larger number of factors and constraints than in our example are increasingly difficult to solve, it would not have been needed. Once a suitable candidate set has been obtained, SAS can be used to find the required design in the usual fashion.

12.8 Other Algorithms and Further Reading

The BLKL exchange algorithm was initially made available as FORTRAN code in Appendix A of Atkinson and Donev (1992); it is currently incorporated in the computer package GENSTAT (`http://www.vsn-intl.com/genstat/`). Miller and Nguyen (1994) also provided a computer code for the Fedorov exchange algorithm. These codes came with limitations as they were programmed to be used only for certain standard classes of problems, but they do provide a relatively easy opportunity to develop extensions for other situations where no standard designs are available. A list of subsequent work of this type includes Atkinson and Donev (1996) who generate D-optimum designs in the presence of a time trend (see Chapter 15); Donev (1997, 1998) who constructs optimum crossover designs; Trinca and Gilmour (2000), Goos and Vandebroek (2001*a*), Goos (2002), and Goos and Donev (2006*a*) who construct response surface designs in random blocks. Goos and Vandebroek (2001*b*, 2003) construct D-optimum split-plot designs. Most of these tasks can be implemented in SAS—see Chapter 13.

All algorithms discussed so far limit their search to a pre-specified list of candidate points. There is however no guarantee that the D-optimum design needs only a subset of these points as support. Donev and Atkinson (1988) investigate the extension of the search to the neighbourhood of the best design found by searching over the candidate list. They thus find the optimum more precisely. This is achieved by considering beneficial perturbations of the co-ordinates of the currently best available design. Candidate points are not used. The application of this *adjustment algorithm* to standard design problems leads to modest improvements of most of the designs found by searching over the candidate list. However, this approach might be more useful in practical situations where the design regions are constrained and where the initial list of candidate points are a poor guess for what the best support of the required design should be.

Meyer and Nachtsheim (1995) extend this idea by iteratively improving all initial designs by searching for beneficial co-ordinate exchanges. Thus, their Co-ordinate Exchange algorithm requires no candidate set of points, making it applicable in cases when the number of factors is so large that a candidate set would be unwieldy, or perhaps even infeasible—for example an optimal response surface design for 30 factors.

Note that the adjustment algorithm can be viewed as a particular method of derivative-free optimization. However, for response surface designs the D-criterion is a smooth function of the factor values, so the derivatives of this relationship can greatly aid in determining the optima. These derivatives can be cumbersome to derive analytically, but our experience is that numerical calculation of the derivatives works well with general optimization methods, and this is how we recommend finding the precisely optimum factor values for an exact design. See §13.4 for an example.

Bohachevsky *et al.* (1986) use simulated annealing to construct an exact design for a non-linear problem in which replication of points is not allowed. The search is not restricted to a list of candidate points. In the application of simulated annealing reported by Haines (1987), D-, G-, and V-optimum exact designs were found for a variety of polynomial models. Atkinson (1992) uses a segmented annealing algorithm to speed the search which is applied to Example 12.2. Payne *et al.* (2001) and Coombes *et al.* (2002) argue that simulated annealing is very slow and inefficient because it does not make use of the past history of the search and compare it to the Tabu and the Reactive Tabu searches. The Tabu search bans moves that would return the current design to one of the recently considered designs. A refinement of this method is the Reactive Tabu search, proposed by Battiti and Tecchiolli (1992). In contrast to the Tabu search, the parameters of the search are tuned as the search progresses, thus the optimum design is found faster.

In principle, it is easy to modify any of the algorithms for criteria other than D-optimality. For example, Welch (1984) describes how DETMAX can be adapted to generate G- and V-optimum designs, while Tack and Vandebroek (2004) found cost-effective designs by incorporating the cost of the observations into the criterion of optimality.

SAS can generate both D- and A-optimum designs. It also offers two space-filling criteria that are not directly related to the information-based criteria that we discuss. The construction of space-filling designs is one of the features of the `gossett` package (Hardin and Sloane 1993).

Neither DETMAX nor exchange algorithms are guaranteed to find the globally optimum design. In rare, and small, cases globally optimum designs can be found, at the cost of increased computation, by the use of standard methods for combinatorial optimization. Another general method of search is that of branch and bound, which is applied by Welch (1982) to finding globally D-optimum designs over a list of candidate points. The method guarantees to find the optimum design. However, the computational requirements increase rapidly with the size of the problem.

13

OPTIMUM EXPERIMENTAL DESIGN WITH SAS

13.1 Introduction

For optimum exact design, the key SAS tool is the OPTEX procedure. For optimum continuous designs, there is no specialized tool in SAS, but with a moderate amount of programming the non-linear optimization tools in SAS/IML software serve this purpose. Finally, the ADX Interface provides a point-and-click approach for constructing optimum designs for many of the kinds of experiments discussed in this book.

For this book, our aim will be to survey the optimum design capabilities of these tools, and to demonstrate their use for the kinds of optimum design problems with which we are concerned. It is not our goal to cover all aspects of how to use these programs. For more complete information on the use of these tools, refer to the SAS documentation (SAS Institute Inc. 2007*b*,*c*).

13.2 Finding Exact Optimum Designs Using the OPTEX Procedure

The OPTEX procedure searches discrete candidate sets for optimum exact designs for linear models. The candidates are defined outside the procedure, using any of SAS's facilities for manipulating data sets. Models are specified using the familiar general linear modelling syntax of SAS analytic procedures, including both quantitative and qualitative effects. OPTEX can find both D- and A-optimum designs. A variety of search algorithms are available, as well as various measures of design efficiency (although these are not necessarily precisely the same as the measures we discuss in this book; see §13.3). The design constructed by OPTEX can be saved to a data set for further processing, including randomization, editing, and eventually, analysis.

OPTEX is a very versatile procedure and much of the material on SAS in the remainder of the book will reveal different features of its use and computational details. In this chapter we will simply introduce the procedure and discuss some general features of its use. OPTEX offers five different search algorithms, which can be combined with various methods of initializing the search. All of the search algorithms are different specializations of KL exchange, discussed in §12.6, and initialization in general begins with some number $0 \leq N_r \leq N$ of randomly chosen candidate points and fills up the rest of the starting design by sequential addition. An arbitrary data set may also be given as the initial design.

The minimum requirements for using OPTEX are a data set of candidate points and a linear model. For example, the following statements first define a 3×3 set of candidates for two quantitative factors and search for a D-optimum design for a quadratic model.

```
data Candidates;
   do x1 = -1 to 1;
      do x2 = -1 to 1;
         output;
         end;
      end;
run;

proc optex data=Candidates;
   model x1 x2 x1*x2 x1*x1 x2*x2;
run;
```

These minimal statements will find an optimum design with a default size of the number of parameters in the model plus 10, which amounts to 16 in this case. But usually the size of the design will be fixed by the constraints of the experiment, requiring use of the `N=` option to set it. Furthermore, we advise almost always using a couple of options—`CODING=ORTHCAN`, in order to make the efficiency measures more interpretable, and `METHOD=M_FEDOROV`, in order to use a more reliable search algorithm than the default one. The default search algorithm is the simple exchange method of Mitchell and Miller (1970), starting with a completely random design. However, increases in computing power have made more computationally intense approaches feasible. We have found that the Modified Fedorov algorithm of Cook and Nachtsheim (1980) is the most reliable at quickly finding the optimum design.

```
proc optex data=Candidates coding=orthcan;
   model x1 x2 x1*x2 x1*x1 x2*x2;
   generate n=16 method=m_fedorov;
run;
```

OPTEX searches for a D-optimum design 10 times with different random initial designs, and displays the efficiencies of the resulting designs in a table (Table 13.1). As is discussed in §13.3, the `CODING=ORTHCAN` option enables

TABLE 13.1. Exact optimum two-factor design: OPTEX efficiencies, as defined in §13.3

Design Number	D-Efficiency	A-Efficiency	G-Efficiency	Average prediction standard error
1	100.9290	97.4223	92.5649	0.6204
2	100.9290	97.4223	92.5649	0.6204
3	100.9290	97.4223	92.5649	0.6204
4	100.9290	97.4223	92.5649	0.6204
5	100.9290	97.4223	92.5649	0.6204
6	100.9290	97.4223	92.5649	0.6204
7	100.9290	97.4223	92.5649	0.6204
8	100.9290	97.4223	92.5649	0.6204
9	100.9290	97.4223	92.5649	0.6204
10	100.9290	97.4223	92.5649	0.6204

the D- and A-efficiencies to be interpreted as true efficiencies relative to a design uniformly weighted over the candidate points. The G-efficiency is relative to a continuous optimum design with support on the given candidate points. Perhaps the most salient feature of this output is the fact that OPTEX finds designs with the same characteristics in all 10 tries, indicating that this likely is truly the D-optimum design. In order to see the chosen points, add an OUTPUT statement to the OPTEX statements above, and print the results, as follows.

```
proc optex data=Candidates coding=orthcan;
   model x1 x2 x1*x2 x1*x1 x2*x2;
   generate n=16 method=m_fedorov;
   output out=Design;
run;

proc print data=Design;
run;
```

SAS Task 13.1. Use PROC OPTEX to find exact D-optimum two-factor designs over the 3×3 grid for a second-order response surface model for $N = 6$, $N = 7$, $N = 8$, and $N = 9$. Compare your results to Figure 12.1.

One final comment about how we advise typically running OPTEX. It is well-known that discrete optimization problems are often best solved by searching many times with different random initializations, and this has

TABLE 13.2. Typical options for PROC OPTEX

Option	Default setting	Suggested setting
N=	$p + 10$	
CODING=	STATIC	ORTHCAN
METHOD=	EXCHANGE	M_FEDOROV
NITER=	10	1000
KEEP=	10	10

been our experience with OPTEX. Moreover, for most of the design situations considered in this book, each search in OPTEX runs very quickly. Therefore, we advise habitually using many more 'tries' than the default 10. For example, increasing the number of tries to 1000 in the two-factor response surface design, as shown in the following code, hardly changes the run-time.

```
proc optex data=Candidates coding=orthcan;
   model x1 x2 x1*x2 x1*x1 x2*x2;
   generate n=16 method=m_fedorov niter=1000 keep=10;
   output out=Design;
run;
```

For this small design the results are the same, but for larger designs the benefits of increasing the number of tries can be appreciable.

Table 13.2 summarizes the OPTEX options we suggest typically employing, with their respective default and suggested settings.

13.3 Efficiencies and Coding in OPTEX

By default, OPTEX prints the following measures, labelled as efficiencies, for the designs found on each of its tries.

$$\text{D-efficiency} = 100 \times \left(\frac{|F^{\mathrm{T}}F|^{1/p}}{N} \right)$$

$$\text{A-efficiency} = 100 \times \left(\frac{p/N}{\mathrm{tr}\{(F^{\mathrm{T}}F)^{-1}\}} \right)$$

$$\text{G-efficiency} = 100 \times \left(\frac{p/N}{\max_{x \in \mathcal{C}} f^{\mathrm{T}}(x)(F^{\mathrm{T}}F)^{-1}f(x)} \right)^{1/2},$$

where p is the number of parameters in the linear model, N is the number of design points, and $\mathcal{C}$ is the set of candidate points. Since these measures are monotonically related to $|F^{\mathrm{T}}F|$, $\mathrm{tr}\{(F^{\mathrm{T}}F^{-1})\}$, and $\max_{x\in\mathcal{C}} f^{\mathrm{T}}(x)(F^{\mathrm{T}}F)^{-1}f(x)$, maximizing them is indeed equivalent to optimizing D-, A-, and G-efficiency, respectively. But the values are not usually the same as the values computed as efficiencies in this book, which are typically relative to the continuous optimum design.

By the General Equivalence Theorem, see §10.7, we know that for a continuous G-optimum design

$$\max_{\mathbf{x}\in\mathcal{C}} f^{\mathrm{T}}(x)M^{-1}(\xi)f(x) = p,$$

implying that the G-efficiency computed by OPTEX accurately reflects how well the exact optimum design found by OPTEX does relative to the theoretically G-optimum design. But for D- and A-efficiency, finding $|M(\xi)|$ and $\mathrm{tr}\{M(\xi)^{-1}\}$ for the continuous optimum design requires a non-linear optimization, which is not implemented in OPTEX.

The formulae above were suggested by Mitchell (1974) in the context where orthogonal designs are reasonable, since they compare $F^{\mathrm{T}}F$ to an information matrix proportional to the identity. This is not generally the case, but we can make it so by *orthogonally coding* the candidate vectors $\{f(x)|x \in \mathcal{C}\}$. The set of candidate points $\mathcal{C}$ generates a matrix F_C of candidate vectors of explanatory variables. When you specify the `CODING=ORTHCAN` option, OPTEX first computes a square matrix R such that $\sum_{x\in\mathcal{C}} f(x)f(x)^{\mathrm{T}} = F_C^{\mathrm{T}}F_C = R^TR$. Then each candidate vector of explanatory variables is linearly transformed as

$$f(x) \rightarrow f(x)R^{-1}\sqrt{N_C}\,, \tag{13.1}$$

where N_C is the number of candidate points. D-optimality is invariant to linear transformations of the candidate points. Since

$$|F^{\mathrm{T}}F| \rightarrow |N_C(R^{\mathrm{T}})^{-1}F^{\mathrm{T}}FR^{-1}| = \frac{|F^{\mathrm{T}}F|}{|F_C^TF_C/N_C|},$$

the D-optimum designs are the same. Note that $F_C^{\mathrm{T}}F_C/N_C = M(\xi_E)$, the information matrix for a continuous design with equal weights for all candidate points. Moreover, the D-efficiency computed above becomes

$$|F^{\mathrm{T}}F|^{1/p}/N \rightarrow \frac{|F^{\mathrm{T}}F/N|^{1/p}}{|M(\xi_E)|^{1/p}}$$

comparing the pth root of the determinant of $F^{\mathrm{T}}F/N$ to that of $M(\xi_E)$.

Unless the optimum continuous design is orthogonal, the efficiencies computed by OPTEX will not match the ones we typically show in this book. The relative efficiency values are the same, so OPTEX's results are still optimum, but the raw efficiency values usually differ by a constant factor. To explain this, first note that all design efficiencies are relative efficiencies, quantifying the theoretical proportional size of the design required to match the relevant information characteristics of a standard design. Usually this standard is the continuous design with optimum weights over the candidate region, or with optimum weights over a discrete grid of candidate points. In contrast, the standard for OPTEX's efficiencies is a design orthogonal for the given coding. As noted above, for orthogonal coding, this is equivalent to a continuous design with equal weights for the discrete candidate points. Note that if the equal-weight design is not a particularly good one, then the efficiencies computed by OPTEX can even be greater than 100%.

In order to compute efficiencies relative to the continuous optimum design, in general a non-linear optimization is required. In the next section it is demonstrated how to do this optimization in the SAS/IML matrix programming language, and then how to compute the efficiencies.

13.3.1 Evaluating Existing Designs

In typical runs, PROC OPTEX displays the efficiency values for the designs that it finds. It can also be used to calculate and display the corresponding values for an existing design. The trick is to supply the N-trial design to be evaluated as the initial design in the sequential construction of an N-trial design. The search procedure `METHOD=SEQUENTIAL` requires no initialization and, in this case, passes the initial design through as the final design; the efficiency calculations for this design are accordingly displayed. For example, for the small response surface design that we found and stored in a data set named Design, above, the following code evaluates the design.

```
proc optex data=Candidates coding=orthcan;
   model x1 x2 x1*x2 x1*x1 x2*x2;
   generate n=16 method=sequential initdesign=Design;
run;
```

13.4 Finding Optimum Designs Over Continuous Regions Using SAS/IML Software

In §9.6 the non-linear optimization capabilities of SAS/IML software were introduced and applied to finding continuous optimum design measures. The non-linear optimization tools in IML can also be used to find optimum exact designs over continuous regions. In this case, instead of fixing F_C for the grid points and optimizing weights as in §9.6, we need to optimize the determinant of $F^{\mathrm{T}}F$ as a function of the factor values. Also, we need to accomodate the constraints on the design region. The following code defines the appropriate function of the factor values and provides constraints for non-linear optimization of a 6-run design. Notice that the 6×2 matrix of factor values is transferred to the non-linear optimization function as a vector of 12 parameters, requiring it to be reshaped for building the design matrix.

```
start DCritX(xx);
   x = shape(xx,nrow(xx)*ncol(xx)/2,2);
   F =      j(nrow(x),1)
       || x[,1]          || x[,2]
       || x[,1]#x[,1]  || x[,2]#x[,1]
                         || x[,2]#x[,2];
   return(det(F`*F));
finish;

Nd = 6;
x0init = 2*ranuni((1:(Nd*2))`)-1;
con    =      shape(-1,1,Nd*2)        /* Lower bounds */
        //  shape( 1,1,Nd*2);        /* Upper bounds */

call nlpqn(rc,            /* Return code              */
           x0,            /* Returned optimum factors */
           "DCritX",      /* Function to optimize     */
           x0init,        /* Initial value of factors */
           1,             /* Specify a maximization   */
           con);          /* Specify constraints      */
x0 = shape(x0,nrow(x0)*ncol(x0)/2,2);
print x0;
```

The resulting design corresponds to the 6-run design as shown in Figure 12.1.

A caveat about this approach is in order. In general, constrained non-linear optimization is a very difficult computational task, with no method guaranteed to find the global optimum. This is also the case for continuous design optimization. Reliability in finding the true global optimum can be much improved by a judicious choice of initial design. For example, instead of the random initial design used in the previous code, one might try using the optimum design over the discrete candidate set.

SAS Task 13.2. Use PROC IML to find exact D-optimum two-factor designs over the continuous region $[-1, 1]^2$ for a second-order response surface model for $N = 7$, $N = 8$, and $N = 9$. Compare your results to Figure 12.1.

SAS Task 13.3. (Advanced). Use PROC IML to solve the same problem as SAS Task 13.2, but replace the constrained optimization with an unconstrained optimization using sine transformation (9.26).

As noted at the end of the previous section, the matrix programming tools of IML also provide a convenient way to compute efficiencies relative to various different designs. For example, the following code evaluates the optimum 6-run design computed above relative to the optimum continuous design.

```
p = ncol(FC);
F0 =     j(nrow(x0),1)
     || x0[,1]          || x0[,2]
     || x0[,1]#x0[,1]   || x0[,2]#x0[,1]
                        || x0[,2]#x0[,2];
print ( (det(F0‘            *F0/Nd)**(1/p))
       /(det(FC‘*diag(sw)*FC    )**(1/p)));
```

The result matches the efficiency calculated for the 6-run design over the continuous square region in Table 12.1.

SAS Task 13.4. Use PROC IML to evaluate the exact optimum designs constructed in SAS Tasks 13.1 and 13.2. Compare your results to Table 12.1.

13.5 Finding Exact Optimum Designs Using the ADX Interface

In addition to constructing many standard designs (see §7.7), the ADX Interface offers a point-and-click front-end to the OPTEX procedure. For any of the broad categories of designs (two-level, response surface, mixture, and mixed level), if a design of the desired size is not available, then there is always an 'Optimal' selection available under the 'Create' menu. In addition, there is a broad category of design called 'Optimal' which allows the user to find a design for a mix of factors, both quantitative and qualitative, with and without blocking.

SAS Task 13.5. Use the ADX Interface to create an exact optimum quadratic design for two factors. This is a type of 'response surface' design in ADX's categorization, so it is easiest to proceed by first selecting to create such a design. On the design selection screen, select a design with two factors and then select 'Optimal' under 'Create' at the top.

13.5.1 Augmenting Designs

The computational formula for building a design sequentially depends only on the relationship between the current form of the information matrix $F^{\mathrm{T}}F$

and the points to be added. For this reason, it is easy to modify optimum design search algorithms for optimum augmentation of a given set of experimental runs. This can be useful in sequential experimentation, when new experimental runs are to be analysed together with previous runs in order to gain information on more effects. The `AUGMENT=` option in OPTEX allows for optimum design augmentation, but we discuss how to use the ADX interface for this.

As an example, consider a screening experiment for seven two-level factors. It is possible to construct an orthogonal 16-run design that has resolution IV, allowing for estimates of all 7 main effects unconfounded with any of the 21 potential two-factor interactions. To gain more information about the interactions, we seek to augment this design with 16 more runs in such a way that the combined 32 runs allow for estimation of all main effects and two-factor interactions. The following steps demonstrate using the ADX interface to create the original 16-run design as well as the 32-run augmented design.

1. Construct a new two-level design.
2. Select the 16-run design for 7 factors.
3. Save the 16-run design and return to the main ADX window. Select the saved design, right-click on it, and choose 'Augment...'.
4. Navigate the resulting 'wizard' to add 16 more runs to the design in order to estimate main effects and two-factor interactions.

SAS Task 13.6. Use the ADX Interface to create an augmented screening design for 7 factors according to the recipe above. Add arbitrary values of response variables to both the 16-run and the 32-run designs and fit the full second-order model to confirm that interactions are not all estimable in the first design but they are in the second.

14

EXPERIMENTS WITH BOTH QUALITATIVE AND QUANTITATIVE FACTORS

14.1 Introduction

This chapter is concerned with the design of experiments when the response depends on both qualitative and quantitative factors. For instance, in Example 1.2 there are two quantitative variables, naphthenic oil and filler level, and one qualitative variable, kind of filler, taking three levels because as many fillers were used in the experiment. Likewise, an experiment on a new drug may involve the quantitative factors of composition and dosage as well as the qualitative factor of mode of administration (orally or by injection).

There is a huge literature on optimum designs when only one kind of factor is present. For example, the papers collected in Kiefer (1985) consider the optimality, over a wide range of criteria, of both block designs and Graeco-Latin squares, and derive designs for regression over a variety of experimental regions. In contrast, relatively little attention has been given to designs that involve both classes of factor.

Example 14.1 Quadratic Regression with a Single Qualitative Factor: Example 5.3 continued In this model the response depends not only on a quadratic in a single quantitative variable x but also on a qualitative factor z at b levels. If the effect of z is purely additive, the model is

$$\mathrm{E}(y_i) = \sum_{j=1}^{b} \alpha_j z_{ij} + \beta_1 x_i + \beta_2 x_i^2 \quad (i = 1, \ldots, N). \tag{14.1}$$

In the matrix notation of (5.16) this was written

$$\mathrm{E}(y) = X\gamma = Z\alpha + F\beta, \tag{14.2}$$

where Z, of dimension $N \times b$, is the matrix of indicator variables z_j, taking the values 0 or 1 for the level of the qualitative factor.

The extension of the design given in (5.17) is to repeat the three-trial D-optimum design for the quadratic at each level of z. As b increases, this design involves an appreciable number of trials for the estimation of rather few parameters. ■

Example 14.2 Second-order Response Surface with One Qualitative Factor
The chief means of illustrating the structure of designs when the two kinds of factor are present will be the extension of the previous example to two quantitative factors. For illustration, in the chemical industry the yield of a process might depend not only on the quantitative factors temperature and pressure, but also on such qualitative factors as the batch of raw material and the type of reactor. Suppose that the effect of the quantitative factors can be described by a quadratic response surface, the mean value depending on the value of a single qualitative factor, the levels of which represent combinations of raw material and reactor type. This formulation implies no interaction between the qualitative and quantitative factors. The model is then

$$\mathrm{E}(y) = \sum_{i=1}^{b} \alpha_i z_i + \beta_1 x_1 + \beta_2 x_2 + \beta_{11} x_1^2 + \beta_{22} x_2^2 + \beta_{12} x_1 x_2, \tag{14.3}$$

when the qualitative factor is at b levels. It is important to recognize that we are making two critical assumptions which may not be valid for all experiments: that all combinations of qualitative and quantitative factors are admissible, and, to repeat, that there are no interactions between qualitative and quantitative factors. ■

Simple examples like this indicate some of the many possibilities and complications. Although the design region for the factors x will usually be the same for all levels of z, there is no reason why this should be the case: certain combinations of x_1 and x_2 might be inadmissible for some levels of the qualitative factor. The resulting restricted design region presents no difficulties for the numerical calculation of exact designs, but theoretical results are unavailable. A second potential complication is that the model (14.3) might be too simple, since there could be interactions between the two groups of factors, causing the shape of the response surface to depend on the level of the qualitative factor.

Nevertheless, under these assumptions, there are some general theoretical results. Kurotschka (1981) shows that, under certain conditions, the optimum continuous design consists of replications of the design for the quantitative factors at each level of the qualitative ones. Such designs are called product designs. This work is outlined in the next section. A disadvantage of these designs is that the number of experimental conditions needed is large compared with the number of parameters. Therefore they will often be impracticable. Accordingly, §14.3 is concerned with exact designs, particularly when the number of trials is not much greater than the number of the parameters. These designs, which are often almost as efficient as the continuous product designs, exhibit several interesting properties when compared with continuous designs. One is that the numbers of trials at the levels of the qualitative factors are often not even approximately equal. A second is that D-efficiency seems to prefer designs with some structure in the quantitative factors at the individual levels of the qualitative factors. A third, less appealing, feature is that for some values of N the addition of one extra trial can cause a rather different design to be optimum. This suggests that care may be needed in the choice of the size of the experiment. $\mathrm{D_S}$-optimum designs when the qualitative factors are regarded as nuisance parameters are considered in Chapter 15. The case when the quantitative factors are the components of a mixture is treated in Chapter 16.

14.2 Continuous Designs

In this section we extend Example 14.1 to a general model that accounts for two kinds of factors

$$\mathrm{E}(y) = \eta(x, z, \gamma), \tag{14.4}$$

where x represents m quantitative factors and z represents B_F qualitative ones, having $b_i,\ i = 1, \cdots, B_F$ levels, respectively. The parameterization of (14.4) can be complicated, even for linear models, if there are interactions between x and z. It is convenient to follow the classification introduced by Kurotschka (1981) who distinguishes three cases:

1. Complete interaction between qualitative and quantitative factors.
2. No interaction between qualitative and quantitative factors, although there may well be interaction within each group of factors.
3. The intermediate case of some interaction between groups.

The model corresponding to Case 1 has parameters for the quantitative factor which are allowed to be different at each combination of the qualitative factors. The design problem then becomes that of a series of distinct design regions $\mathcal{X}_i$ $(i = 1, \cdots, b,\ b = \prod_{B_F} b_i)$. The models need not all be the same. Let the model at the ith level of z have parameter vector β_i of dimension p_i. The D-optimum continuous design for this model over $\mathcal{X}$ is δ_i^*. In order to find the optimum design for the whole experiment, we also need to consider the distribution of experimental effort between the levels of z. Let ν be the measure which assigns mass ν_i to the experiments at level i. Then the measure on the points of $\mathcal{X}_i$ is the product $\nu_i \times \delta_i = \xi_i$. From the Equivalence Theorem, the D-optimum design must be such that the maximum variance is the same at all design points. Therefore ν_i^* must be proportional to p_i and the optimum measure is

$$\xi_i^* = \frac{p_i}{\sum p_i} \times \delta_i^*. \tag{14.5}$$

If the models and design regions are the same for all levels of z while the parameters remain different, the optimum design can be written in the simpler product form as

$$\xi^* = \nu^* \times \delta^*, \tag{14.6}$$

where $\nu^* = \{1/b\}$ is now a uniform measure over the levels of z. Similar conditions can also be found for A-optimality.

For Case 2, in which there is no interaction between x and z, the model has the straightforward form (14.2). In a simple case it may be possible to assume that the change in response when moving from one level of one of the qualitative factors to another level of that factor remains the same for all levels of the remaining qualitative factors. For example, with two qualitative factors we can let $Z\alpha = Z_1\alpha_1 + Z_2\alpha_2$ (i.e. a main-effects-only model in the qualitative factors), with the identifiability constraint that one of the elements of either α_1 or α_2 be set to zero. If the numbers of the levels of the qualitative factors are b_1 and b_2 respectively, the number of extra parameters in the model needed to describe the effect of the qualitative factors is $b_1 + b_2 - 2$.

However, if the structure of the qualitative factors z is complicated it may be necessary to regard the experiment as one with a single qualitative factor acting at b levels which are formed from all possible combinations of the qualitative factors (an interaction model in the qualitative factors). For example, with two qualitative factors this form could represent all $b_1 \times b_2 - 1$ conditions of a full factorial or the smaller number of treatment combinations for a fractional factorial. With more factors the qualitative variables could, for example, represent the cells of a Graeco-Latin square, which again would be treated as one factor at b levels.

For Case 2, with complex structure for the qualitative factors and the same experimental region at each level of the quantitative factors, the product design (14.6) is A- and D-optimum for α and γ in (14.2) with all elements of $\nu^* = 1/b$, although δ^* will of course depend on the design criterion. Case 3, in which there is some interaction between groups, is not susceptible to general analysis.

In a sense, Kurotschka's Case 1 is not very interesting: designs can be found using the general theory of Chapter 11. Our interest will be in Case 2 which covers many models of practical importance.

Example 14.1 Second-order Response Surface with One Qualitative Factor continued If the design region for the quantitative factors in (14.3) is the square for which $-1 \leq x_i \leq 1$ ($i = 1, 2$), the D-optimum continuous design has support at the nine points of the 3^2 factorial for each level of z. The optimum design, which is of product form, has the following design points and weights:

$4\,b$	corner points $(\pm 1, \pm 1)$	$0.1458/b$	
$4\,b$	centre points $(0, \pm 1, \pm 1, 0)$	$0.0802/b$	(14.7)
b	centre points $(0, 0)$	$0.0962/b$	

When $b = 1$ this is the D-optimum second-order design for two factors (12.2). For general b the design has support at $9b$ design points with unequal weights. The number of parameters is only $b + 5$. Even with a good integer approximation to (14.7), such as repetitions of the 13-trial design formed by replicating the corner points of 3^2 factorial, the ratio of trials to parameters rapidly becomes very large as b increases. In the next section we look for much smaller exact designs. ■

14.3 Exact Designs

As usual, there is no general construction for exact designs. The design for each value of N has to be calculated individually, typically by a search algorithm as discussed in Chapters 12 and 13. In this section we give some examples to demonstrate the features of exact designs and their differences from continuous designs.

Example 14.1 Second-order Response Surface with One Qualitative Factor continued To calculate the exact designs for the second-order response surface (14.3) a search was made over the nine points of the 3^2 factorial for x_1 and x_2 at each level of the qualitative factor z. There is no constraint on the number of trials N_i at each level except that $\sum_b N_i = N$. Suitable algorithms for the construction of exact designs are described in Chapter 12 with their SAS implementation in Chapter 13. Interest was mainly in designs when N is equal to, or just greater than, the number of parameters p. This is not only because of the practical importance of such designs, but also because their structure is furthest from that of the product designs of the continuous theory.

Figure 14.1 shows the D-optimum nine-trial design for model (14.3) when $b = 3$. The number of observations at each level of z is the same, that is $N_1 = N_2 = N_3 = 3$, but the design is different for each of the levels. Of course, it does not matter which level of the qualitative factor is associated with which design. One interesting feature is that the projection of the design obtained by ignoring z does not result in the best design when $b = 1$. This, for $N = 9$, is the 3^2 factorial. The best design with such a projection for

$b = 3$ has a value of 0.1806×10^{-3} for the determinant $|M(\xi_9)|$, as opposed to 0.1873×10^{-3} for the optimum design—a difference which, whilst real, is negligible for practical purposes. For the D-optimum design $d_{\text{ave}} = 8.77$ and $d_{\text{max}} = 16.55$, whereas for the design which projects into the 3^2 factorial $d_{\text{ave}} = 8.45$ and $d_{\text{max}} = 19.50$.

A second example, given in Figure 14.2, exhibits some further properties of the optimum designs. Here $b = 2$ and $N = 13$. The optimum design has five trials at one level and eight at the other, rather than the nearly equal, here six to seven, division which would be indicated by the continuous product design.

For the exact design, the designs at each level have a clear structure. It is also interesting that projection of the design yields the 13-trial

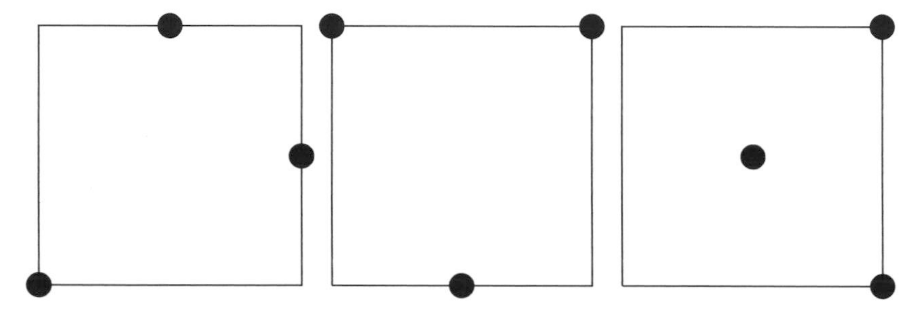

FIG. 14.1. Example 14.2: second-order response surface with one qualitative factor at three levels. D-optimum nine-trial design.

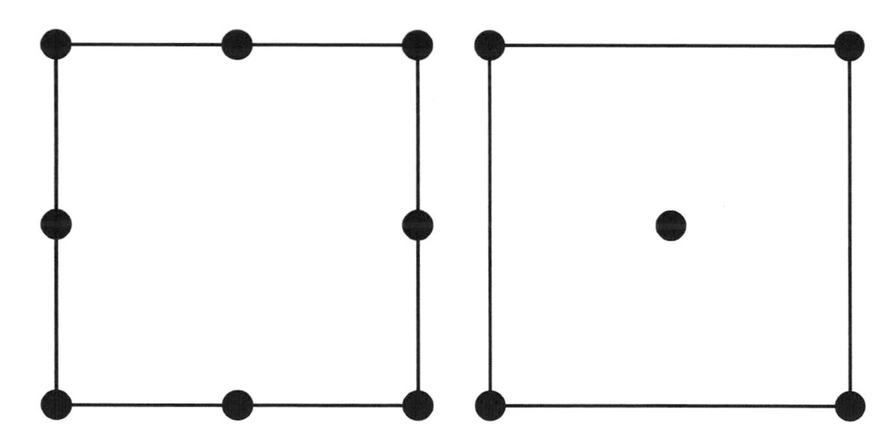

FIG. 14.2. Example 14.2: second-order response surface with one qualitative factor at two levels. D-optimum 13-trial design.

approximation to the continuous design for $b = 1$ mentioned in §14.2 in which the corner points of the 3^2 factorial are replicated.

Table 14.1 gives the values of the determinants of the information matrices of the designs for Example 14.2 for a variety of values of b and N, as well as for designs with a single quantitative factor. Also given are results from the use of the adjustment algorithm, briefly mentioned in §12.8 and the D-efficiencies of the adjusted designs. Perhaps most informative is the division of the number of trials between the levels of the qualitative factor. These results provide further evidence that for small values of N like these, when product designs are inapplicable, the continuous designs provide little guidance to the structure of exact designs. An algorithm like those described in Chapter 12 has to be employed. ■

Example 14.3 Second-order Response Surface with Two Qualitative Factors D-optimum designs for experiments with two or more qualitative factors, possibly with interaction, have similar features that distinguish them from those with one qualitative factor. One common feature is that the reparametrization of the required models can be done in many ways and is more complex than when there is a single blocking variable; now non-identifiabilty cannot be removed just by ignoring the constant term as was the case for Example 14.1. For instance, the 18-trial D-optimum design for two quantitative factors and two qualitative factors acting at two and three levels, respectively, for the quadratic model

$$\mathrm{E}(y) = \alpha_{1.} z_{1.} + \sum_{j=1}^{2} \alpha_{.j} z_{.j} + \beta_0 + \beta_1 x_1 + \beta_2 x_2 + \beta_{11} x_1^2 + \beta_{22} x_2^2 + \beta_{12} x_1 x_2 \quad (14.8)$$

is shown in Figure 14.3. In (14.8), as a result of the adopted reparametrization, β_0 denotes the intercept of the model when the second level of the first qualitative factor and the third level of the second qualitative factor are used. Also, $\alpha_{1.}$ denotes the mean difference between the effects associated with the first and the second level of the first qualitative factor at any level of the second qualitative factor. Similarly, $\alpha_{.j}$ denotes the mean difference between the effects associated with the jth ($j = 1,\ 2$) and the third level of the second qualitative factor at any level of the first qualitative factor. The indicator variable $z_{1.}$ is equal to one when the first level of the first factor is used and it is zero otherwise. Finally, $z_{.j}$ is equal to one when the jth level of the second factor is used and it is zero otherwise. The number of the model parameters is 9. However (14.8) is just one full-rank

TABLE 14.1. D-optimum exact N-trial designs for a second-order model in m quantitative factors with one qualitative factor at b levels. Results labelled 'AA' are from use of the adjustment algorithm, §12.8

m	b	N_1	N_2	N_3	p	N	$\|M(\xi)\|$	$\|M^{AA}(\xi)\|$	D_{eff}	D_{eff}^{AA}
1	2	2	2		4	4	0.156×10^{-1}	0.219×10^{-1}	0.8059	0.8774
1	2	2	3		4	5	0.200×10^{-1}	0.267×10^{-1}	0.9118	0.9212
1	2	3	3		4	6	0.370×10^{-1}	0.370×10^{-1}	1.0000	1.0000
1	3	1	2	2	5	5	0.128×10^{-2}	0.180×10^{-2}	0.7474	0.8000
1	3	2	2	2	5	6	0.309×10^{-2}	0.327×10^{-2}	0.8913	0.9016
1	3	2	2	3	5	7	0.357×10^{-2}	0.394×10^{-2}	0.9176	0.9358
1	3	2	3	3	5	8	0.439×10^{-2}	0.450×10^{-2}	0.9566	0.9610
1	3	3	3	3	5	9	0.549×10^{-2}	0.549×10^{-2}	1.0000	1.0000
2	2	3	4		7	7	0.157×10^{-2}	0.162×10^{-2}	0.9183	0.9225
2	2	4	4		7	8	0.183×10^{-2}	0.188×10^{-2}	0.9384	0.9422
2	2	4	5		7	9	0.193×10^{-2}	0.194×10^{-2}	0.9453	0.9460
2	2	4	6		7	10	0.207×10^{-2}	0.212×10^{-2}	0.9553	0.9582
2	2	5	6		7	11	0.225×10^{-2}	0.229×10^{-2}	0.9667	0.9688
2	2	5	7		7	12	0.236×10^{-2}	0.238×10^{-2}	0.9729	0.9744
2	2	5	8		7	13	0.265×10^{-2}	0.265×10^{-2}	0.9894	0.9894
2	2	6	8		7	14	0.267×10^{-2}	0.267×10^{-2}	0.9905	0.9905
2	2	7	8		7	15	0.256×10^{-2}	0.256×10^{-2}	0.9844	0.9844
2	2	8	8		7	16	0.257×10^{-2}	0.258×10^{-2}	0.9850	0.9855
2	2	8	9		7	17	0.253×10^{-2}	0.253×10^{-2}	0.9826	0.9830
2	2	8	10		7	18	0.258×10^{-2}	0.258×10^{-2}	0.9856	0.9856
2	3	2	3	3	8	8	0.137×10^{-3}	0.159×10^{-3}	0.8688	0.8849
2	3	3	3	3	8	9	0.187×10^{-3}	0.206×10^{-3}	0.9031	0.9137
2	3	3	3	4	8	10	0.246×10^{-3}	0.261×10^{-3}	0.9345	0.9413
2	3	3	3	5	8	11	0.282×10^{-3}	0.295×10^{-3}	0.9506	0.9561
2	3	3	4	5	8	12	0.315×10^{-3}	0.324×10^{-3}	0.9637	0.9673
2	3	5	5	5	8	15	0.363×10^{-3}	0.364×10^{-3}	0.9809	0.9814
2	3	5	5	8	8	18	0.359×10^{-3}	0.359×10^{-3}	0.9795	0.9795

way of writing the model; the optimum design does not depend on what parameterization we choose. SAS, for example, merely requires specification as to which variables are qualitative.

If such a relatively simple structure for the effects corresponding to the levels of the qualitative factors cannot be assumed, the model will have a

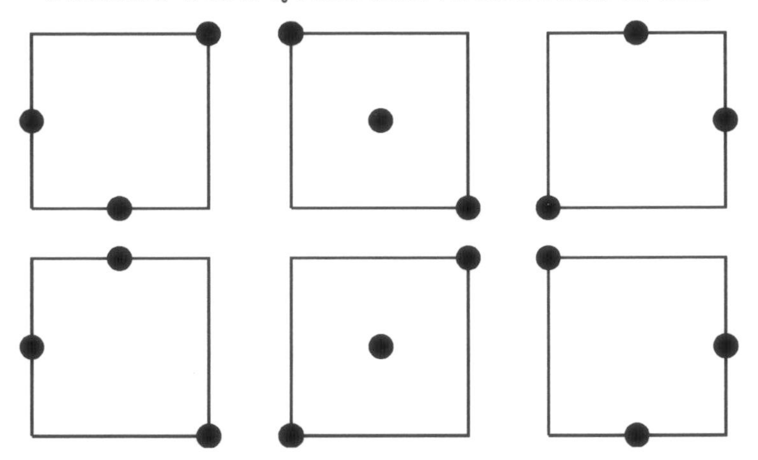

FIG. 14.3. Example 14.3: second-order response surface with two qualitative factors at two and three levels. D-optimum 18-trial design for model (14.8).

larger number of model parameters and therefore the estimate for the variance of the error will be based on a smaller number of degrees of freedom. As discussed earlier, in the most complicated case, though still assuming no interaction between the effects of the quantitative and the qualitative factors, the model (14.3) can be used with $2 \times 3 = 6$ and 11 parameters. ■

In the examples so far the designs have as their support the points of the 3^2 factorial. If the quantitative factors x can be adjusted more finely than this, one possibility is to search for exact designs over the points of factorials with more levels. An alternative, which we have found preferable, is to use an adjustment algorithm. An example is shown in Figure 14.4, where the design of Figure 14.1 is improved by adjustment. However, the increase in D-efficiency is small, from 90.31% to 91.37%. In some environments, particularly when it is difficult to set the quantitative factors with precision, this increase may not be worth achieving at the cost of a design in which the factor levels no longer only have the scaled values -1, 0, and 1.

14.4 Designs with Qualitative Factors in SAS

In Chapter 13 we saw how to use the OPTEX procedure to create optimal exact designs for response surface models over discrete candidate spaces. The facilities for specifying a linear model in OPTEX also allow for qualitative

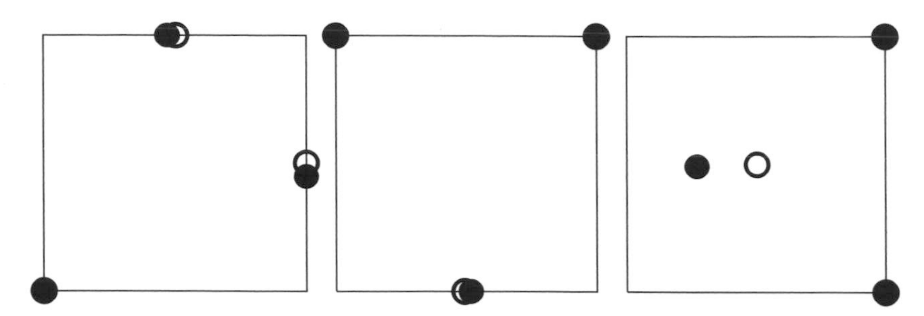

FIG. 14.4. Example 14.2: second-order response surface with one qualitative factor at three levels. D-optimum nine-trial design: effect of the adjustment algorithm; ∘ design of Figure 14.1 on points of 3^2 factorial; • adjusted design points.

factors, in both main effects and interactions. Thus, OPTEX can be applied to designs with qualitative factors in precisely the same way as in Chapter 13, to whit, a data set containing the admissible combinations of factors (both quantitative and qualitative) is defined and supplied to OPTEX, along with the appropriate model.

In SAS, factors are declared to contribute qualitatively to a model by naming them in a CLASS statement before specifying the model. Thus, the second-order response surface model with one qualitative factor of Example 14.1 is specified with the following two statements

```
class Group;
model Group x1 x2 x1*x2 x1*x1 x2*x2;
```

where `Group`, `x1`, and `x2` are variables in the candidate data set. Declaring `Group` to be qualitative by naming it in the CLASS statement means that its term in the model contributes $k-1$ columns to the design matrix, where k is the number of levels it has in the candidate data set. (As usual, a term for the intercept is included by default.) The complete syntax for generating the design of Example 14.1 begins by defining this candidate data set.

```
data Candidates;
   do Group = 1 to 3;
      do x1 = -1 to 1; do x2 = -1 to 1;
         output;
         end; end;
      end;
run;
```

TABLE 14.2. Characteristics for designs found on 10 tries of OPTEX for Example 14.1

Design number	D-Efficiency	A-Efficiency	G-Efficiency	Average prediction standard error
1	92.1182	80.6561	69.5183	1.0498
2	92.1182	80.6561	69.5183	1.0498
3	92.1182	80.6561	69.5183	1.0498
4	92.1182	80.6561	69.5183	1.0498
5	92.1182	80.6561	69.5183	1.0498
6	92.1182	80.6561	69.5183	1.0498
7	92.1182	80.6561	69.5183	1.0498
8	92.1182	80.6561	69.5183	1.0498
9	92.1182	80.6561	69.5183	1.0498
10	92.1182	80.6561	69.5183	1.0498

This sets up the 27 different possible combinations of the 3^2 factorial for x_1 and x_2 with 3 groups in a data set named `Candidates`. Then the following syntax directs the OPTEX procedure to select 9 points from this candidate set D-optimum for estimating all eight parameters of the model, and then prints the resulting design.

```
proc optex data=Candidates coding=orthcan;
   class Group;
   model Group x1 x2 x1*x2 x1*x1 x2*x2;
   generate n=9 method=m_fedorov niter=1000 keep=10;
   output out=Design;
proc print data=Design;
run;
```

The output displayed by the procedure, shown in Table 14.2, gives various measures of design goodness. As discussed in §13.3, these efficiencies are relative to a design with uniform weights over the candidates.

Note that if you run the same code you may not see the very same table of efficiencies, since OPTEX is non-deterministic by default; but a design with a computed D-efficiency of 92.1182% should be at the top. Also, as discussed in §13.3, because of the `CODING=ORTHCAN` option, this D-efficiency is relative to a continuous design uniformly weighted over the cross-product candidate set. The fact that this design was found at least 10 times indicates that the algorithm will probably not be able to improve on it. The resulting design is not necessarily composed of the very same runs depicted in Figure 14.1, but it yields the same value of $|M(\xi)|$.

SAS Task 14.1. Use SAS to find an optimal design for two qualitative factors, `Row` and `Column` with 2 and 3 levels respectively, and two quantitative factors, assuming a model composed of a quadratic response surface in the two quantitative factors and main effects of the qualitative factors.

14.5 Further Reading

The continuous theory leading to product designs was described by Kurotschka (1981) and Wierich (1986). Schwabe (1996) provides a book length treatment of these multifactor designs. Lim, Studden, and Wynn (1988) consider designs for a general model G in the quantitative factors. Each of the qualitative factors is constrained to act at two levels. These factors can interact with each other and with submodels G_i of G. The resulting D-optimum continuous designs involve a weighting similar to that of the generalized D-optimum designs of §10.8. Further discussion and examples of exact designs are given by Atkinson and Donev (1989).

15

BLOCKING RESPONSE SURFACE DESIGNS

15.1 Introduction

This chapter, like Chapter 14, is concerned with the design of experiments when the response depends on both qualitative and quantitative factors. However, we now focus on the case when the experimenter is not interested in the model parameters corresponding to the levels of the qualitative factors. These factors may be included in the study because this is convenient or even unavoidable, or because their inclusion ensures the required experimental space and, hence, range for the validity of the results. Such qualitative variables are called blocking variables and the groups of observations corresponding to their levels, blocks. Examples of experimental situations requiring blocking were given in Chapter 1.

The next section presents models accommodating fixed and random blocking variables and discusses the choice of criteria of optimality for blocked designs. Orthogonal blocking is discussed in §15.3, while §15.4 briefly summarizes related literature. Examples of construction of block designs using SAS are given in §15.5. Blocking of experimental designs for mixture models is discussed in §16.5.

15.2 Models and Design Optimality

The complexity of the blocking scheme depends on the nature of the experiment. Fixed, random, or both types of blocking variables can be required. In this section we discuss the appropriate models for a number of standard situations. The block effects introduce extra parameters into the model, but in all cases these are considered nuisance parameters.

15.2.1 Fixed Blocking Variables

When the observations can be taken in blocks of homogeneous units generated by B_F blocking variables, and the corresponding effects are regarded as fixed, the model becomes

$$\begin{aligned} y &= F\beta + Z_F\alpha_F + \epsilon \\ &= X\gamma + \epsilon. \end{aligned} \tag{15.1}$$

In model (15.1) $\epsilon \sim N(0_N, \sigma_F^2 I_N)$, 0_N is a vector of zeroes, α_F is a vector of f fixed block effects, $X = [\ F\ Z_F\ ]$ and $Z_F = BC$, where B is the $N \times c$ design matrix corresponding to the indicator variables for the fixed blocks and C is an $c \times f$ matrix identifying an estimable reparametrization of the fixed block effects of the model. When the block effects are additive $c = \sum_{i=1}^{B_F} b_i$, where b_i is the number of the levels of the ith blocking variable, but it will be larger when there are interactions between the blocking variables.

Since the block effects are not of interest, one option is simply to ignore them, at the cost of inflating the variance of the experimental error by a factor related to $\alpha_F^{\mathrm{T}} Z_F^{\mathrm{T}} Z_F \alpha_F$ (or its expectation if block effects are random). Whether this is advisable depends on the relative sizes of $\alpha_F^{\mathrm{T}} Z_F^{\mathrm{T}} Z_F \alpha_F$ and σ_F^2, the relative D-efficiencies $|F^{\mathrm{T}}F - F^{\mathrm{T}}Z_F(Z_F^{\mathrm{T}}Z_F)^{-1}Z_F^{\mathrm{T}}F|$ and $|F^{\mathrm{T}}F|$, and the proportion of experimental information lost in estimating block effects.

Least squares estimators of all parameters of the model are given by

$$\hat{\gamma} = (X^{\mathrm{T}}X)^{-1}X^{\mathrm{T}}y. \tag{15.2}$$

For the comparison and construction of designs for this model we can use the particular case of $\mathrm{D_S}$-optimality when only the parameters β in (15.1) are of interest. We call this criterion D_β-optimality; it requires maximization of

$$|M_\beta(N)| = \frac{|X^{\mathrm{T}}X|}{|Z_F^{\mathrm{T}}Z_F|} = |F^{\mathrm{T}}F - F^{\mathrm{T}}Z_F(Z_F^{\mathrm{T}}Z_F)^{-1}Z_F^{\mathrm{T}}F|. \tag{15.3}$$

It is important that if the block sizes are fixed, then so is $|Z_F^{\mathrm{T}}Z_F|$, and thus optimizing $|M_\beta(N)|$ is equivalent to optimizing $|M(N)|$. So, the D- and D_β-optimum designs are the same for fixed block sizes, but in general this is not true when the block sizes are not fixed.

If the block sizes can be chosen, $|M_\beta(N)|$ may be increased by using unequal block sizes which minimize $|Z_F^{\mathrm{T}}Z_F|$. This could provide practical benefits in situations where some blocks of observations are easier or cheaper to obtain than others. However, there is no statistical value in using blocks of size one and they should be omitted.

Example 15.1 Quadratic Regression: Single Factor Generating Three Fixed Blocks Figure 14.1 shows the D-optimum design for $m = 2$, $B_F = 3$, and

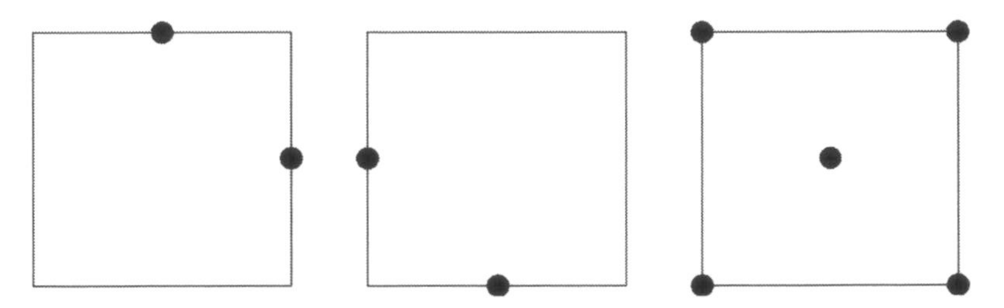

FIG. 15.1. Example 15.1: D_β-optimum design for second-order response surface in three blocks.

$N = 9$. It has a $3 : 3 : 3$ division of the observations between blocks with $|M(9)| = 0.1873 \times 10^{-3}$ and $|M_\beta(9)| = 0.6939 \times 10^{-5}$. The D_β-optimum design for the same parameter values is shown in Figure 15.1. This has the less equal $5 : 2 : 2$ division with $|M(9)| = 0.1427 \times 10^{-3}$ and $|M_\beta(9)| = 0.7136 \times 10^{-5}$. While the optimum block sizes in this case are 2, 2, and 5, the experimenter may still prefer to use equal group sizes or other group sizes as practically convenient. ■

15.2.2 Random Blocking Variables

If there are B_R random blocking variables, the model that has to be estimated is

$$y = F\beta + Z_R\alpha_R + \epsilon, \tag{15.4}$$

where $\epsilon \sim N(0_N, \sigma_R^2 I_N)$, α_R is a vector of r random effects and Z_R is the corresponding design matrix. We assume that the random effects are normally distributed, independently of each other and of ϵ and have zero means and variances $\sigma_i^2, i = 1, 2, \ldots, B_R$, that is,

$$\text{var}(\alpha_R) = G = \text{diag}(\sigma_1^2 I_{b_1}, \sigma_2^2 I_{b_2}, \ldots, \sigma_{B_R}^2 I_{B_R}), \tag{15.5}$$

where b_i is the number of levels of the ith blocking variable. Hence,

$$\begin{aligned} \text{var}(y) = V &= Z_R G Z_R^{\mathrm{T}} + \sigma_R^2 I_N \\ &= \sigma_R^2 (Z_R H Z_R^{\mathrm{T}} + I_N), \end{aligned} \tag{15.6}$$

where

$$H = \text{diag}\,\{\eta_1 I_{b_1}, \eta_2 I_{b_2}, \ldots, \eta_{B_R} I_{B_R}\}$$

and $\eta_i = \sigma_i^2/\sigma^2, i = 1, 2, \ldots, B_R$. In this case $r = \sum_{i=1}^{B_R} b_i$. In general α_R may include random interactions, such as interactions between explanatory variables and random blocking variables, or between fixed and random blocking variables. Hence, the number of the elements of α_R can be larger than $\sum_{i=1}^{B_R} b_i$.

Customarily, σ_R^2 and $\sigma_i^2, i = 1, 2, \ldots, B_R$ are not known and need to be estimated, typically, by residual maximum likelihood (Patterson and Thompson 1971). The generalized least squares estimators for the parameters of interest are

$$\hat{\beta} = (F^{\mathrm{T}} V^{-1} F)^{-1} F^{\mathrm{T}} V^{-1} y, \tag{15.7}$$

with variance–covariance matrix

$$\mathrm{var}(\hat{\beta}) = (F^{\mathrm{T}} V^{-1} F)^{-1} \sigma_R^2. \tag{15.8}$$

The determinant of $\mathrm{var}(\hat{\beta})$ is minimum for a D-optimum design. Note that this criterion depends on the relative variability within and between blocks through the ratios $\eta_i, i = 1, 2, \ldots, B_R$.

15.2.3 Fixed and Random Blocking Variables

Suppose now that the observations are divided in f fixed blocks and r random blocks generated by B_F and B_R fixed and random blocking variables, respectively. Under assumptions and notation similar to those introduced in §§15.2.1 and 15.2.2 the model can be written as

$$\begin{aligned} y &= F\beta + Z_F \alpha_F + Z_R \alpha_R + \epsilon \\ &= F\beta + Z\alpha_{FR} + \epsilon \\ &= X\gamma + Z_R \alpha_R + \epsilon, \end{aligned} \tag{15.9}$$

where α_{FR} is the vector of all block effects, $Z = [\; Z_F \quad Z_R \;]$, and $\epsilon \sim N(0_N, \sigma_{FR}^2 I_N)$. Then

$$\hat{\gamma} = (X^{\mathrm{T}} V^{-1} X)^{-1} X^{\mathrm{T}} V^{-1} y, \tag{15.10}$$

and

$$\mathrm{var}(\hat{\gamma}) = (X^{\mathrm{T}} V^{-1} X)^{-1}, \tag{15.11}$$

where the expression for V is given by (15.6) with σ_R^2 replaced by σ_{FR}^2. As in §15.2.1, $|X^{\mathrm{T}} X|$ can be factored into a part that depends only on Z_F and V and a part that depends on the information matrix for β,

$$M_\beta(N) = F^{\mathrm{T}} V^{-1} F - F^{\mathrm{T}} Z_F (Z_F^{\mathrm{T}} V^{-1} Z_F)^{-1} Z_F^{\mathrm{T}} F. \tag{15.12}$$

As in §15.2.2, finding D_β-optimum designs is made difficult by the dependence of this criterion on $\eta_i, i = 1, 2, \ldots, B_R$. Jones (1976) shows that when there is a single blocking variable and these ratios tend to infinity, the D_β-optimum designs with fixed and random blocks coincide. Goos and Donev (2006*a*) extend this result to an arbitrary number of random blocking factors.

Example 15.2 Second-order Response Surface: Six Blocks Generated by Two Blocking Variables Suppose that the blocking variables act at 2 and 3 levels, respectively. Hence, the problem is similar to that of Example 14.3, though here the effects of the blocking variables are nuisance parameters. Assume that the model

$$\mathrm{E}(y) = \alpha_{1\cdot}z_{1\cdot} + \sum_{j=1}^{2} \alpha_{\cdot j}z_{\cdot j} + \beta_0 + \beta_1 x_1 + \beta_2 x_2 + \beta_{11}x_1^2 + \beta_{22}x_2^2 + \beta_{12}x_1 x_2 \quad (15.13)$$

will be used to explain the data. Figure 15.2 shows that four designs are D_β-optimum for different values of η_1 and η_2. These are Designs I, II, III, and IV shown in Figures 15.3–15.6. Design IV is also the D_β-optimum design when all block effects are fixed and when the first blocking variable is random and the second is fixed, while Design I is D_β-optimum when the first blocking variable is fixed and the second is random. The values of the D_β-optimality criteria of the designs close to the borderlines are very similar. It is therefore clear that precise knowledge of η_1 and η_2 is not crucial for choosing a suitable design. If the block structure of (15.13) appears to be too simple for a particular study, a more complex model using interactions can be used. Using a model similar to (14.3) may be good if the block effects

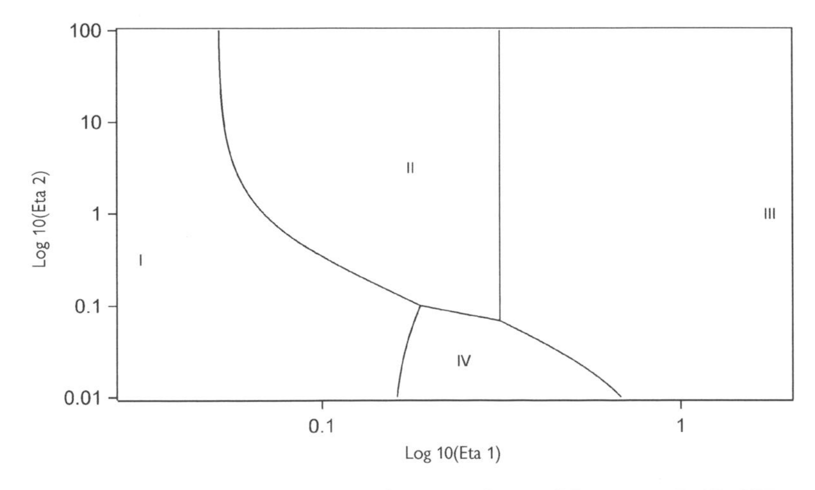

FIG. 15.2. Example 15.2: regions of optimality of Designs I, II, III, and IV.

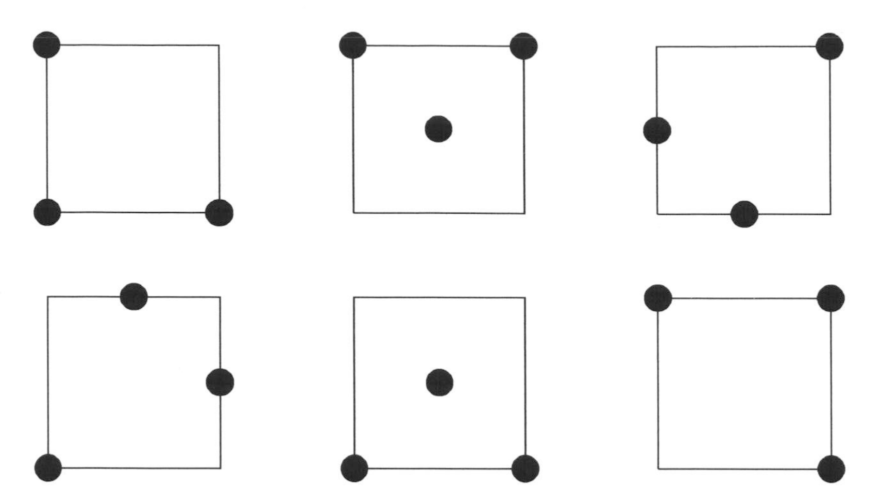

FIG. 15.3. Example 15.2: D-optimum design for a full quadratic model in two explanatory variables in the presence of two blocking variables when the values of η_1 and η_2 fall in the region I shown in Figure 15.2.

are fixed. However, this model could be too restrictive if the block effects are random, as it assumes equal variances for effects generated by the two blocking variables. ■

15.3 Orthogonal Blocking

Suitable blocking of experimental designs introduces a number of desirable features to the studies where it is used. However, it may also lead to difficulty in the interpretation of the results. This problem is considerably reduced when the designs are orthogonally blocked, that is, when the parameters β are estimated independently of the parameters in the model representing the block effects.

If we let $F = \begin{bmatrix} 1_N & \tilde{F} \end{bmatrix}$ and $\beta^{\mathrm{T}} = \begin{bmatrix} \beta_0 & \tilde{\beta}^{\mathrm{T}} \end{bmatrix}$, model (15.9) becomes

$$y = \lambda_0 1_n + \tilde{F}\tilde{\beta} + \tilde{Z}_R\alpha_R + \epsilon,$$

where 1_N is a vector of N ones, $\lambda_0 = \beta_0 + N^{-1} 1_N^{\mathrm{T}} Z_R \alpha_R$ and $\tilde{Z}_R = (I_N - N^{-1} 1_N 1_N^{\mathrm{T}}) Z_R$. A design is orthogonally blocked if the columns of F are orthogonal to those of $\tilde{Z}_R$, that is if

$$F^{\mathrm{T}}\tilde{Z}_R = F^{\mathrm{T}}(I_N - N^{-1} 1_N 1_N^{\mathrm{T}}) Z_R = 0_{p\times(f+r)},$$

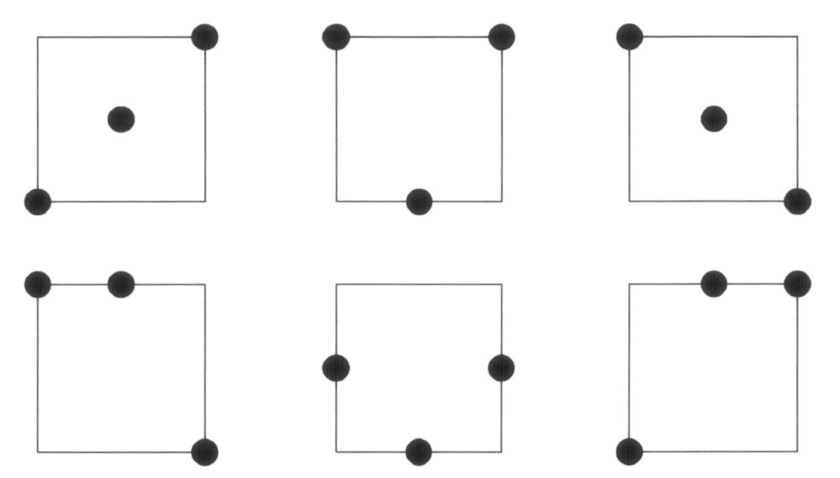

FIG. 15.4. Example 15.2: D-optimum design for a full quadratic model in two explanatory variables in the presence of two blocking variables when the values of η_1 and η_2 fall in the region II shown in Figure 15.2.

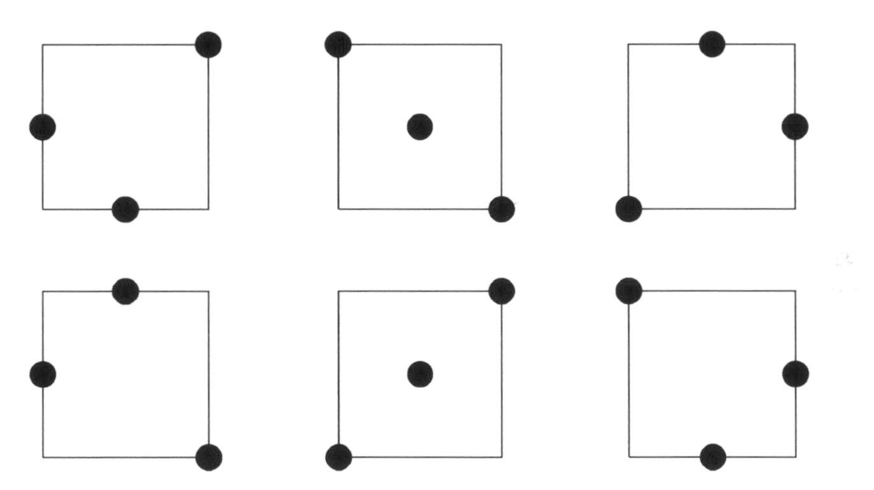

FIG. 15.5. Example 15.2: D-optimum design for a full quadratic model in two explanatory variables in the presence of two blocking variables when the values of η_1 and η_2 fall in the region III shown in Figure 15.2.

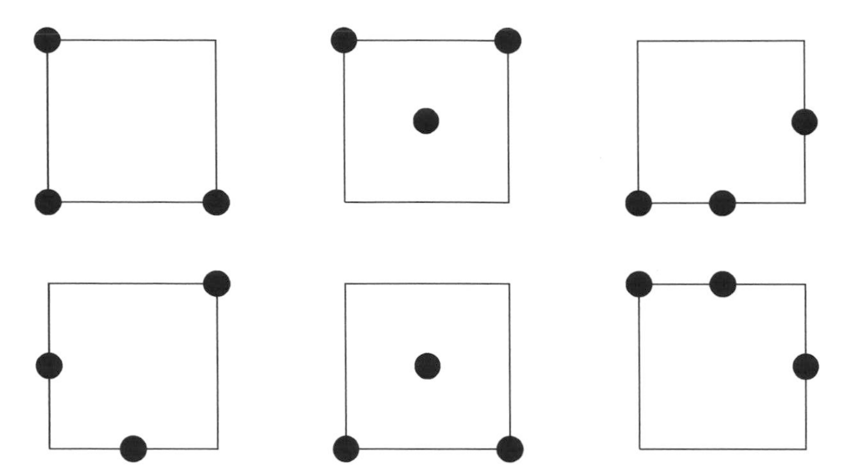

FIG. 15.6. Example 15.2: D-optimum design for a full quadratic model in two explanatory variables in the presence of two blocking variables when the values of η_1 and η_2 fall in the region IV shown in Figure 15.2.

where $0_{p\times(f+r)}$ is a matrix of zeroes. If the block effects are additive, this condition becomes

$$N_{ij}^{-1} F_{ij}^{\mathrm{T}} 1_{N_{ij}} = N^{-1} F^{\mathrm{T}} 1_N, \quad i = 1, 2, \ldots, B; j = 1, 2, \ldots, b_i, \qquad (15.14)$$

where F_{ij} is the part of F corresponding to the jth level of the ith blocking variable and N_{ij} is the number of observations at that level. Thus, the average level of the regressors in an orthogonally blocked design is the same at each level of each blocking variable. Orthogonally blocked designs with equal block sizes have good statistical properties (Goos and Donev 2006*a*). An example of such a design is given in Figure 15.5. However, orthogonality can also be achieved in designs with unequal block sizes.

When an experimental design is orthogonally blocked, its information matrix is block-diagonal. Then the orthogonality measure

$$\left(\frac{|F^{\mathrm{T}} V^{-1} F| |Z_F^{\mathrm{T}} V^{-1} Z_F|}{|X^{\mathrm{T}} V^{-1} X|} \right)^{1/p} \qquad (15.15)$$

will be 1, whilst being less than one for designs that are not orthogonally blocked.

Many D_β-optimum designs are orthogonally blocked. Examples include numerous orthogonally blocked standard two-level factorial and fractional factorial designs which are D_β-optimum for estimable first-order models that may include some interactions between the qualitative factors.

Orthogonally blocked D_β-optimum designs do not exist for all block structures. However, designs that are D_β-optimum usually also perform well with respect to the orthogonality measure (15.15). Notable exceptions are experiments with mixtures where the orthogonality comes at a considerably higher cost in terms of lost D-efficiency; see Chapter 16 for more details.

15.4 Related Problems and Literature

In the previous sections the block sizes were specified or found by an algorithmic search. An important extension is to designs in which the block sizes are not specified but there are upper bounds N_i^U, on the number of trials in at least some of the blocks. The numbers of trials per block N_i then satisfy $N_i \leq N_i^U$ with $\sum N_i^U > N$, with, perhaps, some block sizes specified exactly, that is, $N_i = N_i^U$, for some, but not all, i. Such designs can be constructed by use of an exchange algorithm such as those described in Chapter 14; see §15.5 for discussion of how this can be done in SAS.

Sometimes an alternative to blocking an experiment is to allow for continuous concomitant variables. The partitioned model is

$$\mathrm{E}(y) = F\beta + T\theta, \tag{15.16}$$

where the β are the parameters of interest and T is the design matrix for the variables describing the properties of the experimental unit. If the experimental units were people in a clinical trial, the variables in T might include age, blood pressure, and a measure of obesity, variables which would be called prognostic factors. In a field trial, last year's yield on a unit might be a useful concomitant variable. By division of the continuous variables T into categories, blocks could be created which would, in the medical example, contain people of broadly similar age, blood pressure, and obesity. The methods for design construction described earlier can then be used. However, with several concomitant variables, unless the number of units is large, the blocks will tend either to be small or to be so heterogeneous as to be of little use. In such cases there are advantages in designing specifically for the estimation of β in (15.16). Nachtsheim (1989) describes the theory for approximate designs. Exact designs and algorithms are given by Harville (1974, 1975), Jones (1976), and Jones and Eccleston (1980).

Atkinson and Donev (1996) consider another important case where the response is described by model (15.16). In this case the observations are collected sequentially over a certain period of time, possibly in groups, and it is believed that the response depends on the time as well as the regressors defining F. In this case T defines the dependence of the response on a time trend in terms of linear, quadratic, or other known functions of the time. Clearly the θ are nuisance parameters and it is desirable to estimate β independently of θ. A measure similar to (15.15) is used to compare and construct designs, and for time-trend-free designs it is equal to one. Atkinson and Donev (1996) show that an algorithmic search allows for finding time-trend-free or nearly time-trend-free D-optimum designs in the majority of cases. Such designs can be obtained using SAS—again see §15.5.

Similar, but somewhat more complex, is the case when blocking arises as a result of factors which are hard to change or which are applied to experimental units in discrete stages. The design is therefore not randomized in the usual way, but observations for which the hard-to-change factors are the same can be considered as being run in blocks. For example, Trinca and Gilmour (2001) describe a study where changing the levels of one of the experimental variables is expensive as it involves taking apart and reassembling equipment.

Designs with hard-to-change factors belong to a larger class of split-plot designs. The terminology originates from agricultural experiments where units are, in fact, plots of land, and some factors (such as fertilizer type) can be applied to small *sub-plots* while other factors (such as cultivation method) can only be applied to collections of sub-plots called *whole plots*. Thus, considering sub-plots to be the fundamental experimental unit, whole-plot factors must be constant across all the units within a single whole plot. Such constraints are also very common in chemical and process industries, where factors of interest are often applied at different stages of the production process and the final measurements of interest are made on the finished product. In this case, the different stages of production may give rise to multiple, nested whole-plots.

It is possible but inadvisable to ignore whole-plot constraints when selecting a model with which to analyse a split-plot design. Whole plots are typically subject to their own experimental noise, with the result that sub-plots within the same whole plot are more correlated than those in different whole plots. Thus, a mixed fixed/random effect model of the form (15.4) would typically be appropriate to describe the data from such studies. As usual, the algorithmic approach to the construction of such designs is very effective; see Goos and Vandebroek (2001), and Goos and Donev (2007*b*) for illustrations. Myers *et al.* (2004) provide an extensive general review of the literature devoted to response surface methodology.

15.5 Optimum Block Designs in SAS

In order to use SAS to find the optimum designs discussed in this chapter, we need to introduce a new statement for the OPTEX procedure, the BLOCK statement. Using this statement, OPTEX can be directed to find D_β-optimum designs for fixed blocks, either balanced or unbalanced, as in §15.2.1; for random block effects as in §15.2.2; or for any fixed covariate effects (not necessarily qualitative), as in §15.4. When block sizes are not fixed, the OPTEX procedure cannot by itself find D_β-optimum designs, but with some straightforward SAS programming this, too, is possible.

Up to this point, we have discussed how to use OPTEX to maximize $F^{\mathrm{T}}F$, where the rows of F are selected from a given candidate set, as defined by a data set of feasible factor values and a linear model. In general, the BLOCK statement in OPTEX defines a matrix A such that $|F^{\mathrm{T}}AF|$ is to be maximized instead. For fixed block effects, this matrix is $A = I - Z_F(Z_F^{\mathrm{T}}Z_F)^{-1}Z_F^{\mathrm{T}}$, as in (15.3), while for random block effects $A = V^{-1}$, as in (15.8).

15.5.1 Finding Designs Optimum for Balanced Blocks

The BLOCK statement is especially easy to use for optimum designs in balanced blocks, with the option STRUCTURE=(b)k defining b blocks of size k each. For example, the following statements find an optimum response surface design in two factors for three blocks of size three.

```
data Candidates;
   do x1 = -1 to 1;
      do x2 = -1 to 1;
         output;
         end;
      end;
run;

proc optex data=Candidates coding=orthcan;
   model x1 x2 x1*x2 x1*x1 x2*x2;
   block structure=(3)3 niter=1000 keep=10;
   output out=Design;
run;
```

15.5.2 Finding Designs Optimum for Unbalanced Blocks

If blocks are not balanced, an auxiliary data set containing blocks of the appropriate size needs to be defined, and named in the BLOCK statement option DESIGN=. Subsequent CLASS and MODEL statements define the block model. For example, the following statements use the Candidates data set defined earlier and find an optimum response surface design in two factors for blocks with a $2:2:5$ division.

```
data Blocks; keep Block;
   Block = 1; do i = 1 to 2; output; end;
   Block = 2; do i = 1 to 2; output; end;
   Block = 3; do i = 1 to 5; output; end;
run;

proc optex data=Candidates coding=orthcan;
   model x1 x2 x1*x2 x1*x1 x2*x2;
   block design=Blocks niter=1000 keep=10;
   class Block;
   model Block;
   output out=Design;
run;
```

> **SAS Task 15.1.** Use PROC OPTEX to find exact D-optimum two-factor designs over the 3×3 grid for a second-order response surface model with blocks of size (a) $4:4:4$ and (b) $3:4:5$.

15.5.3 Finding Designs Optimum for Arbitrary Covariates

Notice that there are two MODEL statements in the OPTEX code above. The first refers to variables in the `DATA=Candidates` data set, and the second, which follows the BLOCKS statement, refers to the BLOCK `DESIGN=Blocks` data set. Just as the first MODEL statement can be used to define quite general linear models, with qualitative and quantitative factors, including interactions, so can the MODEL statement following the BLOCK statement. Thus, code with the same general structure can be used to find designs optimum for general covariate models. For example, the following code uses OPTEX to find a 9-run design for a quadratic model in two factors on $[-1,1]^2$ which is time-trend-free: all terms in the model are uncorrelated with the linear effect of time. Again, we use the Candidates data set defined earlier.

```
data Runs;
   do Time = 1 to 9;
      output;
      end;
run;

proc optex data=Candidates coding=orthcan;
   model x1 x2 x1*x2 x1*x1 x2*x2;
   block design=Runs niter=1000 keep=10;
   model Time;
   output out=Design;
run;
```

SAS Task 15.2. Starting with the time-trend-free design constructed above, using either SAS/IML or a DATA step, construct variables for the quadratic and cross-product terms. Then use PROC CORR to confirm that these and the (linear) factor terms are all uncorrelated with Time.

15.5.4 Finding Designs Optimum for Random Block Effects

Finding designs optimum for random block effects, or more generally for a given covariance matrix for the runs, calls for a third way of using the BLOCK statement. Recall from Section 15.2.2 that a D-optimum design in this case maximizes $|F^{\mathrm{T}}V^{-1}F|$, where the matrix V depends on the block structure and the assumed variances of the random effects. In this case, the V matrix can be stored in a data set and supplied to OPTEX as an argument to the COVAR= option of the BLOCK statement. We illustrate with Example 15.2, a second-order response surface for two factors in six blocks generated by two random blocking variables. Assume that $\eta_1 = \eta_2 = 1$ where $\eta_1 = \sigma^2_{\mathrm{Col}}/\sigma^2_R$ and $\eta_2 = \sigma^2_{\mathrm{Row}}/\sigma^2_R$. The following code creates these blocking factors in a data set and then uses IML to construct the corresponding variance matrix, saving it in a data set named Var.

```
data RowCol; keep Row Col;
   do Row = 1 to 2; do Col = 1 to 3;
      do i = 1 to 3; output; end;
      end; end;
run;

proc iml;
   use RowCol; read all var {Row Col};
   Zr = design(Row);
   Zc = design(Col);

   s2Row = 1;
   s2Col = 1;
   s2R = 1;

   V = s2Row * Zr*Zr‘ + s2Col * Zc*Zc‘ + s2R*i(nrow(Zr));

   create Var from V;
   append from V;
```

Having constructed V in a data set, the following code uses OPTEX to find the design that maximizes $|F^{\mathrm{T}}V^{-1}F|$, merging the result with the row/column structure for printing.

```
data Candidates;
   do x1 = -1 to 1; do x2 = -1 to 1; output; end; end;
run;

proc optex data=Candidates coding=orthcan;
   model x1 x2 x1*x2 x1*x1 x2*x2;
   block covar=Var var=(col:) niter=1000 keep=10;
   output out=Design;
run;

data Design; merge RowCol Design;
proc print data=Design noobs;
run;
```

The D-efficiency calculated by OPTEX for the best design is 62.9961. According to Figure 15.2, this design should be equivalent to the one depicted in Figure 15.5, with respect to $|F^{\mathrm{T}}V^{-1}F|$. It is difficult to see that this is necessarily so, but one way to confirm this is to use OPTEX to evaluate those designs and to compare the efficiency values. In §13.3.1 we saw how to evaluate a design when no BLOCK statement was involved. When there is a BLOCK statement, the trick is similar: name the design to be evaluated as the initial design for `METHOD=SEQUENTIAL`, then use BLOCK search options (`INIT=CHAIN`, `NOEXCHANGE`, and `NITER=0`) that guarantee that this design will pass through the search algorithm without change. The following code creates the design of Figure 15.5 in a data set named Design3, and then uses OPTEX to evaluate it.

```
data Design3; input x1 x2 @@; cards;
-1  0    0 -1    1  1   -1  1    0  0    1 -1
-1 -1    0  1    1  0   -1  0    0  1    1 -1
-1 -1    0  0    1  1    0 -1   -1  1    1  0
;
proc optex data=Candidates coding=orthcan;
   model x1 x2 x1*x2 x1*x1 x2*x2;
   generate initdesign=Design3 method=sequential;
   block covar=Var var=(col:) init=chain noexchange niter=0;
run;
```

Again, the D-efficiency calculated by OPTEX is 62.9961.

SAS Task 15.3. Use OPTEX to confirm that each of the designs of Figures 15.3 through 15.6 are indeed optimum in the regions indicated by Figure 15.2.

15.5.5 Finding D_β-optimum Designs

There are no direct facilities in SAS for finding D_β-optimum designs when the sizes of blocks are not fixed—that is, when D_β-optimality is not equivalent to D-optimality, a situation discussed in §15.2.1. However, the BLOCK statement in OPTEX can be combined with SAS macro programming for a more or less brute force solution to the problem. We illustrate with an example in which blocks are unavoidable nuisance factors, with, in addition, limited sizes.

A clinical trial on an expensive new treatment is to be carried out, with as many as seven hospitals available for performing the trial. There is a sufficient budget to study a total of 30 patients, but no hospital will be able to handle more than 10. Considering the hospital to be a fixed block effect, we need to determine a D_β-optimum design for a response surface model in two quantitative factors, Dose and Admin (i.e., the number of administrations of the treatment).

As discussed earlier `DESIGN=` option for the BLOCK statement is the appropriate tool for finding an optimum design for specific block sizes n_1, n_2, $\ldots n_7$. The following code illustrates this approach for the present example, using macro variables `&n1`, `&n2`, ... `&n7` to hold the block sizes.

```
%let n1 = 4; %let n2 = 4; %let n3 = 4; %let n4 = 4;
%let n5 = 4; %let n6 = 5; %let n7 = 5;

/*
/ Create candidates for quantitative factors.
/-----------------------------------------------------*/
data Candidates;
   do Dose = -1 to 1;
      do Admin = -1 to 1;
         output;
         end;
      end;

/*
/ Create blocks.
/-----------------------------------------------------*/
data Hospitals;
   Hospital = 1; do i = 1 to &n1; output; end;
   Hospital = 2; do i = 1 to &n2; output; end;
   Hospital = 3; do i = 1 to &n3; output; end;
   Hospital = 4; do i = 1 to &n4; output; end;
   Hospital = 5; do i = 1 to &n5; output; end;
   Hospital = 6; do i = 1 to &n6; output; end;
   Hospital = 7; do i = 1 to &n7; output; end;
```

```
/*
/  Select quantitative runs D_beta optimum for a
/  response surface model with the given blocks.
/---------------------------------------------------*/
proc optex data=Candidates coding=orthcan;
   model Dose|Admin Dose*Dose Admin*Admin;
   block design=Hospitals niter=1000 keep=10;
   class Hospital;
   model Hospital;
   output out=Design;
run;
```

One way to solve the entire problem, then, is to loop through all feasible block sizes. Macro statements which do so are shown later. Note that these statements themselves do not constitute allowable SAS code. They need to be submitted in the context of a SAS macro program, and 'wrapped around' the OPTEX code above. Additional SAS programming techniques are required to make it easy to tell which block sizes yield the D_β-optimum design.

```
%do n1=0   %to 30;
   %if (&n1<=10) %then %do;
%do n2=&n1 %to 30- &n1;
   %if (&n2<=10) %then %do;
%do n3=&n2 %to 30-(&n1+&n2);
   %if (&n3<=10) %then %do;
%do n4=&n3 %to 30-(&n1+&n2+&n3);
   %if (&n4<=10) %then %do;
%do n5=&n4 %to 30-(&n1+&n2+&n3+&n4);
   %if (&n5<=10) %then %do;
%do n6=&n5 %to 30-(&n1+&n2+&n3+&n4+&n5);
   %if (&n6<=10) %then %do;
      %let n7=  %eval(30-(&n1+&n2+&n3+&n4+&n5+&n6));
      %if ((&n6<=&n7) & (&n7<=10)) %then %do;
         .
         . (Insert code above at this point)
         .
         %end;
   %end; %end;
   %end; %end;
   %end; %end;
   %end; %end;
   %end; %end;
   %end; %end;
```

16

MIXTURE EXPERIMENTS

16.1 Introduction

This chapter is concerned with the design of experiments when there is a constraint on q of the factors

$$\sum_{i=1}^{q} x_i = 1 \quad (x_i \geq 0). \tag{16.1}$$

We shall chiefly consider the case when the remaining factors, if any, do not vary during the experiment. Then (16.1) defines the q components of a mixture. In mixture experiments the response will depend only on the proportions of the components in the mixture, for example an alloy, but not on the total amount. Experiments of this kind occur frequently in such areas as chemistry, medicine, and agriculture. An extensive survey of mixture experiments is given by Cornell (2002).

The general theory of the optimum design of experiments applies to mixture experiments. However, constraint (16.1) introduces some special features. In particular, changes in the values of one of the factors will lead to changes in the value of at least one of the other factors. The design region becomes a $(q-1)$-dimensional regular simplex and the ordinary polynomial models are inappropriate.

The choice of model for a mixture experiment is not always straightforward. Some of the many models in the literature are described in the next section, together with some classical designs. Once an appropriate model has been chosen, the optimum design can be calculated in the standard way. In §16.3 consideration is given to the situation where the experiment is to be performed in a constrained part of the simplex. In the following two sections we discuss the cases when qualitative and quantitative factors have also to be studied and when blocking is required. We show in §16.6 how mixture experiments in which the response depends on both the proportions of the q components and the amount of the mixture can be regarded as a $(q+1)$-component mixture problem. The final section is devoted to the use of SAS to construct designs for experiments with mixtures.

16.2 Models and Designs for Mixture Experiments

The canonical polynomials of Scheffé (1958) have been widely applied because of the flexible family of models they provide. They are obtained by reparameterization of standard polynomials allowing for the relationship between the factors following from (16.1). The first-order Scheffé polynomial is

$$E(y) = \sum_{i=1}^{q} \beta_i x_i, \tag{16.2}$$

which does not explicitly contain a constant term, while the second-order model becomes

$$E(y) = \sum_{i=1}^{q} \beta_i x_i + \sum_{i=1}^{q-1} \sum_{j=i+1}^{q} \beta_{ij} x_i x_j. \tag{16.3}$$

In (16.3) the effect of the constraint is to render redundant the pure quadratic terms. For higher-order models the reparameterization does not lead to such simple expressions. For example, a symmetrical way of writing the third-order model is

$$\begin{aligned} E(y) = &\sum_{i=1}^{q} \beta_i x_i + \sum_{i=1}^{q-1} \sum_{j=i+1}^{q} \beta_{ij} x_i x_j + \sum_{i=1}^{q-1} \sum_{j=i+1}^{q} \gamma_{ij} x_i x_j (x_i - x_j) \\ &+ \sum_{i=1}^{q-2} \sum_{j=i+1}^{q-1} \sum_{k=j+1}^{q} \beta_{ijk} x_i x_j x_k. \end{aligned} \tag{16.4}$$

As designs for estimation of the parameters in these canonical polynomials, Scheffé (1958) proposed 'simplex lattice' designs. For the polynomial of degree d these designs include all possible combinations of trials in which each component takes values 0, $1/d$, $2/d$,..., 1. Such designs are saturated, that is, the number of trials is equal to the number of parameters in the model. This simplifies the expression for $M(\xi)$ and leads to simple formulae for the estimation of the parameters. Like other saturated designs, simplex lattice designs suffer from the disadvantage that they provide no information on lack of fit; additional trials have to be added to check the adequacy of the model. A specific shortcoming of the designs is that they contain a high proportion of trials for which at least one component takes the value zero. If the behaviour of the complete mixture is markedly different from the behaviour of simpler systems lacking one or more components, the information provided by the simplex lattice designs will be seriously inadequate.

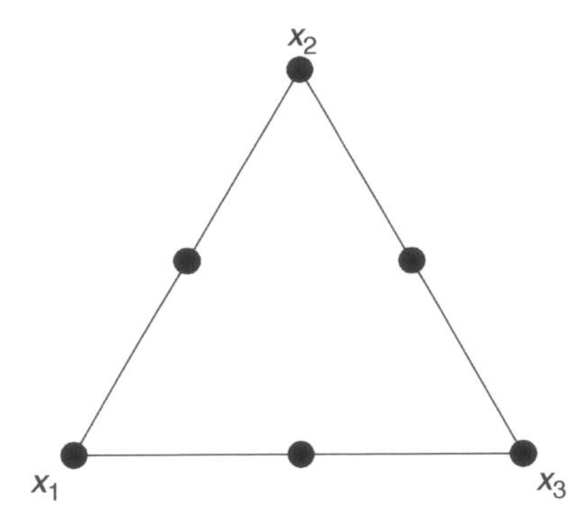

FIG. 16.1. Three-component mixture: second-order simplex lattice design.

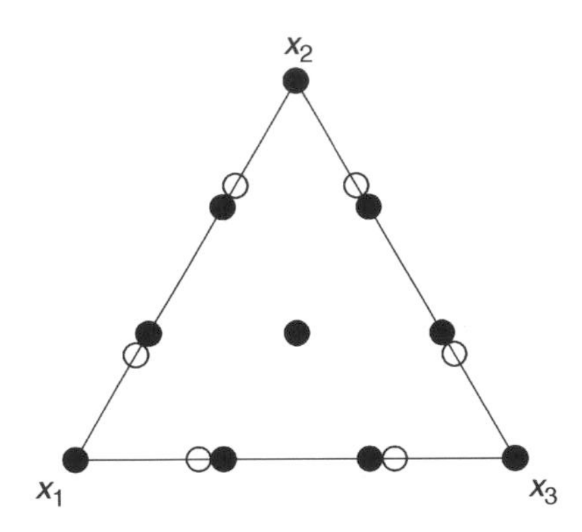

FIG. 16.2. Three-component mixture: • third order simplex lattice design; ○ D-optimum design for third-order model (16.4).

The D-optimality of the simplex lattice designs for the first- and second-order canonical polynomials was established by Kiefer (1961). As an example Figure 16.1 shows the six-trial D-optimum design for the second-order polynomial (16.3) when $q = 3$. This is also the optimum continuous design, the measure putting weight 1/6 at the vertices and middles of the edges of the triangle representing the design region. However, the points of the simplex lattice are not the support of optimum designs for higher-order models.

Figure 16.2 shows the third-order simplex lattice design which has support at the vertices of the design region, at the centre, and on the edges at points corresponding to all two-component mixtures in which one of the components is one-third of the mixture and the other is two-thirds. For this design the determinant of the information matrix $|M(N)|^{1/10} = 0.00667$. The D-optimum design for (16.4) with $q = 3$ has the same structure except that the $(1/3, 2/3)$ mixtures are replaced by those derived from the results on D-optimality for dth-order polynomials in Table 11.3. There, for the third-order polynomial over $\mathcal{X} = [-1, 1]$, the interior design points were at $x = \pm 1/\sqrt{5}$. Rescaling of these points on to $[0, 1]$ yields design points at $(1 \pm 1/\sqrt{5})/2$, which are the proportions for the optimum two-component mixtures, namely $(0.2764, 0.7246)$. These design points are shown in Figure 16.2 as open circles. For the resulting design $|M(\xi^*)|^{1/10} = 0.00701$. a more than 5% improvement over the simplex lattice design when the number of parameters is taken into account.

In order to improve the interpretation of the parameters in polynomial mixture models Cox (1971) suggests reparameterization to represent the change in the response as one factor is varied, with the relative proportions of the remaining factors being held constant. Draper and John (1977) recommend the inclusion of inverse terms into Scheffé polynomials, provided that the levels of the components never reach zero. For example, their first-order model is

$$\mathrm{E}(y) = \sum_{i=1}^{q} \beta_i x_i + \sum_{i=1}^{q} \frac{\beta_{-i}}{x_i},$$

which allows behaviour near to lower-dimensional mixtures to be quite different from that for the q-component mixture.

If one of the components has an additive blending effect, the models proposed by Becker (1968) may be useful. Three possibilities are:

$$\mathrm{E}(y) = \sum_{i=1}^{q} \beta_i x_i + \sum_{i=1}^{q-1} \sum_{j=i+1}^{q} \beta_{ij} \min(x_i, x_j) + \cdots + \beta_{12\ldots q} \min(x_1, x_2, \ldots, x_q),$$

$$\mathrm{E}(y) = \sum_{i=1}^{q} \beta_i x_i + \sum_{i=1}^{q-1} \sum_{j=i+1}^{q} \frac{\beta_{ij} x_i x_j}{x_i + x_j} + \cdots + \frac{\beta_{12\ldots q} x_1 x_2 \ldots x_q}{x_1 + x_2 + \ldots + x_q},$$

$$\mathrm{E}(y) = \sum_{i=1}^{q} \beta_i x_i + \sum_{i=1}^{q-1} \sum_{j=i+1}^{q} \beta_{ij} (x_i x_j)^{1/2} + \cdots + \beta_{12\ldots q} (x_1 x_2 \ldots x_q)^{1/q}.$$

These models are symmetrical in the q factors and so are only appropriate if the extremum of the surface they are modelling is at or near the centre of the simplex. Becker also suggests a decentralized form of the first of these models. A further class of models (Kenworthy 1963; Becker 1969; Becker 1970) incorporates ratios of the components.

The customary designs for several of these models are saturated. Vuchkov (1982) uses the results on saturated designs of §11.1 for design problems for canonical polynomials. In particular, the variance of the predicted response $\hat{y}$ at the support points x_i is

$$\mathrm{var}(\hat{y}_i) = \sigma^2 f(x_i)^{\mathrm{T}}(F^{\mathrm{T}}F)^{-1}f(x_i) = \frac{\sigma^2}{N(x_i)} \quad (i = 1, \ldots, n), \tag{16.5}$$

where $n = p$. In (16.5), $N(x_i) > 0$ is the number of replications at the point x_i. Further, the covariance between the predicted values at the points x_i and x_j is zero. These results are helpful in §16.4 for the construction of designs when there are both mixture and qualitative factors.

16.3 Constrained Mixture Experiments

Often the design region for mixture experiments is restricted to part of the $(q-1)$-dimensional regular simplex. For example, gunpowder is a mixture of sulphur, carbon, and saltpetre, the interesting explosive properties of which only occur for a small part of the triangular region of possible blends. In general, the sub-area of the simplex forming the design region can have any irregular shape. The simplest constrained case is when the design region is again a $(q-1)$-dimensional regular simplex within the larger simplex. Such a situation can occur when additional constraints of the form

$$x_i \geq x_{i,\min} > 0$$

are imposed on some or all the factors. The design region will also again be a regular simplex if the constraints are

$$x_i \leq x_{i,\max} < 1$$

and the sum of the $(q-1)$ largest upper bounds is less than or equal to unity.

For constrained design regions which are a regular simplex, Kuroturi (1966) and Crosier (1984) propose the use of pseudo-components $g_i, (i = 1, \ldots, q)$. These are linear transformations of the original variables which allow use of the standard results for unconstrained mixture experiments. Let the columns of the $q \times q$ matrix A be the co-ordinates of the vertices of the restricted design region. Then the relationship between the co-ordinates of a mixture x in the original components and as a pseudo-component g is

$$x = Ag \tag{16.6}$$

with inverse

$$g = A^{-1}x. \tag{16.7}$$

The design is constructed for the pseudo-components g, with the values of the components in the design calculated from (16.6). Since the vectors x and g are $q \times 1$, the rows of the design matrices are respectively x^{T} and g^{T}. The model can be fitted either in the pseudo-components or in the original mixture components. If pseudo-components are used for fitting, the inverse transformation (16.7) is required to give estimates of the original parameters.

Example 16.1 Modification of an Acrylonitrile Powder Garvanska *et al.* (1992) describe the development of a modified acrylonitrile powder with improved electrophysical properties. The modification to the surface of the powder used a three-component mixture of chemicals with factors the proportions of:

x_1 copper sulphate ($CuS0_4$)
x_2 sodium thiosulphate ($Na_2S_2O_3$)
x_3 glyoxal $(CHO)_2$.

Consideration of the mechanism of the reaction imposed constraints on the factor levels:

$$\begin{aligned} 0.2 &\le x_1 \le 0.8 \\ 0.2 &\le x_2 \le 0.8 \\ 0.0 &\le x_3 \le 0.6. \end{aligned} \tag{16.8}$$

It is straightforward to verify that the resulting constrained design region is again a simplex. The matrix A (16.6) for the pseudo-components is

$$A = \begin{bmatrix} 0.8 & 0.2 & 0.2 \\ 0.2 & 0.8 & 0.2 \\ 0 & 0 & 0.6 \end{bmatrix},$$

where each column defines one of the vertices of the restricted design region.

One response of interest was the electric resistivity W per unit volume. This was measured using a Scheffé lattice design for a second-order model in the pseudo-components. The results are given in Table 16.1. To check this saturated model for lack of fit, additional trials were performed at points, not listed in Table 16.1, which were uniformly spread through the design region. Despite the very different response value for the last trial, there was no evidence of lack of fit: it was expected that the response would change rapidly in this part of the experimental region. ■

Example 16.1 illustrates the use of the pseudo-components in the design of mixture experiments when the constrained design region is again a regular simplex. However, with more general constraints of the form

$$0 \leq x_{i,\min} \leq x_i \leq x_{i,\max} \leq 1,$$

the design region will usually be an irregular simplex, so that the use of pseudo-components is not possible. As an alternative, McLean and Anderson (1966) propose extreme vertices designs which Saxena and Nigam (1973) show may contain clusters of points in the design region. Their alternative symmetric simplex designs overcome the problem of clustering, but both design strategies suffer from an excessive number of trials as the size of the problem increases. In such cases the trials generated by these procedures or by the XVERT algorithm of Snee and Marquardt (1974) which finds the extreme vertices and generalized edge-centroids of any linearly constrained region, can be considered as a list of candidate points. A sensible number of them can then be selected by an optimum design algorithm.

Example 16.2 Linear Constraints The three components x_i of a mixture are subject to the constraints

$$0.2 \leq x_1 \leq 0.7$$

$$0.1 \leq x_2 \leq 0.6$$

$$0.1 \leq x_3 \leq 0.6$$

as well as the usual mixture constraints. The resulting hexagonal design region is shown in Figure 16.3 The figure also shows the points of the extreme vertices design constructed following the method of McLean and Anderson (1966) which consists of all vertices, centres of edges, and the centre point, 13 points in all.

Suppose that the model is the second-order canonical polynomial

$$\mathrm{E}(y) = \beta_1 x_1 + \beta_2 x_2 + \beta_3 x_3 + \beta_{12} x_1 x_2 + \beta_{13} x_1 x_3 + \beta_{23} x_2 x_3$$

with six parameters. The extreme vertices design could be used. However, this design takes no account of the model, merely reflecting the structure of the design region. Further, the design can be expected to be inefficient for the second-order model as the number of design points would be appreciably larger than the number of parameters. We therefore take the points of the

TABLE 16.1. Example 16.1: modification of an acrylonitrile powder. Use of pseudo-components for design in a constrained region

Pseudo-components			Mixture variables			Responses			
g_1	g_2	g_3	x_1	x_2	x_3	$\log W$	$y(z_1)$	$y(z_2)$	$y(z_3)$
1	0	0	0.8	0.2	0	1.25	8	9	9
0	1	0	0.2	0.8	0	1.82	3	4.8	4
0	0	1	0.2	0.2	0.6	0.17	8	8.5	8
0.5	0.5	0	0.5	0.5	0	0.44	7.8	8.2	7.5
0.5	0	0.5	0.5	0.2	0.3	0.03	8	8.8	8
0	0.5	0.5	0.2	0.5	0.3	11.14	0.6	2	1.8

The responses $y(z_i)$ are the electromagnetic damping at the three wavelengths $z_i = 8, 10,$ and 12 GHz. W is the electric resistivity per unit volume.

TABLE 16.2. Example 16.2: mixture experiment with linear constraints. Exact D-optimum designs for the non-standard design region in Figure 16.3 (second-order model)

Point				Number of design points $N(x)$				
number	x_1	x_2	x_3	10	20	30	35	100
1	0.7	0.1	0.2	1	2	3	4	10
2	0.2	0.6	0.2	1	2	3	4	10
3	0.7	0.2	0.1	1	2	3	3	10
4	0.2	0.2	0.6	1	2	3	3	10
5	0.3	0.6	0.1	1	2	3	3	10
6	0.3	0.1	0.6	1	2	3	4	10
7	0.7	0.15	0.15					
8	0.2	0.4	0.4	1	2	3	4	11
9	0.5	0.1	0.4	1	2	3	4	11
10	0.25	0.6	0.15					
11	0.5	0.4	0.1	1	2	3	4	11
12	0.25	0.15	0.6					
13	0.4	0.3	0.3	1	2	3	2	7

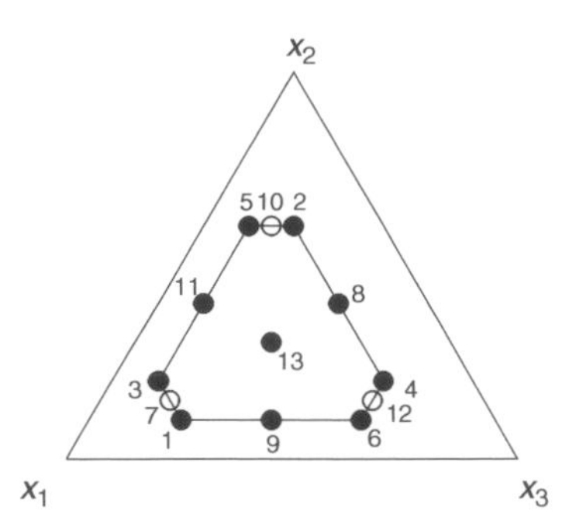

FIG. 16.3. Example 16.2 mixture experiment with linear constraints. D-optimum design for second-order model with points numbered as in Table 16.2: • design points; ○ unused candidate points.

extreme vertices design as a list of candidate points for the construction of exact designs.

The results of using an optimum design algorithm to find moderate-sized exact designs are given in Table 16.2. For $N = 10$ the design puts one trial at each point of the candidate set except for the centres of the short edges. The design is shown in Figure 16.3, where the open symbols represent the unused candidate points. For $N = 20$ and $N = 30$ this pattern is repeated two and three times respectively. For $N = 35$ this repetition is not possible: the design in Table 16.2 puts only two trials at the centre point, as a result of which there is a symmetrical structure in the other points. Points at the centres of the long sides and at one end of the short sides each receive four trials, with three trials at the other end of each short side. Designs with equal efficiency can be found by interchanging the points on short sides receiving three and four trials.

The design for $N = 35$ does not give quite as efficient a design measure as the equi-replicated design, although the difference is less than 0.1%. However, the design does serve as a further reminder that exact designs do depend, sometimes in an unpredictable way, on the specific value of N. To obtain some idea about the continuous D-optimum design we used the first-order algorithm adding design points where the variance $d(x, \xi_N)$ is a maximum. The starting design was the equi-replicated 10-point design of Table 16.2. As we saw in Chapter 11, this algorithm does not converge monotonically towards the optimum but passes through good balanced designs, which are then distorted by the sequential addition of design points. In this example a particularly interesting design was found for 100 trials, with

$\bar{d}(\xi) = 6.306$ when evaluated over the 13 candidate points, as opposed to 6 for the continuous D-optimum design. The structure of the exact weights approximates that of the optimum weights, with the centre point having the lowest weight, the centres of the long sides the highest weight, and with the remaining weight equally distributed over the end-points of the short sides of the design region. (Demonstrating that the optimum weights over these 13 candidate points have this structure is given as SAS Task 16.5.) ■

This example thus illustrates two features of mixture designs for constrained regions. One is that not all the candidate points of the extreme vertices design are required. The second is that even continuous designs for constrained regions can be appreciably more complicated than those for regular regions. Here we have 10 unequally replicated support points rather than the six equally replicated points of the simplex lattice design. As we have seen, the calculation of exact designs introduces further complications.

Example 16.3 Quadratic Constraints The following example is taken from an unpublished report by Z. Iliev. As the result of the first phase of an experiment with a three-component mixture, models were fitted to two responses. In the second phase measurements are to be made of a third response, but only in that portion of the design region where the other two responses have satisfactory values. The requirements $\hat{y}_1 \geq c_1$ and $\hat{y}_2 \geq c_2$ for specified c_1 and c_2 lead to the quadratic constraints

$$
\begin{aligned}
-4.062x_1^2 + 2.962x_1 + x_2 &\geq 0.6075 \\
-1.174x_1^2 + 1.057x_1 + x_2 &\leq 0.5019.
\end{aligned}
$$

These two quadratics define the smile-shaped experimental region shown in Figure 16.4. The D-optimum continuous design for the second-order canonical polynomial was found by searching over a fine grid covering this region. The resulting six-point design is given in Table 16.3 and shown in Figure 16.4. An interesting feature of this design is that, like the D-optimum simplex lattice design, it also has six support points.

This formulation assumes that the constraints defining the design region are known precisely. Allowing for variation in the design region caused by the sampling variability in the coefficients of the estimated constraints is a more complicated design problem. Donev (2004) discusses the design of experiments where there is uncertainty about the region where observations can be taken as well as the possibility of losing observations close to the borders of the design region. ■

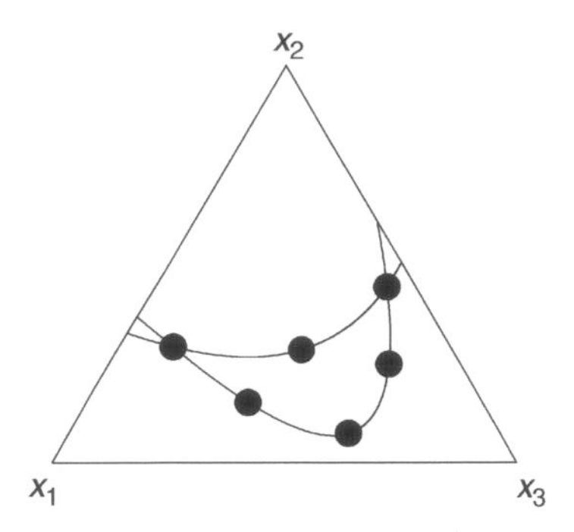

FIG. 16.4. Example 16.3 mixture experiment with quadratic constraints. D-optimum second-order design of Table 16.3.

TABLE 16.3. Example 16.3: quadratic constraints for a mixture experiment. D-optimum design for the non-regular region of Figure 16.4

x_1	x_2	x_3
0.59	0.28	0.13
0.50	0.15	0.35
0.34	0.07	0.59
0.33	0.28	0.39
0.16	0.24	0.60
0.07	0.43	0.50

16.4 Mixture Experiments with Other Factors

In addition to the q mixture variables an experiment may also include one or more qualitative or quantitative factors, so that the design region will often be the product of two regions, one for the mixture variables and the other for the remaining factors. For example, with three mixture variables and one quantitative factor the design region will be a triangular prism, provided that experiments are possible at all combinations of the two sets of variables. Once the required model has been specified, the results presented in Chapter 10 can be used to construct designs that satisfy a chosen criterion of optimality, for instance by using one of the algorithms presented in Chapter 12.

Unexpectedly interesting designs arise when the q components of a mixture are studied simultaneously with a qualitative factor at b levels. If the mixture variables can be modelled using second-order canonical polynomials, the model is expressed

$$\mathrm{E}(y) = \sum_{i=1}^{b-1} \alpha_i z_i + \sum_{i=1}^{q} \beta_i x_i + \sum_{i=1}^{q-1} \sum_{j=i+1}^{q} \beta_{ij} x_i x_j, \tag{16.9}$$

where the identifiability condition $\alpha_b = 0$ is imposed. The continuous D-optimum design for (16.9) is a product design, with the lattice design of Figure 16.1 repeated at each level of b. Product designs have good statistical properties. However, in general they may require an impracticably large number of observations. As in previous sections, we therefore consider reasonably small exact designs. An interesting distinction between these designs for mixtures and those of Chapter 14 is that here it is possible to derive useful theoretical results about the structure and properties of small exact designs.

Example 16.1 Modification of an Acrylonitrile Powder continued For the experimenters a second interesting property of the modified acrylonitrile powder of Table 16.1 was the effect of the modification on the electromagnetic damping y of the material. It was expected that this would depend on the wavelength of the electromagnetic radiation. From previous experience the effect of the wavelength was expected to be additive and representable by a vector of indicator variables Z. Therefore, in this case, the wavelength can be treated as a qualitative variable. The response was measured at three wavelengths (Table 16.4). The simplex lattice design for the three pseudo-components was repeated at each level of the indicator variables. The results are given in Table 16.1, for which the fitted model in terms of the pseudo-components is

$$\hat{y} = 0.983 z_2 + 0.483 z_3 + 8.178 g_1 + 3.444 g_2 + 7.678 g_3 + 6.133 g_1 g_2 - 0.600 g_1 g_3 - 18.333 g_2 g_3.$$

In order to achieve identifiability of the coefficients in the model, there is no term in z_1. This corresponds to the absence of the constant term in the canonical polynomial (16.3). ■

We shall mainly consider designs with support at the points of the optimum product design. Thus, the candidate runs will be the points of the simplex centroid design, as depicted in Figure 16.1. The slight loss in efficiency due to this restriction will usually be off-set by the relative ease with which such designs can be constructed and employed. One important aspect

TABLE 16.4. Example 16.1: modification of an acrylonitrile powder. Indicator variables for the three wavelengths

Wavelength (GHz)	Indicators z_1	z_2	z_3
8	1	0	0
10	0	1	0
12	0	0	1

of such a design will be the properties of the projection of the design at the various levels of the qualitative factor onto one level.

Let this projection design have design matrix F, with $N(s_i)$ trials at support point s_i, where the fitted response is

$$\hat{y}(s_i) = \hat{\beta}^{\mathrm{T}} f(s_i). \tag{16.10}$$

Therefore it follows from (11.4) that

$$\frac{\mathrm{var}\{\hat{y}(s_i)\}}{\sigma^2} = \frac{1}{N(s_i)}$$

$$\mathrm{cov}\{\hat{y}(s_i), \hat{y}(s_j)\} = 0 \quad (i \neq j). \tag{16.11}$$

We now show that the determinant of the information matrix of the design is the product of the determinant of the information matrix for the design at the projection level and the determinant of a symmetric matrix R with elements depending both on the number of trials at the same location on the projection level, but at different levels of the qualitative factor, and on the numbers of replicates at the projection level. R is thus purely a counting matrix. This structure greatly simplifies the search for optimum designs. Since the model can be written as (14.2), it follows that the D-optimum design maximizes

$$|X^{\mathrm{T}}X| = |F^{\mathrm{T}}F| \times |Z^{\mathrm{T}}Z - Z^{\mathrm{T}}F(F^{\mathrm{T}}F)^{-1}F^{\mathrm{T}}Z|. \tag{16.12}$$

This expression emphasizes the importance of the design at the projection level. The determinant of the information matrix of this design is $|F^{\mathrm{T}}F|$, and the elements of $F(F^{\mathrm{T}}F)^{-1}F^{\mathrm{T}}$, which are of the form $f^{\mathrm{T}}(s_i)(F^{\mathrm{T}}F)^{-1}f(s_j)$, are the variances and covariances given by (16.11). Since the covariances are zero, the effect of the indicator variables Z is to form sums of these variances. The information matrix $|Z^{\mathrm{T}}Z|$ for the indicator variables, is a diagonal matrix $\{N_i\}$, where N_i is the number of trials at the ith level of z.

It is simplest to illustrate the resulting structure of (16.12) for specific values of b. Let O_i, denote the set of N_i design points at the ith level of z, let O_{ij} be the subset of points at levels i and j with the same location on the projection level, and let O_{ii} be the set of r_{ii} points at level i with the same location at the projection level as trials from any other level of z. For $b = 3$ with the identifiability condition $\alpha_3 = 0$, $Z^{\mathrm{T}}Z$ is 2×2 and

$$|X^{\mathrm{T}}X| = |F^{\mathrm{T}}F| \times \left| \left\{ \begin{array}{cc} N_1 & 0 \\ 0 & N_2 \end{array} \right\} - \left\{ \begin{array}{cc} \sum_{i \in O_1} N^{-1}(s_i) & \sum_{i \in O_{12}} N^{-1}(s_i) \\ \sum_{i \in O_{21}} N^{-1}(s_i) & \sum_{i \in O_2} N^{-1}(s_i) \end{array} \right\} \right| \tag{16.13}$$

For those design points which are not replicated $N(s_i) = 1$, so that (16.13) becomes

$$\begin{aligned} |M(N)| &= |X^{\mathrm{T}}X| \\ &= \left|F^{\mathrm{T}}F\right| \times \left| \begin{array}{cc} r_{11} - \sum_{i \in O_{11}} N^{-1}(s_i) & -\sum_{i \in O_{12}} N^{-1}(s_i) \\ -\sum_{i \in O_{21}} N^{-1}(s_i) & r_{22} - \sum_{i \in O_{22}} N^{-1}(s_i) \end{array} \right| \\ &= \left|F^{\mathrm{T}}F\right| \times |R| . \end{aligned} \tag{16.14}$$

If $b > 3$, the expression for R extends to include the terms obtained by counting shared design points between trials at pairs of levels of z. Otherwise, the structure of $|M(N)|$ is the same. It is the product of two terms, one of which is a counting matrix for the numbers of shared locations at the different levels and the other has only to do with the determinant of the projection matrix.

For $b = 2$ there is appreciable simplification and

$$|M(N)| = |F^{\mathrm{T}}F| \times \left\{ r - \sum_{i=1}^{r} N^{-1}(s_i) \right\}, \tag{16.15}$$

where r ($r \leq N_1$) is the total number of replicated points between the levels. It is clear from (16.15) that to maximize $|M(N)|$ the number of replicated points should be maximized, along with maximization of $|F^{\mathrm{T}}F|$. Then for N less than the number of trials of the product design, the D-optimum N-trial exact design has no more than two replicates at any point on the projection level.

This general result is important for the construction of experiments with both mixture and qualitative factors. The design criterion is seen to be the product of two terms, one dependent only on the replicates of the trials and the other only on the projection design. Thus the generation of an N-trial D-optimum exact design can be carried out in two steps. The first is

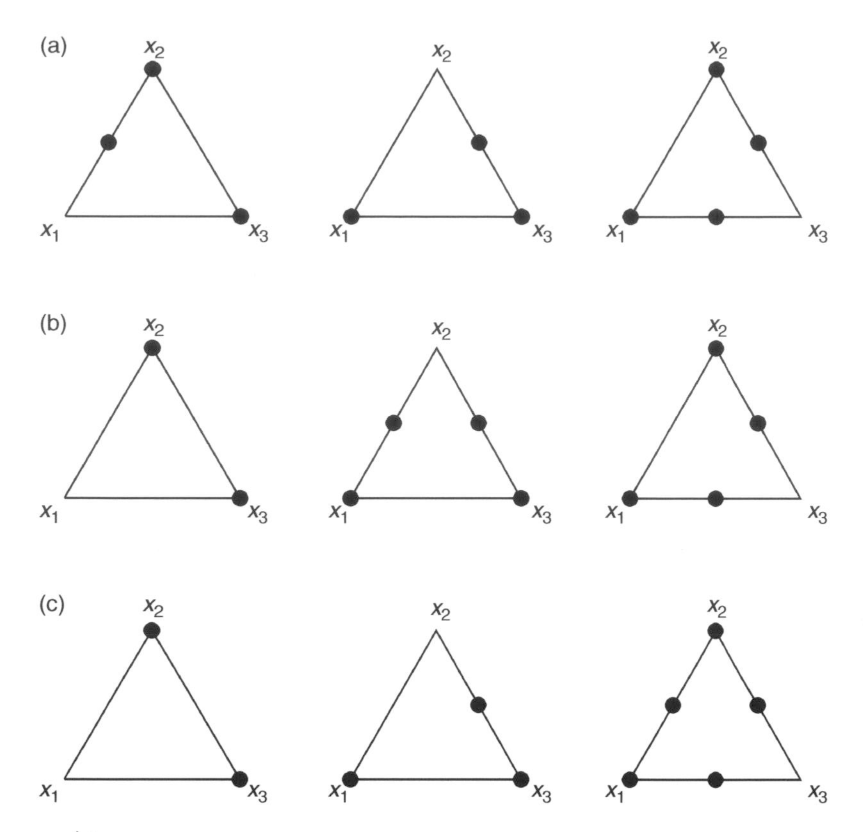

FIG. 16.5. Example 16.4: second-order model for a three-component mixture with a qualitiative factor at three levels. D-optimum 10-trial designs. The divisions of the trials between the levels of the qualitative factor are (a) $3:3:4$, (b) $2:4:4$, and (c) $2:3:5$. All three designs give the same value of $|M(10)|$.

to generate the exact N-trial D-optimum design for the projection level. The second is to obtain the optimum number of replicated points and their location. The distribution of the non-replicated points does not matter and can be made according to criteria other than D-optimality. There is often appreciable freedom in the positions of the design points and usually several different divisions of the experimental conditions between the levels of the qualitative factor which yield the same value for D-optimality. The results also have an important interpretation when the qualitative factor is a blocking variable. In order to illustrate (16.14) we give an example of a mixture design in which there are no individual constraints on the numbers of trials N_i at the levels of the qualitative factor.

Example 16.4 Second-order Model for Three-component Mixture with Qualitative Factor at Three Levels The 10-trial designs in Figure 16.5 are all D-optimum for a three-component mixture with a qualitative factor at three levels. Although the division of trials between the levels is different, being $3:3:4$, $2:4:4$, and $2:3:5$, respectively, all give rise to the same design at the projection level. Furthermore, only one pair of points is replicated between levels 1 and 2. It then follows from (16.15) that $|M(10)|$ is the same for each design. To illustrate this we write

$$|M(10)| = |R| \times |F^{\mathrm{T}}F| = |R| \times |M_s(10)| \qquad (16.16)$$

to stress the dependence of the design at the projection level on the number of trials. The determinant for the design at the projection level is $|M_s(10)| = 0.3906 \times 10^{-2}$. In this case O_1 and O_2 consist of the sets of points $(1,0,0)$, $(0,1,0)$, $(0,0.5,0.5)$ and $(1,0,0)$, $(0.5,0.5,0)$, $(0,0,1)$, respectively, O_{11} includes $(1,0,0)$, $(0,1,0)$, O_{22} includes $(1,0,0)$, $(0,0,1)$, and O_{12} includes only $(1,0,0)$. Then, according to (16.14)

$$|R| = \begin{vmatrix} 2-(0.5+0.5) & -0.5 \\ -0.5 & 3-(0.5+0.5+0.5) \end{vmatrix} = 1.25$$

so that $|M(10)| = 0.4883 \times 10^{-2}$ for all three designs. Because of the identifiably condition $\alpha_3 = 0$, the number of replicates at level 3 does not explicitly appear in this calculation (although it is implicit in the calculations at the projection level). However, relabelling the levels of the qualitative factor can have no effect on the properties of the design. For example, there are two pairs of replicated points between levels 2 and 3 for each of these designs. ■

Goos and Donev (2006*b*) show that (16.12) extends also to the case when there are several qualitative factors each contributing additive effects to the model. For example, suppose there are two qualitative factors with b_1 and b_2 levels respectively, N_{ij} is the number of observations when the first qualitative factor is at its *i*th level while the second is at its *j*th level, $N_{i.}$ is the number of the observations at the *i*th level of the first qualitative factor while $N_{.j}$ is the number of the observations at the *j*th level of the second. Suppose also that the model is

$$\mathrm{E}(y) = \sum_{i=1}^{b_1-1} \alpha_{i.} z_{i.} + \sum_{j=1}^{b_2-1} \alpha_{.j} z_{.j} + \sum_{i=1}^{q} \beta_i x_i + \sum_{i=1}^{q-1} \sum_{j=i+1}^{q} \beta_{ij} x_i x_j.$$

Then (16.12) holds with

$$Z^{\mathrm{T}} = \left[F_{1.}^{\mathrm{T}} 1_{N_{1.}} \quad \cdots \quad F_{b_1-1,.}^{\mathrm{T}} 1_{N_{b_1-1,.}} \quad F_{.1}^{\mathrm{T}} 1_{N_{.1}} \quad \cdots \quad F_{.,b_2-1}^{\mathrm{T}} 1_{N_{.,b_2-1}}\right],$$

where $F_{i.}^{\mathrm{T}} = [F_{i1}^{\mathrm{T}}\, F_{i2}^{\mathrm{T}} \dots F_{ib_2}^{\mathrm{T}}]$, $F_{.j}^{\mathrm{T}} = [F_{1j}^{\mathrm{T}}\, F_{2j}^{\mathrm{T}} \dots F_{b_1 j}^{\mathrm{T}}]$ and

$$Z^{\mathrm{T}}Z = \begin{bmatrix} N_{1.} & \cdots & 0 & N_{11} & \cdots & N_{1,b_2-1} \\ 0 & \cdots & 0 & N_{21} & \cdots & N_{2,b_2-1} \\ \vdots & \ddots & \vdots & \vdots & \ddots & \vdots \\ 0 & \cdots & N_{b_1-1,.} & N_{b_1-1,1} & \cdots & N_{b_1-1,b_2-1} \\ N_{11} & \cdots & N_{b_1-1,1} & N_{.1} & \cdots & 0 \\ N_{12} & \cdots & N_{b_1-1,2} & 0 & \cdots & 0 \\ \vdots & \ddots & \vdots & \vdots & \ddots & \vdots \\ N_{1,b_2-1} & \cdots & N_{b_1-1,b_2-1} & 0 & \cdots & N_{.,b_2-1} \end{bmatrix}.$$

16.5 Blocking Mixture Experiments

The general method described in Chapter 14 can be used to block designs for mixture experiments where the blocking structure is generated by one or more qualitative factors. Depending on the nature of the experiment the block effects can be fixed or random. The results of the previous section indicate that there is great freedom in the distribution of trials between the levels of the qualitative factors. Usually, for a given value of N there will be designs for several different block sizes which all have the same value of $|M(\xi_N)|$.

The most striking consequences occur when the design is in two blocks and the block effects are fixed. If the number of points in the second-order lattice is k, the design should have as a projection the N-trial exact D-optimum design for the mixture variables in which $N-k$ of the points are replicated. As the value of $|M(\xi_N)|$ does not depend on which points are replicated, any distribution of the N trials into two blocks for which $\max(b_1, b_2) < k$ will yield an exact D-optimum design with the same value of the design criterion. For example, the three nine-trial designs in Figure 16.6 are equivalent with respect to D-optimality. Those in Figure 16.6(a) and (b) are chosen to have the same projection and replicated points, but the non-replicated ones are located in a different way. The design in Figure 16.6(c) has different replicated points and a different projection, but equal block sizes, to that of Figure 16.6(b).

Goos and Donev (2006*b*, 2007*a*) give examples of block designs when the blocks are generated by several blocking variables. Figures 16.7 and 16.8 show 16-trial designs in four mixture components divided in six blocks. The blocks are generated by two factors, acting at two and three levels respectively. The element of choice in the position of the replicated points in the

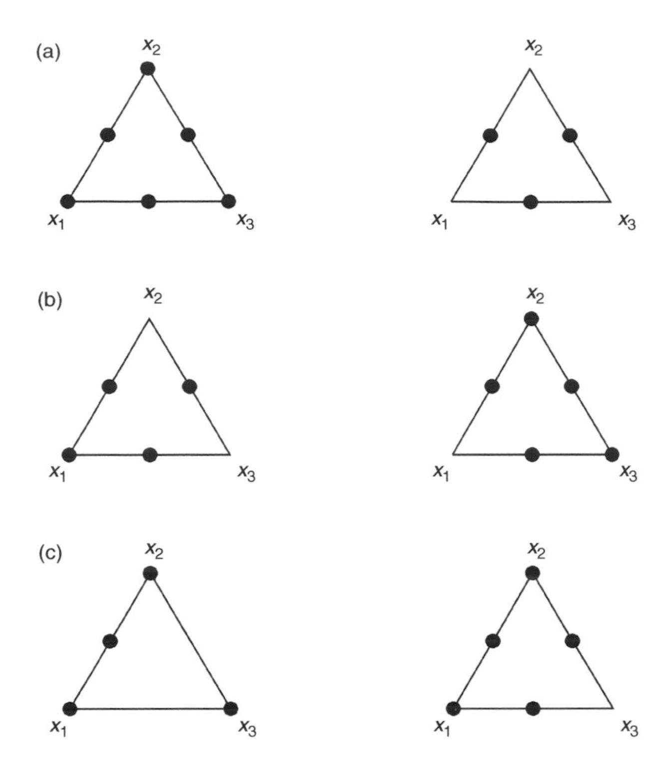

FIG. 16.6. Three-component mixture in two blocks. Some D-optimum designs: (a) $b_1 = 6$, $b_2 = 3$; (b) $b_1 = 4$, $b_2 = 5$, but with the same projection and replication as (a); (c) $b_1 = 4$, $b_2 = 5$ with replicated points and projection different from (b). All designs have the same value of $|M(10)|$.

projection block and of the division of non-replicated points between the blocks makes it possible to achieve improved designs with regard to other criteria, for example, D_β-optimality.

As the block effects are regarded as nuisance parameters, it may be desirable to be able to estimate the effects associated with the mixture components independently from the block effects. Conditions for orthogonal blocking of mixture experiments have been investigated by Nigam (1976), John (1984), and others. Lewis *et al.* (1994) and Prescott (2000) propose methods for constructing such designs. Examples of the use of orthogonally blocked mixture designs are given in Draper *et al.* (1993) and Cornell (2002).

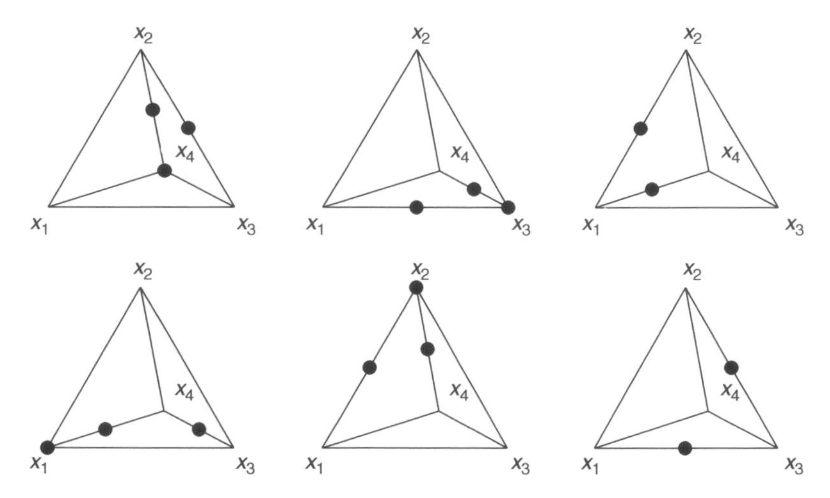

FIG. 16.7. A 16-trial D-optimum design in six blocks for a four-component mixture: approximately even allocation of the observations between the blocks.

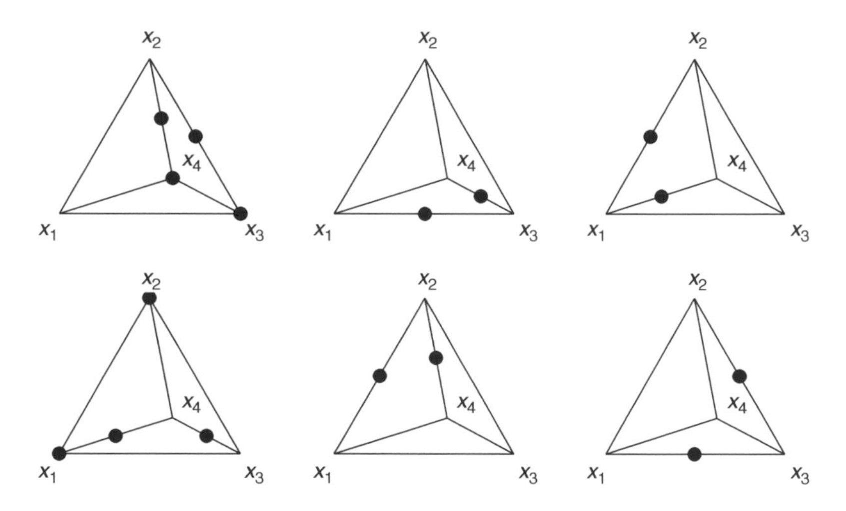

FIG. 16.8. A 16-trial D-optimum design in six blocks for a four-component mixture: larger blocks for the first level of one of the blocking factors.

However, unlike some factorial and response surface designs that can be divided in blocks with little or no loss of efficiency, the orthogonally blocked mixture experiments usually perform badly with respect to the D-optimality

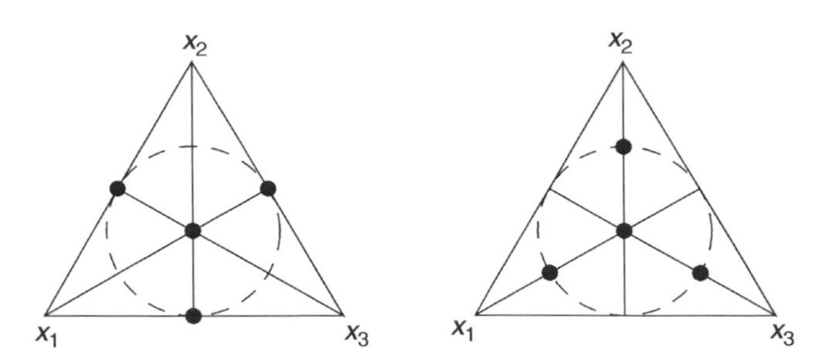

FIG. 16.9. An eight-trial orthogonally blocked design in two blocks for a three-component mixture.

criterion. For example, the eight-trial orthogonally blocked design for a second order Scheffé polynomial in three mixture components in two blocks in Figure 16.9 has $|M|^{1/p} = 0.1667$, while for the eight-trial D-optimum minimum support design in Figure 16.10 it is 0.4529, resulting in relative efficiency of 2.72. The orthogonality measure (15.15) for the D-optimum minimum support design is 0.8908. Hence the sacrifice of orthogonality allowed us to obtain an experimental design that was nearly three times more efficient. When the number of observations is larger the gain in relative D-efficiency as well as the loss in orthogonality for designs obtained this way is smaller; see examples in Goos and Donev (2007*a*).

The blocked designs of this section are minimum support designs and have been found by searching over a design region obtained by repeating the D-optimum design for the mixture components at each block. As was shown in Chapter 15, a slight improvement in design properties can sometimes be found by the perturbation of moving a small way from such a grid. This extra complication may yield a negligible, or no, increase in optimality. For example, the design in Figure 16.11 has a projection with 8 points and the pth root of the determinant of its information matrix is equal to 0.4118, so that its relative D-efficiency against the minimum support D-optimum design is 1.032. In most cases this difference will not have practical importance. The designs shown in Figures 16.9–16.11 are listed in Table 16.5.

16.6 The Amount of a Mixture

So far we have followed the convention, which applies to many mixture experiments, that the response depends only on the proportions of the

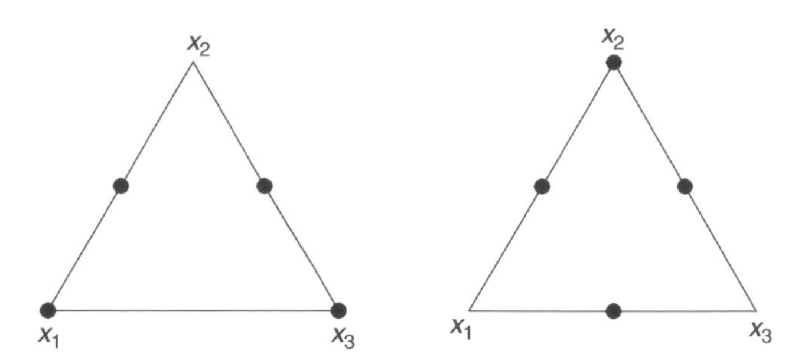

FIG. 16.10. An eight-trial minimum support D-optimum design for estimating a second-order Scheffé polynomial in three mixture components in two blocks.

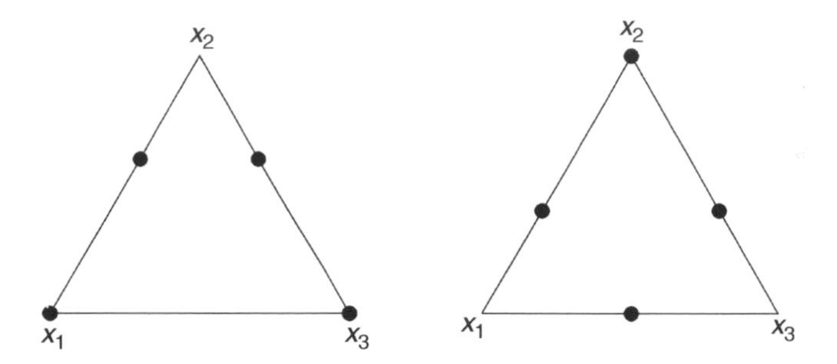

FIG. 16.11. An eight-trial D-optimum design for estimating a second-order Scheffé polynomial in three mixture components in two blocks.

mixture components. But suppose that the response also depends on the amount A of the mixture. Let this be defined to vary in the experiment between some minimum and maximum values so that

$$0 \leq A_{\min} \leq A \leq A_{\max} \leq 1.$$

When $A_{\min}$ is strictly positive and the amount can be regarded as a qualitative factor at two levels, we can apply the results of Claringbold (1955), Vuchkov *et al.* (1981), and Piepel and Cornell (1983), who in addition suggest models and designs for this case. The results given in the following section on designs when both qualitative and mixture variables are present can also be applied. However, when $A_{\min} = 0$, this approach is not possible.

TABLE 16.5. Three-component mixture designs with two blocks of size four for estimating quadratic Scheffé polynomials

	Orthogonal			D-optimum			Minimum support		
Block	x_1	x_2	x_3	x_1	x_2	x_3	x_1	x_2	x_3
1	2/3	1/6	1/6	0.6	0.4	0	0.5	0.5	0
1	1/6	2/3	1/6	0.5	0	0.5	0.5	0	0.5
1	1/6	1/6	2/3	0	1	0	0	0	1
1	1/3	1/3	1/3	0	0.4	0.6	0	0.5	0.5
2	1/2	1/2	0	0.4	0.6	0	1	0	0
2	1/2	0	1/2	0	0.6	0.4	0	1	0
2	0	1/2	1/2	1	0	0	0.5	0	0.5
2	1/3	1/3	1/3	0	0	1	0	0.5	0.5

In the general case when the amount is a qualitative factor it is possible to define $q+1$ pseudo-components

$$g_{ij} = \begin{cases} \frac{x_{ij}}{A\text{max}} & (i = 1, \ldots, q) \\ \frac{A\text{max} - A_j}{A\text{max}} & (i = q+1) \end{cases} \tag{16.17}$$

which form a $(q+1)$-component mixture including the total mixture amount as well as the q actual components of the mixture (Donev 1989). In (16.17) x_{ij}, is the amount of the ith component in the mixture with total mixture amount A_j, where

$$A_j = \sum_{i=1}^{q} x_{ij}.$$

The pseudo-component g_{q+1} corresponds to the total mixture amount and takes the value unity when the amount is zero, decreasing to zero when the total mixture amount increases to its maximum possible value. In this way the design and analysis of experiments for mixtures, where the total amount is a factor, can be regarded as a $(q+1)$-dimensional mixture problem, although the interpretation of the results is different.

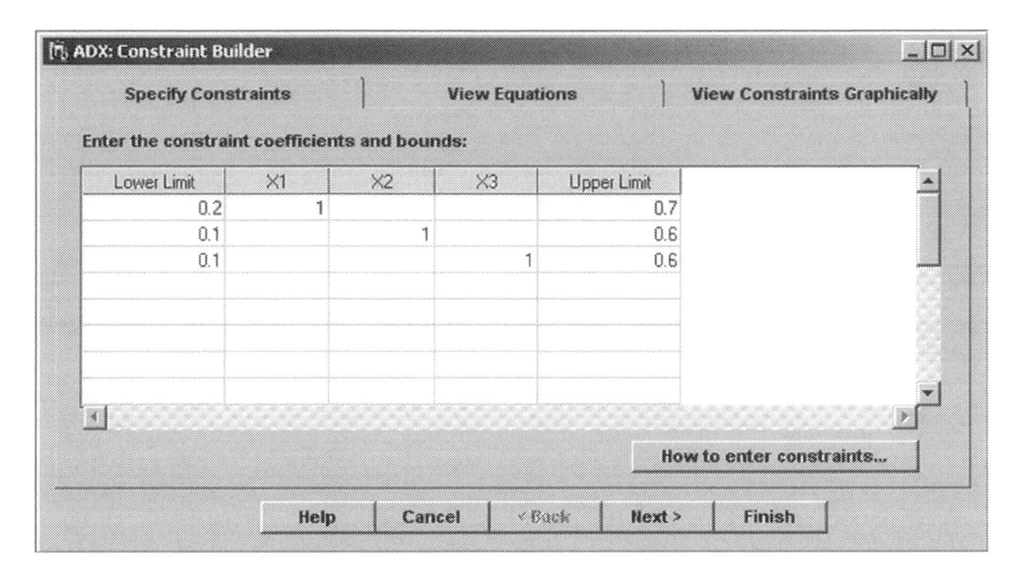

FIG. 16.12. ADX constraint definition screen.

16.7 Optimum Mixture Designs in SAS

Generating optimum designs in SAS for mixture experiments is not very different from generating optimum designs for the response surface experiments discussed in previous chapters. Mixture designs without blocks can be constructed by the ADX Interface. As with other designs in ADX, once the factors are defined, the user has the option of either selecting a standard design (see Chapter 7 for options) or searching for an optimum design with a custom number of runs or for a custom model. Since simple upper or lower constraints on the factors can make the candidate space irregular, the numerical calculation of optimum designs is required relatively more often for mixture experiments than for other kinds of experiments. Accordingly, ADX provides a special interface for entering and viewing linear constraints for mixture factors (Figure 16.12).

> **SAS Task 16.1.** Use the ADX Interface to construct a three-factor simplex-centroid design as shown in Figure 16.1.

> **SAS Task 16.2.** Use the ADX Interface to find an optimum constrained mixture design in 20 runs for Example 16.2.

Blocked mixture designs and designs with non-standard models call for direct SAS programming with the OPTEX procedure. As with the designs

TABLE 16.6. Syntax for full polynomial models in three mixture factors

Order	Syntax
1	`model x1-x3 / noint;`
2	`model x1|x2|x3@2 / noint;`

3	`model x1 x2`
	`x1*x1 x1*x2 x2*x2`
	`x1*x1*x1 x1*x1*x2 x1*x2*x2 x2*x2*x2;`

discussed in previous chapters, OPTEX requires minimally a candidate data set and an appropriate model. First- and second-order models are particularly easy to define, remembering to use the NOINT option on the model statement to direct OPTEX not to automatically add the intercept. For higher-order polynomial models, it is easiest to define an equivalent model in one less mixture factor. Syntax for several full polynomial models in three mixture factors is exhibited in Table 16.6.

Mixture experiments also require techniques for making candidate sets that are different from the simple loops for full factorial data sets of previous chapters. For small, unconstrained mixture experiments, a relatively easy modification of a factorial loop yields a simplex lattice. The following code demonstrates the relevant techniques.

```
data Grid; keep x1-x3;
   do ix1 = 0 to 100;
   do ix2 = 0 to 100;
   do ix3 = 0 to 100;
      if (ix1 + ix2 + ix3 = 100) then do;
         x1 = ix1 / 100;
         x2 = ix2 / 100;
         x3 = ix3 / 100;
         output;
         end;
      end; end; end;
run;
```

The code above begins with an integer lattice $(0, 1, \ldots, 100)^3$, only retaining the points of the lattice which sum to 100. Then dividing these points by 100 yields a simplex-lattice with an increment of $1/100$.

SAS Task 16.3. Use PROC OPTEX to confirm the D-optimality with respect to the simplex lattice defined above of the third-order design depicted in Figure 16.2.

For irregularly constrained designs, such as that depicted in Figure 16.3, one technique is to start with a simplex-lattice as above and use a WHERE statement to retain only the subset of points which satisfy the constraints, as demonstrated in the following code.

```
data Grid; set Grid;
   where (  (0.2 <= x1 <= 0.8)
          & (0.2 <= x2 <= 0.8)
          & (0   <= x3 <= 0.6));
run;
```

SAS Task 16.4. Use PROC OPTEX with the appropriate subset of a simplex lattice as a candidate set to confirm the D-optimality of the linearly constrained second-order design depicted with the filled-in circles in Figure 16.3.

SAS Task 16.5. (Advanced). Use nonlinear optimization in PROC IML, as discussed in §13.4 to confirm the structure of the optimal design weights for Example 16.2.

SAS Task 16.6. Use PROC OPTEX with the appropriate subset of a simplex lattice as a candidate set to confirm the D-optimality of the non-linearly constrained second-order design depicted in Figure 16.4.

When there are many mixture factors, simplex lattice designs may have too many points to provide useful candidate sets, or they may fail to have enough points that satisfy mixture factor constraints. As discussed in §16.3, the extreme vertices and generalized edge centroids of the constrained region often provide an excellent candidate set from which to select the points of an optimum design; and the XVERT algorithm of Snee and Marquardt (1974) can be used to find these points. The ADX interface employs XVERT when the mixture factors are subject to constraints. For direct SAS programming, the macro subroutine `%ADXXVERT`, which is shipped with SAS, implements XVERT. The arguments for this macro are:

- the data set in which the extreme vertices and generalized edge centroids are to be created;
- a list of mixture factors for the design, with lower and upper bounds, and
- the maximum order m of centroid to be generated.

If the argument for m is left blank, centroids of all orders are generated.

The following code demonstrates how to use `%ADXXVERT` to generate the candidate points depicted in Figure 16.3.

```
%adxgen;
%adxmix;
%adxxvert(Candidates,x1 0.2-0.7 /
                     x2 0.1-0.6 /
                     x3 0.1-0.6 );
```

The first two statements above load the required macro library. The %ADXXVERT statement first names the data set of vertices and generalized edge centroids to be created, and then names the factors, separated by slashes, each followed by their respective bounds.

> **SAS Task 16.7.** Use PROC OPTEX with the %ADXXVERT macro to confirm the D-optimality of the linearly constrained second-order design depicted with the filled-in circles in Figure 16.3.

Two other macro subroutines included with SAS are the %ADXSLD macro, for simplex-lattice designs, and the %ADXSCD macro, for simplex-centroid designs. They are invoked similarly to the %ADXXVERT macro, except that they do not allow bounds on the factors.

Finally, it is usually transparent in SAS to include other factors in a mixture design, in addition to the mixture factors, by (1) building such factors into the candidate data set and (2) naming their effects in the model. For example, the following code finds a D-optimum second-order mixture design in 10 runs with a three-level qualitative factor.

```
%adxsld(Candidates,x1 x2 x3,2);
data Candidates; set Candidates;
   do a = 1 to 3; output; end;
run;

proc optex data=Candidates coding=orthcan;
   class a;
   model a x1|x2 x1*x1 x1*x2 x2*x2;
   id x3;
   generate n=10 method=m_fedorov niter=1000 keep=1;
   output out=Design2;
run;
```

The code above employs the %ADXSLD macro mentioned previously to make a simplex-lattice design with increment 1/2, which is the minimum support design for this problem. Each of these simplex-lattice points is combined with three different levels of the qualitative factor A to make the final candidate set. In OPTEX, we declare A as a qualitative factor in the CLASS statement; and in order to avoid ambiguity about the intercept, instead of a second-order mixture model in all three mixture factors, we use an equivalent full second-order response surface model in just the first two. The ID statement specifies that even though the factor X3 is not included in the model, its value should be written to the output design data set.

SAS Task 16.8. Use PROC OPTEX or PROC IML to evaluate the three designs depicted in Figure 16.5, confirming that they all have the same value of the determinant of the information matrix $|M|$ as the optimum design found by OPTEX. If you use PROC OPTEX with `CODING=ORTHCAN` to compute the D-criterion, be careful to supply the same candidate set for all three designs.

SAS Task 16.9. (Advanced). Use PROC OPTEX to solve again the problem of a D-optimum second-order mixture design in 10 runs with a three-level qualitative factor, but this time use a finer grid that includes the simplex-lattice with increment 1/2. For example, use the simplex-lattice with increment 1/10. Compute the D-efficiency of your new design relative to the three designs depicted in Figure 16.5. Why do the results seem to contradict the discussion following equation 16.12?

16.8 Further Reading

Cornell (1988) points out that in many mixture experiments, quantitative or process factors are difficult to change. Therefore, split-plot designs have to be constructed so that the dependence of the observations introduced by the way the experiment is organized is taken into account. Examples are also given by Kowalski *et al.* (2000) and Kowalski *et al.* (2002). Goos and Donev (2007*b*) demonstrate the merits of the algorithmic search for optimum designs in such cases.

Martin, Bursnall, and Stillman (2001) give examples of the use of mixture designs over the constrained regions of §16.3 that arise in glass manufacture and discuss properties of the optimum designs. A comprehensive survey of fifty years of research work on mixture experiments is given by Piepel (2006).

17

NON-LINEAR MODELS

17.1 Some Examples

The experimental designs derived in the previous chapters for linear models do not depend on the values of the parameters β of the linear model. On the other hand, an important feature of designs for non-linear models is that they do depend on the values of the parameters. In general the values of these parameters will be unknown. In this chapter we describe locally optimum designs that depend on a point prior estimate of the unknown parameter. In the next chapter we extend the designs to include more general prior distributions of the parameters.

In this section we give four examples of non-linear models. Locally D-optimum designs are calculated in §17.2 with a discussion of experimental strategies for local optimality in §17.3. We then briefly consider designs of bounded efficiency when the optimum conditions may not be exactly adhered to; an example is the administration of treatments in a clinic. Locally c-optimum designs are in §17.5. Some comments on the analysis of non-linear experiments are in §17.6. Because of the dependence of the designs on the unknown parameter values, sequential designs alternating design, experimentation, and analysis are often of great importance with non-linear models. An example is in §17.7. We then find optimum designs when the models are defined by differential equations without an analytical solution. An example of a D-optimum design for a multivariate response is in §17.9. The primary SAS tool for fitting non-linear models is the NLIN procedure in SAS/STAT software. In §17.10 we will demonstrate how to use this procedure to fit the models we discuss in this chapter, along with SAS strategies for creating optimum designs for non-linear modelling.

Example 17.1 Exponential Decay Many non-linear models arise in the study of pharmacokinetics and chemical kinetics. One of the simplest is the first-order reaction, in which the transformation of a compound A into a compound B at a constant rate θ is denoted

$$A \xrightarrow{\theta} B. \tag{17.1}$$

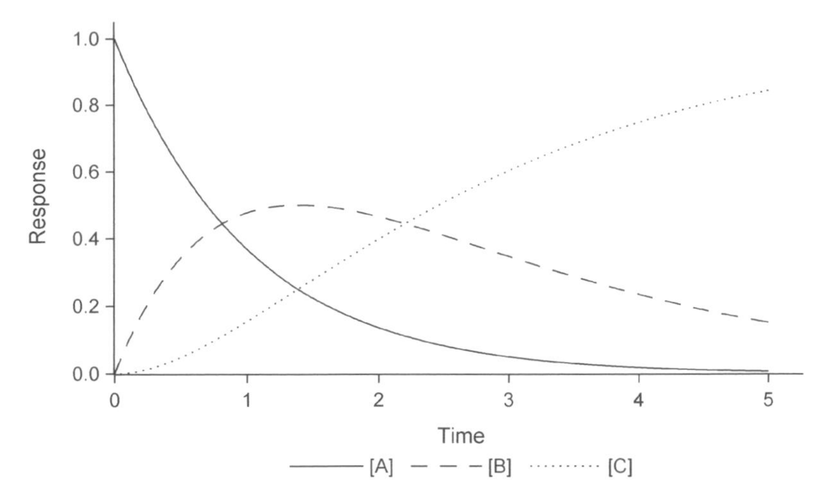

FIG. 17.1. Responses against time for Examples 17.1 and 17.3.

Measurement is made at time t of the concentration of A, sometimes written $[A]$. If the initial concentration of A is $[A_0]$

$$\eta(t, \theta) = [A] = [A_0] \exp(-\theta t) \quad (t \geq 0, \theta > 0). \tag{17.2}$$

The solid black curve in Figure 17.1 shows a plot of this response when both $[A_0]$ and θ equal one. When $t = 1$ the response $\eta(1) = e^{-1} = 0.3679$. So, after one time unit, the concentration of A is reduced to 36.79% of the initial value. For larger values of θ the reaction proceeds more rapidly and the response decreases at a greater rate.

This model is closely related to that used in §4.2 to introduce the idea of a non-linear model. If instead of measuring the concentration of A, that of B is measured, when $[A_0]$ is to be estimated as well as θ, we obtain

$$\eta(t, \theta) = [B] = [A_0]\{1 - \exp(-\theta t)\},$$

the two-parameter model of §4.2 plotted in Figure 4.4. In this model the initial concentration $[A_0]$ enters linearly. A consequence is that, when the errors satisfy the second-order assumptions introduced in §5.1, the optimum design to estimate θ and $[A_0]$ will not depend on the value of A_0. Most of the models discussed in this chapter have this feature. For this reason, we will usually ignore the initial concentration and write the exponential decay model as

$$\eta(t, \theta) = [A] = \exp(-\theta t) \quad (t \geq 0, \theta > 0). \tag{17.3}$$

■

Example 17.2 Inverse Polynomial A model with very similar properties to (17.3) is the first-order inverse polynomial

$$\eta(t, \theta) = \frac{1}{1 + \theta t} \quad (t \geq 0, \theta > 0). \tag{17.4}$$

At $t = 0$ the response has a value of unity. As t increases, the response decays to zero. For large t (17.4) decays more slowly than (17.3). However, when measurements are made with error, it may be hard to detect the difference between the two curves. ■

Example 17.3 Two Consecutive First-order Reactions If the first-order reaction scheme of Example 17.1 is extended to two consecutive first-order reactions

$$A \xrightarrow{\theta_1} B \xrightarrow{\theta_2} C,$$

then the concentration of B at time t, when $[A_0]$ is taken as one, is

$$\eta(t, \theta) = [B] = \frac{\theta_1}{\theta_1 - \theta_2}\{\exp(-\theta_2 t) - \exp(-\theta_1 t)\} \quad (t \geq 0), \tag{17.5}$$

provided that $\theta_1 > \theta_2 > 0$ and that the initial concentration of A is one. The grey curve in Figure 17.1 is a plot of this response for $\theta_1 = 1$ and $\theta_2 = 0.5$. Initially, the concentration of B is zero. It rises to a maximum, for these parameter values at $t = 2.506$, and then decreases more slowly to zero as t increases. ■

Example 17.4 A Compartmental Model Example 17.3 is a special case of a class of models that arises frequently in the study of pharmacokinetics, in which the response function is a weighted sum of exponentials with differing exponents. As an example we take

$$\eta(t, \theta) = \theta_3\{\exp(-\theta_2 t) - \exp(-\theta_1 t)\} \quad (t \geq 0), \tag{17.6}$$

with the exponential parameters in the same order as in (17.5) so that again $\theta_1 > \theta_2$. All three parameters are positive. This model, discussed more fully in §17.5, can be thought of as arising from a reaction scheme similar to that yielding (17.5) with the difference that the initial concentration of A is unknown.

The exponential structure of such models reflects an underlying scheme of first-order chemical reactions. Other orders of reaction give more complicated models, for which explicit solutions of the type shown here are rarely, if at all, available. Design under such conditions is discussed in §17.8. ■

A feature common to all four models is that the optimum design will depend upon the value of θ; it is said to be *locally* optimum. For example, it is clear from Figure 17.1 that measurements taken when t is near zero, or very large, will not be informative about θ; for a large range of values of θ the response will either be close to unity or close to zero at such values of t. Measurements where the expected response is close to 0.5 might be expected to be more informative, but the value of t for which $\eta(t,\theta) = 0.5$ will depend upon θ. Since the purpose of the design is to estimate θ, the dependence of the design on the value of the parameter is unfortunate, but unavoidable for efficient designs with non-linear models. The effect of a bad choice of θ on the design can be reduced by the use of sequential designs or by designs for several prior values of θ. We discuss these and other possibilities in §17.3, with an example of a sequential design in §17.7. However, in this chapter we mainly focus on the construction of designs for linearized models using a point prior estimate θ_0. The resulting locally optimum designs would be optimum if θ_0 were the true parameter value.

17.2 Parameter Sensitivities and D-optimum Designs

To begin, suppose that θ is scalar, as it is in Examples 17.1 or 17.2. Then Taylor expansion of the model about the value θ_0 yields the linearization

$$\begin{aligned} \mathrm{E}\,(y) = \eta(t,\theta) &= \eta(t,\theta_0) + (\theta - \theta_0) \left.\frac{\partial\eta(t,\theta)}{\partial\theta}\right|_{\theta=\theta_0} + \cdots \\ &\doteq \eta(t,\theta_0) + (\theta-\theta_0) f(t,\theta_0), \end{aligned} \tag{17.7}$$

where the derivative $f(t,\theta_0)$ is often called the parameter sensitivity, here evaluated at $\theta = \theta_0$. The relationship with linear models is emphasized by rewriting (17.7) as

$$\mathrm{E}\,(y) - \eta(t,\theta_0) = \beta f(t,\theta_0), \tag{17.8}$$

with $\beta = \theta - \theta_0$. Linearization has thus resulted in a linear regression model through the origin in which the explanatory variable is the parameter sensitivity $f(t,\theta_0)$. The design for which β and, equivalently, θ is estimated with minimum variance is therefore that for which the information $\int f^2(t,\theta_0)\xi(dt)$ is maximized. This is achieved by putting all experimental effort at the point where $f(t,\theta_0)$ is a maximum.

Example 17.1 Exponential Decay continued For the exponential decay response function (17.3) the parameter sensitivity is

$$f(t,\theta) = \frac{\partial\eta(t,\theta)}{\partial\theta} = -t\exp(-\theta t). \tag{17.9}$$

Figure 17.2 is a plot of $-f(t,\theta)$ evaluated at $\theta = 1$. This plot starts at zero, rises to a maximum and then decreases to zero, a shape that makes explicit

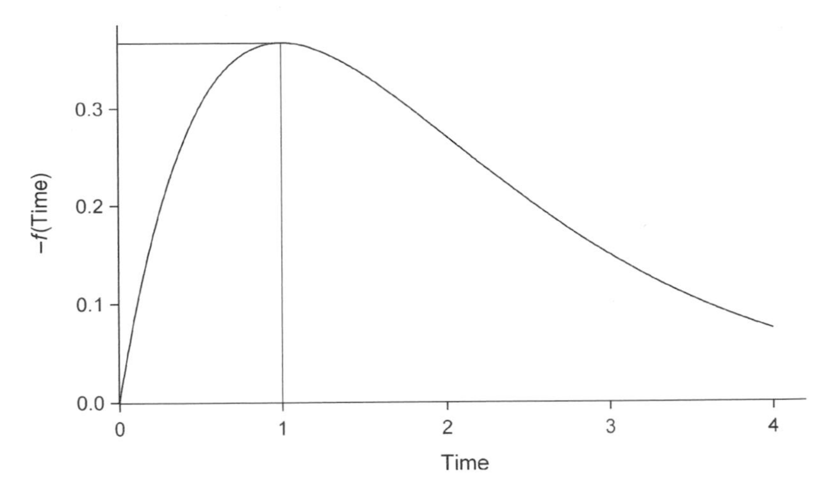

FIG. 17.2. Example 17.1: exponential decay, negative derivative (parameter sensitivity) $-f(x, \theta)$ for prior parameter value $\theta_0 = 1$.

the idea that experiments at very small or very large values of t will not be informative.

The maximum of (17.9) is found by differentiation with respect to t to satisfy the relationship $t\theta = 1$. Thus, to obtain an estimate of θ with minimum variance, all experiments should be performed at $t = 1/\theta$, at which value of t the expected yield is e^{-1}. For the guess $\theta_0 = 1$, all measurements should therefore be taken at $t = 1$. For a single parameter model like (17.3) this is the D-optimum design as well, of course, as being A- and E-optimum. It shares with many locally D-optimum designs for non-linear models with $m = 1$ the characteristic of having the number of support points equal to p. Therefore there are no degrees of freedom available for checking the model. Some approaches to checking non-linear models are mentioned in §§20.4.4 and 21.6. A second potential defect of this one point design is that it may be inefficient if the assumed value θ_0 is far from the true θ. Methods of avoiding over dependence on a single parameter value are outlined in §17.3.

That the design with all trials at $t = 1$ is locally D-optimum can be checked by applying the General Equivalence Theorem using (17.8). For the linearized one-parameter model, the required variance can be written as

$$d(t, \xi, \theta) = \frac{f^2(t, \theta)}{\int f^2(t, \theta)\xi(dt)}. \tag{17.10}$$

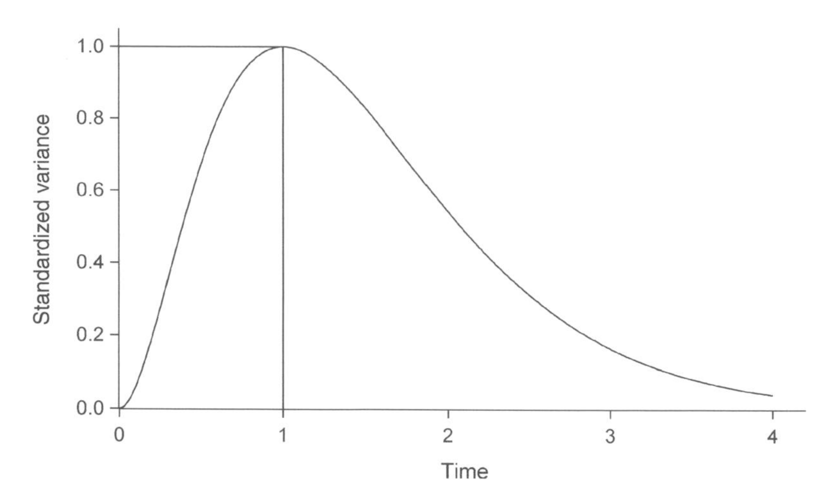

FIG. 17.3. Example 17.1: exponential decay, variance $d(t, \xi^*)$ for linearized model: reference line at locally D-optimum design.

For the first-order decay model $f(t, \theta)$ is given by (17.9). Evaluation of (17.10) with $\theta = 1$ for the optimum design that puts all trials at $t = 1$ yields

$$d(t, \xi^*) = t^2 \exp\{2(1-t)\}, \tag{17.11}$$

plotted in Figure 17.3. The maximum value $\bar{d}(\xi^*) = p$, here one, occurs at $t = 1$. The design is indeed locally D-optimum.

This design has been derived under the assumption, common to all designs found so far, of additive independent errors of constant variance. In this example, independent errors could arise from repeating the experiment several times, each time taking a single measurement of y at time t_i. An alternative form of experiment is to monitor the evolution of the concentration of A over time, obtaining for each experiment the value of y at a series of values of t. The resulting time series of y values might have a correlation structure that should be allowed for in both design and analysis. Design for correlated observations is the subject of Chapter 24.

It may also be that the errors are not of constant variance. In particular, as $\eta \to 0$, the observations will usually remain positive, so that a normal model with additive errors of constant variance cannot hold. Two possibilities, discussed in Chapter 24 are to design for transformation of y or to model the change of variance with E (y) and to design for non-homogeneous variance. ■

The extension of locally D-optimum designs to models with more than one non-linear parameter is straightforward. Let θ be of dimension $p \times 1$,

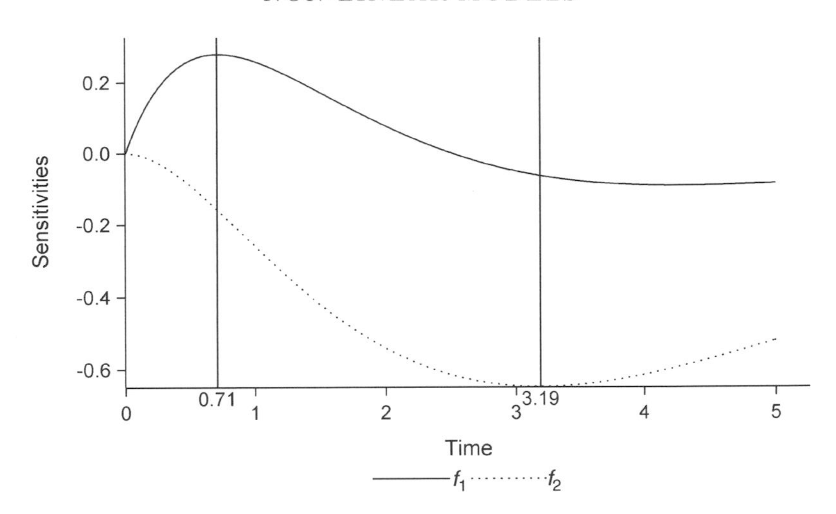

FIG. 17.4. Example 17.3: two consecutive first-order reactions, sensitivities against time.

with θ_0 the vector of p prior values of θ. Then the multivariable version of the Taylor expansion that led to (17.8) yields the vector of parameter sensitivities

$$f^{\mathrm{T}}(t, \theta_0) = \left\{ \frac{\partial \eta(t, \theta)}{\partial \theta_1} \; \frac{\partial \eta(t, \theta)}{\partial \theta_2} \; \dots \; \frac{\partial \eta(t, \theta)}{\partial \theta_p} \right\} \Bigg|_{\theta=\theta_0} \tag{17.12}$$

for the terms of the p variable linearized multiple regression model. The vector of sensitivities for observation i is written $f^{\mathrm{T}}(t_i, \theta)$ and forms row i of the extended design matrix F for this linearized model, for which the information matrix is $F^{\mathrm{T}}F$.

Example 17.3 Two Consecutive First-order Reactions continued For this two-parameter model, differentiation of (17.5) followed by substitution of the parameter values $\theta_1 = 1$ and $\theta_2 = 0.5$ yields the sensitivities

$$\begin{aligned} f_1(t) &= (2 + 2t)\exp(-t) - 2\exp(-0.5t) \\ f_2(t) &= (4 - 2t)\exp(-0.5t) - 4\exp(-t). \end{aligned} \tag{17.13}$$

Both are functions of the single design variable t against which they are plotted in Figure 17.4. The maximum value of $|f_1(t)|$ is at $t = 0.71$ and of $|f_2(t)|$ at 3.19.

The locally D-optimum design maximizes $|M(\xi, \theta_0)|$ with the rows of the $N \times 2$ extended design matrix given by (17.13). The optimum design can

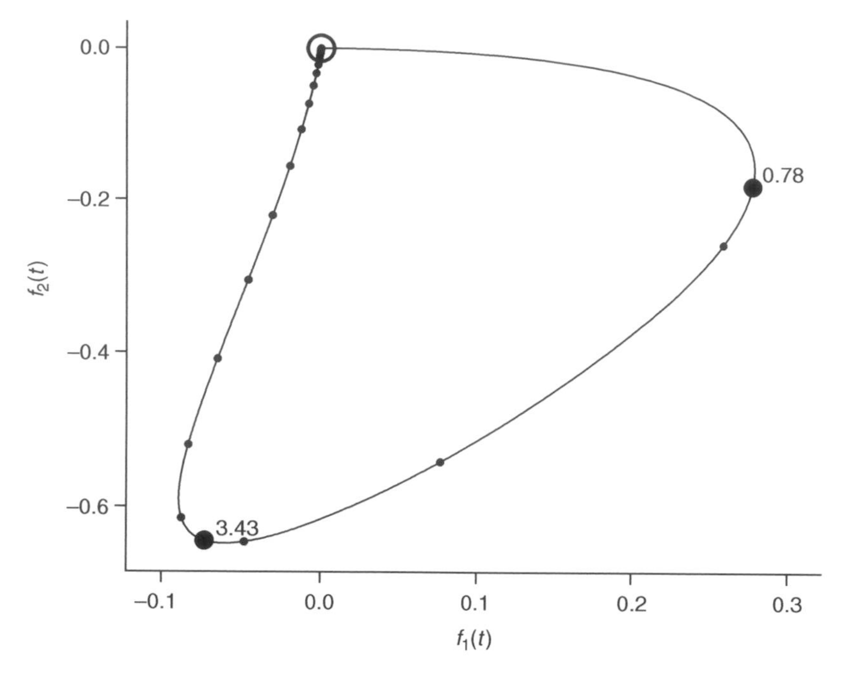

FIG. 17.5. Example 17.3: two consecutive first-order reactions, derivatives $f_1(t)$ and $f_2(t)$; 'design locus'.

be found using the methods described in Chapter 9. We now consider some geometrical properties of this optimum design.

We shall see in a moment that this optimum design has $n = 2$ points of support. As with other D-optimum designs with $n = p$, the optimum design puts weight $1/p$ at each of the support points; in this case weight $1/2$. For any such equi-weighted design F is a square matrix and

$$|F^{\mathrm{T}}F| = |F|^2. \tag{17.14}$$

For $p = 1$ the optimum design had all weight concentrated where the absolute value of $f(t, \theta_0)$ was a maximum. The generalization of this result is, from (17.14), that the D-optimum design will maximize $|F|$, which is equivalent to maximizing the volume, in the space of the sensitivities, of the simplex formed by the p points $f(t_i, \theta_0)$ $(i = 1, \ldots, p)$ and the origin.

This geometrical interpretation is illustrated in Figure 17.5 which shows the values of the sensitivities plotted against each other. A point on this 'design locus' gives the values of the two sensitivities for the particular value of t. At $t = 0$ and $t = \infty$ both sensitivities are zero (see Figure 17.5) and this origin is marked by a circle. As t increases possible values of $f_1(t, \theta_0)$

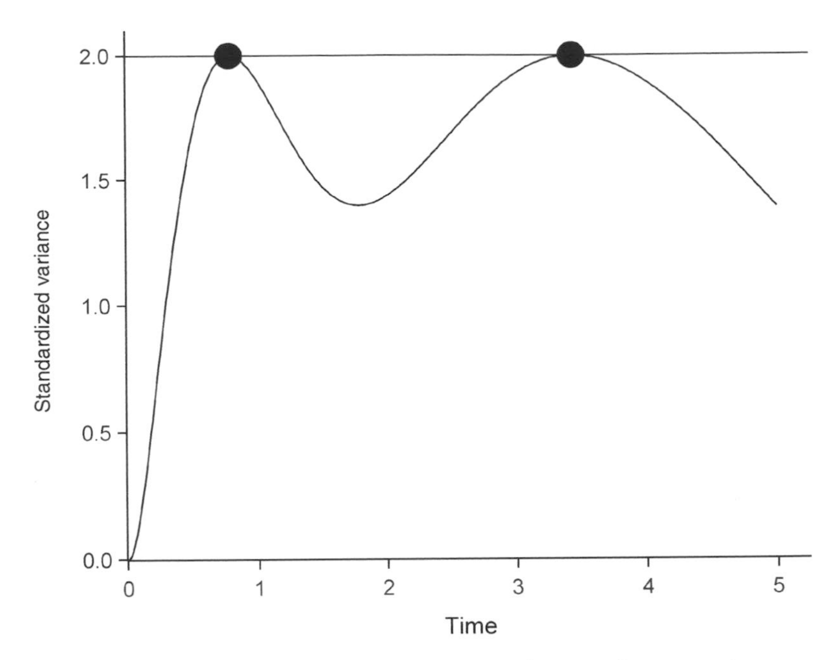

FIG. 17.6. Example 17.3: two consecutive first-order reactions, variance $d(t, \xi^*, \theta_0)$ for linearized model, • locally D-optimum design.

and $f_2(t, \theta_0)$ lie on the curve in a clockwise direction. The optimum design consists of trials at times 0.78 and 3.43 marked with large dots in the figure.

Together with the origin the design points in Figure 17.5 form a triangle of maximum area on the design locus. Since the triangle has maximum value, the tangent to the design locus at t_2 must be parallel to the line joining the origin to the design point on the locus at t_1. The same property holds with t_1 and t_2 reversed. This geometrical argument makes it possible to check whether the proposed design is indeed D-optimum, given that it requires only two support points.

It is generally easier to use the equivalence theorem to confirm the optimality of a proposed design. In this case Figure 17.6 shows the plot of $d(t, \xi^*, \theta_0)$ for the design with equal support at $t = 0.78$ and $t = 3.43$. for this linearized model $\bar{d}(\xi^*, \theta_0) = 2$ with the maxima at the support points of the design.

As Figure 17.1 indicates the maximum yield occurs at $t = 1.39$. The design we have found has one trial at a smaller value of t than this and one at a larger value. The expected responses at the design points are respectively 29.5% and 43.7% of the maximum. These design points are not quite those at the maxima of the two sensitivities plotted in Figure 17.4. These occur at times of 0.71 and 3.19. A design with equal weight at these two points has a D-efficiency of 99%, very close to one. Whether the design putting trials at the maximum values of each sensitivity individually is efficient depends on the shape of the design locus, itself a function of the model and of the parameter values. For linear models including first-order terms, the design locus is in the p-dimensional space of $\mathcal{X}$ and the terms forming the columns of the extended design matrix, including the constant term, if any.

17.3 Strategies for Local Optimality

This section briefly describes several methods of overcoming the dependence of optimum designs on the unknown parameter θ_1. The methods are quite general, not being restricted to non-linear regression models, but also being appropriate, for example, for generalized linear models (Chapter 22).

17.3.1 Sequential Designs

Because of the dependence of the optimum design on the unknown parameter, sequential experimentation is more important for non-linear than for linear models. The usual procedure consists of several steps:

1. Start with a preliminary estimate, or guess, of the vector parameter values. This may be either a prior point estimate θ_0 or a prior distribution for θ based on past experience, including the analysis of related experiments.
2. The model is linearized by Taylor series expansion.
3. The optimum design is found for the linearized model.
4. One or a few trials of the optimum design for the linearized model are executed and analysed. If the new estimate of θ is sufficiently accurate, the process stops. Otherwise, step 2 is repeated for the new estimate and the process continued until sufficient accuracy is obtained or the experimental resources are exhausted.

We give an example of the most cautious variant of this scheme in §17.7 in which single trials are added with the model being refitted after each trial. If, as in agricultural field trials, the experiments take a long time to perform or analyse, such a scheme is clearly unsatisfactory. The sequential scheme furthest from the addition of single trials is, when resources for n experiments are available, to devote about $\sqrt{n}$ of the effort to a preliminary study, with the resulting parameter estimates used in the design of the larger final experiment. In group sequential designs, groups of fixed, or increasing, numbers of trials are added and analysed.

The construction of an optimum sequential scheme depends on the relative costs of experimentation and of the time delay in obtaining an accurate answer. In all but the simplest cases the calculations are rarely performed.

17.3.2 Bayesian Designs

Locally D-optimum designs maximize $\log |M(\xi, \theta_0)|$. If a prior distribution $p(\theta)$ is available for θ, D-optimum designs can be found to maximize

$$\Phi(\xi) = \mathrm{E}_\theta \log |M(\xi, \theta)| = \int_\theta \log |M(\xi, \theta)| p(\theta) d\theta. \qquad (17.15)$$

An equivalence theorem then applies to the variances

$$d(x, \xi) = \mathrm{E}_\theta d(x, \xi, \theta) = \int_\theta d(x, \xi, \theta) p(\theta) d\theta. \qquad (17.16)$$

Designs maximizing expectations such as (17.15) ignore the additional effect of the prior information about θ on the information matrix, which is shown, for a linear model, in (18.1). The designs are accordingly sometimes called pseudo-Bayesian. They can be found in a manner similar to those used in this chapter for locally optimum designs, the integration being performed analytically or numerically, perhaps by sampling the prior distribution. These topics are explored more fully in Chapter 18.

17.3.3 Maximin Designs

Another approach to avoid dependence of designs on the unknown value of θ is to develop maximin designs in which the parameter θ belongs to a set Θ. Then the design is found for which

$$\Phi(\xi^*) = \max_\xi \min_{\theta \in \Theta} \log |M(\xi, \theta)|.$$

That is, the design ξ^* is found that maximizes $\log |M(\xi, \theta)|$ for that value of θ for which the determinant is a minimum. A potential objection to these designs is that the maximin design occurs at the edges of the parameter space. If a prior distribution is available, such points may have a very low probability; their importance in the design criterion may therefore be being over-emphasized by the maximin criterion. Such designs may also have a large number of support points.

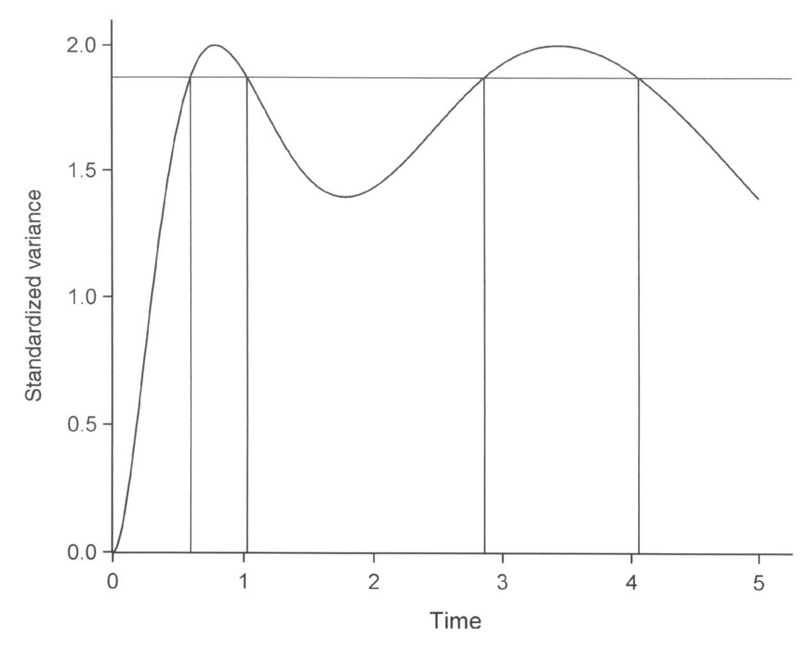

FIG. 17.7. Example 17.3: two consecutive first-order reactions, variance $d(x, \xi^*, \theta)$ for linearized model, with intervals around the locally D-optimum design.

17.4 Sampling Windows

The D-optimum design for Example 17.3 specifies that measurements should be made at times 0.78 and 3.43. Although such precise times may be possible in some technological experiments, they are not likely to be adhered to in clinical experiments where patients have either to attend a clinic for treatment or treat themselves. In this section, following Bogacka *et al.* (2007), we therefore describe a method of deriving intervals within which observations should be taken for the experiment to have a specified minimum efficiency.

It is clear from the two peaks in the plot of $d(t, \xi^*, \theta)$ in Figure 17.6 that the optimum design is more sensitive to deviations of the same absolute amount in the optimum time t_1^* than in t_2^*. To quantify this idea we have repeated Figure 17.6 as Figure 17.7 with the addition of a horizontal line at a value κp, $\kappa < 1$. Let the points of intersection of this line with $d(t, \xi^*)$ be t_1^L and t_1^U, for lower and upper, around t_1^* and, likewise t_2^L and t_2^U around t_2^*. We propose these two intervals as regions in which observations should be taken to obtain a design with a specified minimum efficiency.

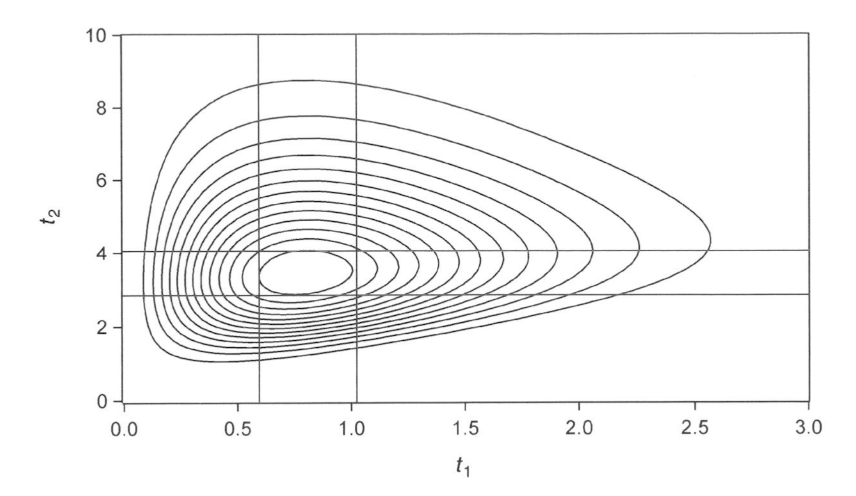

FIG. 17.8. Example 17.3: two consecutive first-order reactions, contour plot of $|M(\xi, \theta)|$.

Figure 17.8 is a contour plot of $|M(\xi, \theta)|$ as a function of t_1 and t_2 on which we have included lines at t_1^L, t_1^U, t_2^L, and t_2^U. The resulting rectangular region forms the design region $\mathcal{X}_W$ for pairs of values t_1 and t_2 for which we require to find the minimum efficiency of the design.

The value of $|M(\xi)|$ is, of course, a maximum at t^* which lies within the region. The minima over the region occur at one, or more, of the four vertices of the region $\mathcal{X}_W$, that is for experiments with observations at

$$(t_1^L, t_2^L) \;\; (t_1^U, t_2^L) \;\; (t_1^L, t_2^U) \;\text{ and }\; (t_1^U, t_2^U).$$

Calculations at these four vertices determine the minimum values of $|M(\xi)|$ over $\mathcal{X}_W$ and so the minimum efficiency for a specified value of κ. Numerical search over values of κ will thus lead to the value for a specified minimum efficiency. For Example 17.4, specifying that the minimum efficiency should have a value of 0.9 leads to a value of 0.936 for κ. The corresponding values of the windows for the design points are (0.594, 1.025) and (2.855, 4.057).

This procedure provides a simple method for obtaining a design with minimum efficiency in which the range of values for the second time does not depend on the first time. We now outline a rule in which the width of the sampling window for t_2 depends on the actual value t_1^0 at which the first measurement was taken.

Because of the rectangular form of $\mathcal{X}_W$ all designs consisting of pairs of boundary points, other than those for which the minimum value is obtained, will be of higher efficiency. The loci of pairs of points for which the efficiency is constant are the contours of $|M(\xi)|$, plotted in Figure 17.8. For a given t_1, say t_1^0, the range of values of t_2 yielding a design with a specified minimum efficiency is bounded by the two points where the contour for that efficiency intersects with $t_1 = t_1^0$. If t_1^0 is close to t_1^* the range of values for t_2 will be large, decreasing as t_1^0 moves away from t_1^*. However, such a rule loses the practical appeal of the sampling window described here in which the window for t_2 does not depend on t_1^0.

17.5 Locally c-optimum Designs

The locally D-optimum designs of §17.2 are appropriate if interest is in precise estimation of all the parameters of the model. Interest in a subset of the parameters would indicate the use of locally D_S-optimum designs, developed in an analogous manner. In this section we consider instead designs when estimation of a specified function of the parameters is of interest. We begin with an example.

Example 17.4 A Compartmental Model continued The three-parameter model (17.6) was used by Fresen (1984) to analyse the data introduced in Chapter 1 as Example 1.5 on the concentration of theophylline in the blood of a horse. Fresen used an 18-point design. The focus here is not whether, by use of optimum design methods, it is possible to do better than this 18-point design. Rather we shall be concerned with how the optimum design depends upon the aspect of the model that is of interest.

The least squares estimates of the parameters for these results are

$$\theta_1^0 = 4.29 \quad \theta_2^0 = 0.0589 \quad \theta_3^0 = 21.80. \tag{17.17}$$

The fitted curve is similar to Figure 17.2. There is a rapid rise of the response from zero to a maximum at a time of 1.01 minutes. Thereafter the exponential decay of the response is governed almost entirely by the value of θ_2, with θ_3 determining the height of the curve.

If the main interest is in estimation of the parameters, the prior estimates (17.17) can be used to find the locally D-optimum design. However, in biological work, derived functions of the parameters are often of interest. We consider three.

1. *The Area Under the Curve (AUC).* Model (17.6) is being used rather than (17.5) because not all the theophylline administered appears in the bloodstream. Some is metabolized after intragastric administration. The total area under the curve is

$$\mathrm{AUC} = \int_0^\infty \eta(t,\theta)\mathrm{d}t = \frac{\theta_3}{\theta_2} - \frac{\theta_3}{\theta_1} = \theta_3\left(\frac{1}{\theta_2} - \frac{1}{\theta_1}\right) = g_1(\theta). \tag{17.18}$$

This function is linear in θ_3, which is a multiplier of all terms in (17.6) and depends on only two ratios of the three parameters.

2. *Time to Maximum Concentration.* The time to maximum concentration is found by differentiation of $\eta(t, \theta)$ to be

$$t_{\max} = \frac{\log \theta_1 - \log \theta_2}{\theta_1 - \theta_2} = g_2(\theta). \tag{17.19}$$

For the parameter values (17.17) $t_{\max} = 1.01$, a value that does not depend on θ_3.

3. *The Maximum Concentration.* The maximum concentration is found by substituting $t_{\max}$ in $\eta(t, \theta)$. The explicit expression for the maximum concentration is more complicated than informative, so we leave it for any interested reader to derive. ■

We now consider designs to estimate functions such as (17.18) with minimum variance. Let the general function be $g(\theta)$. For linear models and linear functions the c-optimum designs of §10.5 minimize

$$\text{var}\, g(\hat{\theta}) = \text{var}\,(c^{\mathrm{T}}\hat{\beta}) = c^{\mathrm{T}} M^{-1}(\xi) c, \tag{17.20}$$

where the $p \times 1$ vector c is of known constants. To estimate a linear combination of the parameters of a non-linear model we again find designs to minimize (17.20) but the information matrix is now $M(\xi, \theta)$ for the linearized model. To extend (17.20) to the non-linear function $g(\theta)$ we expand the function in a Taylor series to give the coefficient of θ_j as

$$c_j(\theta) = \frac{\partial g(\theta)}{\partial \theta_j}, \tag{17.21}$$

estimated at the prior value θ_0. The resulting locally optimum design will depend on θ both through the coefficients (17.21) which enter the design criterion (17.20) and through the information matrix $M(\xi)$. To stress this dependence the criterion can therefore be called c_θ-optimality.

For linear models c-optimum designs can be singular, in which case the equivalence theorem must be adapted slightly (see, for example, Silvey 1980, p. 27). When the design is singular it is not possible to estimate all the parameters in the model, although it must be possible to estimate the linear combination of interest. For non-linear models, or for non-linear functions of the parameters in a linear model, the c_θ-optimum design may again be singular with the property of interest now not estimable. We first consider some examples and then discuss the implications for design.

Example 17.4 A Compartmental Model continued Let the three quantities of interest be written as $g_i(\theta)$ $(i = 1, 2, 3)$. To represent the linearized combinations of the parameters let $c_{ij} = \partial g_i(\theta)/\partial\theta_j$ $(j = 1, 2, 3)$. Then the quantities and coefficients are

1. The area under the curve:

$$\begin{aligned} g_1(\theta) &= \theta_3/\theta_2 - \theta_3/\theta_1 \\ c_{11}(\theta) &= \theta_3/\theta_1^2 \\ c_{12}(\theta) &= -\theta_3/\theta_2^2 \\ c_{13}(\theta) &= 1/\theta_2 - 1/\theta_1. \end{aligned}$$

2. The time to maximum concentration:

$$g_2(\theta) = \frac{\log\theta_1 - \log\theta_2}{\theta_1 - \theta_2}.$$

Let $a = \theta_1 - \theta_2$ and $b = \log(\theta_1/\theta_2)$. Then $t_{\max} = b/a$ and

$$\begin{aligned} c_{21}(\theta) &= (a/\theta_1 - b)/a^2 \\ c_{22}(\theta) &= (b - a/\theta_2)/a^2 \\ c_{23}(\theta) &= 0. \end{aligned}$$

3. The maximum concentration:

$$g_3(\theta) = \eta(t_{\max}, \theta).$$

Let $e_1 = \exp(-\theta_1 t_{\max})$, $e_2 = \exp(-\theta_2 t_{\max})$, and $f = \theta_1 e_1 - \theta_2 e_2$. Then

$$\begin{aligned} c_{31}(\theta) &= \theta_3\{e_1 t_{\max} + f c_{21}(\theta)\} \\ c_{32}(\theta) &= \theta_3\{-e_2 t_{\max} + f c_{22}(\theta)\} \\ c_{33}(\theta) &= e_2 - e_1. \end{aligned}$$

Table 17.1 shows the locally D- and locally c-optimum designs calculated for the parameter values (17.17). The D-optimum design for this three-parameter model has three support points, each with weight 1/3. The plot of the variance is similar to Figure 17.6 except that there are now three local maxima, each of height three. This plot introduces no new ideas and so is not reproduced here.

The D-optimum design, with three support points, allows estimation of the three parameters. The c_θ-optimum designs, with only two points of support, or even with only one, are singular. In order to calculate the designs the singularity of $M(\xi)$ was overcome by use of the ridge type regularization procedure (10.10) in which a small quantity ϵ is added to the diagonal of $M(\xi)$ before inversion. An ϵ value of 10^{-5} was found to be adequate. With this regularization it is possible to check the equivalence theorem that, for

TABLE 17.1. Example 17.4: a compartmental model. D_θ- and c_θ-optimum designs

Criterion	Time t	Design weight	Criterion value
D-optimum	0.2292	1/3	7.3908
	1.3907	1/3	
	18.4164	1/3	
AUC	0.2331	0.0135	2193
	17.6322	0.9865	
Time to maximum concentration	0.1796	0.6061	0.02822
	3.5678	0.3939	
Maximum concentration	1.0137	1	1.000

each optimum design,

$$\{f^{\mathrm{T}}(x)M^{-1}(\xi^*)c(\theta)\}^2 \leq c^{\mathrm{T}}(\theta)M^{-1}(\xi^*)c(\theta) \tag{17.22}$$

for all $x \in \mathcal{X}$.

The c_θ-optimum design for estimating the area under the curve has only two points of support. This makes some sense, as the criterion is a function (the difference) of the two ratios θ_3/θ_1 and θ_3/θ_2. The reading at the low time of 0.23 allows efficient estimation of the ratio θ_3/θ_1 whereas that at $t = 17.6$ is for the ratio θ_3/θ_2. When the response (17.6) is plotted it is seen that the curve rises very rapidly to the maximum at $t = 1.10$, declining slowly thereafter. The relationship between θ_3 and θ_2 is therefore of greater importance in determining the area under the curve, an importance that is reflected in the design putting over 98% of the experimental effort at the higher value of t.

The c_θ-optimum design for the time to maximum concentration again has two points of support. In comparison with the design for the area under the curve, the experimental effort is much more evenly spread over the two design points. In addition, these points are relatively close to the calculated time of maximum concentration.

The last of the three functions, the maximum concentration, yields a c_θ-optimum design concentrated on one point; all measurements are taken at $t_{\max}$, the time at which the maximum is believed to occur. This is an extreme example of a c_θ-optimum design for which the quantity of interest is not estimable. If this design were to be used, so that measurements were

TABLE 17.2. Example 17.4: a compartmental model. Efficiencies of the D_θ- and c_θ-optimum designs of Table 17.1

	Efficiency for			
Design	D-optimum	AUC	$t_{\max}$	$y_{\max}$
D-optimum	100.0	34.31	65.94	36.10
AUC	0	100.0	0	0
$t_{\max}$	0	0	100.0	0
$y_{\max}$	0	0	0	100.0
18-point	67.65	24.00	28.61	36.77

$t_{\max}$, time to maximum concentration; $y_{\max}$, maximum concentration.

taken at only one point, it would be impossible to tell where, in fact, the response was a maximum.

These results demonstrate that, whichever criterion of optimality is used, the optimum design has far fewer points of support than the 18-point design of Table 1.4. As with the one-point design for exponential decay, these optimum designs are derived on the assumption of independent observations, an assumption made in the customary analysis of such data by least squares. ∎

Locally optimum designs for non-linear models depend upon the prior value θ_0. For locally D-optimum designs this information seems not to overwhelm the design criterion; all parameters can be estimated from the design. It may however sometimes be desirable to check the efficiency of the locally D-optimum design for a range of values around θ_0. For the locally c-optimum designs, on the other hand, the prior value in this example seems to provide too much information, leading to designs that rely so much on the specified value θ_0 that the quantities of interest are not estimable without including the prior information in the analysis. There are at least three possible solutions:

1. General Purpose Designs. A general purpose design, such as the locally D-optimum design or the 18-point design, can be used. If the purpose is to estimate a specific quantity, the efficiency of the design is calculated for that purpose. Table 17.2 gives the efficiencies of the D_θ-optimum and 18-point designs for Example 18.4, as well as the efficiencies of the c_θ-optimum designs. These last three efficiencies are all zero apart from the quantities for which the designs are optimum. The efficiency of the D-optimum design for the three quantities varies from 34–66%, whereas the range for the 18-point design is 24–36%. This design also has a relatively low D-efficiency of 67%. However, it is the only design that would allow for detection of lack of fit. If, however, detection of lack of fit is important, the methods of Chapter 20 would provide more efficient designs for specified departures.

2. Compound Designs. Several or all of the quantities of interest may be designed for at the same time. As the results of Table 17.1 show, the magnitudes of the design criteria for the three quantities are very different. Minimization of the sum of the variances would therefore lead to a design in which the numerically largest variance, that for the area under the curve, was given undue prominence. A more satisfactory criterion would be to minimize the product of the variances. This is identical to minimizing

$$\Psi\{M(\xi)\} = \sum_{j=1}^{3} \log\{c_j^{\mathrm{T}}(\theta)M^{-1}(\xi)c_j(\theta)\}, \tag{17.23}$$

a special form of compound generalized D-optimality (10.9).

A second family of compound designs is obtained by combining efficiencies for the D-optimum design with those for one or more of the quantities of interest. The resulting CD-optimum designs, in which the efficiencies are combined in a flexible parametric manner are calculated for this example in §21.9.

3. Bayesian Designs. If the purpose of the design really is to estimate one quantity with minimum variance, the preceding proposals will not provide the best design. The seeming paradox of singular designs arises because of the prior specification of θ_0 as a single value. If θ really is so well known, experimentation is unnecessary. A more realistic specification is to design with a prior specification for θ that allows for the uncertainty in the parameter values. The resulting Bayesian optimum designs are the subject of Chapter 18, with Example 18.4 continued in §18.4.

17.6 The Analysis of Non-linear Experiments

The principles of the analysis of experiments do not depend on whether the models fitted are linear or non-linear. However, some details differ between the two classes of model.

The least squares estimates of the parameters minimize the sum of squares

$$S(\theta) = \sum_{i=1}^{N} \{y_i - \eta(x_i, \theta)\}^2. \qquad (17.24)$$

For linear models the minimum of (17.24) is found explicitly as the set of values $\hat{\theta} = (F^{\mathrm{T}}F)^{-1}F^{\mathrm{T}}y$ (5.20). For non-linear models the values $\hat{\theta}$ minimizing (17.24) have to be found using an iterative numerical method. Whatever method is used, the vector of sensitivities from linearizing the model is as in (17.12), but now evaluated at $\hat{\theta}$, to give an extended design matrix $\hat{F}$.

An approximate $100(1-\alpha)\%$ region for θ is given by those values for which

$$(\theta - \hat{\theta})^{\mathrm{T}}\hat{F}^{\mathrm{T}}\hat{F}(\theta - \hat{\theta}) \leq ps^2 F_{p,\nu,\alpha}, \qquad (17.25)$$

where $F_{p,\nu,\alpha}$ is a percentage point of the F distribution, seemingly as in the linear case (5.22). However, there are two shortcomings of this approximate region.

1. The content of the region is only asymptotically $100(1-\alpha)\%$. For experiments of a practical size a nominal 95% region might have a content appreciably greater or less than this. If the exact content is important, it can be found by simulation.

2. For linear models the region (17.25) corresponds to a contour of the sum of squares $S(\theta)$ that lies on a plane in the space of the parameters. For non-linear models the region is found by first linearizing the model; (17.25) then lies on the tangent plane to $S(\theta)$ at $\hat{\theta}$. The resulting ellipsoids may be a poor approximation to the true contours of $S(\theta)$, which are often asymmetric about $\hat{\theta}$ and may be twisted into p-dimensional shapes sometimes likened to bananas.

References are given in §17.11 to ways in which improvements can be made to the elliptical contours given by (17.25) and to the implications for experimental design. An advantage of (17.25) is ease of calculation. The calculation of exact contours of $S(\theta)$ is a daunting task, particularly for multi-parameter problems.

17.7 A Sequential Experimental Design

Optimum designs for non-linear models may depend strongly on the values of the parameters θ. In §17.3.1 we described a sequential design scheme consisting of cycles of non-linear optimum design, experimentation, and parameter

TABLE 17.3. Example 17.5: the dehydration of n-hexanol. The initial factorial experiment

Observation number	x_1	x_2	y
1	1	1	0.126
2	2	1	0.219
3	1	2	0.076
4	2	2	0.126

estimation, the estimates to be used in determining the optimum conditions for the next experiment, until sufficiently accurate parameter estimates are obtained. We illustrate this procedure for an experiment in which one trial is added at a time. The description is adapted from Box and Hunter (1965).

Example 17.5 The Catalytic Dehydration of n-Hexanol The response is the rate of catalytic dehydration of n-hexanol at 555°F given by

$$\eta(x, \theta) = \frac{\theta_1 \theta_3 x_1}{1 + \theta_1 x_1 + \theta_2 x_2}, \tag{17.26}$$

so that there are two explanatory variables and three parameters. The variables are the partial pressures of the product and of water and the parameters are rate and equilibrium constants. The design region $\mathcal{X}$ is the square

$$0 \le x_1 \le 3, 0 \le x_2 \le 3. \tag{17.27}$$

In the absence of any knowledge of the parameters, or of the methods of optimum design, a plausible starting design is a 2^2 factorial shrunk away from the edges of the design region at values of 1 and 2. The results of these four trials are given in Table 17.3. The resulting least squares estimates of the parameters are

$$\hat{\theta}_1^4 = 7.1687 \ \hat{\theta}_2^4 = 33.6446 \ \hat{\theta}_3^4 = 0.7444. \tag{17.28}$$

For optimum sequential design we need the parameter sensitivities. Differentiation of (17.26) with respect to the components of θ yields

$$f_1(x, \theta) = \frac{\partial \eta(x, \theta)}{\partial \theta_1} = \frac{\theta_3 x_1 (1 + \theta_2 x_2)}{(1 + \theta_1 x_1 + \theta_2 x_2)^2},$$

with similar expressions for $f_2(x, \theta)$ and $f_3(x, \theta)$. To find the conditions for the fifth observation we evaluate the sensitivities (17.28) at $\hat{\theta}^4$ and find the

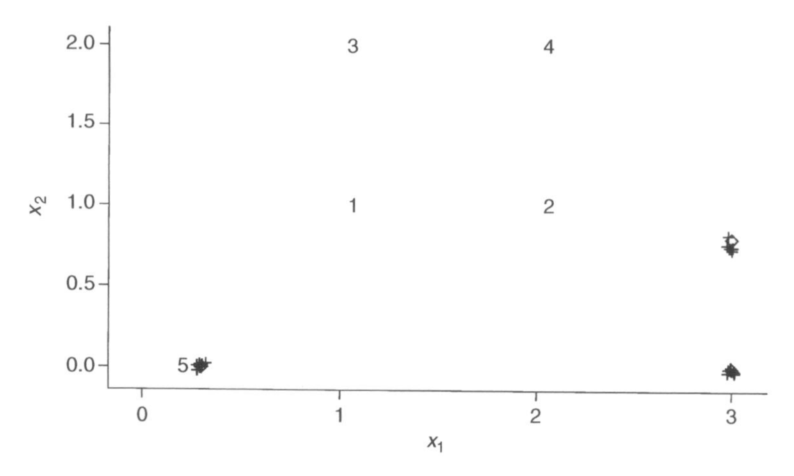

FIG. 17.9. Example 17.5: catalytic dehydration, initial points (1–4), sequentially added points (5 and crosses), and locally optimum points (diamonds).

maximum of $d(x, \xi)$ over $\mathcal{X}$ by a grid search, the candidate set consisting of the 3721 discrete pairs (x_1, x_2) in $\mathcal{X}$ at intervals of 0.05. The maximum is at $x_1 = 0.15, x_2 = 0.00$ which form the conditions for the fifth experiment. To simulate the sequential design we take

$$\theta_1 = 2.9 \;\; \theta_2 = 12.2 \;\; \theta_3 = 0.69, \tag{17.29}$$

with the observations having additive independent errors with variance $\sigma^2 = 0.01$. We iterate the above procedure—estimating θ based on all data so far, determining the point of maximum $d(x, \xi)$, and observing the reaction at that point—until the width of the approximate confidence intervals for the parameters are all less than 10% of the values of the estimated parameter values.

For a particular random sequence of simulated responses, this iterative procedure requires 22 additional points in order to achieve confidence intervals of the required relative width. Figure 17.9 shows a scatter plot of the initial and sequentially generated points, along with the locally optimum three-point design. Even though the parameter values $\hat{\theta}^4$ (17.28) are far from the true, but unknown, values (17.29), the design quickly settles down to being close to the optimum. The optimum design for the true parameter values puts equal weight of $1/3$ at the three points represented as diamonds in the figure. The D-efficiency of the initial 2^2 factorial design for these parameter values is only about 0.8%. In tangible terms, it would take hundreds of times as many replicate observations of the initial factorial design to achieve parameter estimates that are as accurate as those of the sequentially generated points. Further, this low efficiency explains the great distance between the preliminary parameter estimates (17.28) and those generating the data (17.29). These remarks illustrate how inefficient designs based on intuitive ideas in the factor space can be when the model is non-linear. The design also emphasizes how important it is to use sequential procedures. ■

The sequential updating of the design after each new observation is obtained is the most efficient sequential procedure for information per trial. However, it may not always be possible or convenient and the experiment may well be designed in larger groups of two or three, or more trials. Alternatively, the number of observations taken between updating of the parameter estimates and design calculations can increase with N as the parameter become better estimated.

17.8 Design for Differential Equation Models

Kinetic models of chemical reactions are in the form of sets of differential equations. Sometimes, as in the case of Example 17.3 with two consecutive first-order reactions, the differential equations can be solved analytically to give the expression for $\eta(t, \theta)$, in this example (17.5). Differentiation of $\eta(t, \theta)$ then gives the sensitivities, (17.13) for example. In this section we consider the more frequent and complicated case when the differential equations cannot be solved analytically. Numerical solution of the differential equations of the kinetic model also requires the use of numerical methods to find the sensitivities. We use the 'direct' method, described by Valko and Vajda (1984), in which the sensitivities are found from solution of a second set of differential equations.

Example 17.6 A General Consecutive Reaction The model for the two consecutive reactions is, as in Example 17.3,

$$A \overset{\theta_1}{\rightarrow} B \overset{\theta_2}{\rightarrow} C, \tag{17.30}$$

but now the reactions are not first-order. In the general consecutive reaction the differential equations determining the concentrations of the reactants are

$$\begin{aligned}
\frac{d[A]}{dt} &= -\theta_1[A]^{\lambda_1} \\
\frac{d[B]}{dt} &= \theta_1[A]^{\lambda_1} - \theta_2[B]^{\lambda_2} \\
\frac{d[C]}{dt} &= \theta_2[B]^{\lambda_2}.
\end{aligned} \tag{17.31}$$

where λ_1 and λ_2 are the orders of reaction. In Example 17.3, both reactions were first-order, that is $\lambda_1 = \lambda_2 = 1$. Chemical theory shows that the orders will be small integers, or ratios of small integers. Experimental results yielding other orders are often an indication that some parts of the reaction scheme have not been correctly modelled.

For the general consecutive reaction the vector of parameters is conveniently written

$$\psi = (\theta_1, \theta_2, \lambda_1, \lambda_2),$$

so that four sensitivities will be required. ■

We now consider a general model $y = \eta(t, \psi)$ with y a h-dimensional response. The experiment consists of measuring some or all of the h responses at a set of times $t \in \mathcal{T}$. Numerical solution of the set of h differential equations defining the response requires a prior value ψ_0 of the parameters. The auxiliary equations satisfied by the sensitivities come from the rule of total differentiation applied to these h model equations.

More specifically let $\eta(t, \psi) = (\eta_1(t, \psi), \ldots, \eta_h(t, \psi))^{\mathrm{T}}$ be a set of h functions representing the expected multivariate response at time $t \in \mathcal{T}$ with common parameters $\psi = (\psi_1, \ldots, \psi_p)^{\mathrm{T}}$. The functions η_i, $i = 1, \ldots, h$, are solutions of a set of differential equations

$$\frac{d\eta_1(t, \psi)}{dt} = g_1(\eta, t, \psi)$$

$$\vdots$$

$$\frac{d\eta_h(t, \psi)}{dt} = g_h(\eta, t, \psi) \tag{17.32}$$

with initial conditions given by

$$\eta(t_0, \psi) = \eta_0(\psi), \tag{17.33}$$

where $g_i(\cdot)$, $i = 1, \ldots, h$, are known differentiable functions of η, t, and ψ. Furthermore, let f_j be the vector of sensitivities for the jth parameter, that is the vector of derivatives of the response functions with respect to ψ_j:

$$f_j(t, \psi) = (f_{1j}(t, \psi), \ldots, f_{hj}(t, \psi))^{\mathrm{T}} = \left(\frac{\partial \eta_1(t, \psi)}{\partial \psi_j}, \ldots, \frac{\partial \eta_h(t, \psi)}{\partial \psi_j} \right)^{\mathrm{T}}, \quad j = 1, \ldots, p. \tag{17.34}$$

In matrix notation the sensitivity equations may be written as

$$\frac{df_j(t, \psi)}{dt} = \frac{\partial g(\eta, t, \psi)}{\partial \eta} f_j(t, \psi) + \frac{\partial g(\eta, t, \psi)}{\partial \psi_j}, \quad j = 1, \ldots, p, \tag{17.35}$$

or, in an extended notation, we get the sets of equations

$$\left[\begin{array}{c} \frac{df_{1j}(t,\psi)}{dt} \\ \vdots \\ \frac{df_{hj}(t,\psi)}{dt} \end{array}\right] = \left[\begin{array}{ccc} \frac{\partial g_1(\eta,t,\psi)}{\partial \eta_1} & \cdots & \frac{\partial g_1(\eta,t,\psi)}{\partial \eta_h} \\ & \vdots & \\ \frac{\partial g_h(\eta,t,\psi)}{\partial \eta_1} & \cdots & \frac{\partial g_h(\eta,t,\psi)}{\partial \eta_h} \end{array}\right] \left[\begin{array}{c} f_{1j}(t,\psi) \\ \vdots \\ f_{hj}(t,\psi) \end{array}\right] + \left[\begin{array}{c} \frac{\partial g_1(\eta,t,\psi)}{\partial \psi_j} \\ \vdots \\ \frac{\partial g_h(\eta,t,\psi)}{\partial \psi_j} \end{array}\right]. \tag{17.36}$$

These have to be solved together with equations (17.32), all for some *a priori* given values of the parameter vector ψ.

Once numerical values of the sensitivities have been calculated, the optimum experimental design can be found in the same way as the others in this chapter. If the errors follow the multivariate normal distribution and a locally D-optimum design is required for all p parameters, the multivariate design criterion of 10.10 is appropriate. An example of such a design is in §17.9.

Example 17.6 A General Consecutive Reaction continued We now illustrate the exact method on the general consecutive reaction. For comparison with the design for first-order reactions we assume again that only $[B]$ is measured.

In this example the first of the three equations (17.31) can be solved analytically to give the concentration of chemical A at time t as

$$[A] = \{1 - (1 - \lambda_1)\theta_1 t\}^{1/(1-\lambda_1)} \quad (\lambda_1, \theta_1, t \geq 0; \lambda_1 \neq 1), \tag{17.37}$$

if it is assumed that the initial concentration of A is 1. If λ_1 is less than one, the reaction is complete at $t = 1/\theta_1$, after which $[A] = 0$. If $\lambda_1 = 1$ we obtain exponential decay (17.3). Substitution of $[A]$ from (17.37) into the expression for $[B]$ (17.31) yields a differential equation from which analytical expressions for $[B]$ can only be found for a few values of λ_1 and λ_2. However, even if an explicit expression is available for the concentration for these particular values, the expression will not provide the derivatives of the concentration with respect to the orders λ_1 and λ_2 which we require in order to design the experiment.

Substitution of (17.37) in the second equation gives the differential equation for the concentration of the compound B

$$\frac{d[B]}{dt} = \theta_1\{1 - (1 - \lambda_1)\theta_1 t\}^{\lambda_1/(1-\lambda_1)} - \theta_2[B]^{\lambda_2}. \tag{17.38}$$

This is a single differential equation ($m = 1$) and the system (17.36) simplifies to one equation for each parameter. There are four parameters, that is

$$\psi = (\theta_1, \theta_2, \lambda_1, \lambda_2)^{\mathrm{T}},$$

TABLE 17.4. Example 17.6 a generalized consecutive reaction: D-optimum designs for both rate and order. The weights are 0.25 at each design point

Prior orders of reaction $(\lambda_1^0, \lambda_2^0)$	Time t_1	t_2	t_3	t_4
(1,1)	0.47	1.66	3.81	7.66
(2,1)	0.29	1.23	3.47	8.35
(1,2)	0.54	1.90	5.00	20.00
(2,2)	0.34	1.45	4.74	20.00

and so four sensitivities f_j when one response is measured: f_1 and f_2 are for θ_1 and θ_2, with f_3 and f_4 for λ_1 and λ_2. We need to solve (17.38) together with each of the following

$$\frac{df_1}{dt} = -\theta_2\lambda_2[B]^{\lambda_2-1}f_1 + (1-\theta_1 t)\{1-(1-\lambda_1)\theta_1 t\}^{\lambda_1/(1-\lambda_1)-1}$$

$$\frac{df_2}{dt} = -\theta_2\lambda_2[B]^{\lambda_2-1}f_2 - [B]^{\lambda_2}$$

$$\frac{df_3}{dt} = -\theta_2\lambda_2[B]^{\lambda_2-1}f_3 + \frac{\theta_1}{(1-\lambda_1)^2}\{1-(1-\lambda_1)\theta_1 t\}^{\lambda_1/(1-\lambda_1)}$$
$$\times \log\{1-(1-\lambda_1)\theta_1 t\} + \frac{\lambda_1\theta_1^2 t}{1-\lambda_1}\{1-(1-\lambda_1)\theta_1 t\}^{\lambda_1/(1-\lambda_1)-1}$$

$$\frac{df_4}{dt} = -\theta_2\lambda_2[B]^{\lambda_2-1}f_4(t,\psi) - \theta_2[B]^{\lambda_2}\log[B]. \qquad (17.39)$$

Equation (17.38) together with the sensitivity expressions forms four systems of two equations to be solved numerically.

For our four-parameter model the D-optimum design has four points of support, with weight 0.25 at each. However, the values of the design points depend on the prior values of the parameters. Table 17.4 gives the optimum design points for the four values of λ^0: (1,1), (1,2), (2,1), and (2,2). The design region is $\mathcal{T} = [0,20]$. The values of the rate constants were $\theta_1^0 = 1$ and $\theta_2^0 = 0.5$ as in Example 17.3. These designs were found by searching

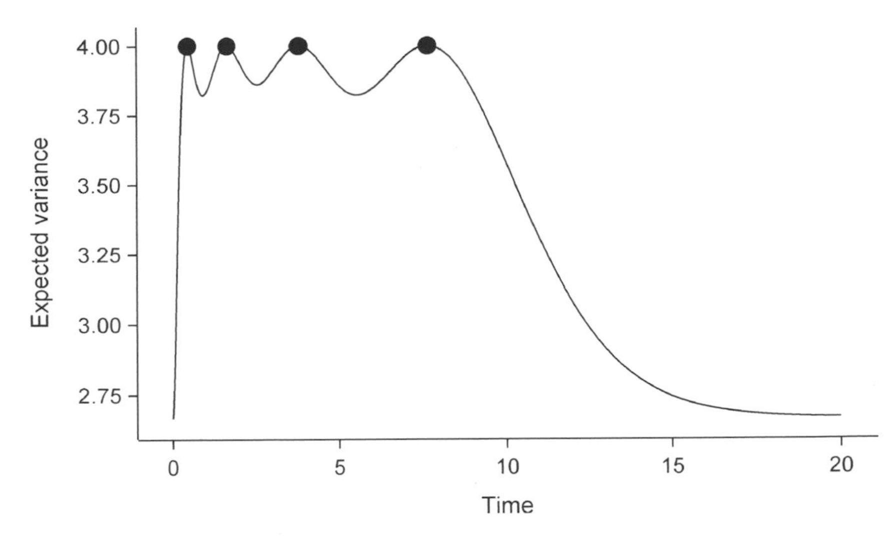

FIG. 17.10. Example 17.6 a generalized consecutive reaction: $d(t, \xi^*, \psi)$ for the D-optimum design $\lambda_1^0 = \lambda_2^0 = 1$. The design satisfies the equivalence theorem since the maxima of the variance are at the design points and equal four.

over the four continuous values of time, but with the weights held known at 0.25.

This table shows that the two lower design points depend principally on the prior value of λ_1. However, the most spectacular effect is the dependence of t_4 on the prior value of λ_2. When $\lambda_2^0 = 2$ both experimental designs have values of t_4 of 20, the upper limit, whereas for $\lambda_2^0 = 1$ the values are below 10.

The plots of Figures 17.10 and 17.11 show the usefulness of the variance $d(t, \xi^*, \psi)$ in checking optimum designs and exploring their structure. We optimized assuming that there were four points of support with equal weights. Figure 17.10 shows the variance function for $\lambda^0 = (1, 1)$. There are indeed four maxima, one at each optimum design point, at each of which $d(t, \xi^*, \psi) = 4$. The same is true for the D-optimum design for $\lambda^0 = (2,2)$ in Figure 17.11 except that now the fourth design point is on the boundary of the design region and the derivative of the variance is not zero. If the design region were increased in size a design with a larger value of t_4 would be obtained, which would have an increased value of the design criterion. The designs were constructed with weights w_i taken as 0.25. Since the designs satisfy the conditions for an optimum design, the figures also confirm that the weights for the optimum designs are indeed 0.25. ■

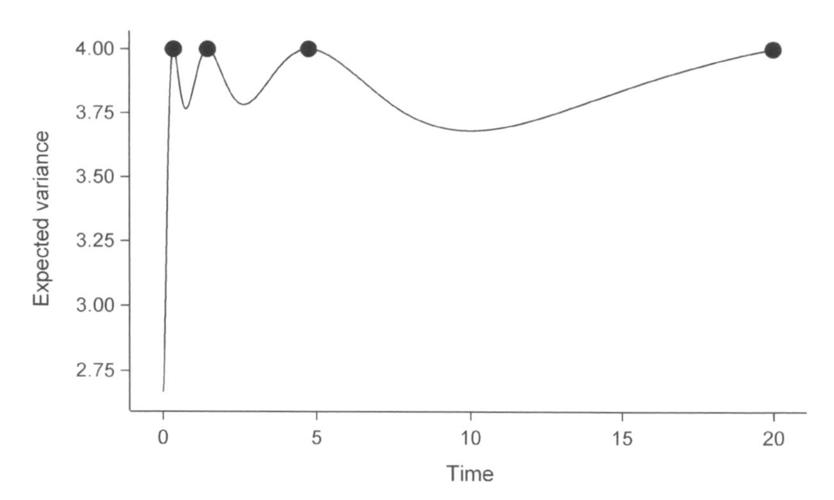

FIG. 17.11. Example 17.6 a generalized consecutive reaction: $d(t, \xi^*, \psi)$ for the D-optimum design $\lambda_1^0 = \lambda_2^0 = 2$. The design like that of Figure 17.10 also satisfies the equivalence theorem, but now the maximum at $t = 20$ indicates that measurements could with advantage be taken at a larger time.

17.9 Multivariate Designs

In the consecutive chemical reactions of Example 17.6 there are three chemical species, but we have found optimum designs under the assumption that only one of them is measured. We now use the results on multivariate D-optimum designs of §10.10 to find optimum designs when more than one species is measured. For this illustration we use a four-parameter model that is an extension of Example 17.3 and similarly has all reactions first-order. We can therefore write down explicit expressions for the responses and so find the sensitivities by analytical differentiation. The important property of the example is that the D-optimum design for this four-parameter model when three of the responses are measured has only three points of support.

Example 17.7 A Four-Parameter Model Seemingly simple chemical reactions may have a more complicated structure. For example, the first reaction in Example 17.3 may consist not of A going directly to B, but through some intermediate species D. It may also be that the reaction $B \rightarrow C$ is reversible, so that the concentration of B does not go to zero as $t \rightarrow \infty$. Rather there is an equilibrium between the concentrations of B and C. With these extensions the reaction scheme becomes

$$A \overset{\theta_4}{\rightarrow} D \overset{\theta_1}{\rightarrow} B \underset{\theta_3}{\overset{\theta_2}{\rightleftharpoons}} C. \tag{17.40}$$

If all reactions are first-order and the initial concentration of A is one with those of B, C, and D all zero, $[A] = \eta_A(t, \theta)$ follows exponential decay (17.3) with $\theta = \theta_4$. The other concentrations are given by

$$\begin{aligned}
[D] = \eta_D(t,\theta) &= \frac{\theta_4}{\theta_4 - \theta_1}\left\{e^{-\theta_1 t} - e^{-\theta_4 t}\right\} \\
[B] = \eta_B(t,\theta) &= \frac{\theta_3}{\theta_2 + \theta_3}\left\{1 - e^{-(\theta_2+\theta_3)t}\right\} \\
&\quad - \frac{(\theta_1 - \theta_3)\theta_4}{(\theta_1 - \theta_2 - \theta_3)(\theta_4 - \theta_1)}\left\{e^{-\theta_1 t} - e^{-(\theta_2+\theta_3)t}\right\} \\
&\quad - \frac{(\theta_4 - \theta_3)\theta_1}{(\theta_2 + \theta_3 - \theta_4)(\theta_4 - \theta_1)}\left\{e^{-\theta_4 t} - e^{-(\theta_2+\theta_3)t}\right\} \\
[C] = \eta_C(t,\theta) &= 1 - \eta_A(t,\theta) - \eta_B(t,\theta) - \eta_D(t,\theta).
\end{aligned} \tag{17.41}$$

If $\theta_1 >> \theta_4$, the overall rate of reaction will be determined by θ_4 (the rate determining step) and the first two reactions effectively become one, as they do if, conversely, $\theta_4 >> \theta_1$.

These equations show that, as $t \to \infty$, $[B] \to \theta_3/(\theta_2+\theta_3)$ with $[C] \to 1 - [B]$. Measurements at large values of t will therefore only provide information about the ratio θ_2/θ_3. Measurements at lower times are needed to obtain estimates of all parameters.

To illustrate we assume that $[A]$, $[B]$, and $[C]$ are measured, but not $[D]$. Differentiation of the three measured responses with respect to the four parameters gives the set of parameter sensitivities (17.36) to be used in calculating optimum designs. We take the prior values of the parameters as $\theta_1^0 = 0.8$, $\theta_2^0 = 0.2$, $\theta_3^0 = 0.15$, and $\theta_4^0 = 1.4$, with the initial concentration $A = 1$. We assume that the three responses are uncorrelated and with equal variance so that the covariance matrix Σ (§10.10) is a multiple of the identity matrix. The resulting three-point optimum design when $\mathcal{X} = [0, 20]$ puts a quarter of the weight at $t = 20$, nearly half ($w = 0.4836$) at $t = 1.1528$, and the rest at $t = 4.8258$.

The D-optimum design depends, of course, on which responses are measured, although all designs put a weight close to 0.25 at $t = 20$. Measuring just $[A]$ or $[D]$, or both of these responses, results in a singular design. The designs for the single responses $[B]$ or $[C]$ have four points of support, with weights one quarter. The six designs in which $[D]$ is measured along with one or more of $[A]$, $[B]$, or $[C]$ also have four points of support, but with unequal weights. In addition to the design for $[A]$, $[B]$, and $[C]$, the three designs for any pair of these responses also have three points of support. Finally, all the designs can be well approximated by designs with three or four points of support with weights 0.25, or 0.5 on a particular one of the three points of support, which are not quite the same as those for the optimum designs with unrestricted weights. All except one of these restricted designs has an efficiency greater than 99%. ■

The assumptions about the error structure are important. Although the true values of the concentrations of the species given by (17.41) must sum to one, the observed concentrations are assumed to have additive normal errors, here taken as independent. If all four species were measured the total would be distributed around a mean of one with a normal distribution. Other error structures may be appropriate. If $h - 1$ components are measured but the hth is also recorded so that the observed concentrations seem to sum to one, then there are only $h - 1$ observed values, although it may not be easy to determine which are the true observations. If all h measurements are made, but then scaled to add to one, (10.26) does not apply; one approach is that of compositional data (Aitchison 2004).

17.10 Optimum Designs for Non-linear Models in SAS

There is a variety of procedures in SAS for fitting non-linear models. In this section we shall focus on three:

- the NLIN procedure for fitting models with non-linear least squares, as discussed in §17.6;
- the NLMIXED procedure, which can estimate and test functions of the parameters; and
- the MODEL procedure, a somewhat more general non-linear modelling procedure with a variety of estimation methods, which we can use to fit the differential equation models of §17.8.

For a typical non-linear model, including any of the examples introduced in this chapter, the estimated parameters using any of these three procedures are identical to within differences of algorithmic convergence. The procedures have slightly different input and output corresponding to the different kinds of non-linear models they are designed to address. We will first describe how to use these procedures for fitting non-linear models, then discuss how to use them for computing optimum designs.

17.10.1 Fitting Non-linear Models with SAS

The NLIN procedure provides the most direct implementation of the non-linear least squares procedure outlined in §17.6. Minimal required input includes a data set with the observations to be fitted, a MODEL statement defining the non-linear model, and a PARAMETERS statement giving initial values for the parameters of the model. For example, for the three-parameter compartment model of §17.5, the following statements define the data and fit the model with PROC NLIN, using starting values of $\theta_i^0 = 1$, $i = 1, \ldots, 3$.

```
data Theophylline;
   input Time Concentration;
   Time = round(60*Time)/60;
datalines;
 0.166 10.1
 0.333 14.8
 0.5   19.9
 0.666 22.1
 1     20.8
 1.5   20.3
 2     19.7
 2.5   18.9
 3     17.3
 4     16.1
 5     15.0
 6     14.2
 8     13.2
10     12.3
12     10.8
24      6.5
30      4.6
48      1.7
;

proc nlin data=Theophylline;
   parameters t1=1 t2=1 t3=1;
   model Concentration = t3*(exp(-t1*Time) - exp(-t2*Time));
run;
```

The PARAMETERS statement names the parameters of the model and gives initial values, and the MODEL statement defines the model for the response variable Concentration. The default output includes a table of the parameters and the sum of squares as the iterative least squares procedure proceeds, an overall analysis of variance for the significance of the final model, and estimates, confidence limits, and correlations for the parameters. In this case, the parameters are estimated with reasonable precision (Table 17.5).

TABLE 17.5. Example 17.4: a compartmental model. Parameter estimates from PROC NLIN

Parameter	Estimate	Approx Std Error	Approximate 95% confidence limits	
t1	0.0589	0.00489	0.0484	0.0693
t2	4.2899	0.4702	3.2878	5.2920
t3	21.7965	0.6040	20.5091	23.0839

SAS Task 17.1. The data in Table 17.6 give 29 observations on the amount of nitrogen in US lakes. The variables are:

x_1: average influent nitrogen concentration;
x_2: water retention time; and
y: the mean annual nitrogen concentration.

The data were analysed by Stromberg (1993) using the model

$$y_i = \frac{x_{1i}}{1 + \theta_1 x_{2i}^{\theta_2}} + \epsilon_i. \tag{17.42}$$

Use PROC NLIN to fit this model to the data of Table 17.6. Atkinson and Riani (2000, p. 164) find that observations 10 and 23 are outliers. Do you find any evidence of this? How do the parameter estimates change when one or both of these observations are deleted?

The NLMIXED procedure provides an interface similar to that of the NLIN procedure. The NLMIXED procedure is primarily intended for fitting non-linear *mixed effect* models using the method of maximum likelihood, but it can also be used to fit the same models that NLIN fits. The maximum likelihood setting additionally provides a natural way to estimate and test arbitrary functions of the parameters, as discussed in §17.5. The following statements use PROC NLMIXED to fit the same three-parameter model to the theophylline data, and also to estimate the area under the curve.

```
proc nlmixed data=Theophylline;
   parameters t1=1 t2=1 t3=1;
   model Concentration
            ~ normal(t3*(exp(-t1*Time) - exp(-t2*Time)),s2);
   estimate "AUC" t3*(1/t1 - 1/t2);
run;
```

The MODEL statement for NLMIXED has a somewhat different form than the one for NLIN. Instead of describing just the model for the mean of the response, it explicitly parameterizes the distribution of the response.

TABLE 17.6. Nitrogen concentration in American lakes

Number	x_1	x_2	y
1	5.548	0.137	2.590
2	4.896	2.499	3.770
3	1.964	0.419	1.270
4	3.586	1.699	1.445
5	3.824	0.605	3.290
6	3.111	0.677	0.930
7	3.607	0.159	1.600
8	3.557	1.699	1.250
9	2.989	0.340	3.450
10	18.053	2.899	1.096
11	3.773	0.082	1.745
12	1.253	0.425	1.060
13	2.094	0.444	0.890
14	2.726	0.225	2.755
15	1.758	0.241	1.515
16	5.011	0.099	4.770
17	2.455	0.644	2.220
18	0.913	0.266	0.590
19	0.890	0.351	0.530
20	2.468	0.027	1.910
21	4.168	0.030	4.010
22	4.810	3.400	1.745
23	34.319	1.499	1.965
24	1.531	0.351	2.555
25	1.481	0.082	0.770
26	2.239	0.518	0.720
27	4.204	0.471	1.730
28	3.463	0.036	2.860
29	1.727	0.721	0.760

x_1: average influent nitrogen concentration; x_2: water retention time; y: mean annual nitrogen concentration.

TABLE 17.7. Example 17.4: a compartmental model. Parameter estimates from PROC NLMIXED

Parameter	Estimate	Approx Std Error	Approximate 95% confidence limits	
t_1	0.0589	0.00468	0.0490	0.0687
t_2	4.2899	0.3886	3.4736	5.1062
t_3	21.7965	0.5491	20.6429	22.9501
s_2	0.9988	0.3329	0.2993	1.6983

TABLE 17.8. Example 17.4: a compartmental model. Additional estimates from PROC NLMIXED

Label	Estimate	Approx Std Error	Approximate 95% confidence limits	
AUC	365.27	23.7157	315.44	415.09

In this case, we are using the normal distribution for which non-linear least squares is appropriate, additionally naming the variance parameter s2. Accordingly, the table of parameter estimates contains an additional row for the variance (Table 17.7). Notice that although the estimated parameters are the same as in Table 17.5, the standard errors and confidence limits are slightly different. This is because NLMIXED uses the maximum likelihood estimate of the variance $S(\hat{\theta})/N$ rather than the approximately unbiased $S(\hat{\theta})/(N-p)$. The ESTIMATE statement also causes PROC NLMIXED to produce similar statistics for the area under the curve (Table 17.8).

SAS Task 17.2. Use PROC NLMIXED to estimate the time to maximum concentration and the maximum concentration for the theophylline experiment.

For fitting differential equation models, the appropriate SAS tool is the MODEL procedure. MODEL is a general non-linear modeling tool, and as such it can also handle the simple non-linear models discussed so far. But with its versatility comes a syntax that is a little more difficult to use than NLIN and NLMIXED. As an example, consider the following simulated data for an experiment on a three-component reaction, with measured concentrations given by the variables C1, C2, and C3.

```
data Experiment;
   input Time @@;
   do Rep = 1 to 5;
      input C1 C2 C3 @@;
      output;
      end;
datalines;
0.5 0.643 0.343 0.057    0.585 0.389 0.036
    0.617 0.343 0.037    0.607 0.330 0.044
    0.620 0.328 0.034
1.0 0.352 0.484 0.149    0.341 0.486 0.181
    0.396 0.469 0.187    0.347 0.458 0.174
    0.376 0.476 0.179
1.5 0.211 0.486 0.272    0.222 0.518 0.254
    0.247 0.492 0.300    0.220 0.471 0.232
    0.215 0.512 0.286
2.0 0.126 0.482 0.381    0.161 0.436 0.407
    0.134 0.441 0.402    0.119 0.463 0.382
    0.151 0.466 0.438
4.0 0.011 0.266 0.727    0.013 0.238 0.731
    0.028 0.203 0.763    0.036 0.241 0.754
    0.001 0.223 0.759
6.0 0.017 0.122 0.883    0.001 0.111 0.930
    0.001 0.082 0.900    0.017 0.113 0.895
    0.001 0.115 0.883
8.0 0.001 0.066 0.951    0.001 0.027 0.945
    0.001 0.019 0.960    0.026 0.041 0.978
    0.001 0.040 0.933
;
```

This data set needs to be massaged slightly to make it appropriate for PROC MODEL. PROC MODEL assumes that the data are a well-ordered time series, with no duplicate time values in individual series, and with the initial time as the first observation. For the data above, this is easiest to accomplish by appending an initial time value for each replicate, and then sorting by replicate and time. The following statements perform these data manipulations.

```
data InitialTime;
   Time = 0; do Rep = 1 to 5; output; end;
data ModelData; set Experiment InitialTime;
proc sort data=ModelData;
   by Rep Time;
run;
```

When the data are not naturally organized in identifiable replicates, an alternative approach is to 'jitter' the time values by a small random amount, in order to make them well-ordered in a single long series, add a single initial time value, and then sort by time, as in the following statements.

```
data ModelData; set Experiment;
   Time = Time + 1e-6*ranuni(1);
data InitialTime;
   Time = 0; output;
data ModelData; set ModelData InitialTime;
proc sort data=ModelData;
   by Time;
run;
```

However an appropriate form of the data is constructed, the following statements use PROC MODEL to define and fit the general consecutive reaction model (17.31) for these data. Notice that λ_1 and λ_2 have been redefined slightly, so that the default tests against $\lambda_i = 0$ will apply to inferences about whether a first-order model is sufficient. The code being a bit more complicated than what we have seen thus far, we use SAS macros to make the parts of the PROC MODEL syntax more transparent.

```
%let InitialParameters = Theta1 1 Theta2 1 Lambda1 0 Lambda2 0;
%let InitialConditions = C1 0.99999 C2 0.00001 C3 0;
%macro DefineModel;
   dert.C1 = - Theta1*(C1**(1+Lambda1));
   dert.C2 =   Theta1*(C1**(1+Lambda1))
             - Theta2*(C2**(1+Lambda2));
   dert.C3 =   Theta2*(C2**(1+Lambda2));
%mend;

proc model data=ModelData;
   parameters &InitialParameters;
   dependent  &InitialConditions;
   %DefineModel;
   fit C2 / dynamic;
run;
```

As before, the PARAMETERS statement names the parameters and specifies their initial values. The new DEPENDENT statement names the responses and their initial conditions; we use values slightly different from the theoretical ones. The model is defined by three differential equations, assumed to be with respect to time. Finally, the FIT statement produces the fit of this model to the observed values of C2. The resulting parameter estimates are shown in Table 17.9. The indication is that a first-order model is sufficient for both reactions.

17.10.2 Finding Optimum Designs for Non-linear Models with SAS

Finding optimum designs for non-linear models is in principle not much different from finding them for linear models, the difference being that instead of dealing with the actual model matrix we deal with the approximate model matrix $F = (\partial/\partial\theta)\eta(t,\theta)$. Thus, in order to find the optimum design for

TABLE 17.9. Example 17.6: a general consecutive reaction. Parameter estimates from PROC MODEL

Parameter	Estimate	Approx Std Error	t	P(|T| > t)
Theta1	1.019026	0.0591	17.25	<.0001
Theta2	0.51716	0.0754	6.86	<.0001
Lambda1	0.05425	0.1207	0.45	0.6563
Lambda2	0.043823	0.1230	0.36	0.7241

modelling exponential decay with $\theta = 1$, we compute the sensitivities of the model with respect to θ as in (17.9) over a suitable grid of candidate time values, and then run PROC OPTEX on the resulting (one-column) design matrix, as shown in the following code.

```
data Jacobian;
   do Time = 0 to 50 by 0.01;
      f = -Time*exp(-1*Time);
      output;
      end;
run;

proc optex data=Jacobian coding=orthcan;
   model f / noint;
   generate n=5 method=m_fedorov niter=1000 keep=10;
   output out=Design;
   id Time;
proc print data=Design;
run;
```

The results confirm that the single point Time=1 is the optimum design.

SAS Task 17.3. Starting with the sensitivities defined in (17.13), confirm that the optimum design for the two consecutive first-order reaction model with $\theta_1 = 1$ and $\theta_2 = 0.5$ is the pair of points (0.78, 3.43).

SAS tools for fitting non-linear models can also be used to compute the sensitivities automatically, so that the only user input required is the model itself and the initial guess for the parameter values. There is a bit of trickery involved: although the computation is possible, the procedures were not designed for it, so the syntax is a little complex and the printed output rather cryptic. The idea is to fit the model to arbitrary response values, fixing the parameters at the initial guess and storing in an output data set the sensitivities at those fixed values. The following statements demonstrate the method.

```
data Candidates;
   do Time = 0 to 50 by 0.01;
      output;
      end;
proc nlin data=Candidates maxiter=0 noprint;
   parameters Theta 1;
   _y = 0;
   Predicted = exp(-Theta*Time);
   model _y = Predicted;
   f = getder(Predicted,Theta);
   id f;
   output out=Jacobian(drop=_y);
run;
```

In the PROC NLIN code above, the variable _y is the arbitrary response and the subsequent two statements fit the exponential decay model to it. The `MAXITER=0` option on the PROC MODEL statement fixes the parameter as the initial guess $\theta = 1$ specified in the PARAMETERS statement. Finally, the variable `f` is defined as the derivative of the model with respect to θ, and subsequent statements save it to the data set `Jacobian`. This data set is numerically identical to the result of computing the derivative function directly in the previous code.

SAS Task 17.4. Perform SAS Task 17.3 again, this time using PROC NLIN to compute the parameter sensitivities.

Finally, the sensitivities for differential equation models can also be computed automatically in SAS. The extra complication is that the method requires the numerical solution for the model at the initial guess for the parameters θ as input for computing the sensitivities. Thus, two invocations of PROC MODEL are involved, as shown in the following code, which computes the sensitivities for the general three-component consecutive reaction model (17.31). As in the example of fitting this reaction model with PROC MODEL, we employ SAS macros to make the parts of the PROC MODEL syntax more transparent.

```
%let InitialParameters = Theta1 1 Theta2 0.5
                         Lambda1 1 Lambda2 1;
%let InitialConditions = C1 0.99999 C2 0.00001 C3 0;
%macro DefineModel;
   dert.C1 = - Theta1*(C1**Lambda1);
   dert.C2 =   Theta1*(C1**Lambda1)
             - Theta2*(C2**Lambda2);
   dert.C3 =   Theta2*(C2**Lambda2);
%mend;

data Candidates;
   do Time = 0 to 50 by 0.01;
      output;
      end;
```

```
proc model data=Candidates noprint;
   parms     &InitialParameters;
   dependent &InitialConditions;
   %DefineModel;
   solve C1 C2 C3 / out=Solution time=time;
proc model data=Solution noprint;
   parms     &InitialParameters;
   dependent &InitialConditions;
   %DefineModel;
   f1 = getder( C2, Theta1  );
   f2 = getder( C2, Theta2  );
   f3 = getder( C2, Lambda1 );
   f4 = getder( C2, Lambda2 );
   outvar f1 f2 f3 f4;
   restrict Theta1=1, Theta2=0.5, Lambda1=1, Lambda2=1;
   fit C2 / out=Jacobian dynamic;
run;
```

In this code, the first PROC MODEL invocation computes the solution to the differential equations defining the model at the initial guess for every time point in the candidate data set and stores the results in a data set named `Solution`. Then the second PROC MODEL invocation takes both this solution and the model definition as input and computes the sensitivities `f1`, ..., `f4`, storing them in a data set named `Jacobian`.

> **SAS Task 17.5.** Confirm the D-optimum designs shown in Table 17.4 by using PROC MODEL to compute the relevant Jacobians and PROC OPTEX to select $N = 4$ points.

17.11 Further Reading

The methodology of non-linear regression is discussed in detail by Ratkowsky (1989), Bates and Watts (1988), and Seber and Wild (1989). A method for obtaining and presenting intervals of various sizes for a single parameter in a multi-parameter non-linear model is presented by Cook and Weisberg (1990). The methods for fitting non-linear models with the NLIN and MODEL procedures in SAS are described in the chapters for those procedures in SAS Institute Inc. (2007*d*) and SAS Institute Inc. (2007*a*), respectively.

Many of the criteria of Chapter 10 have been applied to design for non-linear models. For example, Box and Lucas (1959), Ford, Titterington and Kitsos (1989), Haines (1993), and Hedayat, Zhong and Nie (2003) obtain D-optimum designs. Minimax design are studied by Dette and Sahm (1998). Dette and O'Brien (1999) propose a new criterion of optimality, which they call I-L-optimality. Based on the predicted variance, this criterion is invariant with respect to different parameterizations of the model. G- and D-optimality can be seen as its special cases. The use of E-optimality is discussed in Dette and Haines (1994) and Dette, Melas, and Pepelyshev (2004). The c_θ-optimum designs of §17.5 are described by Kitsos, Titterington, and Torsney (1988) and Atkinson *et al.* (1993).

Ratkowsky (1990) provides an introduction to non-linear models, with an emphasis on the properties of a variety of models for one or a few factors. Some non-linear models have received special attention because of the important role they play in some scientific investigations. For instance, design for the model proposed by Michaelis and Menten (1913) (see Exercise 20) is discussed by Dette, Melas, and Pepelyshev (2003), Dette and Biedermann (2003) and Dette, Melas, and Wong (2005). The optimum design of experiments when the models are exponential is investigated by Mukhopadhyay and Haines (1995), Dette and Neugebauer (1997), and Han and Chaloner (2003). D-optimum maximin designs are found by Biedermann, Dette, and Pepelyshev (2004) for a compartmental model.

The Bayesian approach (see Chapter 18) for constructing experimental designs for non-linear models has been applied by many researchers, see for instance Chaloner (1993), Chaloner and Verdinelli (1995), and Dette (1996). Han and Chaloner (2004) give an example when a non-linear mixed-effects model is required. We discuss this further in §25.5. A functional approach to optimum design for non-linear models is proposed by Melas (2005, 2006).

Often several non-linear models are suitable for explaining a particular type of data, for example when the response asymptotically converges to a constant. A non-sequential procedure for parameter estimation and model discrimination for a collection of such models is proposed by O'Brien and Rawlings (1996). The problem of model selection is discussed in detail in Chapter 22.

The most numerically demanding of the design problems discussed in this chapter are those of §17.8, where the responses are defined by sets of differential equations for which there are no analytical solutions. For this class of design problems Uciński (1999) compared the direct method with two others, a finite-difference method and an adjoint method. He found the direct method to be most appropriate for a wide range of applications in engineering, chemistry, physics, and other experimental sciences.

All of these approaches to design for non-linear models are based on the first-order Taylor expansion of the model introduced in §17.2 in which the N-dimensional curved non-linear surface of the model as θ varies is replaced by the tangent plane approximation at θ_0. More refined statistical methods then have to take account of higher-order derivatives. Beale (1960) and

Bates and Watts (1980) introduce curvature measures for non-linear models that lead to the design procedures of Hamilton and Watts (1985) and of O'Brien (1992). Walter and Pronzato (1997, p. 329) remark that it seems hard to evaluate the accuracy of the approximation to the volume of confidence regions on which these procedures are based. Accordingly, Pázman and Pronzato (1992) and Pronzato and Pázman (2001) base designs for non-linear models on, hopefully improved, approximations to the distribution of the estimates of the parameters. The context of this work and further results on experimental design for non-linear models are given booklength treatment by Pázman (1993).

18

BAYESIAN OPTIMUM DESIGNS

18.1 Introduction

The design of the experiments considered so far has depended in an informal way on prior information. In the discussions in Chapters 2 and 3 the purpose of the experiment, the range of the variables and possible models, among other aspects, all reflect the prior knowledge and experience of the experimenters. However, the only aspect of this knowledge that has been susceptible to mathematical analysis is the information matrix $M(\xi)$. In this chapter we extend the design criteria to allow formal inclusion of prior information about the parameters of the model. In the resulting Bayesian designs the mathematical analysis again reduces to the study of the implications for design of various forms of information matrix.

For the normal theory linear model with known variance σ^2, suppose that the prior information is that β is normally distributed with mean β_0 and covariance matrix $\sigma^2 M^{-1}(N_0)$. The information matrix $M(N_0)$ formalizes the prior information as being equivalent to the results of N_0 observations, where N_0 need not be an integer. The design for the prior observations can be written $N_0\xi_0$, when $M(N_0) = N_0 M(\xi_0)$. The experiment consists of N trials with design measure ξ. The information matrix for the posterior distribution of β is then

$$\widetilde{M}(\xi) = N_0 M(\xi_0) + N M(\xi). \tag{18.1}$$

Bayesian D-optimum designs for the linear model maximize $|\widetilde{M}(\xi)|$, which does not depend on the prior value β_0. These designs maximize the expected gain in Shannon information between the prior and posterior of β. For a thorough review of Bayesian experimental design see Chaloner and Verdinelli (1995). As the prior information becomes less important, that is as $N_0/N \to 0$, the Bayesian criterion (18.1) tends to standard D-optimality.

The form of (18.1) is the same as that used in the frequentist augmentation of designs, when the N_0 trials form the design for a previous experiment and the results of all $N_0 + N$ trials are to be analysed together. These augmentation designs are the subject of §19.2.

The theory for non-linear models $\eta = f(x, \theta)$, where the prior distribution of θ is $p(\theta)$, is more complicated and requires the use of approximations to the distribution of the posterior estimator $\tilde{\theta}$. If the prior information matrix in the equivalent of (18.1) is ignored and the least squares estimator $\hat{\theta}$ is used instead of $\tilde{\theta}$, the estimator will be asymptotically normal with a distribution depending only on the information matrix of the experiment. Then, approximately, the design maximizing expected Shannon information will maximize the Bayesian D-optimality criterion

$$\Phi(\xi) = \mathrm{E}_\theta \log |M(\xi, \theta)| = \int_\theta \log |M(\xi, \theta)| p(\theta) d\theta, \tag{18.2}$$

introduced in §17.3.2. Unlike (18.1), this criterion for the non-linear model does depend on the prior distribution of θ, not just on the prior information matrix. Chaloner and Verdinelli (1995) derive Bayesian versions of other criteria from different utility functions. For example if, as in §17.5, interest is in a function $g(\theta)$, squared error loss leads to a Bayesian version of c-optimality.

In Chapter 17 the dependence of the design on θ was resolved by replacing the distribution $p(\theta)$ by a point prior giving mass one to the value θ_0. Locally optimum designs resulted. In this chapter we consider instead designs incorporating the prior distribution for θ, which may be either discrete or continuous. The extra information is incorporated into the design, as in (18.2) for D-optimality, by taking expectations of the design criterion over the prior distribution. A general equivalence theorem for such criteria is given in §18.2. The principal difference between the designs satisfying this equivalence theorem and those satisfying the theorem of §9.2 is that Carathéodory's Theorem no longer provides an upper bound on the number of trials in the experimental design. Thus the designs exhibit the common-sense property that the number of support points increases as the prior for θ becomes more dispersed. This behaviour is illustrated in §18.3 which is concerned with several extensions of D-optimality which incorporate prior information.

Section 18.4 describes the parallel extension of c-optimality: as the prior information in this case becomes more dispersed, the singular designs of §17.5 for the compartmental model are replaced by designs with sufficient support points to allow estimation of all parameters. In §18.5 we describe a Monte-Carlo method for the simple calculation of the expectations involved in Bayesian designs. But the chapter begins with three examples of non-linear models, two of which are continued from Chapter 17. In all examples it is assumed that the second-order error assumptions hold and that estimation is by least squares.

Note that this chapter contains no section devoted to using SAS to find the designs discussed. It is possible to do so; indeed, all of the specific designs presented, in both tables and figures, were constructed with SAS. But there are no specialized SAS tools to faciliate such constructions, beyond the general optimization capabilities of SAS/IML software, which are illustrated in other chapters.

Example 18.1 Truncated Quadratic Model For the first example in this chapter, the expected value of the response y is related to the single explanatory variable x by the truncated quadratic relationship

$$\begin{aligned} \mathrm{E}(y) = \eta(x,\beta,\theta) = \beta x\theta(1 - x\theta) &= \beta f(x,\theta) \quad (0 \leq x \leq 1/\theta) \\ &= 0 \qquad\qquad\quad \text{otherwise.} \end{aligned} \tag{18.3}$$

For known θ this is a standard linear model with a single parameter β, except that the expected value of the response is constrained to be non-negative. The model is related to those used in pharmacokinetic studies to describe the flow of a drug through a subject, although such models usually involve linear combinations of exponential terms, as does the compartmental model of Example 18.3. In the pharmacokinetic interpretation $1/\theta$, which would vary between subjects, is the time to complete elimination of the drug. The maximum of the curve, corresponding to the maximum concentration of the drug, is $\beta/4$; interest is in estimation of β. Observations for which $x\theta > 1$ are not informative about the value of β, although they may be about θ.

This simple model demonstrates clearly the properties of designs in the presence of prior information. For a given value of θ the variance of $\hat{\beta}$, the least squares estimate of β, is minimized by putting all trials at the point where the response is maximum, that is, at $x = 1/2\theta$. If the value of θ is not known a priori, but is described by a prior distribution, this locally optimum design could be used with θ replaced by its expected value. But there may be values of θ within the prior distribution for which concentration of the design on one value of x will give estimates of β with large, or even infinite, variance. The designs derived in this chapter are intended to provide, for example, a small value of the expected variance of $\hat{\beta}$ taken over the distribution of θ. As we shall see, such designs can be very different from those which maximize the expected information about β. ■

Example 18.2 Exponential Decay; Example 17.1 continued A simple model arising from chemical kinetics, which was discussed in Chapter 17, is that for first-order or exponential decay in which

$$\mathrm{E}(y) = \eta(t,\theta) = \exp(-\theta t) \quad (t,\ \theta \geq 0). \tag{18.4}$$

Linearization about the prior value θ_0 gave the model

$$\mathrm{E}(y) = \eta(t,\theta_0) + (\theta - \theta_0)f(t), \tag{18.5}$$

where $f(t) = f(t, \theta_0) = -t\exp(-\theta_0 t)$.

The relationship between (18.5) and (18.3) is expressed more forcefully by writing

$$\begin{aligned} \mathrm{E}(y) - \eta(t, \theta_0) &= (\theta - \theta_0) f(t) \\ &= \beta f(t). \end{aligned} \tag{18.6}$$

The variance of $\hat{\beta}$ is again minimized by performing all trials where $f(t)$ is a maximum, that is, at $t = 1/\theta_0$ when $\eta = e^{-1}$. As in Example 17.1, if the true value of θ is far from θ_0, the variance of $\hat{\beta}$ will be large because experiments will be performed where $f(t)$ is small. In such regions the value of the response is near zero or unity, providing little information about θ. ■

Example 18.3 A Compartmental Model; Example 17.4 continued The properties of the three-parameter model

$$\mathrm{E}(y) = \theta_3\{\exp(-\theta_2 t) - \exp(-\theta_1 t)\} \tag{18.7}$$

were discussed in §17.5 and a variety of designs were derived. The locally D-optimum design found by linearization of the model required three equally replicated design points. The c_θ-optimum designs for estimating the area under the curve and for finding the time to maximum concentration had only two design points, in neither case equally replicated, whereas the design for the maximum concentration required trials at only one value of time. It was argued in §17.5 that use of a single prior value of θ_0 for these designs seemed to provide so much information that singular designs resulted. The Bayesian procedure of §18.4 remedies this defect and leads to non-singular designs from which the properties of interest may be estimated. ■

18.2 A General Equivalence Theorem Incorporating Prior Information

In this section the General Equivalence Theorem of §9.2 is extended to include dependence of the information matrix on a vector parameter θ. For linearized non-linear models dependence on θ is through the vector of p partial derivatives

$$f^{\mathrm{T}}(x, \theta) = \left\{ \frac{\partial \eta(x_i, \theta)}{\partial \theta_j} \right\} \quad j = (1, \ldots, p).$$

For c_θ-optimum designs there is a further dependence through the coefficients $c_{ij}(\theta)$ in (17.21). We write the information matrix as

$$M(\xi, \theta) = \int_{\mathcal{X}} f(x, \theta) f^{\mathrm{T}}(x, \theta) \xi(\mathrm{d}x) = \int_{\mathcal{X}} M(\bar{\xi}, \theta) \xi(dx).$$

TABLE 18.1. Equivalence theorem for Bayesian versions of D-optimality: design criteria and derivative functions

Criterion	$\Psi\{M(\xi)\}$	Derivative function $\phi(x,\xi)$
I	$\mathrm{E}\log\|M^{-1}\|$	$p - \mathrm{E}\{\mathrm{tr}M^{-1}M(\bar{\xi},\theta)\}$
II	$\log\mathrm{E}\|M^{-1}\|$	$p - \mathrm{E}\{\|M^{-1}\|\mathrm{tr}M^{-1}M(\bar{\xi},\theta)\}/\mathrm{E}\|M^{-1}\|$
III	$\log\|\mathrm{E}M^{-1}\|$	$p - \mathrm{E}\{\mathrm{tr}M^{-1}(\mathrm{E}M^{-1})M^{-1}M(\bar{\xi},\theta)\}$
IV	$\log\{\mathrm{E}\|M\|\}^{-1}$	$p - \mathrm{E}\{\|M\|\mathrm{tr}M^{-1}M(\bar{\xi},\theta)\}/\mathrm{E}\|M\|$
V	$\log\|\mathrm{E}M\|^{-1}$	$p - \mathrm{tr}(\mathrm{E}M)^{-1}M(\bar{\xi},\theta)$

EM is short for $E_\theta M(\xi,\theta)$ etc. and $M(\bar{\xi},\theta)$ is the elementary information matrix corresponding to a design with one support point.

The generalization is to consider design criteria of the form

$$\Psi\{M(\xi)\} = E_\theta\Psi\{M(\xi,\theta)\}.$$

For the one-parameter example of §18.1 reasonable extensions of D-optimality would be to find designs to maximize the expected information about the parameter or to minimize the expected variance of the parameter estimate. The results of §18.3.2 show that these designs are not the same.

Similarly, there are several generalizations of D-optimality when θ is a vector. The obvious generalization is to take

$$\Psi_{\mathrm{I}}\{M(\xi)\} = -E_\theta\log|M(\xi,\theta)| = E_\theta\log|M^{-1}(\xi,\theta)|, \qquad (18.8)$$

for which a Bayesian justification was mentioned in §18.1. Another possibility is

$$\Psi_{\mathrm{II}}\{M(\xi)\} = \log E_\theta|M^{-1}(\xi,\theta)|. \qquad (18.9)$$

which, when $p = 1$, reduces to minimizing the expected variance of the parameter estimate. Firth and Hinde (1997) show that for $p > 1$ this criterion may be non-concave for dispersed prior distributions, leading to local maxima in the criterion function. Five possible generalizations of D-optimality are listed in Table 18.1, together with their derivative functions, for each of which an equivalence theorem holds (Dubov 1971; Fedorov 1981). These criteria are compared in §18.3.2. Although Criterion I arises naturally from a Bayesian analysis incorporating a loss function, the other criteria do not have this particular theoretical justification.

These results provide design criteria whereby the uncertainty in the prior estimates of the parameters is translated into a spread of design points. In the standard theory the criteria are defined by matrices $M(\xi)$, which are linear combinations, with positive coefficients, of elementary information matrices $M(\bar{\xi})$ corresponding to designs with one support point. But in the extensions of D-optimality, for example, dependence is on such functions of matrices as $E_\theta M^{-1}(\xi, \theta)$ or $E_\theta|M(\xi, \theta)|$, the non-additive nature of which precludes the use of Carathéodory's Theorem. As a result the number of support points is no longer bounded by $p(p+1)/2$. The examples of the next two sections show how the non-additive nature of the criteria leads to designs with an appreciable spread of the points of support.

18.3 Bayesian D-optimum Designs

18.3.1 Example 18.1 The Truncated Quadratic Model Continued

As a first example of design criteria incorporating prior information we calculate some designs for the truncated model (18.3), concentrating in particular on Criterion II given by (18.9). In this one-parameter example this reduces to minimizing the expected variance of the parameter estimate. We contrast this design with that maximizing the expected information about β.

The derivative function for Criterion II is given in Table 18.1. It is convenient when referring to this derivative to call $d(x, \xi) = p - \phi(x, \xi)$ the expected variance. Then, for Criterion II,

$$d(x, \xi) = \frac{E_\theta|M^{-1}(\xi, \theta)|d(x, \xi, \theta)}{E_\theta|M^{-1}(\xi, \theta)|}, \tag{18.10}$$

where $d(x, \xi, \theta) = f^{\mathrm{T}}(x, \theta)M^{-1}(\xi, \theta)f(x, \theta)$. The expected variance is thus a weighted combination of the variance of the predicted response for the various parameter values. In the one-parameter case the weights are the variances of the parameter estimates. It follows from the equivalence theorem that the points of support of the optimum design are at the maxima of (18.10), where $d(x, \xi^*) = p$.

Suppose that the prior for θ is discrete with weight p_m on the value θ_m. The design criterion (18.9) to be minimized is

$$E_\theta M^{-1}(\xi, \theta) = \sum_m \frac{p_m}{f^2(x, \theta_m)}, \tag{18.11}$$

with $f(x, \theta)$ given in (18.3). To illustrate the properties of the design let the prior for θ put weight 0.2 on the five values 0.3, 0.6, 1, 1.5, and 2. Trials at values of $x > 1/\theta$ yield a zero response. Thus for $\theta = 2$ a reading at

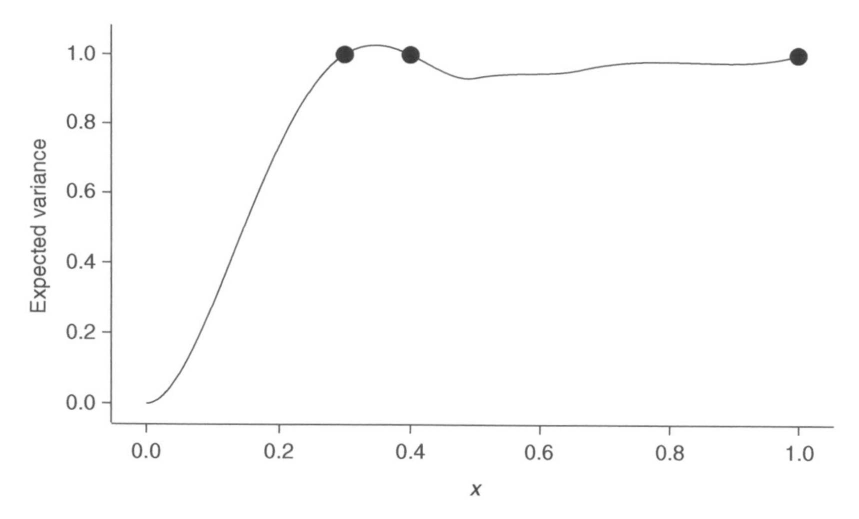

FIG. 18.1. Example 18.1: truncated quadratic model. Expected variance $d(x, \xi^*)$ for the three-point Bayesian optimum design from searching over the grid $x = 0.1, 0.2, \ldots, 1.0$: • design points. Five point prior for θ.

any value of x above 0.5 will be non-informative. Unless the design contains some weight at values less than this, the criterion (18.11) will be infinite. Yet, for the three smallest parameter values, the locally optimum designs, at $x = 1/2\theta$, all concentrate weight on a single x value at or above 0.5.

The expected values required for the criterion (18.11) are found by summing over the five parameter values. Table 18.2 gives three optimum continuous designs for Criterion II. The first design was found by searching over the convex design space [0, 1], and the second and third were found by grid search over respectively 20 and 10 values of x. The designs have either two or three points of support, whereas Carathéodory 's Theorem indicates a single point for the locally optimum design. The design for the coarser grid has three points; the others have two. That the three-point design is optimum can be checked from the plot of $d(x, \xi^*)$ in Figure 18.1. The expected variance is $p = 1$ at the three design points and less than 1 at the other seven points of the discrete design region. However, it is 1.027 at $x = 0.35$, which is not part of the coarse grid. Searching over a finer grid leads to the optimum design in which the weights at 0.3 and 0.4 are almost combined, yielding a two-point design, for which the expected variance of $\hat{\theta}$ is slightly reduced. It is clear why the number of design points has changed. But such behaviour is impossible for the standard design criteria when the additivity property holds and the model contains a single parameter.

TABLE 18.2. Example 18.1: truncated quadratic model. Continuous optimum designs ξ^* minimizing the expected variance of the parameter estimate (Criterion II) over three different grids in $\mathcal{X}$

Region				Criterion value
(a) Convex [0,1]				32.34
x	0.3430	1		
w^*	0.6951	0.3049		
(b) 20-point grid				32.37
x	0.35	1		
w^*	0.7033	0.2967		
(c) 10-point grid				32.95
x	0.3	0.4	1	
w^*	0.4528	0.2405	0.3066	

The design ξ^* puts weight w_i^* at point x_i.

The effect of the spread of design points is to ensure that there is no value of θ for which the design is very poor. The appearance of Figure 18.1 indicates that it is the sum of several rather different curves arising from the various values of θ. However, not all design criteria lead to a spread of design points. If we use instead a criterion like V of Table 18.1 in which the expected information about β is maximized, maximization of (18.11) is replaced by maximization of

$$E_\theta M(\xi, \theta) = \sum_m p_m f^2(x, \theta_m). \tag{18.12}$$

For the coarse grid the optimum design is at the single point $x = 0.3$. The effect of little or no information about β for a specific θ value may well be outweighed by the information obtained for other θ values. This is not the case for designs using (18.11), when variances can be infinite for some parameter values, whereas the information is bounded at zero.

18.3.2 A Comparison of Design Criteria

The results of 18.3.1 illustrate the striking difference between designs which minimize expected variance and those which maximize expected information. In this section we use the exponential decay model, Example 18.2, to compare the five generalizations of D-optimality listed in Table 18.1.

When, as here, $p = 1$ the five criteria reduce to the three listed in Table 18.3, in which the expectation of integer powers of the information matrix, in

TABLE 18.3. Equivalence theorem for Bayesian versions of D-optimality: reduction of criteria of Table 18.1 for single-parameter models

Criterion	$\Psi\{M(\xi)\}$	Power parameter	Expected variance weight a(θ)
I	$-E_\theta \log M(\xi,\theta)$	0	1
II,III	$E_\theta M^{-1}(\xi,\theta)$	-1	$M^{-1}(\xi,\theta)$
IV,V	$-E_\theta M(\xi,\theta)$	1	$M(\xi,\theta)$

this case a scalar, are maximized or minimized as appropriate. The values of the power parameter are also given in Table 18.3. The equivalence theorem for these criteria involves an expected variance of the weighted form

$$d(x,\xi) = \frac{E_\theta\{a(\theta)d(x,\xi,\theta)\}}{E_\theta\{a(\theta)\}},$$

where the weights $a(\theta)$ are given in Table 18.3. For Criterion I, $a(\theta) = 1$, so that the combination of variances is unweighted.

For a numerical comparison of these criteria we use Example 18.2 with, again, five equally probable values of θ, now $1/7, 1/\sqrt{7}$, 1, $\sqrt{7}$ and 7. For each parameter value the locally D-optimum design is at $x = 1/\theta$, so that the design times for these individual locally optimum designs are uniformly spaced in logarithmic time.

The designs for the three one-parameter criteria are given in Table 18.4. The most satisfactory design arises from Criterion I in which $E_\theta \log|M(\xi,\theta)|$ is maximized. This design puts weights in the approximate ratio of 2:1:1 within the range of the optimum designs for the individual parameter values. By comparison, the design for Criterion II, in which the expected variance is minimized, puts 96.69% of the weight on $x = 0.1754$. This difference arises because, in the locally D-optimum design for the linearized model, $\text{var}(\hat{\theta}) \propto \theta^2 e^2$. Large parameter values, which result in rapid reactions and experiments at small values of x, are therefore estimated with large variances relative to small parameter values. Designs with Criterion II accordingly tend to choose experimental conditions in order to reduce these large variances. The reverse is true for the design with Criterion V, in which the maximization of expected information leads to a one-point design dominated by the smallest parameter value, for which the optimum design is at $x = 7$: all the weight in the design of Table 18.4 is concentrated on $x = 6.5218$.

The numerical results presented in this section indicate that Criterion I is most satisfactory. We have already mentioned the Bayesian justification for this criterion. A third argument comes from the equivalence theorem.

TABLE 18.4. Example 18.2: exponential decay. Comparisons of optimum designs satisfying criteria of Table 18.3

Criterion	Power	t	w^*
I	0	0.2405	0.4781
		1.4868	0.2708
		3.9915	0.2511
II,III	−1	0.1755	0.9669
		2.5559	0.0331
IV,V	1	6.5218	1

For each value of θ the locally optimum design will have the same maximum value for the variance, in general p. The results of Table 18.3 show that the weight $a(\theta)$ for Criterion I is unity. Therefore, the criterion provides an expected variance which precisely reflects the importance of the different θ values as specified by the prior distribution. In other criteria the weights $a(\theta)$ can be considered as distorting the combination of the already correctly scaled variances.

Despite these arguments, there may be occasions when the variance of the parameter estimates is of prime importance and Criterion II is appropriate. For Example 18.1 this criterion produced an appealing design in §18.3.1 because the variance of $\hat{\beta}$ for the locally optimum design does not depend on θ. But the results of this section support the use of the Bayesian criterion in which $E_\theta \log M^{-1}(\xi, \theta)$ is minimized. In Example 18.1 a property of the design using Criterion I is that a close approximation to the continuous design is found by replacing the weights in Table 18.4 by two, one, and one trials.

18.3.3 The Effect of the Prior Distribution

The comparisons of criteria in §18.3.2 used a single five-point prior for θ. In this section the effect of the spread of this prior on the design is investigated together with the effect of more plausible forms of prior. Criterion I is used throughout with Example 18.2.

The more general five-point prior for θ puts weight of 0.2 at the points $1/\nu$, $1/\sqrt{\nu}$, 1, $\sqrt{\nu}$, and ν. In §18.3.2 taking $\nu = 7$ yielded a three-point design. When $\nu = 1$ the design problem collapses to the locally optimum design with all weight at $t = 1$. Table 18.5 gives optimum designs for these and three other values of ν, giving one-, two-, three-, four-, and five-point designs as ν increases. The design for $\nu = 100$ almost consists of weight 0.2 on each of the separate locally optimum designs for the very widely spaced parameter values. A prior with this range but more parameter values might be expected to give a design with more design points. As one example, a nine-point uniform prior with support ν^{-1}, $\nu^{-3/4}$, $\nu^{-1/2}$, $\ldots, \nu^{-3/4}$, ν, with ν again equal to 100, produces an eight-point design. Rather than explore this path any further, we let Table 18.5 demonstrate one way in which increasing prior uncertainty leads to an increase in the number of design points. In assessing such results, although it may be interesting to observe the change in the designs, it is the efficiencies of the designs for a variety of prior assumptions that is of greater practical importance.

An alternative to these discrete uniform priors in $\log\theta$ is a normal prior in $\log\theta$, so that the distribution of θ is lognormal. This corresponds to a prior assessment of θ values in which $k\theta$ is as likely as θ/k and θ has a log-normal distribution. An effect of continuous priors such as these on the design criteria is to replace the summations in the expectations by integrations. However, numerical routines for the evaluation of integrals reduce once more to the calculation of weighted sums.

The normal distribution used as a prior was chosen to have the same standard deviation τ on the $\log\theta$ scale as the five-point discrete prior with $\nu = 7$, which gave rise to a three-point design. The normal prior was truncated to have range -2.5τ to 2.5τ, and this range was then divided into seven equal intervals on the $\log\theta$ scale to give weights for the values of θ. To assess the effect of this discretization the calculation was repeated with the prior divided into 15 intervals. The two optimum designs are given in Table 18.6. There are slight differences between these five-point designs. However, the important results are the efficiencies of Table 18.7, calculated on the assumption that the 15-point normal prior holds. The optimum design for the seven-point prior has an efficiency of 99.92%, indicating the irrelevance of the kind of differences shown in Table 18.6. More importantly, the three-point design for the five-point uniform prior has an efficiency of 92.51%. The four-trial exact design derived from this by replacing the weights in Table 18.4 with two, one, and one trials is scarcely less efficient. The only poor design is the one-point locally optimum design. In §18.5 we describe a sampling algorithm that provides a simple way of approximating prior distributions, including the log-normal prior of this section.

18.3.4 Algorithms and the Equivalence Theorem

Results such as those of Table 18.6 suggest that there is appreciable robustness of the designs to mis-specification of the prior distribution. A related interpretation is that the optima of the design criteria are flat for Bayesian designs. This interpretation is supported by plots of the expected variance for some of the designs of Table 18.6.

TABLE 18.5. Example 18.2: exponential decay. Dependence of design on range of prior distribution: optimum designs for Criterion I with five-point prior distributions over $1/\nu$, $1/\sqrt{\nu}$, 1, $\sqrt{\nu}$ and ν

ν	t	w^*
1	1	1
3	0.6506	0.7690
	1.5749	0.2310
7	0.2405	0.4781
	1.4858	0.2706
	3.9897	0.2513
13	0.1109	0.3371
	0.4013	0.1396
	1.2841	0.1954
	6.1463	0.3279
100	0.0106	0.2137
	0.1061	0.1992
	1.0610	0.2000
	10.6490	0.2009
	100.000	0.1862

TABLE 18.6. Example 18.2: exponential decay. Optimum designs for discretized log-normal priors

Prior	t	w^*
7	0.1013	0.0947
	0.2295	0.1427
	0.6221	0.3623
	1.6535	0.2549
	4.2724	0.1454
15	0.1077	0.1098
	0.3347	0.2515
	0.7480	0.2153
	1.4081	0.2491
	3.7769	0.1743

The plot of $d(t, \xi^*)$ for the locally optimum design for the exponential decay model, putting all weight at $t = 1$, was given in Figure 17.3. The curve is sharply peaked, indicating that designs with trials far from $t = 1$ will be markedly inefficient. However, the black curve for the design for the

TABLE 18.7. Example 18.2: exponential decay. Efficiencies of optimum designs for various priors using Criterion I when the true prior is the 15-point log-normal

Prior used in design	Efficiency %
One point	22.67
5-point uniform, $\nu = 7$	92.51
Exact design with $N = 4$ for $\nu = 7$	92.11
7-point log-normal	99.92
15-point log-normal	100

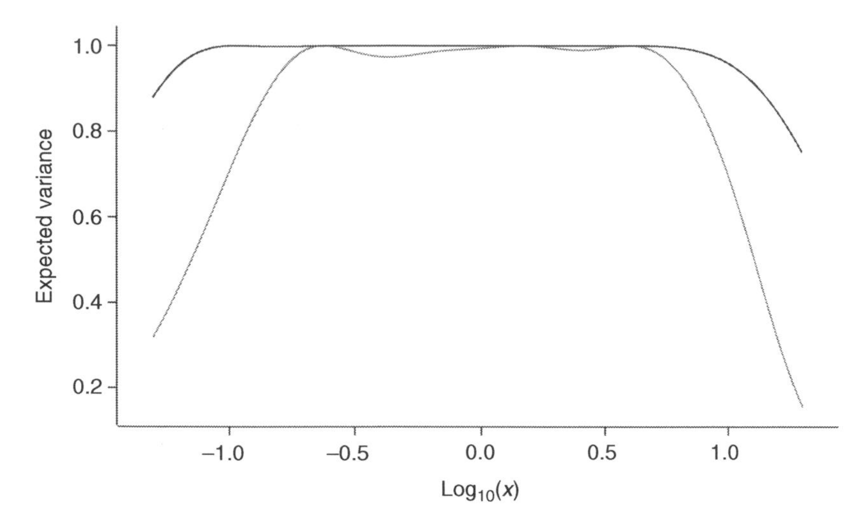

FIG. 18.2. Example 18.2: exponential decay. Expected variance $d(t, \xi^*)$ for Criterion I; 5-point uniform prior, $\nu = 1$—black line, 15-point log-normal prior—grey line.

five-point uniform prior with $\mu = 7$ (Figure 18.2) is appreciably flatter, with three shallow peaks at the three design points. The grey curve for the five-point design for the 15-point normal prior Figure 18.2 is sensibly constant over a 100-fold range of t, indicating a very flat optimum.

The flatness of the optima for designs with prior information has positive and negative aspects. The positive aspect, illustrated in Table 18.7, is the near optimum behaviour of designs quite different from the optimum design; the negative aspect is the numerical problem of finding the precisely optimum design, if such is required.

The standard algorithms of optimum design theory are described in §9.4. They consist of adding weight at the point at which $d(t, \xi)$ is a maximum. For the design of Figure 17.3, with a sharp maximum, the algorithms converge, albeit relatively slowly, since convergence is first-order. For flat derivative functions, such as the grey curve of Figure 18.2, our limited experience is that these algorithms are useless, an opinion supported by the comments of Chaloner and Larntz (1989). One difficulty is that small amounts of weight are added to the design at numerous distinct points; the pattern to which the design is converging does not emerge.

The designs described in this section were found using both numerical optimization and a step we call 'consolidation'. There are two parts to the consolidation step:

1. Points with small weights ($\xi_i < \epsilon_\xi$) are dropped.
2. Nearby points ($|x_i - x_j| < \epsilon_x$) are replaced by their average with weight equal to the sum of the weights of the averaged points.

Initially, weights ξ were optimized for a log-uniform design with 20-points. After that, rounds of consolidation and optimizing x and ξ simultaneously were repeated until x and ξ ceased changing. In all examples, this automated approach led to designs in which the maximum expected variance was equal to 1 ± 0.0001, so that the equivalence theorem was sensibly satisfied.

18.4 Bayesian c-optimum Designs

The two examples of §18.3 are both one-parameter non-linear models. In this section Bayesian designs are considered for the three-parameter compartmental model.

Example 18.3 A Compartmental Model continued Although model (18.7) contains three parameters, θ_3 enters the model linearly and so the value of θ_3 does not affect the D-optimum design. In general, c-optimum designs, even for linear models, can depend on the values of the parameters. However, in this example, the coefficients $c_{ij}(\theta)$ (§17.5) either depend linearly on θ_3 or are independent of it, so that, again, the design does not depend on θ_3. In the calculation of Bayesian optimum designs we can therefore take θ_3 to have a fixed value which, as in §17.5, is 21.80. For comparative purposes, two prior distributions are taken for θ_1 and θ_2. These are both symmetric, centred at $(\theta_1, \theta_2) = (0.05884, 4.298)$, the values given in §17.5, and are both uniform over a rectangular region. The calculation of derivatives and the numerical integration required for both c- and D-optimum designs is thus only over these two dimensions of θ_1 and θ_2.

Prior distribution I takes θ_1 to be uniform on 0.05884 ± 0.01 and, independently, θ_2 to be uniform on 4.298 ± 1.0. These intervals are, very approximately, the maximum likelihood estimates for the data of Table 1.4 plus or minus twice the asymptotic standard errors. For prior distribution II the limits are ± 0.04 and ± 4.0, that is, approximately eight asymptotic standard errors on either side of the maximum likelihood estimator. Prior distribution II thus represents appreciably more uncertainty than prior distribution I. Both priors are such that, for all θ_1 and θ_2 in their support, $\theta_2 > \theta_1$ which is a requirement for the model to be of a shape similar to the concentration $[B]$ plotted in Figure 17.1.

To find the Bayesian c-optimum designs we minimize $E_\theta c^{\mathrm{T}}(\theta)M^{-1}(\xi,\theta)c(\theta)$, the expected variance of the linear contrasts and the analogue of criterion II of §18.2. An alternative would be the minimization of the expected log variance, which would be the analogue of Criterion I.

For the area under the curve, the Bayesian c-optimum design with prior distribution I is similar to the c_θ-optimum design, but has three design points, not two, as is shown in Table 18.8. However, nearly 95% of the design measure is concentrated at $t = 18.5$. Prior distribution II (Table 18.9) gives an optimum design with four design points, the greatest weight being nearly 70%. These two designs are quite different; increased uncertainty in the parameter values leads to an increased spread of design points. The optimum value of the criterion, the average over the distribution of the asymptotic variance, is also much larger under prior distribution II than under I by a factor of almost 3.

Tables 18.8 and 18.9 also give results for the other two contrasts of interest. For the time to maximum concentration, prior distribution I gives a design with three support points and distribution II gives an optimum design with five. Again, these designs are different from each other and from the c_θ-optimum design. The designs for the maximum concentration also change with the prior information. The c_θ-optimum design has one support point. Prior distribution I gives an optimum design with three points and distribution II gives a design with five points. For prior distribution I the optimum criterion value is only slightly larger than that for c_θ-optimum and for II it is about twice as large. An advantage of designs incorporating a prior distribution is that all three parameters can be estimated.

In addition to the c-optimum designs, Tables 18.8 and 18.9 also include the D-optimum designs found by maximizing $E_\theta \log|M(\xi,\theta)|$, that is, Criterion I of §18.2. These designs behave much as would be expected from the results of §18.3.3, with the number of support points being three for prior distribution I and five for prior distribution II. The efficiencies of all

TABLE 18.8. Example 18.3: a compartmental model. Optimum Bayesian designs for prior distribution I

Criterion	Time t	Design weight	Criterion value
D-optimum	0.2288	0.3333	7.3760
	1.4169	0.3333	
	18.4488	0.3333	
Area under the curve	0.2451	0.0129	2463.3
	1.4952	0.0387	
	18.4906	0.9484	
Time to maximum concentration	0.1829	0.6023	0.0303
	2.4638	0.2979	
	8.8561	0.0998	
Maximum concentration	0.3608	0.0730	1.1143
	1.1446	0.9094	
	20.9201	0.0176	

designs are given in Tables 18.10 and 18.11. The Bayesian c-optimum designs are typically very inefficient for estimation of a property other than that for which they are optimum. This is particularly true under distribution I where, for example, the Bayesian c-optimum design for estimating the area under the curve has an efficiency of about 3% for estimating the time of maximum yield. Both the various D-optimum designs and the original 18-point design are, in contrast, quite robust for a variety of properties. Although it is hard to draw general conclusions from this one example, it is clear that if the area under the curve is of interest, then that should be taken into account at the design stage.

We return to this example in §21.9 where we use compound optimality to find designs with both good c- and D-efficiencies. ■

18.5 Sampled Parameter Values

To construct Bayesian optimum designs for continuous prior distributions requires the evaluation of the integral of the criterion over the prior distribution; for Bayesian D-optimality using Criterion I we have to evaluate (18.2). In §18.4 for the compartmental model we, rather unrealistically, took θ_1 and

TABLE 18.9. Example 18.3: a compartmental model. Optimum Bayesian designs for prior distribution II

Criterion	Time t	Design weight	Criterion value
D-optimum	0.2033	0.2870	7.1059
	1.1999	0.2346	
	2.9157	0.1034	
	5.9810	0.0612	
	6.5394	0.0022	
	20.2055	0.3116	
Area under the curve	0.2914	0.0089	6925.0
	1.7302	0.0366	
	13.1066	0.2571	
	39.5900	0.6974	
Time to maximum concentration	0.2515	0.2917	0.1910
	0.9410	0.2861	
	2.7736	0.1464	
	8.8576	0.2169	
	24.6613	0.0588	
Maximum concentration	0.3698	0.0972	1.9867
	1.1390	0.3588	
	2.4379	0.3166	
	6.0684	0.1632	
	24.0678	0.0641	

TABLE 18.10. Example 18.3: a compartmental model. Efficiencies of Bayesian D-optimum and c-optimum designs of Table 18.8 under prior distribution I

	Efficiency for			
Design	D-optimum	AUC	$t_{\max}$	$y_{\max}$
D-optimum	100.0	37.0	67.2	39.9
AUC	23.3	100.0	3.2	4.5
$t_{\max}$	57.4	5.1	100.0	19.6
$y_{\max}$	28.2	1.9	12.5	100.0
18-point	68.4	26.0	30.2	41.0

TABLE 18.11. Example 18.3: a compartmental model. Efficiencies of Bayesian D-optimum and c-optimum designs of Table 18.8 under prior distribution II

	Efficiency for			
Design	D-optimum	AUC	$t_{\max}$	$y_{\max}$
D-optimum	100.0	28.8	64.7	53.8
AUC	23.3	100.0	7.3	10.8
$t_{\max}$	87.6	13.3	100.0	64.3
$y_{\max}$	59.5	10.8	58.2	100.0
18-point	82.9	31.3	77.5	73.8

θ_2 to have uniform distributions. For the single parameter exponential decay model in §18.3.3 we more realistically took θ to be lognormal, but replaced this prior distribution with a discretized form. We now describe a sampling algorithm which allows the straightforward calculation of an approximation to the required expectation for an arbitrary prior distribution.

Instead of numerical integration, Atkinson *et al.* (1995) sample parameter values from the prior and then replace the integral in, for example, (18.2) with a summation. The design criterion then becomes maximization of the approximation

$$\Phi_{\text{APX}}(\xi) = \sum_{i=1}^{n(\theta)} \log |M(\xi, \theta_i)|, \tag{18.13}$$

where $n(\theta)$ is the number of values sampled from the prior distribution $p(\theta)$. Equivalence theorems, such as those in Table 18.1, apply with the expectations E_θ calculated with the summation in (18.13).

In this approximation one sample of values is taken from $p(\theta)$ and the optimum design calculated. Atkinson *et al.* (1995) take $n(\theta) = 100$, but also investigate other values. For a particular problem the value of $n(\theta)$ can be checked by comparing optimum designs for several samples from one value of $n(\theta)$ and then repeating the process for other values. However, even if there is some variation in the designs, the criterion values for Bayesian optimum designs are often very flat, so that different samples from $p(\theta)$ may give slightly different optimum designs with very similar properties; Figure 18.2 illustrates a related aspect of the insensitivity of Bayesian designs.

The method is of particular advantage for multivariate prior distributions. It is straightforward to sample from prior distributions which are multivariate normal. Let this distribution have vector mean μ_θ and variance

Σ_θ and let $SS^T = \Sigma_\theta$, where S is a triangular matrix which can be found by the Choleski decomposition. Then if Z_i is a vector of independent standard normal random variables, the random variables θ_i given by

$$\theta_i = \mu_\theta + SZ_i, \tag{18.14}$$

will have the required multivariate normal distribution. Sampling from other multivariate priors can use transformation of the distribution to near-normality with Box–Cox transformations (Atkinson *et al.* 2004, Chapter 4).

Example 18.4 A Reversible Reaction As an example of the use of sampling in generating Bayesian D-optimum designs this section considers a model with two consecutive first-order reactions with the second reaction reversible which can be written

$$A \overset{\theta_1}{\rightarrow} B \underset{\theta_3}{\overset{\theta_2}{\rightleftharpoons}} C, \tag{18.15}$$

where θ_3 is the rate of the reverse reaction and all $\theta_j > 0$. The kinetic differential equations for $[A], [B]$, and $[C]$ are

$$\begin{aligned}\frac{d[A]}{dt} &= -\theta_1[A]\\ \frac{d[B]}{dt} &= \theta_1[A] - \theta_2[B] + \theta_3[C]\\ \frac{d[C]}{dt} &= \theta_2[B] - \theta_3[C].\end{aligned} \tag{18.16}$$

Since no material is lost during the reaction, if the initial concentration of A is one and those of B and C are zero, $[A] + [B] + [C] = 1$, although the observed concentrations will not obey this relationship. The concentration of A, $\eta_A(t, \theta)$, follows exponential decay and the concentrations are given by

$$\begin{aligned}\eta_A(t,\theta) &= e^{-\theta_1 t}\\ \eta_B(t,\theta) &= \frac{\theta_3}{\theta_2+\theta_3}\left\{1 - e^{-(\theta_2+\theta_3)t}\right\} - \frac{\theta_1-\theta_3}{\theta_1-\theta_2-\theta_3}\left\{e^{-\theta_1 t} - e^{-(\theta_2+\theta_3)t}\right\}\\ \eta_C(t,\theta) &= 1 - \eta_A(t,\theta) - \eta_B(t,\theta).\end{aligned} \tag{18.17}$$

As $t \to \infty$, $[B] \to \theta_3/(\theta_2+\theta_3)$ so that $[B]$ and $[C]$ have informative values for large t. Figure 18.3 shows the responses as a function of time when $\theta_1 = 0.7$, $\theta_2 = 0.2$, and $\theta_3 = 0.15$: the asymptotic value of $[B]$ is therefore 3/7. This model is both an extension of exponential decay, Example 18.2 and a special case of the four-parameter model, Example 17.7.

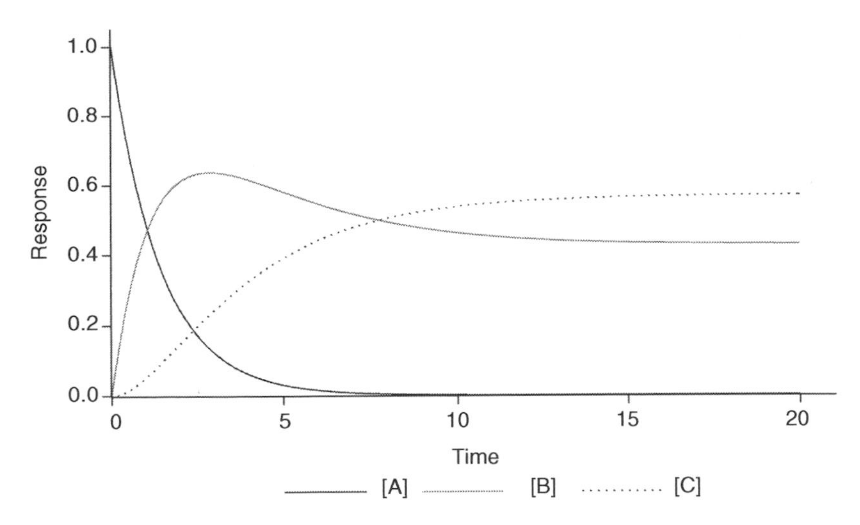

FIG. 18.3. Reversible reaction: concentrations of reactants over time. Reading upwards for large t: [A], [B], and [C].

The parameter sensitivities are found by differentiation of (18.17). Since there are three parameters, single-response designs will have at least three points of support, the third being at the maximum value of t, here taken as 20. The locally D-optimum design when only $[B]$ is measured has three points of support at times $\{1.1666, 4.9554, 20\}$ with, of course, weight one third at each time.

To incorporate uncertainty about the parameter values in the design criterion let the θ_j in (18.17) independently have log-normal distributions generated by the exponentiation of normal random variables with mean $\log \theta_j$ and standard deviation τ. When $\tau = 0.5 \log 2$ just over 95% of the prior distribution lies between $\theta_j/2$ and $2\theta_j$. Because the θ_j are mutually independent, we do not need to use (18.14). However, some values of θ have to be excluded. As $\theta_1 - (\theta_2 + \theta_3) \to 0$, the expression for $\eta_B(t, \theta)$ (18.17) becomes indeterminate and needs to be replaced by

$$\eta_B(t, \theta) = \frac{\theta_3}{\theta_2 + \theta_3} \left\{ 1 - e^{-(\theta_2 + \theta_3)t} \right\} - (\theta_1 - \theta_3) t e^{-\theta_1 t}, \tag{18.18}$$

with a consequent effect on the parameter sensitivities. Such complications were avoided by rejecting any set of simulated values for which $1.1(\theta_2 + \theta_3) > \theta_1$.

Table 18.12 gives Bayesian D-optimum designs maximizing (18.13) for $\tau = 0.5 \log \nu$, $\nu = 1$, 2, and 4. The value $\nu = 1$ corresponds to the locally optimum design for the point prior with $\theta_j = \theta_j^0$. The value of $n(\theta)$ was taken as 50 at both $\nu = 2$ and $\nu = 4$. To indicate the sampling variability we took

TABLE 18.12. Example 18.4: a reversible reaction—locally D-optimum ($\tau = 0$) and Bayesian D-optimum ($\tau > 0$) designs

τ	$n(\theta)$								
$0.5 \log 1$	1	$t = \{$	1.1666	4.9555				20	$\}$
$= 0$		$w = \{$	0.3333	0.3333				0.3333	$\}$
$0.5 \log 2$	50	$t = \{$	1.0606	4.7585				20	$\}$
		$w = \{$	0.3333	0.3333				0.3333	$\}$
$0.5 \log 4$	50	$t_1 = \{$	0.2196	0.7882	2.8455	5.7177		20	$\}$
		$w_1 = \{$	0.0111	0.2860	0.1829	0.2050		0.3150	$\}$
$0.5 \log 4$	50	$t_2 = \{$	0.7116	2.1649	4.8814			20	$\}$
		$w_2 = \{$	0.2969	0.0864	0.2904			0.3263	$\}$
$0.5 \log 4$	50	$t_3 = \{$	0.5496	1.3171	4.8394			20	$\}$
		$w_3 = \{$	0.1754	0.1956	0.3059			0.3231	$\}$
$0.5 \log 4$	500	$t_0 = \{$	0.5783	0.8332	2.4873	2.6964	5.2313	20	$\}$
		$w_0 = \{$	0.0451	0.2594	0.0030	0.1239	0.2477	0.3209	$\}$

three different random samples of size 50 when $\nu = 4$; an optimum design with a sample of size $n(\theta) = 500$ at $\nu = 4$ is also shown. As these results indicate, for small τ the designs are close to the locally optimum design. As τ grows, the optimum designs move away from the locally optimum design. Note in particular that the supports of the optimum designs for $\tau = 0.5 \log 4$ have more than three points. However, the precise location of these support points depends on the prior sample of θ values. This dependence is clarified by Figure 18.4, which shows the expected variance for the three different designs for $\nu = 4$ and $n(\theta) = 50$. The upper support point at $t = 20$ is unambiguous, but the number as well as the values of the other support points vary rather widely.

Although the designs differ, their properties differ less dramatically. With locally optimum designs we can compare efficiencies of designs at the parameter value θ_0. Here we are interested in the expected value of the efficiency of a design over $p(\theta)$. We approximated this by evaluating the efficiencies over the sampled prior with $n(\theta) = 500$. We found that the efficiencies of

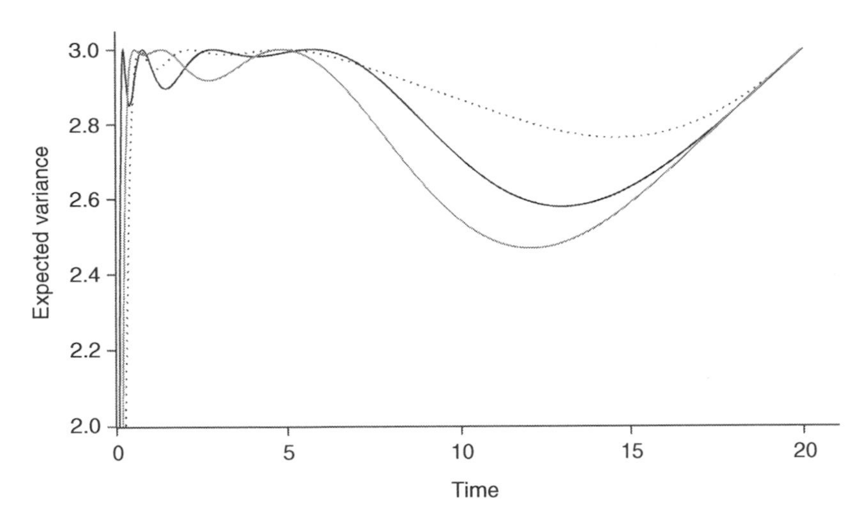

FIG. 18.4. Reversible reaction: expected variance $d(t, \xi^*)$ for optimum design with $\tau = 0.5 \log 4$ for three different samples with $n(\theta) = 50$. The black curve corresponds to the last of the three designs with $n(\theta) = 50$ in Table 18.12.

the three designs for $\nu = 4$ and $n(\theta) = 50$ relative to the optimum design for the 500-point prior were all greater than 99%. ■

18.6 Discussion

The main result of this chapter is the extension of the standard equivalence theorem of §9.2 to incorporate prior information, yielding the General Equivalence Theorem of §18.2. This theorem has then been exemplified by extensions to the familiar criteria of D- and c-optimality. The equivalence theorem for these expectation criteria has a long implicit history. The earliest proof seems to have been due to Whittle (1973), but the implications, particularly for the number of design points, are not clearly stated. The first complete discussion, including examples of designs, is due to Chaloner and Larntz (1989) who consider logistic regression. Chaloner (1988) briefly treats the more general case of design for generalized linear models. Earlier work does not consider either the number of design points, nor the properties of the derivative function, which are of importance in the construction of designs. Läuter (1974, 1976) prove the theorem in the generality required but only gives examples of designs for composite criteria for linear models.

Atkinson and Cox (1974) use the theorem for Criterion I of Table 18.1 with linear models. Cook and Nachtsheim (1982) are likewise concerned with designs for linear models. Pronzato and Walter (1985) calculate numerical optimum designs for some non-linear problems, but do not mention the equivalence theorem. Fedorov and Atkinson (1988) give a more algebraic discussion of the properties of the designs for the criteria of Table 18.1. The example of §18.4 is described in greater detail by Atkinson *et al.* (1993) who also give a more complete discussion of the independence of the optimum design from the value of θ_3. For a more general analysis of such independence for D- and D_S-optimum designs, see Khuri (1984). In all applications, if the prior information used in calculating the designs is also to be used in the analysis of the experiments, the information matrices used in this chapter require augmentation by prior information. Pilz (1983,1991) provide surveys.

A further example of Bayesian optimum designs is given in the next chapter, the subject of which is the design of experiments for discrimination between regression models. The resulting optimum designs, like those of this chapter, depend upon the values of the unknown parameters. In §20.8.2 the Bayesian technique of this chapter is used to define optimum designs maximizing an expectation criterion.

19

DESIGN AUGMENTATION

19.1 Failure of an Experiment

There are many possible reasons for disappointment or dissatisfaction with the results of an experiment. Four common ones are:

1. The model is inadequate.
2. The results predicted from the experiment are not reproducible.
3. Many trials failed.
4. Important conditions, often an optimum, lie outside the experimental region.

One cure for several of these experimental shortcomings is to augment the design with some further trials. The remainder of the chapter discusses the addition of extra trials to an existing design. But first we discuss the four possibilities in greater detail.

Inadequacies of the model should be revealed during the analysis of the data using the methods described in Chapter 8. If the model is inadequate, the investigated relationship may be more complicated than was expected. Systematic departures from the model are often detected by plots of residuals against explanatory variables and by the use of added and constructed variable plots. These can suggest the inclusion of higher-order polynomial terms in the model, whereas systematic trends in the magnitude of the residuals may suggest the need for transformations (Atkinson 1985; Atkinson and Riani 2000, Chapter 4). Other patterns in the residuals may be traced to the effect of omitted or ignored explanatory variables. Examples of the latter, sometimes called 'lurking' variables, are batches of raw materials or reagents, different operators or apparatus, and trends in experimental conditions such as ambient temperature or humidity. These should properly have been included in the experiment as blocking factors or as concomitant observations (Chapter 15). Adjustment for these variables after the experiment may be possible. However, there may be some loss of efficiency in estimation of the parameters of interest. When the design becomes far from orthogonal, it may not be possible to distinguish the effects of some factors from those of the omitted variables.

Another set of possibilities that may be suggested by the data is that the ranges of some factors are wrong. Excessive changes in the response might suggest smaller ranges for some variables in the next part of the experiment, whereas failure to observe an effect for a variable known to be important would suggest that a larger range be taken. Both the revised experimental region that these changes imply and the augmented model following from the discovery of specific systematic inadequacies suggest the design of a new experiment. For this, the decisions taken at each of the stages in §3.2 should be reconsidered. When the new experiment is an augmentation of the first one, the methods of this chapter apply.

The situation is different when a model, believed to be adequate, fails correctly to predict the results of new experiments. This may arise because of systematic differences between the new and old observations, for example an unsuspected blocking factor or other lurking variable. Or, particularly for experiments involving many factors, it may be due to the biases introduced in the process of model selection; it is frequent that models provide appreciably better predictions for the data to which they are fitted than they do for independent sets of data (see, for example, Miller 2002). A third reason for poor predictions from an apparently satisfactory model is that the experimental design may not permit stringent testing of the assumed model. Parsimonious designs for model checking are the subject of §§20.2 – 20.5.

If many individual trials fail there may not be sufficient data to estimate the parameters of the model. It is natural to think of repeating these failed trials. It is important to find out if they are missing because of some technical mishap or whether there is something more fundamentally amiss. Technical mishaps could include accidentally broken or randomly failing apparatus, or a failure of communication in having the correct experiment performed. In such cases the missing trials can be completed, perhaps with augmentation or modification due to anything that has been learnt from the analysis of the complete part of the experiment. On the other hand, the failure may be due to unsuspected features of the system being investigated. For example, the failed trials may all lie in a definable subregion of $\mathcal{X}$, in which case the design region should be redefined, perhaps by the introduction of constraints. Example 19.3 in §19.3 illustrates the construction of optimum designs for non-regular design regions generated by constraints on $\mathcal{X}$.

Finally, the aim of the experiment may be to define an optimum of the response or of some performance characteristic. If this seems to lie appreciably outside the present experimental region, experimental confirmation of this prediction will be necessary. The strategy of §3.3, together with the design augmentation of the next section, provide methods for moving towards the true optimum.

'Failure of an Experiment' is probably too pessimistic a title for this section. That an experiment has failed to achieve all the intended goals does not constitute complete failure. Some information will surely have been obtained that will lead either to abandonment of the project before further resources are wasted or to the planning of a further stage in the experimental programme. As was emphasized in Chapter 3, an experimental study may involve several stages as the solution to the problem is approached. Information gathered at one stage should be carefully incorporated in planning the next stage.

19.2 Design Augmentation and Equivalence Theory

19.2.1 Design Augmentation

We now consider the augmentation of a design by the addition of a specified number of new trials. Examples given in §19.1 which lead to augmentation include the need for a higher-order model than can currently be fitted to the data, a different design region, or the introduction of a new factor. In general, we wish to design an experiment incorporating existing data. The new design will depend on the trials for which the response is known, although not usually on the values of the responses. An exception is for non-linear models, where the sequential design scheme of §17.7 involved re-estimation of the parameters as each new observation was obtained. In this chapter interest is in augmentation by N observations, $N > 1$. The case $N = 1$ corresponds to one step in a sequential design construction, illustrated for D-optimality in §11.2.

Augmentation of a design of size N_0 to one of size $N + N_0$ can use any of the criteria described in Chapter 10. We illustrate only D-optimality, when the algorithms of Chapter 12 can be used. There are however two technical details that require attention. If the design region has changed, the old design points have, of course, to be rescaled using (2.1). These points are then not available for exchange by the algorithm. Second, if the model has been augmented to contain $p > p_0$ parameters, the design for N_0 trials may be singular even for $N_0 > p$.

19.2.2 Equivalence Theory

We first consider the augmentation of a design in which there are N_0 existing or prior observations

$$\xi_0 = \left\{ \begin{array}{ccc} x_1^0 & \dots & x_q^0 \\ w_1^0 & \dots & w_q^0 \end{array} \right\} \tag{19.1}$$

with information matrix $N_0 M_0$ where

$$M_0 = \sum_{i=1}^{q} w_i^0 f(x_i^0) f^{\mathrm{T}}(x_i^0). \tag{19.2}$$

In (19.1) the weights w_i^0 are therefore multiples of $1/N_0$.

The new information in the experiment comes from an N-trial design ξ with information matrix in the usual form $NM(\xi)$. Combining the previous or prior information with that from the experiment yields the posterior information matrix

$$\widetilde{M}(\xi) = N_0 M_0 + NM(\xi). \tag{19.3}$$

The D-optimum designs with which we are concerned maximize $|\widetilde{M}(\xi)|$.

We can find either exact designs, for which the design weights w_i in (19.3) are multiples of $1/N$ or continuous designs in which the weights are not so constrained. In this case it may seem forced to talk of augmentation with N trials and we introduce weights

$$\alpha = \frac{N}{N_0 + N} \quad \text{and} \quad 1 - \alpha = \frac{N_0}{N_0 + N}.$$

For stating the equivalence theorem we then use the normalized information matrix for a continuous design ξ

$$M_\alpha(\xi) = (1 - \alpha) M_0 + \alpha M(\xi). \tag{19.4}$$

Maximizing $\Phi\{M_\alpha(\xi)\}$ for given α is equivalent to maximizing $\Phi\{\widetilde{M}(\xi)\}$ (19.3) for given N_0 and N.

We can now state the Equivalence Theorem for continuous D-optimum augmentation designs ξ^*. From Theorem 11.6 and Lemma 6.16 of Pukelsheim (1993), it follows that ξ^* is D-optimum if

$$f^{\mathrm{T}}(x) \left\{M_\alpha(\xi^*)\right\}^{-1} f(x) \leq \mathrm{tr}\left[M(\xi^*) \left\{M_\alpha(\xi^*)\right\}^{-1}\right] \tag{19.5}$$

for all $x \in \mathcal{X}$, with equality at the support points of ξ^*. For this optimum design

$$d_\alpha(x, \xi^*) = \alpha f^{\mathrm{T}}(x) \left\{M_\alpha(\xi^*)\right\}^{-1} f(x) + (1 - \alpha)\mathrm{tr}\left[M_0 \left\{M_\alpha(\xi^*)\right\}^{-1}\right] \leq p, \tag{19.6}$$

where p is the number of parameters in the model for the augmentation design, that is the dimension of the information matrix $M_\alpha(\xi)$.

A useful re-expression of the condition for the Equivalence Theorem is obtained by substituting M_0 from (19.2) in (19.6) when

$$d_\alpha(x, \xi^*) = \alpha f^{\mathrm{T}}(x) \{M_\alpha(\xi^*)\}^{-1} f(x) + (1-\alpha) \sum_{i=1}^{q} w_i^0 f^{\mathrm{T}}(x_i^0) \{M_\alpha(\xi^*)\}^{-1} f(x_i^0) \leq p. \qquad (19.7)$$

This condition for the optimum design has the informative statistical interpretation

$$d_\alpha(x, \xi^*) = (N/\sigma^2) \text{ var } \{\hat{y}(x)\} + (N_0/\sigma^2) \sum_{i=1}^{q} w_i^0 \text{ var } \{\hat{y}(x_i^0)\} \leq p.$$

The first variance term is the posterior variance at a point in $\mathcal{X}$ and the second a weighted sum of posterior variances at the points of the prior design. If the initial design is D-optimum, the standardized posterior variances in (19.7) are all equal to p, so that the optimum augmentation design is the D-optimum design in the absence of prior information, that is a replicate of the prior design. Usually this will not be the case. As we show, the augmentation design can be very different from the D-optimum design found in the absence of prior information.

We find D-optimum designs for one, or several, values of α. An advantage of the formulation (19.4) is that we find continuous designs which, unlike the exact design of Example 19.1 below, can be checked using the Equivalence Theorem. We can also see how the structure of the optimum designs changes with α. For $N = 1$, that is $\alpha = 1/(N_0 + 1)$, the augmentation is, as stated above, equivalent to one step in the sequential construction of optimum designs illustrated in §11.2. The sequential addition of one trial at a time is thus the same as the sequential algorithm for the construction of D-optimum designs of Wynn (1970). Such designs are optimum as $N \to \infty$ but may be far from optimum for small N. Example 19.2 illustrates this point.

19.3 Examples of Design Augmentation

We start with an example of the numerical calculation of an exact design, as described in §19.2.1 without reference to equivalence theory.

Example 19.1. Augmentation of a Second-order Design to Third Order A second-order model is fitted to the results of a 3^2 factorial. For this design and model $p = 6$ and $N_0 = 9$, so that the model can be tested for adequacy.

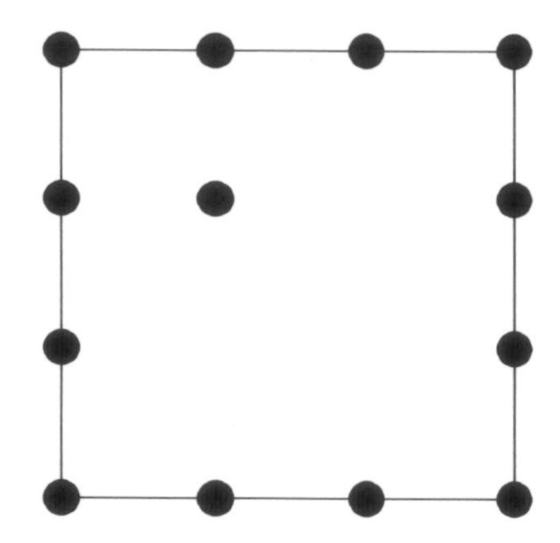

FIG. 19.1. Example 19.1: 13-trial D-optimum third-order design found by searching over the points of the 4^2 factorial.

We leave to Chapter 20 a discussion of efficient designs for testing goodness of fit.

Suppose that the test shows the model to be inadequate and we would like to extend the experiment so that a third-order model can be fitted. The model is thus augmented by the inclusion of terms in x_1^3, $x_1^2x_2$, $x_1x_2^2$, and x_2^3. For illustration we compare two strategies for 13 trial designs, leaving to later a discussion of whether augmentation with $N = 4$ is a good choice.

One possibility is to start again with a D-optimum design for the third-order model. This will require trials at four values of each x. Figure 19.1 shows a 13-trial D-optimum exact design for the cubic model found by searching over the points of the 4^2 factorial with values of $x_i = -1, -1/3, 1/3,$ and 1. Only four trials of this design, those of the 2^2 factorial, coincide with those of the original design from which the data were collected. Thus nine new trials would be indicated. The other possibility is to augment the existing design by the addition of four further points, bringing $N + N_0$ up to 13. Figure 19.2 shows that the second-order design is augmented by the addition of four symmetrically disposed points to yield a 13-trial third-order design. The D-efficiency of the augmented design is 89.7% relative to the 13-trial design of Figure 19.1. In this example, proceeding in two stages and using design augmentation, has resulted in a loss of efficiency equivalent to the information in just over one trial, more precisely 10.3% of 13 trials. The saving is that only four new trials are needed instead of the nine introduced by the design of Figure 19.1.

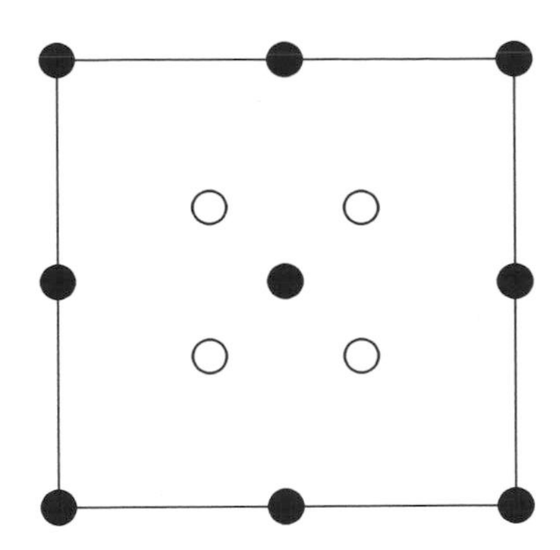

FIG. 19.2. Example 19.1: augmentation of a second-order design to a 13-trial third-order design: • original second-order design; ∘ additional trials. The symmetric structure suggests that $N = 4$ is an efficient choice of augmentation size.

Two general principles are raised by this example. The less general is that third-order models are usually not necessary in practice. Transformation of the response, the subject of Chapter 23, is often a more parsimonious and satisfactory elaboration of the model. More general is that augmentation of the design leads to two groups of experiments run at different times, which it may be prudent to treat as coming from different blocks. The augmentation procedure can be extended straightforwardly to handle the blocking variables of Chapter 15. However, in the present example, the design was unchanged by the inclusion of this extra parameter.

The procedure also raises questions about the design that it is hard to answer in the absence of application of the theory of §19.2.2. In particular, a plot of $d_\alpha(x, \xi)$ (19.7) over $\mathcal{X}$ for $\alpha = 4/13$ would indicate whether the design of Figure 19.2 could be appreciably improved by moving away from the points of the 4^2 factorial with $x_i = -1, -1/3, 1/3$, and 1 and might suggest good exact designs. Similar plots for other values of α, such as 3/12 and 5/14 would likewise indicate the support of exact designs for $N =$ 3 and 5. Comparison of the values of $|\widetilde{M}(\xi)|$ for these values of α would indicate whether appreciably greater information, on a per trial basis, can be obtained by augmenting the design with a different value of N. This seems unlikely given the symmetry of the augmented design in Figure 19.2. However, it does seem likely that the 13-trial design of Figure 19.1, which lacks symmetry, has a low efficiency relative to the D-optimum continuous

TABLE 19.1. Response surface—regular design region: augmentation of the second-order design of Figure 19.3 ($N_0 = 6$). Continuous optimum designs and integer approximations $[Nw_i]$ are given for each value of α. The last two columns give the optimum design for a second-order response surface and an integer approximation

		$\alpha = 1/3$		$\alpha = 3/5$		$\alpha = 13/19$		$\alpha = 19/25$		$\alpha = 1$	
x_1	x_2	w	$[3w]$	w	$[9w]$	w	$[13w]$	w	$[19w]$	w	$[13w]$
−1	−1	—	—	0.082	1	0.107	1	0.122	2	0.146	2
0	−1	—	—	—	—	—	—	—	0	0.080	1
1	−1	0.290	1	0.227	2	0.209	3	0.194	4	0.146	2
1	−0.1	0.056	0	—	—	—	—	—	—	—	—
−1	0	—	—	0.020	0	0.046	1	0.059	1	0.080	1
0	0	—	—	—	—	0.004	0	0.036	1	0.096	1
1	0	—	—	0.114	1	0.112	1	0.104	2	0.080	1
−1	1	0.284	1	0.223	2	0.203	3	0.187	4	0.146	2
−0.1	1	0.067	0	—	—	—	—	—	—	—	—
0	1	—	—	0.109	1	0.113	1	0.110	2	0.080	1
1	1	0.303	1	0.225	2	0.205	3	0.188	4	0.146	2

design for the third-order model. Section 11.5 discusses designs for second-order models for general m; these have a symmetric structure. The highly symmetric design for the cubic model for $m = 2$ is in §3.2 of Farrell *et al.* (1968). ■

Example 19.2. Second-order Response Surface: Augmentation of Design Region As a first example of the use of the equivalence theorem of §19.2.2 we continue with designs for the second-order polynomial in two variables over the square design region $\mathcal{X} = \{-1 \leq x_1 \leq 1, -1 \leq x_2 \leq 1\}$. But now we suppose that the initial six-trial design was concentrated in the lowest quarter of the experimental region. It is still required to fit the six-parameter second-order model. The augmentation will then provide points that span the whole region.

The D-optimum design for the second-order model with $\mathcal{X}$ the unit square, when no prior information is available ($\alpha = 1$), is the well-known design supported on the points of the 3^2 factorial with weights as given in the penultimate column of Table 19.1. The good integer approximation to this design is one replicate of the full factorial with one extra replicate of each corner point, making 13 trials in all. It is given in the last column of Table 19.1.

The six-trial starting design, shown by circles in Figure 19.3 is far from this design. The six points are all in the lower left-hand quarter of the design region: no values of x_1 or of x_2 are greater than zero.

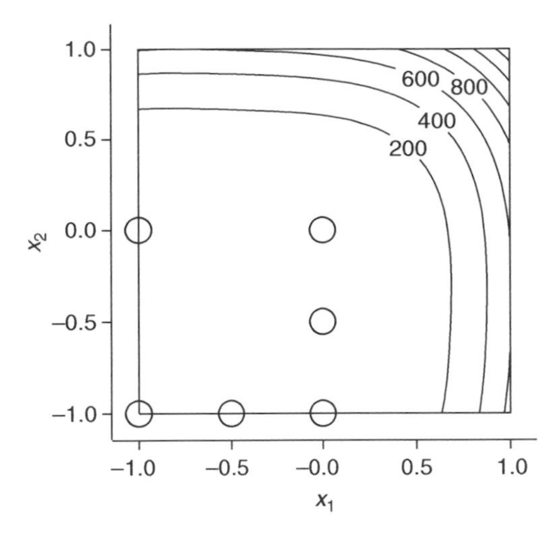

FIG. 19.3. Regular design region: the six points of the second-order design which is to be augmented, with contours of the standardized variance $d(x, \xi)$.

It is clear that any scheme for design augmentation will extend the design over the whole region. Since there are six points, the second-order model can be fitted. The contours of the variance function of the prediction from the initial design are included in Figure 19.3. The maximum value is at (1,1), the corner of the region most remote from the initial design. Augmentation one trial at a time, equivalent to the sequential construction of the D-optimum design one trial at a time, would add a trial at this point. The next point to be added would be at $(1, -0.9)$, not a point of the optimum second-order design. Such perpetuation of distortions introduced by the initial design is one of the drawbacks of the sequential approach. We instead find the optimum continuous design for a specified α and then calculate exact designs from approximations to our optimum continuous measures. In our response surface examples we search over grids of size 0.1 in x_1 and x_2.

Some optimum augmentation designs are given in Table 19.1 for a range of values of α, together with integer approximations. For $\alpha = 1/3$ the optimum measure ξ^* for the augmentation design has five points of support. The contours of the variance function $d_\alpha(x, \xi^*)$ (19.7) given in Figure 19.4 show maximum values of 6 at these five points, the corners most remote from the initial design, and at points near the centres of the remote sides of the region. The continuous design is indeed optimum. The weights on the three corners of the region at which there were no prior experiments

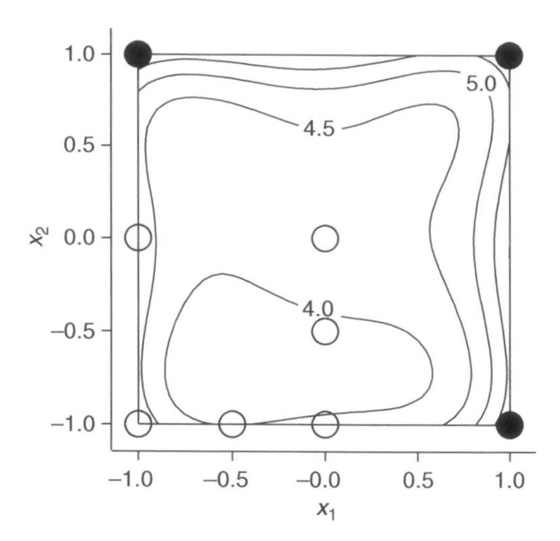

FIG. 19.4. Regular design region: augmentation of the second-order design. Contours of the standardized variance $d(x, \xi^*)$ for the optimum design for $N = 3$ of Table 19.1.

account for 88% of the total, and are nearly equal. When, as here, $N_0 = 6$ and $\alpha = N/(N_0 + N) = 1/3$, we require an augmentation design for $N = 3$. A good integer approximation for the three-trial design is to put one trial at each of the three corners of the region. This integer approximation is given in the fourth column of Table 19.1 with the number of replicates $r_i = [3w_i]$, the integer closest to $3w_i$. These three points are also plotted in Figure 19.4. Since two points with small weights in column 3 of the table were dropped from the design, the resulting nine-trial design is formed by an augmentation which is not quite the optimum continuous design ξ^*. As a result the variance for the nine-trial design at these dropped points is greater than 6.

The best integer approximation to the optimum continuous design for $\alpha = 1/3$ depends on the value of N_0. If N_0 were 60, rather than 6, then, for $\alpha = 1/3$ we would have $N = 30$. The appropriate integer design in column 4 of Table 19.1 would then have $r_i = [30w_i]$, so that all five design points would be included, the extra two with only two replicates.

The remaining columns of the table give continuous designs for $\alpha = 3/5$, 13/19, and 19/25 and exact designs with $r_i = [Nw_i]$. Larger values of α correspond to increasing importance of the N-trial augmentation design relative to the prior design of N_0 trials. The designs should therefore tend towards the unequally weighted 3^2 factorial discussed earlier. And, indeed, the continuous designs in Table 19.1 for increasing α have seven, then eight points of support, all at the points of the 3^2 factorial. For $\alpha = 3/5$ a good integer approximation has six unequally replicated points of support whereas, for $\alpha = 13/25$, the discrete design formed by rounding Nw_i has seven support points, and for $\alpha = 19/25$, it has eight points. However, note that for $\alpha = 19/25$, this integer approximation does not lead to a design with $N = 19$. This points up the fact that in practical situations, for exact prior and augmented designs, tools for exact augmentation should be employed, as discussed in §§13.5.1 and 19.5.

The design weights in Table 19.1 for the points of the 3^2 factorial also show a smooth progression from $\alpha = 1/3$ to $\alpha = 1$. The weights for the three corner points included when $\alpha = 1/3$ decrease steadily to the final values while the other weights increase with α once their support points have been included in the design. ■

Example 19.3. Second-order Response Surface: Constrained Design Region
We now return to the problem of Example 19.1, that of augmenting a design to allow fitting of a higher-order model. But in this example we use the equivalence theory of §19.2.2 to augment a first-order design for a second-order model. In addition, we use an irregular design region that confounds intuition as to what a good design might be.

Background. Often, in chemical or biological experiments, high levels of all factors can lead to conditions which are so severe that inorganic molecules decompose or plants wither. Avoidance of such conditions leads to a constrained design region as in Example 12.2 where unsatisfactory conditions for running an internal combustion engine were avoided by use of a pentagonal design region. Since some of the structure of the preceding augmentation designs depended on the symmetries of the design region, we now consider an example of this kind in which constraints make the design region less regular.

With x_1 and x_2 both scaled so that $-1 \leq x_i \leq 1$, $i = 1, 2$, we add the linear constraints that

$$2x_1 + x_2 \leq 1$$

$$x_1 + x_2 \geq -1$$

$$x_2 - x_1 \leq 1.5.$$

The resulting irregularly hexagonal design region $\mathcal{X}$ is shown in Figure 19.5.

To get a feel for the variety of designs that might be encountered we calculate the D-optimum designs for the first- and second-order models when no prior information is available. The resulting designs are given in Table 19.2 and Figure 19.5. Given an appropriate choice of the N_0 trials of the initial design, the augmentation designs will lie between these two D-optimum designs. The first-order design has three points of support with

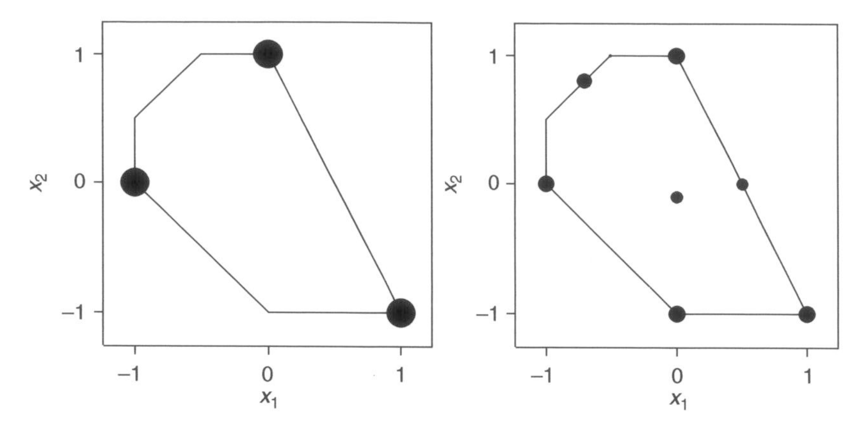

FIG. 19.5. Constrained design region: D-optimum designs of Table 19.2: (a) first-order ($\alpha = 0$) and (b) second-order ($\alpha = 1$). Dot diameter $\propto w_i^{0.8}$.

TABLE 19.2. Second-order response surface—constrained design region: D-optimum first-order and second-order designs

x_1	x_2	First–order w	Second–order designs 8 points w	Exact: $N = 19$ $[19w]$
0	−1	–	0.1595	3
1	−1	1/3	0.1626	3
0	−0.1	–	0.1010	2
−1	0	1/3	0.1648	3
0.5	0	–	0.1010	2
−0.7	0.8	–	0.1479	3
−0.5	1	–	0.0061	0
0	1	1/3	0.1571	3
D-efficiencies			100%	99.95%

weights 1/3. To check the optimality of the design, the variance of the prediction was calculated over the hexagonal region. Not only was the value of the variance three at the design points, it was also three at (0,−1); however this point had a weight of zero and so was not included in the optimum design.

The optimum design for the second-order model has eight points of support; see Table 19.2 and Figure 19.5(b). In order to give some visual impression of the design weights, the diameter of the dots in Figure 19.5 (and also in Figure 19.6) is proportional to $w_i^{0.8}$. The weight for the second-order design on (−0.5, 1) is a negligible 0.0061, and so is hardly visible in the figure. If this point is dropped, the seven-point design for the six-parameter model has weights in the approximate ratio 3:2 and can be well approximated by the 19-trial integer design of Table 19.2. The D-efficiency of this exact design, relative to the eight-point continuous design, is 99.95%.

Design Augmentation. We now use the theory of §19.2.2 for design augmentation starting from a design for model checking derived in Chapter 20. A good approximation to the continuous optimum design for model checking with four support points is given in Table 19.3. This ten-point design has three trials at each of the points of the first-order design of Table 19.2 and one at the fourth support point $(0, -1)$ at which the variance $d(x, \xi^*)$ for the first-order model was equal to three. Suppose that, as a result of this experiment, it seems that a second-order model is needed and so the design is to be augmented. Suppose also that a further five trials are required. If we take as the support points those of the seven-point optimum second-order design of Table 19.2, we find that trials are required at only four support points, those that are not part of the first-order design. The resulting design is in Table 19.3 and Figure 19.6. That this design is optimum is checked by calculating $d_\alpha(x, \xi^*)$, not only at the design points but also over a grid of points in $\mathcal{X}$. The variance is indeed 6 at the design points and less elsewhere. The five trials can be approximated by weights proportional to 1,1,1, and 2 at the support points, as shown in the table. This augmented design has a D-efficiency of 97.224% relative to the eight-point continuous optimum design for the second-order model of Table 19.2.

If, instead, nine trials are to be added, the optimum weights are as shown in the last columns of Table 19.3. Now the weights on the three points of the first-order design, previously exactly zero, are close to that value, with a maximum of 1.46%. An integer approximation to the augmentation design has either two or three trials at each of the other points of support of the second-order design. The combination of this design with the initial design gives the approximation to the second-order non-augmentation design of Table 19.2. However, the continuous augmentation design found in the table is not quite D-optimum for the second-order model: $d\{(-0.5, 1), \xi^*\} = 6.0292$, a further reminder that the optimum design of Table 19.2 has eight, not seven, points of support.

In this example the designs found by augmentation of the model-checking design have efficiencies that are perhaps surprisingly high. These arise because the initial design has many points of support in common with the

TABLE 19.3. Second-order response surface—constrained design region: augmentation of design for checking first-order model

x_1	x_2	Check first-order	$\alpha = 0.333$			$\alpha = 0.474$		
			w	$[5w]$	Total	w	$[9w]$	Total
0	−1	1	0.1757	1	2	0.2254	2	3
1	−1	3	—	—	3	0.0097	0	3
0	−0.1	—	0.2309	1	1	0.2130	2	2
−1	0	3	—	—	3	0.0143	0	3
0.5	0	—	0.2335	1	1	0.2131	2	2
−0.7	0.8	—	0.3600	2	2	0.3130	3	3
−0.5	1	—	—	—	—	—	—	—
0	1	3	—	—	3	0.0116	0	3
D-efficiencies					97.224%			99.95%

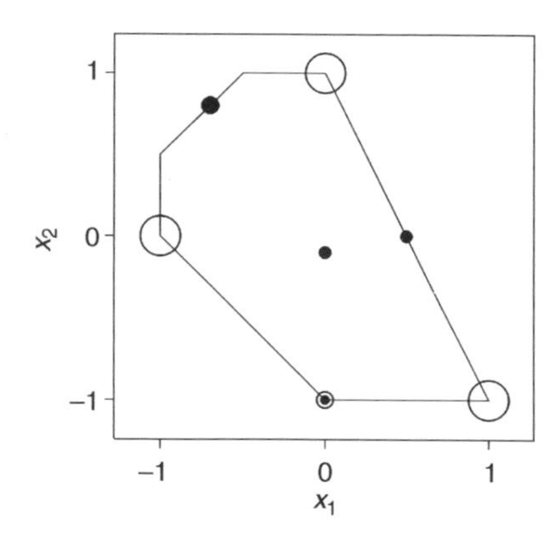

FIG. 19.6. Exact augmentation design of Table 19.3 for $N = 5$: ○ original points, • augmentation.

optimum design for the second-order model. The situation is different in Example 19.2 where two of the support points of the initial design are not present in the D-optimum design for the second-order model; the efficiencies of augmentation designs for comparable values of α are accordingly lower. ■

19.4 Exact Optimum Design Augmentation

It is worth noting that the exchange algorithms of Chapter 12 for exact D-optimum design construction can be applied to the design augmentation problem almost without change. Recall that all of these algorithms search for an optimum design by moving sequentially from design to design by the addition or deletion of points, updating the information matrix as they go. The same sequential approaches can be applied to optimum design augmentation by simply initializing the information matrix to $M(N_0)$ and adding $N - N_0$ more points. Sequential exchange proceeds from that point, except that only the additional points after the first N_0 are considered for deletion.

As noted in §19.2, the information matrix M_0 for the prior design used in applying the theory of §19.2.2 needs to be computed using the augmented design's model, with the same scaling as is applied to the region of interest $\mathcal{X}$ for the augmented design. These considerations are handled by default when you use SAS to augment designs optimally, as presented in the next section.

19.5 Design Augmentation in SAS

As discussed in §13.5.1, the OPTEX procedure finds exact optimum augmented designs through the use of the `AUGMENT=` option. The argument for this option is a data set containing the prior design points. Given this data set, OPTEX takes care of scaling its factors in the same way as with the candidate points, and applying the model for the augmented design.

To demonstrate, consider Example 19.1, the augmentation of a second-order design to enable fitting a third-order model. The following code creates the 3^2 second-order factorial design in a data set named `Prior` and the candidates for augmenting this design in a data set named `Candidates`. A variable named `Source` is added to each data set to distinguish its points in the resulting augmented design.

```
data Prior;
   do x1 = -1,0,1;
      do x2 = -1,0,1;
         Source = "Prior      ";
         output;
         end;
      end;
```

```
data Candidates;
   do x1 = -1,-0.333,0.333,1;
      do x2 = -1,-0.333,0.333,1;
         Source = "Candidates";
         output;
         end;
      end;
run;
```

The following statements find a 13-point design D-optimum for the third-order model, as in Figure 19.1.

```
proc optex data=Candidates;
   model x1 x2
         x1*x1 x1*x2 x2*x2
         x1*x1*x1 x1*x1*x2 x1*x2*x2 x2*x2*x2;
   generate n=13 method=m_fedorov niter=1000 keep=10;
   output out=DOptimumDesign;
run;
```

In order to find a 13-point design by augmenting the 3^2 factorial, the only change required is to name this design as the argument to the AUGMENT= option, as in the following statements.

```
proc optex data=Candidates;
   model x1 x2
         x1*x1 x1*x2 x2*x2
         x1*x1*x1 x1*x1*x2 x1*x2*x2 x2*x2*x2;
   generate n=13 method=m_fedorov niter=1000 keep=10 augment=Prior;
   id Source;
   output out=AugmentedDesign;
run;
```

These statements produce the design shown in Figure 19.2.

SAS Task 19.1. For the example above, use the Source variable to demonstrate, as mentioned in §19.2.2, that modelling the augmenting points as coming from a second block does not change which four points are selected.

SAS Task 19.2. Use OPTEX to find exact augmented designs of size 9, 15, 19, and 25 for Example 19.2, the augmented design region. Compare your answers to the approximate designs computed by rounding multiples of the continuous optimum augmented designs, shown in Table 19.1.

SAS Task 19.3. Use OPTEX to explore the questions raised in the final paragraph of Example 19.1. In particular, compare the supports for exact augmented designs of size 12, 13, and 14; and decide whether a particular number of augmenting points provides appreciably greater information, on a per trial basis.

19.6 Further Reading

The D-optimum designs for design augmentation are described by Dykstra (1971*a*), who considers the sequential addition of one trial at a time. He later comments (Dykstra 1971*b*), as we do in §19.2, that this method is equivalent to the sequential algorithm for the construction of D-optimum designs of Wynn (1970). Evans (1979) finds exact optimum designs for augmentation with specified N. Heiberger, Bhaumik, and Holland (1993) also calculate exact optimum augmentation designs, but for a large and flexible family of criteria, which includes D-optimality. Since these papers describe exact designs, they do not use equivalence theory; derivations and proofs of the results of §19.2.2 are given by Atkinson, Bogacka, and Zocchi (2000).

20

MODEL CHECKING AND DESIGNS FOR DISCRIMINATING BETWEEN MODELS

20.1 Introduction

So far we have assumed that we know the model generating the data. Although this 'known' model may be quite general, for example a second-order polynomial in m factors, any of the terms in the model may be needed to explain the data. Experiments have therefore been designed to allow estimation of all the parameters of the model. In this chapter we find design for two related problems when there is some uncertainty about the model. We begin in §20.2 with a form of Bayesian D-optimum design for parsimonious model checking; typically we want to know whether a simple model is adequate, or whether we need to include at least some of a set of specified further terms. The second part of the chapter begins in §20.6 when we introduce T-optimality for the choice between two or more models; initially these can be non-linear, with neither a special case of the other. In §20.9 we consider the special, but important, case of partially nested linear models and make some comparisons between T- and D_S-optimality for discrimination between nested models.

20.2 Parsimonious Model Checking

20.2.1 General

In the description of the sequential nature of many experiments in §3.3 we suggested augmenting 2^m factorials with a few centre points in order to provide a check of the first-order model against second-order terms. It is not obvious how to extend this procedure to more complicated models and design regions. Further, although it seems intuitively sensible to add centre points to the factorial design, it is not clear how many such points should be included. In this chapter we use the Bayesian method of DuMouchel and Jones (1994) to provide a flexible family of designs for parsimonious model checking and give examples of its application in three situations of increasing complexity. Because the procedure incorporates prior information the algebra is similar to that of Chapter 19 and we are again able to provide an equivalence theorem. This makes clear the non-optimum properties for model checking of the addition of several centre points to factorial designs and provides a quantitative measure of that non-optimality. More importantly, we provide numerical procedures for the construction of model checking designs.

In the formulation followed here, the terms in the model are divided into two sets: the *primary* terms form the model to be checked while the *secondary* terms include those which may have to be included after checking. The method assumes that there is no prior information about the primary terms; the values of the coefficients of these terms are to be determined solely from the experiment. However, some prior information is available about the coefficients of the secondary terms; increasing this prior information has the effect of increasing knowledge about the secondary terms and so reducing the proportion of the experimental effort that is directed towards their estimation. The model checking designs depend on the relative importance of this prior information through the parameter α. For consistency with Chapter 19, we let $1 - \alpha$ reflect the strength of the prior information about the secondary terms. As $\alpha \to 0$, there is appreciable information about the secondary terms and the design tends to that for the primary terms only. Conversely, as $\alpha \to 1$, information on the secondary terms decreases and the design tends to the non-parsimonious design from which all terms can be estimated.

The formulation of the design problem, leading to an equivalence theorem, is in §20.2.2. Because of the parameterization in terms of α, the algebra is similar to that in §19.2.2. However, in that section the prior information came from a previous experiment involving N_0 trials, whereas here the prior information is merely a conceptual device for reducing experimental effort directed towards estimating the secondary terms. Some details of implementation are in §20.2.3. Three examples are in §20.3; we start with the motivating example of this section—checking a first-order model in two factors. Then we continue with Example 19.3, a first-order model over a constrained design region that is to be checked for the presence of second-order terms. Finally we consider a complicated non-linear model; complicated in part because the design depends upon the parameters of the model.

20.2.2 Formulation of Prior Information

We partition the terms in the linear model (or a linear approximation of a non-linear model) into the two groups

$$\mathrm{E}(y) = \theta_r^{\mathrm{T}} f_r(x) + \theta_s^{\mathrm{T}} f_s(x), \tag{20.1}$$

where θ_r is the vector of r primary parameters and θ_s is the vector of s secondary parameters. As usual, $f_r(x)$ and $f_s(x)$ denote vectors of functions of experimental conditions. The terms in $f_r(x)$ are those that it is believed are required in the model, for example first-order terms in a polynomial model. However, as well as designing the experiment to estimate θ_r, it is also required to check that none of the terms of $f_s(x)$ are required. These will typically be higher-order polynomial terms. The parsimonious design problem is to find a design which allows such checking without necessarily allowing estimation of all $r + s$ parameters.

In the Bayesian formulation of DuMouchel and Jones (1994) the absence of specific prior information about the primary parameters θ_r is represented by using a diffuse prior distribution. Let θ_r be normally distributed,

$$\theta_r \sim \mathcal{N}_r(\theta_r^0, \gamma^2 I_r),$$

where interest is in the limit as $\gamma^2 \longrightarrow \infty$. However, we assume that there is some prior information for the secondary parameters, which have distribution

$$\theta_s \sim \mathcal{N}_s(0_s, \tau^2 I_s), \tag{20.2}$$

independently of the distribution of θ_r, where τ^2 is a small positive value. Then the joint prior distribution of all the $p = r + s$ parameters is

$$\theta \sim \mathcal{N}_p\left(\begin{bmatrix} \theta_r^0 \\ 0_s \end{bmatrix}, \begin{bmatrix} \gamma^2 I_r & 0_{r\times s} \\ 0_{s\times r} & \tau^2 I_s \end{bmatrix}\right) = \mathcal{N}_p\left(\begin{bmatrix} \theta_r^0 \\ 0_s \end{bmatrix}, D(\theta)\right). \tag{20.3}$$

To design experiments we require the prior information matrix for θ, that is we require the inverse of the dispersion matrix $D(\theta)$. As $\gamma^2 \longrightarrow \infty$

$$\{D(\theta)\}^{-1} \longrightarrow \frac{1}{\tau^2} K,$$

where

$$K = \begin{pmatrix} 0_{r\times r} & 0_{r\times s} \\ 0_{s\times r} & I_s \end{pmatrix}.$$

Hence the posterior information matrix for θ, given ξ, is

$$\widetilde{M}(\xi) = N_0 K + N M(\xi), \tag{20.4}$$

where $N_0 = \sigma^2/\tau^2$. As we did for design augmentation in §19.2.2, we calculate D-optimum designs using the normalized form of the information matrix

$$M_\alpha(\xi) = (1-\alpha)K + \alpha M(\xi).$$

But now α can be expressed in terms of N, τ^2, and σ^2 as

$$\alpha = \frac{N}{N_0 + N} = \frac{N\tau^2}{\sigma^2 + N\tau^2}. \tag{20.5}$$

Since the variance of the observations, σ^2, is constant, increasing values of τ^2, which mean less precise prior information about the secondary parameters, lead to larger values of α. As $\alpha \to 1$ the design tends to the D-optimum design when the model with all $r+s$ parameters is of interest. Conversely, decreasing α implies more prior knowledge about the secondary terms. As $\alpha \to 0$ the design tends to the D-optimum design for the model with just r parameters.

We can now state an equivalence theorem that parallels that of §19.2.2 with $p = r+s$ and $M_0 = K$. If ξ^* is the D-optimum design maximizing $|M_\alpha(\xi)|$,

$$d_\alpha(x,\xi^*) = \alpha f^{\mathrm{T}}(x)\left\{M_\alpha(\xi^*)\right\}^{-1} f(x) + (1-\alpha)\mathrm{tr}\left[K\left\{M_\alpha(\xi^*)\right\}^{-1}\right] \leq r+s, \tag{20.6}$$

where $r+s$ is the total number of primary and secondary parameters in the model, that is the dimension of the information matrix $M_\alpha(\xi)$.

We can find a simple expression for the function $d(x,\xi^*)$ in (20.6). Let

$$(M_\alpha^*)^{-1} = \{(m_\alpha^*)^{ij}\}_{i,j=1,\ldots,r+s},$$

when

$$\mathrm{tr}K(M_\alpha^*)^{-1} = \sum_{j=r+1}^{r+s}(m_\alpha^*)^{jj}$$

and

$$d(x,\xi^*) = \alpha f^{\mathrm{T}}(x)\left(M_\alpha^*\right)^{-1} f(x) + (1-\alpha)\sum_{j=r+1}^{r+s}(m_\alpha^*)^{jj} \leq r+s, \tag{20.7}$$

which is easily calculated.

20.2.3 Implementation

In the formulation in §20.2.2 it was stated that there was a defined amount of prior information about the secondary terms, but none about the primary terms. To apply this specification requires an easily calculated transformation of the terms of the model.

Since information is available on only one group of terms, the secondary terms should be orthogonal to the primary ones. The prior information for the two groups of terms then acts independently. This orthogonality is achieved by regressing each secondary term on the primary terms. The regression is over a list of candidate points in $\mathcal{X}$. The residuals of the secondary terms are then scaled and used in the construction of the design. The details, for the more complicated case of a non-linear model, are in §20.4.2. Since the design uses only a few of the candidate points, exact orthogonality for each design is obtained by weighted regression over the support points of the design, using the design weights. The residuals then have to be updated for each design. Atkinson and Zocchi (1998) find little difference between the designs obtained from such a procedure and those using approximate orthogonality from regressing over the candidate points. Here we regress over the candidate points.

A second detail of implementation is that DuMouchel and Jones (1994) suggest two scalings of the model terms. The first is that of the primary terms. However, D-optimum designs are invariant to this scaling, which can therefore be omitted.

The second scaling is of the residuals of the secondary terms after regression on the primary terms, so that they all have the same range, or the same variance. If the scales of the residuals of the secondary terms are very different, interpretation of τ^2 in (20.2) as a common variance is strained. In our response surface examples the range of the variables is similar and we do not make this adjustment, calculating designs using the unscaled residuals of the $f_s(x)$. However, we do use this scaling in our non-linear example in §20.4.2.

20.3 Examples of Designs for Model Checking

In this section we find parsimonious model-checking designs for two response surface examples from Chapter 19. An appreciably more complicated non-linear example is described in detail in §20.4.2.

20.3.1 Example 20.1: Second-Order Response Surface: Regular Design Region

We begin with the motivating example of §20.2.1 in which we want to check for second-order departures from a first-order model over a regular design region. We consider only two factors so the full model with $r+s$ parameters is

$$E(y) = \theta_1 + \theta_2 x_1 + \theta_3 x_2 + \theta_4 x_1 x_2 + \theta_5 x_1^2 + \theta_6 x_2^2 \qquad (20.8)$$

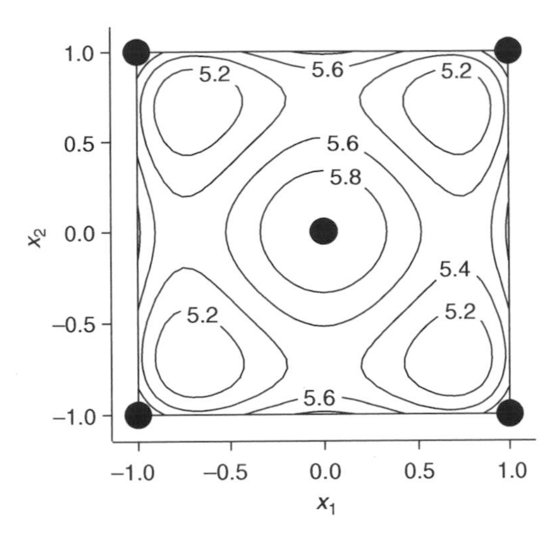

FIG. 20.1. Second-order response surface, regular design region: design of §20.3.1 for checking the first-order model for $\alpha = 2/3$ together with contours of the standardized variance $d(x, \xi^*)$.

over the square design region $\mathcal{X} = \{-1 \leq x_1 \leq 1, -1 \leq x_2 \leq 1\}$. The simplest problem in checking the model is to treat the three second-order terms as secondary, that is the two quadratic terms and the interaction. We thus divide the terms as

$$\begin{array}{rrcllll} & f_r^{\mathrm{T}}(x) & = & (1 & x_1 & x_2) \\ \text{and} & f_s^{\mathrm{T}}(x) & = & (x_1x_2 & x_1^2 & x_2^2), \end{array} \tag{20.9}$$

so that $r = s = 3$. Designs for this problem have a very simple structure. For small α, that is large prior information, the 2^2 factorial is optimum—since the prior is overwhelming, there is no need for experimental evidence to check the model. For larger α, the optimum design is the 2^2 factorial with some weight on a centre point. For example, if $\alpha = 2/3$ the optimum design has weight 0.2343 on each factorial point and the remaining 0.0629 on the centre point. The contours of the variance function for this design are plotted in Figure 20.1. This has a maximum of six at the five design points. But the plot also shows local maxima at the centres of the sides of the region. These are the four remaining points of the 3^2 factorial, at which $d(x, \xi^*) = 5.843$. If α is increased slightly to $8/11 = 0.7273$ this design is no longer optimum, since the variances at the centres of the sides are now 6.025: these points should be included in the design.

Exact designs for this problem are found using SAS in §20.5. For such exact designs it is natural to specify α (20.5) in terms of N and the fictitious number of observations N_0 representing the prior information about the secondary terms.

These designs have one expected feature—that the points of the 2^2 factorial plus centre point can be optimum for checking the first-order model against the second. But it is surprising that such a small weight goes on the centre point. For $\alpha = 0.7188$ the ratio of factorial weights to that for the centre point is $0.2276/0.0894 = 2.546$. For slightly higher values of α the five-point design is not optimal. This is very different from the customary advice on this problem where the factorial might be augmented by two or three centre points, giving a ratio of design weights of 0.5 or less. However, such advice may also incorporate the desire to obtain a rough estimate of the error variance.

20.3.2 Example 20.2: Second-Order Response Surface: Constrained Design Region

We now return to Example 19.3 in which an irregular design region made it difficult to guess efficient experimental designs.

The model is the same as that in §20.3.2, that is the full second-order polynomial in two factors (20.8) with the primary terms again the three first-order terms and secondary terms the second-order terms as in (20.8). In Example 19.3 we found designs to augment a 10-trial design for efficient estimation of (20.8). We now find parsimonious designs for checking whether the second-order terms are needed.

The constraints forming the design region are given in §19.3 with the resulting hexagonal design region plotted in Figure 19.5. Optimum continuous designs for the first- and second-order models are in Table 19.2. The D-optimum continuous design for the second-order model has eight points of support, although one support point has a very small weight. We want designs for checking the first-order model which require fewer than this number of design conditions.

When α is small, so that much of the information is coming from prior knowledge about the secondary terms, the first-order design is augmented by extra trials at the point $(0, -1)$, which was noted in §20.3.2 as having a high variance for the first-order design. Designs with these four points of support are optimum up to $\alpha = 0.318$. The weights are given in Table 20.1. For this value of α, the largest value of $d(x, \xi^*)$ at a point not in the design is 5.9923 at $(-1, 0.5)$. As α increases, this point is included in the design which now has five points. These design points are optimum, with weights changing with α, until $\alpha = 0.531$ when the maximum variance, apart from

TABLE 20.1. Second-order response surface—constrained design region: parsimonious designs for checking first-order model, showing the number of support points of the design increasing with α. Designs are given for values of α such that further increase will augment the design by x_{next}.

		$\alpha = 0.318$		$\alpha = 0.531$	$\alpha = 0.647$	$\alpha = 0.949$
x_1	x_2	w	$[10w]$	w	w	w
0	−1	0.1105	1	0.1691	0.1849	0.1893
1	−1	0.2871	3	0.2625	0.2344	0.1932
−1	0	0.2871	3	0.1893	0.1522	0.1403
0.5	0	—	—	—	0.0659	0.1660
−1	0.5	—	—	0.0919	0.1223	0.0786
−0.5	1	—	—	—	—	0.0989
0	1	0.3153	3	0.2871	0.2468	0.1337
Next to enter		(−1, 0.5)		(0.5, 0)	(−0.5, 1)	(0, −0.1)
$d(x_{\text{next}}, \xi^*)$		5.992		5.987	5.970	5.991

the values for points in the design, is $d\{(0.5, 0), \xi^*\} = 5.9871$. Unlike (−1, 0.5), this new point (0.5, 0) is one of the points of support of the optimum second-order design. Four model-checking designs are shown in Table 20.1 and in Figure 20.2. Each one arises from a value of α at which another support point almost needs to be added to the design measure. For each design the point about to be entered is given in the table, along with the value of $d(x, \xi^*)$, which is just less than six. The ability to determine which points will be included in the design if α increases modestly is a further useful application of the equivalence theorem. There is also a steady change in the design weights as α increases. Those for the three points of the first-order design decrease from their initial values of 1/3, whereas that for (−1, 0.5) increases and then decreases. In the limit for large α this point will not be present in the design.

It is interesting to continue the process of finding optimum designs for model checking for larger values of α. However, the design that is of most practical importance is the four-point design, which we used as the starting point for design augmentation in Chapter 19. An intriguing feature of this 10-point design is that all four points are on the edge of the design region, a design which is unlikely to be found by unaided intuition. The designs we have found for large α are optimum for the model-checking criterion. They, however, have seven points of support and so are no more parsimonious than

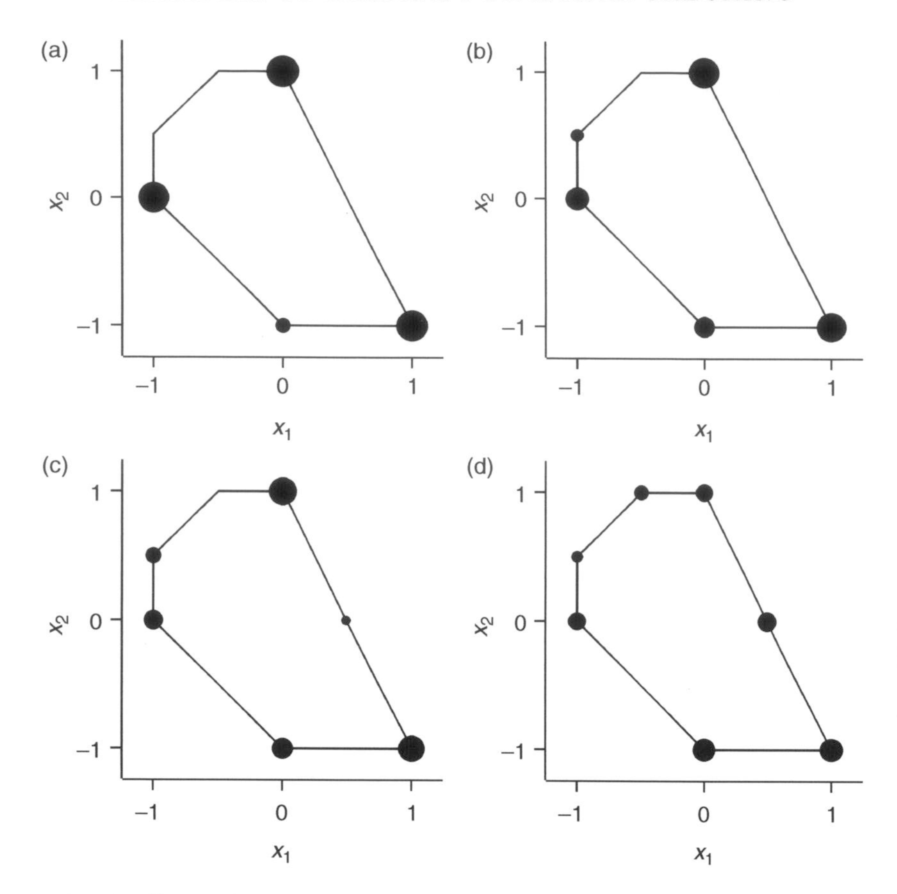

FIG. 20.2. Second-order response surface—constrained design region. Designs for model checking of Table 20.1: (a) $\alpha = 0.318$, (b) $\alpha = 0.531$, (c) $\alpha = 0.647$, and (d) $\alpha = 0.949$. Dot diameter $\propto w_i^{0.8}$.

the seven-point design for the second-order model. Continuing the process of design construction with increasing α will, of course, finally lead to this optimum design for the second-order model.

Several of the continuous designs in Table 20.1 have at least one small design weight. It is therefore to be expected that exact designs for these values of α and small N may have rather different support from the designs in the table. The construction of such exact designs in SAS is the subject of §20.5.

20.4 Example 20.3. A Non-linear Model for Crop Yield and Plant Density

20.4.1 Background

As a last example of parsimonious model checking designs we have a non-linear model which serves to make more explicit some details of the procedure. Because the model is non-linear we require prior information about the values of the model parameters. We also have to rescale the residuals of the secondary terms for the prior variance τ^2 (20.2) to have a common interpretation for all terms. These and other implementation details are the subject of §20.4.2.

Models quantifying the relationship between crop yield and plant density are of great importance in agriculture. Plants compete for resources. The yield per plant therefore tends to decrease as plant density increases although the yield per unit area often continues to increase until appreciably higher plant densities are reached. The yield then either reaches an asymptote or slowly decreases. Because of this behaviour non-linear models are often used. Seber and Wild (1989, §7.6) present a compact review of the subject.

Our example is the yield of soya beans grown in regularly spaced rows—there is a regular distance both between the rows and between plants within the rows. Although the simplest models assume that it is only the area per plant that matters, rather than the shape of the area for each plant, we want a general model which allows for different effects of the two distances, which are of different magnitudes. The equivalent of the general second-order polynomial response surface model (20.8) is the general seven-parameter model

$$\mathrm{E}(y) = \left(\theta_1 + \theta_2 \frac{1}{x_1} + \theta_3 \frac{1}{x_2} + \theta_4 \frac{1}{x_1 x_2} + \theta_5 \frac{1}{x_1^2} + \theta_6 \frac{1}{x_2^2}\right)^{-(1/\theta_7)} \qquad (20.10)$$

where:

$\mathrm{E}(y)$ is the expected yield per plant or biologically definable part of the plant;

$\theta_1, \ldots, \theta_7$ are positive parameters and $0 < \theta_7 \leq 1$;

x_1 is the intra-row spacing, that is, the spacing between plants within a row and

x_2 is the inter-row spacing, that is, the spacing between rows.

The area per plant is therefore $x_1 x_2$ with $1/x_1 x_2$ the density, that is, the number of plants per unit area.

This general model is a second-order polynomial with terms in $1/x_1$ and $1/x_2$. Although the parameters enter linearly, the model is made non-linear by the presence of the power θ_7. Instead of (20.10) we work with the related model for expected yield per unit area

$$E\,(y^*) = \frac{1}{x_1 x_2}\left(\theta_1 + \theta_2\frac{1}{x_1} + \theta_3\frac{1}{x_2} + \theta_4\frac{1}{x_1 x_2} + \theta_5\frac{1}{x_1^2} + \theta_6\frac{1}{x_2^2}\right)^{-(1/\theta_7)} \tag{20.11}$$

obtained by dividing both sides of (20.10) by $x_1 x_2$.

Several simpler models have been proposed, which are special cases of (20.11). We take as our primary model the simplest, the three-parameter model of Shinozaki and Kira (1956) for expected yield per unit area

$$E\,(y^*) = \frac{1}{x_1 x_2}\left(\theta_1 + \theta_4\frac{1}{x_1 x_2}\right)^{-(1/\theta_7)} \tag{20.12}$$

which is obtained by putting $\theta_2 = \theta_3 = \theta_5 = \theta_6 = 0$ and so depends only on the area per plant, ignoring the shape of that area.

To estimate the parameters in the general model (20.11) requires experiments at at least seven combinations of x_1 and x_2. Such a design will usually be inefficient if the experimental purpose is to check the three-parameter model (20.12). We now find an optimum model-checking design which requires trials at only four treatment combinations, rather than seven.

20.4.2 Implementation

Since the models we are using are non-linear, we now need prior estimates of the parameters to be able to design the experiment. For this purpose we use data from Lin and Morse (1975) on the effect of spacing on the yield of soya beans. The data are in Table 20.2. The four levels of the inter-row spacing factor (0.18, 0.36, 0.54, and 0.72 m) and four levels of the intra-row spacing factor (0.03, 0.06, 0.09, and 0.12 m) were used to study the optimum spacing for maximum yield per unit area. The maximum observed yield is near the centre of the region.

When we tried to fit the simple model (20.12) to these data, we found that the fit was improved, as judged by residual plots, if we took logarithms of both sides, so that the primary model for expected yield becomes

$$\eta_0(x,\theta) = E(\log y^*) = -\frac{1}{\theta_7}\log\left(\theta_1 + \theta_4\frac{1}{x_1 x_2}\right) - \log(x_1 x_2).$$

Since yield cannot be negative, a model such as this, which gives a lognormal distribution for y^*, is more plausible than one with a normal distribution of errors on the untransformed scale. The gamma models suggested by McCullagh and Nelder (1989, p. 291) have a similar justification. An interesting

TABLE 20.2. Yield–density relationship: the mean grain yield, in g/m^2, for 16 spacing treatments of the soya bean variety Altona (from Lin and Morse 1975)

Inter-row spacing (m)	Intra-row spacing (m)			
	0.03	0.06	0.09	0.12
0.18	260.0	344.7	279.9	309.2
0.36	305.3	358.3	312.2	267.8
0.54	283.9	342.0	269.0	253.9
0.72	221.8	287.9	230.9	196.9

feature of the logged model is that the models for yield per plant and yield per unit area only differ by the subtraction of the parameterless term $\log(x_1x_2)$, a form of term which McCullagh and Nelder call an offset. The parameter estimates for the logged model were $\hat{\theta}_1 = 0.07469$, $\hat{\theta}_4 = 0.003751$, and $\hat{\theta}_7 = 0.7363$.

The extended seven-parameter logged model is likewise found by taking the logarithm of (20.11). The two groups of parameters are thus the primary parameters

$$\theta_r = \left[\ \theta_1 \quad \theta_4 \quad \theta_7\ \right]^{\mathrm{T}},$$

with prior values

$$\theta_r^0 = \left[\ \hat{\theta}_1 \quad \hat{\theta}_4 \quad \hat{\theta}_7\ \right]^{\mathrm{T}},$$

and the secondary parameters

$$\theta_s = \left[\ \theta_2 \quad \theta_3 \quad \theta_5 \quad \theta_6\ \right]^{\mathrm{T}},$$

with priors θ_s^0 taken as their expected value zero. Note that, particularly in non-linear models, the prior value of the secondary parameters need not be zero. Here we have $r = 3$ and $s = 4$.

The elements of $f_r(x)$ and $f_s(x)$ are the parameter sensitivities, that is the derivatives of $\eta(x, \theta)$ with respect to $\theta_j, j = 1, \ldots, 7$, evaluated at the prior

$$[\theta^0]^{\mathrm{T}} = \left[\ (\theta_r^0)^{\mathrm{T}} \quad (\theta_s^0)^{\mathrm{T}}\ \right],$$

namely

$$\begin{aligned} f_1(x) &= -\left\{\hat{\theta}_7\left(\hat{\theta}_1 + \hat{\theta}_4\frac{1}{x_1x_2}\right)\right\}^{-1}, \\ f_j(x) &= z_j f_1(x) \quad j = 2, \ldots, 6, \\ f_7(x) &= \hat{\theta}_7^{-2}\log\left(\hat{\theta}_1 + \hat{\theta}_4\frac{1}{x_1x_2}\right). \end{aligned} \qquad (20.13)$$

In (20.13) z_j is the coefficient of θ_j in (20.11).

The vectors of the derivatives defined in (20.13) form the rows of the design matrices $F_r(\xi)$ and $F_s(\xi)$ of primary and secondary terms. For design region we let $\mathcal{X} = \{(x_1, x_2) : 0.15 \leq x_1 \leq 0.8 \text{ and } 0.03 \leq x_2 \leq 0.2\}$, slightly larger than that used by Lin and Morse (1975). The prior of §20.2.2 for model checking makes sense if the secondary terms are orthogonal to the primary ones and if the columns of $F_s(\xi)$ are so scaled that the coefficients θ_s have a common prior variance τ^2. In the response surface examples of §20.3 these conditions were satisfied by the scaling of the variables from -1 to 1. We satisfy these conditions for our non-linear model by using scaled residuals from regression to provide the necessary orthogonality. For the regression we use a design measure ξ_c which is uniform over the 14×14 set of points:

$$\{0.15, 0.2, \ldots, 0.8\} \times \{0.03, 0.04307, 0.05615, \ldots, 0.2\},$$

the results not being sensitive to the number of points in $\mathcal{X}$ which are used. The steps of the *scaling procedure* are:

1. Perform the regression of the extra terms on the primary terms computing $B = \{F_r^T(\xi_c)F_r(\xi_c)\}^{-1}\{F_r^T(\xi_c)F_s(\xi_c)\}$ and the residual matrix $R = F_s(\xi_c) - F_r(\xi_c)B$;
2. Calculate the range of each column of R, that is, compute range$(r_j) = \max(r_j) - \min(r_j)$ $(j = 1, \ldots, s)$ where r_j is the jth column of R;
3. Compute $W_s = \text{diag}\{\text{range}^{-1}(r_1), \ldots, \text{range}^{-1}(r_s)\}$;
4. Then the scaled residuals used in constructing the information matrix of the design measure ξ are

$$F_s^*(\xi) = \{F_s(\xi) - F_r(\xi)B\}W_s.$$

20.4.3 Parsimonious Model Checking

In Figure 20.3 and Tables 20.3 and 20.4 we give designs for three different values of α. These were found by numerical maximization of the design criterion, the equivalence theorem being used to check the optimality of our designs. The first, for α in (20.5) equal to zero, is the family of D-optimum designs for estimating the primary parameters. Since the primary model depends on x_1 and x_2 only through the product x_1x_2, the optimum design specifies three values of x_1x_2, without, for one design point, specifying individual values for x_1 or x_2. Specifically

$$\begin{aligned} x_1x_2 &= 0.0045 \quad \text{for } (x_1, x_2) = (0.15, 0.03), \\ x_1x_2 &= 0.16 \quad \text{for } (x_1, x_2) = (0.8, 0.2), \\ \text{and} \quad x_1x_2 &= 0.02205 \quad \text{for any } (x_1, x_2) \text{ giving this value.} \end{aligned}$$

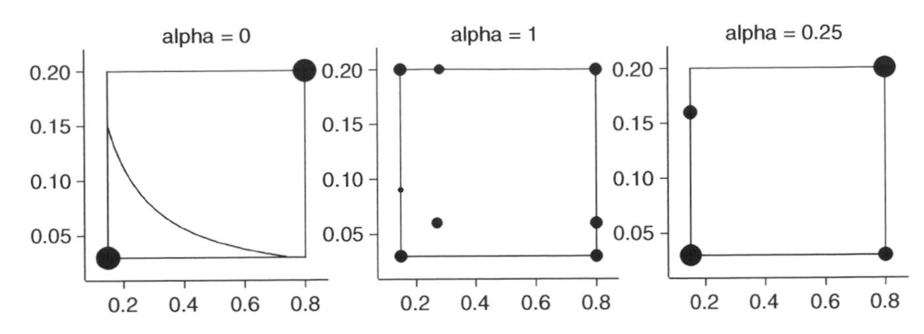

FIG. 20.3. Yield–density relationship. Designs for model checking: $\alpha = 0$: D-optimum design for primary terms. This design depends only on the value of $x_1 x_2$ and the third design point can be anywhere on the line; $\alpha = 1$: D-optimum design for the seven-parameter model with dot diameter $\propto w_i^{0.8}$; $\alpha = 0.25$: optimum design for checking the three-parameter model.

TABLE 20.3. Yield–density relationship: design for $\alpha = 1$: D-optimum design for the seven-parameter model

x_1	x_2	w	$[7w]$
0.15	0.03	0.143	1
0.15	0.0895	0.041	0
0.15	0.2	0.134	1
0.266	0.0633	0.115	1
0.284	0.2	0.138	1
0.421	0.03	0.003	0
0.8	0.03	0.142	1
0.8	0.0613	0.142	1
0.8	0.2	0.142	1
Efficiency			97.0%

The optimum design is therefore not unique. It puts equal weights (1/3) at low, high, and intermediate values of $x_1 x_2$. The graphical representation of the design thus consists of two points at edges corners of the design region and a curve of possible third values, all of which give the same value of the optimality criterion, regardless of the shape of the experimental plot. The specific value for the experiment, if the model were known to be true, could be chosen with respect to an auxiliary criterion.

TABLE 20.4. Yield–density relationship: design for $\alpha = 0.25$, a parsimonious D-optimum design for model checking

Area/plant	x_1	x_2	w	$[6w]$
0.0045	0.15	0.03	0.324	2
0.0237	0.15	0.158	0.176	1
0.0240	0.8	0.03	0.176	1
0.1600	0.8	0.2	0.324	2
Efficiency				99.8%

For α equal to one we obtain the D-optimum design for the full model with all seven terms. This design, like the design for the second-order response surface on a rectangular design region, has nine support points. But now, because the model is non-linear and the range of the two factors is not the same, the design has, as the second panel of Figure 20.3 shows, an approximate symmetry about one-diagonal of the design region. It also has very uneven weights on the support points which are plotted with the dots having diameter proportional to $w_i^{0.8}$. As a result two design points are hard to see. In Table 20.3 we give a seven-point approximation to the design, which has an efficiency of 97.0%.

Finally we use the method of this chapter for model checking to obtain a design with only four support points. We tried several values of α but here give only the results for $\alpha = 0.25$. This design again is not symmetrical, nor does it have equal weight on the four points. However, a good approximation can be found which requires only six trials. The resulting design, plotted in the third panel of Figure 20.3, is close to the D-optimum design for $\alpha = 0$ in the first panel: its two middle points give areas close to the former 0.02205 and the other points are at the same corners of the design region as before. This parsimonious design is therefore highly efficient both for checking the model and for estimating the parameters in the model if it holds.

A design with four support points cannot pick up all departures from the three-parameter model which are possible when the general model has seven parameters. The situation is similar to that of using trials at the centre of a two-level factorial to check that quadratic terms are not needed. If the coefficients of the quadratic terms are not zero, but sum to zero, the quadratic terms will not be detected. Protection against such an unlikely happening for the non-linear model could be obtained by using larger values of α to generate designs with more support points. Such designs would reflect the increased emphasis on model checking implied by larger α.

20.4.4 Departures from Non-linear Models

The secondary terms for linear models are usually higher-order polynomial terms in the factors. The analogue for non-linear models is not obvious. Three methods of forming a more general non-linear model are:

1. Add to the non-linear model a low-order polynomial in the m factors. A systematic pattern in the residuals from the non-linear model might be explained by these terms.

2. Add squared and interaction terms in the parameter sensitivities.

3. Embed the non-linear model in a more general model that reduces to the original model for particular values of some of the non-linear parameters.

These generalizations are identical for linear models. The first embedding depends heavily on the design region and is likely to detect departures in those parts of the region that provide little information about the original model. The addition of higher-order terms in the partial derivatives is, on the contrary, invariant even under non-linear transformation of the factors. Both this method and the third, that of embedding the model in a more general non-linear model, are suitable if the model is already reasonably well established. For linear models, when extra polynomial terms are added, the original model is recovered when all extra parameters are zero. But this is not necessarily the case with non-linear models. For example, in the model for two consecutive first-order reactions (17.5) we get different limiting models depending upon the values towards which the parameters tend. Since we have $\theta_1 > \theta_2$, one possibility is $\theta_1 \to \infty$, when the first reaction becomes increasingly fast; the limiting model is that of first-order decay (17.3) with rate θ_2. On the other hand, if $\theta_2 \to 0$, we obtain (4.13), first-order growth with $\beta_0 = 1$ since, in the limit, none of the B that is formed is decomposed into the third component.

20.5 Exact Model Checking Designs in SAS

For finding exact designs optimal for model checking, OPTEX implements the Bayesian formulation of DuMouchel and Jones (1994) in terms of equation (20.4), assuming that K is diagonal. Primary terms are separated from secondary terms in the MODEL statement simply by inserting a comma between them, and then the `PRIOR=` option gives the value to be added to each set of terms. From equation (20.5) we have

$$N_0 = N \times \frac{1-\alpha}{\alpha}.$$

To demonstrate, consider Example 20.1, in which we want a 5-run design to check for second-order departures from a first-order model over a regular design region. As usual, OPTEX requires a discrete set of candidate points, which the following code creates as the points in the $[-1,1]^2$ square at increments of 1/10.

```
data Grid;
   do ix1 = 1 to 21;
      do ix2 = 1 to 21;
         x1 = -1 + (ix1-1)/10;
         x2 = -1 + (ix2-1)/10;
         output;
         end;
      end;
run;
```

Now, in the following code, the only difference from previous examples using PROC OPTEX is that the linear terms in the MODEL statement (plus the intercept, implicitly) are separated from the quadratic terms by a comma, and the arguments for the `PRIOR=` option give the N_0 value for each group of terms. The primary terms have no prior information, and thus their `PRIOR=` value is 0; while for the secondary terms we use $2.5 = 5 \times (1-2/3)/(2/3)$.

```
proc optex data=Grid;
   model x1 x2, x1*x1 x1*x2 x2*x2 / prior = 0,2.5;
   generate n=5 method=m_fedorov niter=100 keep=10;
   output out=Design;
proc print data=Design;
run;
```

The resulting design has the same five points of support as the continuous optimal design depicted in Figure 20.1.

SAS Task 20.1. For Example 20.1, demonstrate that as the number of points in the design grows, the optimum exact design converges to the optimum continuous one.

SAS Task 20.2. For Example 20.2, confirm that the optimum exact design in 10 runs for $\alpha = 0.318$ is the same as the one given in Table 20.1, obtained by rounding 10 times the weights of the optimum continuous design.

As in §20.3, we have thus far ignored rescaling the residuals of the secondary terms, but this rescaling is more important for non-linear designs, such as Example 20.3 and should be implemented. To show how to compute the rescaling in SAS, we begin by defining a candidate set over the region of interest for Example 20.3, and computing the columns of the Jacobian.

```
%let t1 = 0.07469;
%let t2 = 0;
%let t3 = 0;
%let t4 = 0.003751;
%let t5 = 0;
%let t6 = 0;
%let t7 = 0.7363;
data Can;
   do x1 = 0.15 to 0.8 by 0.01;
   do x2 = 0.03 to 0.2 by 0.01;
      f1 = -(&t7*(&t1+&t4/(x1*x2)))**(-1);
      f2 = f1/x1;
      f3 = f1/x2;
      f4 = f1/(x1*x2);
      f5 = f1/(x1*x1);
      f6 = f1/(x2*x2);
      f7 = ((&t7)**(-2))*log(&t1+&t4/(x1*x2));
      output;
      end; end;
run;
```

In the code above, the two (nested) DO-loops define the region in terms of the row spacing factors x_1 and x_2, and the assignment statements define the Jacobian of the model, as given in (20.13). There are many ways in SAS to perform the regression and rescaling of these Jacobian terms, discussed in §20.3.2. The following code uses the REG procedure to compute the residuals from regressing the secondary terms on the primary terms, then uses the SAS/IML matrix programming language to rescale these residuals, and finally merges these rescaled residuals back into the candidate data set.

```
proc reg data=Can noprint;
   model f2 f3 f5 f6 = f1 f4 f7;
   output out=rCan r = rf2 rf3 rf5 rf6;
proc iml;
   use rCan;
   read all var { f1  f4  f7}      into Fr;
   read all var {rf2 rf3 rf5 rf6} into Rs;
   Ws = diag(1/(Rs[<>,] - Rs[><,]));
   Fs = Rs*Ws;
   create Fs var {rf2 rf3 rf5 rf6};
   append from Fs;
data rJac; merge Can Fs;
run;
```

SAS Task 20.3. For Example 20.3, use the rescaled terms defined above to find exact Bayesian D-optimum designs for

- $\alpha = 1$ in 7 runs, and
- $\alpha = 0.25$ in 6 runs,

and compare them to the exact designs given in the last columns of Tables 20.3 and 20.4.3, respectively.

SAS Task 20.4. (Advanced). Use PROC OPTEX to find an exact Bayesian D-optimum design for 8 two-level factors in 16 runs, where the main effects of the factors are considered to be primary terms and the two-way interactions are considered to be secondary terms. This is a case when the Bayesian D-optimum design does not admit estimation of all parameters, since there are 37 terms in the model but only 16 runs. Use the `ALIASING` option in PROC GLM to confirm that the design has resolution IV: main effects are simultaneously orthogonally estimable; and although two-factor interactions are not all simultaneously estimable, none of them are confounded with main effects.

SAS Task 20.4 illustrates some general features of Bayesian D-optimum designs. In particular, the putative prior information on the secondary terms allows direct information on them to be sacrificed in favor of (1) better information on the primary terms and (2) independence between the primary and secondary terms. Philosophically, the value of the prior is best viewed as quantifying the relative importance of having the design provide information on the primary and secondary terms. When it comes to analysing data from such a design, the value of the prior for which it is Bayesian D-optimum does not come into play. Rather, techniques such as analysis of the alias structure are employed which cope with the lack of information on some terms.

20.6 Discriminating Between Two Models

In the model checking designs considered so far in this chapter the two models were of very different status. The terms of the primary model were certainly needed, but it might be necessary also to include some of the terms of the secondary model. The primary model would be nested within the more general model containing in addition some secondary terms. Now, instead we look at designs when the two models have an equal status; they may be alternative ways of describing the same phenomenon, often only one of which can be true.

Example 20.4. Two Models for Decay In Chapter 17 locally D-optimum designs were found for the exponential decay model

$$\eta_1(x, \theta_1) = \exp(-\theta_1 x) \quad (x, \theta_1 \geq 0). \tag{20.14}$$

It was mentioned that it might be hard to discriminate between this model and the inverse polynomial

$$\eta_2(x, \theta_2) = \frac{1}{1 + \theta_2 x} \quad (x, \theta_2 \geq 0). \tag{20.15}$$

At what values of x should measurements be taken in order to determine which of these two one-parameter models better explains the data? ■

The structure of Example 20.4 is common to the problems investigated in the rest of this chapter. We assume that one model is true and then design experiments to determine which it is.

It is clear for Example 20.4 that both models cannot be true. However, this is not so for nested models.

Example 20.5. Constant or Quadratic Model? Suppose that

$$\eta_1(x, \theta_1) = \gamma \tag{20.16}$$

and

$$\eta_2(x, \theta_2) = \theta_0 + \theta_1 x + \theta_2 x^2. \tag{20.17}$$

Here the constant model (20.16) is a special case of the quadratic (20.17). Therefore if (20.16) is true, so is the degenerate form of (20.17) in which $\theta_1 = \theta_2 = 0$. One way of treating such nested examples is to introduce a constraint on the parameter values, for example to insist that $\theta_1^2 + \theta_2^2 \geq \delta > 0$. This ensures that the two models are separate, with the consequence that (20.17) is an interesting alternative to (20.16) for all parameter values. Under the second model the response is now constrained to be a function of x rather than being allowed to be constant. ■

In this section we describe optimum continuous designs for discriminating between two models, which we call T-optimum. In the next section we demonstrate an algorithm for the practically important problem of the sequential generation of the designs. Designs for more than two models are mentioned briefly in §20.8.1. Whether there are two, or more than two, models, the resulting optimum designs depend upon unknown parameter values. In §20.8.2 we describe Bayesian designs, similar to those of Chapter 18 in which prior information about parameters is represented by a distribution rather than by a point estimate. Finally in §20.9.2, discrimination between nested models is related to D_S-optimality and to designs for checking goodness of fit. A brief note on the analysis of T-optimum designs is in §20.11.

The optimum design for discriminating between two models will depend upon which model is true and, usually, on the values of the parameters of the true model. Without loss of generality let this be the first model and write

$$\eta_{\mathrm{t}}(x) = \eta_1(x, \theta_1). \tag{20.18}$$

A good design for discriminating between the models will provide a large lack-of-fit sum of squares for the second model. When the second model is fitted to the data, the least squares parameter estimates will depend on the experimental design as well as on the value of θ_1 and the errors. In the absence of error the parameter estimates are

$$\hat{\theta}_2(\xi) = \underset{\theta_2}{\operatorname{argmin}} \int_{\mathcal{X}} \{\eta_{\mathrm{t}}(x) - \eta_2(x, \theta_2)\}^2 \xi(dx), \tag{20.19}$$

yielding a residual sum of squares

$$\Delta_2(\xi) = \int_{\mathcal{X}} [\eta_{\mathrm{t}}(x) - \eta_2\{x, \hat{\theta}_2(\xi)\}]^2 \xi(dx). \tag{20.20}$$

For linear models $\Delta_2(\xi)$ is proportional to the non-centrality parameter of the χ^2 distribution of the residual sum of squares for the second model and design ξ. Designs which maximize $\Delta_2(\xi)$ are called T-optimum, to emphasize the connection with testing models; the letter D, which might have served as a mnemonic for **D**iscrimination, having been introduced by Kiefer (1959) for what is now the standard usage for **D**eterminant optimality. The T-optimum design, by maximizing (20.20) provides the most powerful F test for lack of fit of the second model when the first is true. If the models are non-linear in the parameters, the exact F test is replaced by asymptotic results, but we still design to maximize (20.20).

For linear models and exact designs with extended design matrices F_1 and F_2 and parameter vectors θ_1 and θ_2, the expected value of the least squares estimator $\hat{\theta}_2$ minimizing (20.19) is

$$\hat{\theta}_2 = (F_2^{\mathrm{T}} F_2)^{-1} F_2^{\mathrm{T}} F_1 \theta_1. \tag{20.21}$$

Provided the models do not contain any terms in common, (20.20) shows that the non-centrality parameter for this exact design is

$$N\Delta_2(\xi_N)/\sigma^2 = \theta_1^{\mathrm{T}} \{F_1^{\mathrm{T}} F_1 - F_1^{\mathrm{T}} F_2 (F_2^{\mathrm{T}} F_2)^{-1} F_2^{\mathrm{T}} F_1\} \theta_1. \tag{20.22}$$

We call $\Delta_2(\xi_N)$ the standardized non-centrality parameter, since it corresponds to the standard case with both σ^2 and $N = 1$ and so excludes quantities that do not influence the continuous design.

Equation (20.22) makes explicit the dependence of $\Delta_2(\xi_N)$ on the parameters θ_1 of the true model. Common terms in the two models do not contribute to the non-centrality parameter. The algebra for such partially nested models is given in §20.9.1. For the moment we continue with general models that may be linear or non-linear.

The quantity $\Delta_2(\xi)$ is another example of a convex function to which an equivalence theorem applies. To establish notation for the derivative function, let the T-optimum design yield the estimate $\theta_2^* = \hat{\theta}_2(\xi^*)$. Then

$$\Delta_2(\xi^*) = \int_{\mathcal{X}} \{\eta_t(x) - \eta_2(x, \theta_2^*)\}^2 \xi(dx). \tag{20.23}$$

For this design the squared difference between the true and predicted responses at x is

$$\psi_2(x, \xi^*) = \{\eta_t(x) - \eta_2(x, \theta_2^*)\}^2, \tag{20.24}$$

with $\psi_2(x, \xi)$ the squared difference for any other design. We then have the equivalence of the following conditions:

1. The T-optimum design ξ^* maximizes $\Delta_2(\xi)$.
2. $\psi_2(x, \xi^*) \leq \Delta_2(\xi^*)$ for all $x \in \mathcal{X}$.
3. At the points of support of the optimum design $\psi_2(x, \xi^*) = \Delta_2(\xi^*)$.
4. For any non-optimum design, that is one for which $\Delta_2(\xi) < \Delta_2(\xi^*)$,

$$\sup_{x \in \mathcal{X}} \psi_2(x, \xi) > \Delta_2(\xi^*),$$

These results are similar in structure to those we have used for D-optimality and lead to similar methods of design construction and verification.

Example 20.6. Two Linear Models As a first example of T-optimum designs we look at discrimination between the two models

$$\eta_1(x, \theta_1) = \theta_{10} + \theta_{11}e^x + \theta_{12}e^{-x} \tag{20.25}$$

$$\eta_2(x, \theta_2) = \theta_{20} + \theta_{21}x + \theta_{22}x^2 \tag{20.26}$$

for $-1 \leq x \leq 1$. Both models are linear in three parameters and so will exactly fit any three-point design in the absence of observational error. Designs for discriminating between the two models will therefore need at least four support points.

To illustrate the equivalence theorem we find the continuous T-optimum design. As before we assume the first model is true. Then the T-optimum design depends on the values of the parameters θ_{11} and θ_{12}, but not on the

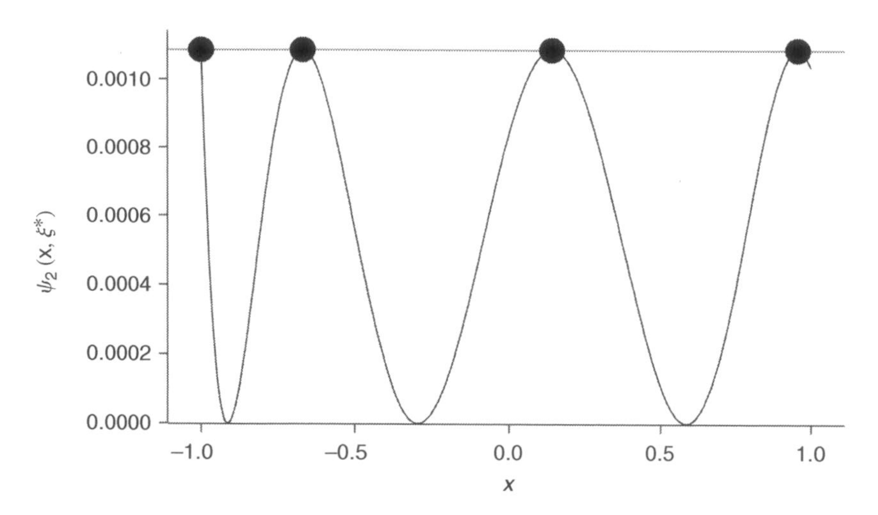

FIG. 20.4. Example 20.6: discrimination between two linear models. Derivative function $\psi_2(x, \xi^*)$ for the T-optimum design: • design points.

value of θ_{10}, since both models contain a constant. We consider only one pair of parameter values, taking the true model as

$$\eta_t(x) = 4.5 - 1.5e^x - 2e^{-x}. \tag{20.27}$$

This function, which has a value of -1.488 at $x = -1$, rises to a maximum 1.036 at $x = 0.144$ before declining to -0.131 at $x = 1$. It can be well approximated by the polynomial model (20.26). The T-optimum design for discriminating between the two models is found by numerical maximization of $\Delta_2(\xi)$ to be

$$\xi^* = \left\{ \begin{array}{cccc} -1 & -0.6694 & 0.1441 & 0.9584 \\ 0.2531 & 0.4280 & 0.2469 & 0.0720 \end{array} \right\} \tag{20.28}$$

for which $\Delta_2(\xi^*) = 1.087 \times 10^{-3}$. A strange feature of this design is that half the weight is on the first and third points and half on the other two.

For the particular parameter values in (20.27) neither is the design symmetrical, nor does it span the experimental region. It has only four support points, the minimum number for discrimination between these two three-parameter models. As an illustration of the equivalence theorem, $\psi_2(x, \xi^*)$ (20.24) is plotted in Figure 20.4 as a function of x. The maximum values of $\psi_2(x, \xi^*)$ is indeed equal to $\Delta_2(\xi^*)$, the maximum occurring at the four points of the optimum design. ■

Example 20.4. Two Models for Decay continued Both models in the preceding example were linear in their parameters. We now find the continuous T-optimum design for discriminating between two non-linear models, using as an example the two models for decay (20.14) and (20.15).

Let the first model be true with $\theta_1 = 1$, so that

$$\eta_t(x) = e^{-x} \quad (x \geq 0).$$

The T-optimum design again maximizes the non-centrality parameter $\Delta_2(\xi)$ (20.20). A complication introduced by the non-linearity of $\eta_2(x, \theta)$ is the iterative calculation required for the non-linear least squares estimates $\hat{\theta}_2(\xi)$. The iterative numerical maximization of $\Delta_2(\xi)$ thus includes an iterative fit at each function evaluation.

The T-optimum design when $\theta_1 = 1$ is

$$\xi^* = \left\{ \begin{array}{cc} 0.327 & 3.34 \\ 0.3345 & 0.6655 \end{array} \right\}, \tag{20.29}$$

a two-point design allowing discrimination between these one-parameter models. We have already seen in §17.2 that the locally D-optimum design for θ_1 when $\theta_1^0 = 1$ puts all trials at $x = 1$. The design given by (20.29) divides the design weights between support points on either side of this value. That this is the T-optimum design is shown by the plot of $\psi_2(x, \xi^*)$ in Figure 20.5 which has two maxima with the value of $\Delta_2(\xi^*) = 1.038 \times 10^{-2}$. ■

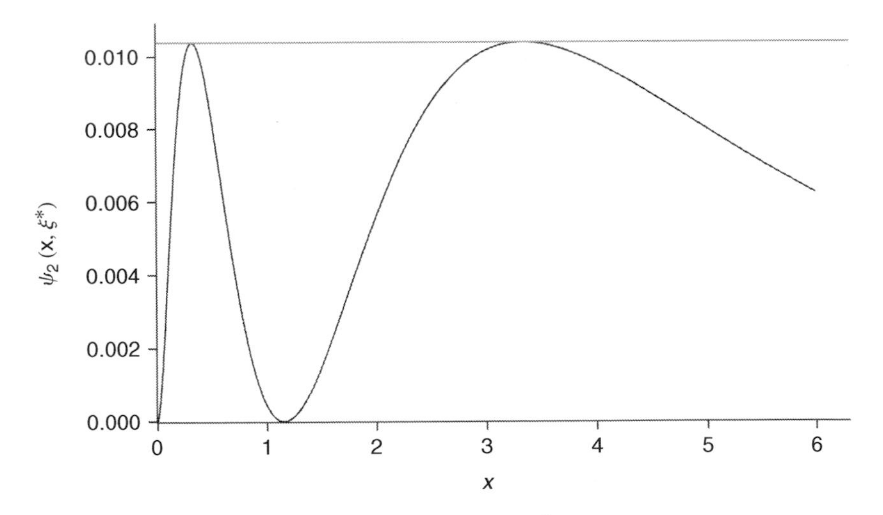

FIG. 20.5. Example 20.6: discrimination between two non-linear models for decay. Derivative function $\psi_2(x, \xi^*)$ for the T-optimum design.

Example 20.5. Constant or Quadratic Model continued In the two preceding examples the parameters of both models can be estimated from the T-optimum design. However, with the nested models of Example 20.5 the situation is more complicated.

Suppose that the quadratic model (20.17) is true. Then disproving the constant model (20.16) only requires experiments at two values of x that yield different values of the response. To find such an optimum design let

$$z = \theta_1 x + \theta_2 x^2,$$

the terms in model 2 not included in model 1. Then from (20.20) and (20.22)

$$\Delta_1(\xi) = \int_{\mathcal{X}} \left\{ z - \int_{\mathcal{X}} z\xi(dx) \right\}^2 \xi(dx),$$

since both models include a constant term. The optimum design thus maximizes the sum of squares of z about its mean. This is achieved by a design that places half the trials at the maximum value of z and half at the minimum, that is at the maximum and minimum of $\eta_2(x, \theta_2)$. The values of x at which this occurs will depend on the values of θ_1 and θ_2. But, whatever these parameter values, the design will not permit their estimation.

In order to accommodate the singularity of the design, extensions are necessary to the equivalence theorem defining T-optimality. The details are given by Atkinson and Fedorov (1975*a*), who provide an analytic derivation of the T-optimum design for this example. The numerical calculation of these singular designs can be achieved using the regularization (10.10). Atkinson and Fedorov also derive the T-optimum design when the constant model is true, with the alternative the quadratic model constrained so that $\theta_1^2 + \theta_2^2 \geq 1$. ■

20.7 Sequential Designs for Discriminating Between Two Models

The T-optimum designs of the previous section depend both on which of the two models is true and, in general, on the parameters of the true model. They are thus only locally optimum. The Bayesian designs described in §20.8.2 provide one way of designing experiments that are not so dependent on precise, but perhaps, erroneous, information. In this section we consider another of the alternatives to locally optimum designs discussed in §17.3, that of the sequential construction and analysis of experiments that converge to the T-optimum design if one of the two models is true.

The key to the procedure is the estimate of the derivative function $\psi(x, \xi_k)$ after k readings have been obtained and analysed, yielding parameter estimates $\hat{\theta}_{1k} = \hat{\theta}_1(\xi_k)$ and $\hat{\theta}_{2k} = \hat{\theta}_2(\xi_k)$. The corresponding fitted models are then $\eta_1(x, \hat{\theta}_{1k})$ and $\eta_2(x, \hat{\theta}_{2k})$. Since the true model and parameter values are not known, the estimate of the derivative function (20.24) for the design ξ_k is

$$\psi(x, \xi_k) = \{\eta_1(x, \hat{\theta}_{1k}) - \eta_2(x, \hat{\theta}_{2k})\}^2. \qquad (20.30)$$

In the iterative construction of T-optimum designs using the first-order algorithm, analogous to that for D-optimum designs in §11.2, trials augmenting ξ_k would be added where $\psi_2(x, \xi_k)$ was a maximum. This suggests the following design strategy:

1. After k experiments let the estimated derivative function be $\psi(x, \xi_k)$, given by (20.30).
2. The point $x_{k+1} \in \mathcal{X}$ is found for which
$$\psi(x_{k+1}, \xi_k) = \sup_{x \in \mathcal{X}} \psi(x, \xi_k).$$
3. The $(k + 1)$st observation is taken at x_{k+1}.
4. Steps 2 and 3 are repeated for $x_{k+2}, x_{k+3}, \ldots$ until sufficient accuracy has been obtained.

Provided that one of the models is true, either $\eta_1(x, \hat{\theta}_{1k})$ or $\eta_2(x, \hat{\theta}_{2k})$ will converge to the true model $\eta_t(x)$ as $k \to \infty$. The sequential design strategy will then converge on the T-optimum design. In order to start the process a design ξ_0 is required, non-singular for both models. In some cases the sequential design will converge to a design singular for at least one of the models. In practice this causes no difficulty as ξ_k will be regularized by the starting design ξ_0.

Example 20.6. Two Linear Models continued The sequential procedure is illustrated by the simulated designs of Figure 20.6 using the two linear models with exponential and polynomial terms (20.25) and (20.26); the true exponential model is again given by (20.27). In all simulations ξ_0 consisted of trials at -1, 0, and 1 and, at each stage, the sequential design was found by searching over a grid of 21 values of x in steps of 0.1. The efficiency of the sequential design is measured by the ratio of $\Delta_2(\xi_k)$ to $\Delta_2(\xi^*)$ for the T-optimum design, which is proportional to the residual sum of squares in the absence of error.

Figure 20.6 shows the efficiencies for designs for four increasing values of the error standard deviation σ. For the first design $\sigma = 0$, corresponding

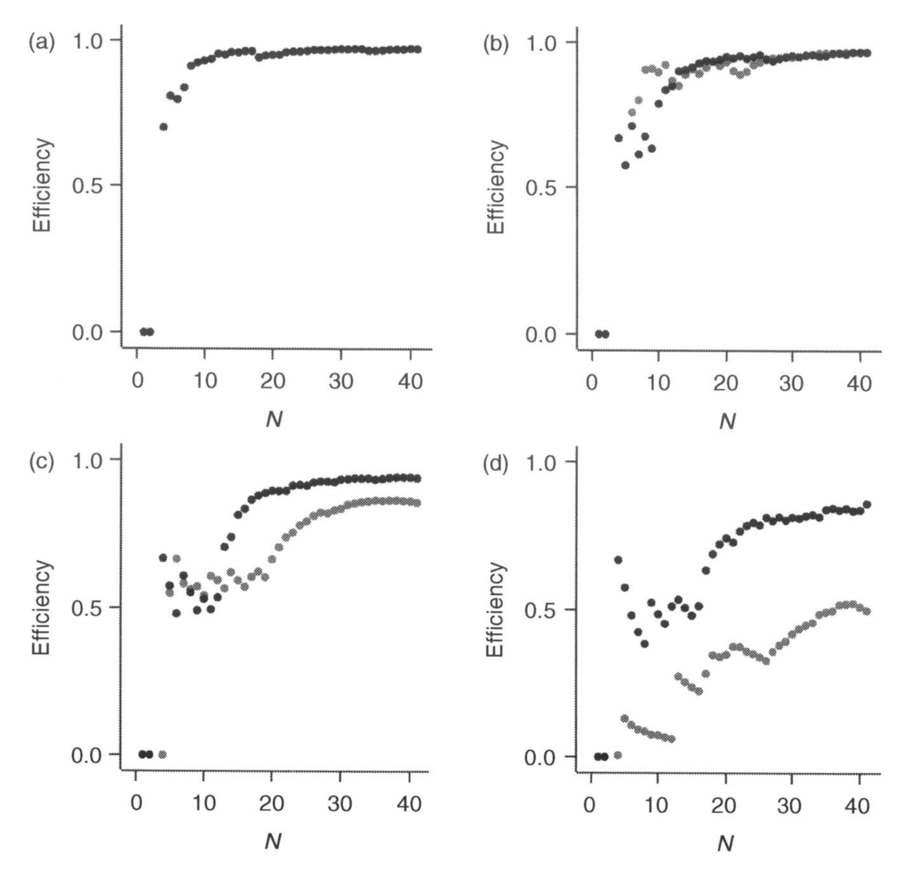

FIG. 20.6. Example 20.6: discrimination between two linear models. Efficiencies of simulated sequential T-optimum designs for increasing error variance: (a) $\sigma = 0$; (b) $\sigma = 0.5$; (c) $\sigma = 1$; (d) $\sigma = 2$.

to the iterative construction of the optimum design with step length $\alpha_k = 1/(k+1)$. Two simulated designs are shown for each of the other values of σ. In general, for increasing values of σ, the effect of random fluctuations takes longer and longer to die out of the design. When $\sigma = 0.5$ the designs start to move rapidly towards the optimum after around 10 trials, whereas, for $\sigma = 1$, they move to the optimum after 20 or so trials. However, when $\sigma = 2$ one design after 40 trials has an efficiency as low as 50%. Since, for the T-optimum design, the maximum difference between the responses is 0.033; even the case of smallest standard deviation ($\sigma = 0.5$) corresponds to an error standard deviation 15 times the effect to be detected. The occasional periods of decreasing efficiency in the plots correspond to the sequential construction of designs for markedly incorrect values of the parameters.

It remains merely to stress that these are examples of sequential designs. As in the non-linear example of §17.7, the results of each trial are analysed and the parameter estimates updated before the conditions for the next trial are selected. In the iterative algorithms for the majority of the continuous designs found in this book, the observed values y_k play no role; iterative design procedures depend only on the values of the factors x. ■

20.8 Developments of T-optimality

20.8.1 More Than Two Models

If there are more than two models the design problem rapidly becomes more complicated. With v models there are $v(v-1)/2$ functions

$$\psi_{ij}(x, \xi_k) = \{\eta_i(x, \hat{\theta}_{ik}) - \eta_j(x, \hat{\theta}_{jk})\}^2, \tag{20.31}$$

which could enter into the sequential construction of a design, instead of the unique function (20.30) when there are two models.

A heuristic solution is to rank the models and to design to discriminate between the two best fitting models. The steps would then be:

1. After k observations with the design ξ_k the residual sums of squares $S_j(\xi_k)$ are calculated for each of the $j = 1, \ldots, v$ models and ranked

$$S_{[1]}(\xi_k) < S_{[2]}(\xi_k) < \cdots < S_{[v]}(\xi_k).$$

2. One step of the sequential procedure for two models of §20.7 is used to find the design point x_{k+1} that maximizes $\psi_{[1][2]}(x, \xi_k)$, that is with the functions for the two best-fitting models substituted in (20.31).

A difficulty is that the two best fitting models may change during the course of the experiment. Although this does not affect the sequential procedure, it does affect the design to which it converges.

Suppose that we augment the two linear models of Example 20.6 with a third model. The set of models is then

$$\begin{aligned}
\eta_1(x, \theta_1) &= \theta_{10} + \theta_{11}e^{x} + \theta_{12}e^{-x} \\
\eta_2(x, \theta_2) &= \theta_{20} + \theta_{21}x + \theta_{22}x^2 \\
\eta_3(x, \theta_2) &= \theta_{30} + \theta_{31}\sin(\pi x/2) + \theta_{32}\cos(\pi x/2) + \theta_{33}\sin(\pi x).
\end{aligned} \tag{20.32}$$

As before, let the first model be true with parameter values given by (20.27). The third model is relatively easily disproved; T-optimum designs solely for discriminating between models 1 and 3 have a non-centrality parameter over five times as large as those for discriminating between models 1 and 2. The focus of the heuristic sequential procedure for three models is on designs for discriminating between models 1 and 2; we saw in (20.28) that a four-point design is optimum for this purpose. But the trigonometric polynomial (20.32) has four parameters, so designs with four points of support provide no evidence against this model. As the proposed heuristic sequential procedure continues and evidence against the second model accumulates, the third model becomes the second best fitting. One trial is then included at a fifth design point informative about the lack of fit of model 3. This then again becomes the worst fitting model and sequential discrimination continues for a while between models 1 and 2 until model 3 again fits second best. This see-saw procedure is not optimum, although it may not be far from being so.

In this example there are three models, one of which is true. Either of the other two, depending on the design, can be closest to the true model. Under these conditions the optimum design will have the two non-centrality parameters equal; this common value should then be maximized. In the more general situation of $v \geq 3$ models, let $J^*(\xi)$ be the set of closest models, which will depend upon the design ξ. For all models j in $J^*(\xi)$, $\Delta_j(\xi) = \Delta^*_{\mathrm{J}}(\xi)$. The T-optimum design maximizes the non-centrality parameter $\Delta^*_{\mathrm{J}}(\xi)$.

Although it is straightforward to define the T-optimum design in this more general situation, numerical construction of the design is complicated both by the need to find the set $J^*(\xi)$ and by the requirement to maximize $\Delta^*_{\mathrm{J}}(\xi)$, which maximization may cause changes in the membership of $J^*(\xi)$. Atkinson and Fedorov (1975*b*) and Atkinson and Donev (1992, §20.3) describe numerical methods for overcoming these problems, including the incorporation of a tolerance in the ordering of the residual sum of squares $S_j(\xi_k)$ to give a set of approximately closest models. Maximization is then of a weighted sum of the non-centrality parameters of the models in this expanded set, with the weights found by numerical search. Table 20.1 of Atkinson and Donev (1992) compares several sequential procures for the three-model example (20.32). In this example there is a slight, but definite, improvement in using the sequential procedure with a tolerance-defined subset, as opposed to using the heuristic sequential procedure outlined at the beginning of this section. Atkinson and Fedorov (1975*b*) give an example in which ignoring the tolerance set gives a design with only 69% efficiency.

20.8.2 Bayesian Designs

The T-optimum designs of the previous sections depend on which model is true and on the parameter values for the true model. The sequential procedures such as that of §20.7 and that with a tolerance outlined above, overcome this dependence by converging to the T-optimum designs for the true model and parameter values. But, if sequential experiments are not possible, we are left with a design that is only locally optimum. One possibility, as in §17.3.2, is to specify prior distributions and then to take the expectation of the design criterion over this distribution. In this section we revert to the two-model criterion of §20.6. We first assume that it is known which model is true, taking expectations only over the parameters of the true model. Then we assign a prior probability to the truth of each model and also take expectations of the design criterion over this distribution. In both cases, straightforward generalizations of the equivalence theorem of §20.6 are obtained.

To begin we extend our notation, to make explicit the dependence of the design criterion on the values of the parameters. If, as in (20.20), model 1 is true, the standardized non-centrality parameter is $\Delta_2(\xi, \theta_1)$ with the derivative function $\psi_2(x, \xi, \theta_1)$. For every design and parameter value θ_1 the least squares estimates of the parameters of the second model (20.19) are $\hat{\theta}_2(\xi, \theta_1)$.

Let E_1 denote expectation with respect to θ_1. Then if we write

$$\begin{aligned} \Delta_2(\xi) &= \mathrm{E}_1\, \Delta_2(\xi, \theta_1) \\ \psi_2(x, \xi) &= \mathrm{E}_1\, \psi_2(x, \xi, \theta_1), \end{aligned} \tag{20.33}$$

the equivalence theorem of §20.6 applies to this compound criterion.

Example 20.4. Two Models for Decay continued The two models (20.14) and (20.15) are respectively exponential decay and an inverse polynomial. We showed in §20.6 that if the exponential model is true with $\theta_1 = 1$, the T-optimum design (20.29) puts design weight at the two points 0.327 and 3.34.

As the simplest illustration of the Bayesian version of T-optimality, we now suppose that the prior distribution of θ_1 assigns a probability of 0.5 to the two values 1/3 and 3, equally spaced from 1 on the logarithmic scale. The T-optimum design is

$$\xi^* = \left\{ \begin{array}{ccc} 0.1160 & 1.073 & 9.345 \\ 0.1608 & 0.4014 & 0.4378 \end{array} \right\}, \tag{20.34}$$

with three points of support. That this is indeed the Bayesian T-optimum design can be shown in the usual way by plotting $\psi_2(x, \xi^*)$ against x. Figure 20.7 of Atkinson and Donev (1992), in which the plot is against $\log x$, shows that there are three maxima at the design points, all of which are equal in value to $\Delta_2(\xi^*)$. ■

A similar approach can be used when it is not known which of the models is true. Let the prior probability that model j is true be π_j with $\sum \pi_j = 1$. Then the expected value of the standardized non-centrality parameter, taken over models and over parameters within models is, by extension of (20.33),

$$\Delta(\xi) = \pi_1 \mathrm{E}_1\, \Delta_2(\xi, \theta_1) + \pi_2 \mathrm{E}_2\, \Delta_1(\xi, \theta_2),$$

with the expected squared difference in responses given by

$$\psi(x, \xi) = \pi_1 \mathrm{E}_1\, \psi_2(x, \xi, \theta_1) + \pi_2 \mathrm{E}_2\, \psi_1(x, \xi, \theta_2). \tag{20.35}$$

That is, for each model assumed true, the calculation is of the expected value of the quantity disproving the other model, combined according to the prior probabilities π_j. The equivalence theorem applies to this more general criterion as it did to its special case (20.33).

Numerical results from the application of the extended Bayesian criterion (20.35) to Example 20.6 are presented by Atkinson and Donev (1992, p. 245). They take a grid of 10 parameter values for the exponential model (20.25) and of five values for the polynomial model (20.26) and obtain a design that, unlike (20.28), has five points of support and spans the experimental region. In finding this design independence was assumed for the prior distributions of the parameters between models. It might be more realistic, in some cases, to consider priors that give equal weights to similarly shaped response curves under the two models.

The extension of (20.35) to three or more models is straightforward when compared with the difficulties encountered in §20.8.1. There the existence of more than one model closest to the true model led to a design criterion constrained by the equality of non-centrality parameters. Here expectations can be taken over all $v(v-1)$ non-centrality parameters, yielding a comparatively smooth and well-behaved design criterion. The price is that experimental effort will be dispersed over disproving a wider variety of models than is the case with T-optimality, with some consequent loss of power.

20.9 Nested Linear Models and D_S-optimum Designs

20.9.1 Partially Nested Linear Models

In this section we derive explicit formulae for the non-centrality parameters of two linear models containing some terms in common and establish the relationship between T-optimality and D_S-optimality. We then discuss some D_S-optimum designs for model checking. Finally we derive the explicit form of the non-centrality parameter in Example 20.1 when a centre point is added to the 2^2 factorial. Throughout interest is in comparisons of only two models.

The two models, linear in the parameters, are

$$\text{Model 1:} \quad \mathrm{E}(y) = F_1\theta_1; \quad \text{Model 2:} \quad \mathrm{E}(y) = F_2\theta_2.$$

When the second model is fitted to data generated by the first, the expectation of the least squares estimator of θ_2 is

$$\mathrm{E}\{\hat{\theta}_2(\xi)\} = M_{22}^{-1}(\xi)M_{21}(\xi)\theta_1, \quad \text{where} \quad M_{ij}(\xi) = F_i^{\mathrm{T}}WF_j. \tag{20.36}$$

The standardized non-centrality parameter when the first model is not a special case of the second model is

$$\Delta_2(\xi) = \theta_1^{\mathrm{T}}M_{1(2)}(\xi)\theta_1, \quad \text{with} \tag{20.37}$$

$$M_{1(2)}(\xi) = M_{11}(\xi) - M_{12}(\xi)M_{22}^{-1}(\xi)M_{21}(\xi).$$

If the two models have terms in common, the elements of $M_{1(2)}(\xi)$ corresponding to these elements are zero. Let the combined model with duplicate terms eliminated be

$$\mathrm{E}(y) = F\theta = F_1\theta_1 + \tilde{F}_2\tilde{\theta}_2 = \tilde{F}_1\tilde{\theta}_1 + F_2\theta_2, \tag{20.38}$$

where $\tilde{F}_j\tilde{\theta}_j$ represents the complement of model not j in the combined model $F\theta$. For example, if $j = 1$, we have the complement of model 2, that is model 1 with terms in common omitted. Then the standardized non-centrality parameter (20.37) can be replaced by

$$\Delta_2(\xi) = \tilde{\theta}_1^{\mathrm{T}}\tilde{M}_{1(2)}(\xi)\tilde{\theta}_1, \tag{20.39}$$

$$\text{where} \quad \tilde{M}_{1(2)}(\xi) = \tilde{M}_{11}(\xi) - \tilde{M}_{12}(\xi)M_{22}^{-1}(\xi)\tilde{M}_{21}(\xi),$$

$$\text{with} \quad \tilde{M}_{11}(\xi) = \tilde{F}_1^{\mathrm{T}}W\tilde{F}_1 \quad \text{and} \quad \tilde{M}_{12}(\xi) = \tilde{F}_1^{\mathrm{T}}WF_2.$$

When the two models differ by a single term, $\tilde{\theta}_1$ is scalar and

$$\Delta_1(\xi) = \tilde{\theta}_1\{\tilde{M}_{11}(\xi) - \tilde{M}_{12}(\xi)M_{22}^{-1}(\xi)\tilde{M}_{21}(\xi)\}\tilde{\theta}_1. \tag{20.40}$$

Then the value of ξ maximizing (20.40) does not depend on $\tilde{\theta}_1$ and the T-optimum design is also the $\mathrm{D_S}$-optimum for the term $\tilde{f}_1$. In general, for vector $\tilde{\theta}_1$, the $\mathrm{D_S}$-optimum design maximizes the determinant

$$|\tilde{M}_{11}(\xi) - \tilde{M}_{12}(\xi)M_{22}^{-1}(\xi)\tilde{M}_{21}(\xi)|.$$

Unlike T-optimality, this criterion does not depend on the value of $\tilde{\theta}_1$.

20.9.2 D_S-optimality and Model Checking

In the parsimonious model checking designs of §20.2 there were r primary terms and s secondary terms. The designs were particularly effective when it was desired to check for the existence of the secondary terms with fewer than $r+s$ trials. As prior information on the secondary terms decreases, the designs tend towards the D-optimum design for all $r+s$ parameters. On the contrary, D_S-optimum designs always permit estimation of all s secondary terms, so the design will often have more points of support than those of §20.2. In addition, emphasis is on precise estimates of only the secondary terms.

Example 20.7. Simple Regression To detect quadratic departures from the simple regression model

$$\mathrm{E}(y) = \eta_2(x, \theta_2) = \theta_{20} + \theta_{21}x.$$

we take as the true model the quadratic

$$\eta_1(x, \theta_1) = \theta_{10} + \theta_{11}x + \theta_{12}x^2. \tag{20.41}$$

Then model 1 is not a special case of model 2, although the converse is true. In the notation of (20.38) $\tilde{F}_1\tilde{\theta}_1$ is the column of values of $\theta_{12}x^2$ at the design points. The D_S-optimum design for θ_{12} when $\mathcal{X} = [-1, 1]$ puts half the trials at $x = 0$ and divides the other half equally between -1 and 1.

As the results of §20.9.1 show, this is also the T-optimum design for discriminating between the linear and quadratic models. The parsimonious model checking design of §20.2 tends, as the prior information on the secondary terms decreases, to the D-optimum design for the quadratic model with weights 1/3 at the support points of the D_S-optimum design.

The D_S-optimum design has a D-efficiency of $1/\sqrt{2} = 70.7\%$ for the first-order model, which may be too great an emphasis on model checking to be acceptable. The efficiency, for the first-order model, of the D-optimum design for the quadratic model is $\sqrt{(2/3)} = 81.6\%$, a higher value. The parsimonious model checking designs will therefore have efficiencies for the first-order model upwards from this value. In Chapter 21 we use compound optimality to generate designs with good efficiency properties for both model checking and parameter estimation. Like D_S-optimum designs these have at least as many support points as there are parameters in the model. ■

A generalization of Example 20.7 is to take $\eta_2(x, \theta_2)$ to be a polynomial of order $d-1$ in one factor. In order to check the adequacy of this model the D_S-optimum design is required for the coefficient of x^d in the dth order polynomial when the design region is again $[1, -1]$. The optimum design

(Kiefer and Wolfowitz 1959) was given in §11.4 and has design points

$$x_j = -\cos\left(\frac{j\pi}{d+1}\right) \quad (j = 0, 1, \ldots, d+1),$$

with weights

$$w_j = \begin{cases} \frac{1}{2(d+1)} & (j = 0, d+1) \\ \frac{1}{d+1} & (j = 1, \ldots, d). \end{cases}$$

Example 20.8. 2^2 Factorial Plus Centre Point Example 20.1 considered a family of parsimonious model checking designs for the detection of lack of fit from a first-order model in two factors. The designs were augmented 2^2 factorials. As a last example of model checking designs, we obtain an explicit expression for the non-centrality parameter when the true model is a quadratic, but the fitted model is first-order including interaction. The design is a 2^2 factorial with one centre point; a similar procedure with one centre point added to a two-level factorial design is often followed for the detection of departures from multifactor first-order models. To be specific we assume that

$$\begin{aligned} \eta_1(x, \theta_1) &= \theta_{10} + \theta_{11}x_1 + \theta_{12}x_2 + \theta_{13}x_1x_2 + \theta_{14}x_1^2 + \theta_{15}x_2^2 \\ \eta_2(x, \theta_2) &= \theta_{20} + \theta_{21}x_1 + \theta_{22}x_2 + \theta_{23}x_1x_2 \end{aligned} \tag{20.42}$$

Let the design be a 2^2 factorial with $x_j = \pm 1$ to which a single centre point is added in order to detect departures from the first-order model. Since the interaction term can be estimated from this design we have included the interaction term in the first model, a different division of terms from that in §20.9.

Calculation of the non-centrality parameter (20.22) requires the matrices

$$\begin{aligned} F_2^{\mathrm{T}} F_2 &= \operatorname{diag}\left[\ 5 \quad 4 \quad 4 \quad 4\ \right] \\ \tilde{F}_1^{\mathrm{T}} \tilde{F}_1 &= \begin{bmatrix} 4 & 4 \\ 4 & 4 \end{bmatrix} \\ \tilde{F}_1^{\mathrm{T}} F_2 &= \begin{bmatrix} 4 & 0 & 0 & 0 \\ 4 & 0 & 0 & 0 \end{bmatrix}, \end{aligned}$$

when

$$\begin{aligned} 5\Delta_2(\xi_5) &= \begin{bmatrix} \theta_{14} & \theta_{15} \end{bmatrix} \begin{bmatrix} 16/5 & 16/5 \\ 16/5 & 16/5 \end{bmatrix} \begin{bmatrix} \theta_{14} \\ \theta_{15} \end{bmatrix} \\ &= \frac{16}{5}(\theta_{14} + \theta_{15})^2. \end{aligned} \tag{20.43}$$

Thus addition of the single centre point will not lead to the detection of departures for which $\theta_{14} + \theta_{15}$ is small relative to the observational error. This result also follows from comparison of the expected average response at the factorial points and at the centre point. In (20.43) $\Delta_2(\xi_5)$ is the average per trial contribution to the non-centrality parameter when $\sigma^2 = 1$. ■

In general, trials at the centre of 2^m factorials will fail to detect departures for which $\sum_{j=1}^{m} \beta_{jj}$ is small. This condition implies that either all β_{jj} are individually small, or that there is a saddle point in the response surface. Although uncommon, saddle points do sometimes occur in the exploration of response surfaces. The parsimonious model checking designs introduced in §20.2 provide an efficient method for investigating such phenomena.

20.10 Exact T-optimum Designs in SAS

We begin our discussion of how to compute T-optimum designs in SAS with the sequential procedure given in §20.7. The first step in that procedure is to compute the derivative function $\psi_2(x, \xi_k)$, the squared difference between the predicted values of the two models, both fitted to the observed data. For Example 20.6, assuming observed values of the factor x and the response y at step k are collected in a data set named Design, the following code computes $\psi_2(x, \xi_k)$ over a set of candidate points between -1 and 1 at intervals of 0.1.

```
data Can;
   do x = -1 to 1 by 0.1; output; end;
data DesignCan; set Design Can;
   x11 = exp( x); x12 = exp(-x);
   x21 = x;       x22 = x*x;
proc reg data=DesignCan; /* Predict eta1 */
   model y = x11 x12;
   output out=P1 p=Eta1;
proc reg data=DesignCan; /* Predict eta2 */
   model y = x21 x22;
   output out=P2 p=Eta2;
data Psi2; merge P1 P2;
   where (y = .);
   Psi2 = (Eta1 - Eta2)**2;
run;
```

After defining the candidate points, the above statements combine them with the design points. It is crucial that the variable y *not* be defined in the candidate set. This has the effect of giving y missing values for the corresponding observations in the combined `DesignCan` data set. When the predicted values $\eta_i(x, \hat{\theta}_i)$ are computed in the two PROC REG steps, such observations do not contribute to estimating the parameters, but predicted values are still computed for them. The WHERE statement in the final DATA step above retains the $\psi_2(x, \xi_k)$ values for only these observations.

The second and third steps in the sequential procedure identify the point at which $\psi_2(x, \xi_k)$ is maximized and add it to the design. The following statements accomplish this by first sorting the $\psi_2(x, \xi_k)$ values in descending order and then appending the x value for the first one to the design.

```
proc sort data=Psi2;
   by descending Psi2;
data Design; set Design Psi2(obs=1 keep=x);
run;
```

SAS Task 20.5. Use the code above to find the next point to be added to the initial data given by the following data step.

```
data Design;
   input x y;
datalines;
-1 0.31644
 0 0.92008
 1 0.08340
;
```

SAS Task 20.6. (Advanced). Continuing with SAS Task 20.5, set the value of y for the next point using equation (20.27). Iterate this procedure until the design has 100 points, and compare the resulting design to the optimal continuous design given in equation (20.28). (NB: The macro programming facility of SAS will make this task much easier.)

SAS Task 20.6 essentially implements for T-optimality the forward sequential procedure defined for D-optimality in §12.4. PROC OPTEX has no features to find exact T-optimum designs, but this sequential procedure can be used to find approximate ones.

Finally, the SAS code for the sequential procedure for non-linear models is nearly identical to that for linear models. The only difference is that the linear modelling procedure used to compute the predicted values $\eta_i(x, \hat{\theta}_i)$ (PROC REG in the code above) is replaced by a non-linear modelling procedure. For example, in the context of Example 20.4, the following two PROC NLIN steps would replace the PROC REG steps in the code above (with corresponding changes to the candidates).

```
proc nlin data=DesignCan; /* Predict eta1 */
   parms t1 1;
   model y = exp(-t1*x);
   output out=P1 p=Eta1;
proc nlin data=DesignCan; /* Predict eta2 */
   parms t2 1;
   model y = 1/(1 + t2*x);
   output out=P2 p=Eta2;
run;
```

20.11 The Analysis of T-optimum Designs

T-optimum designs were introduced in §20.6 with the aim of providing a large lack of fit sum of squares for the incorrect model. As with many other optimum designs, the resulting designs, sequential or not, will contain appreciable replication. These replicate observations provide an estimate of pure error, uncontaminated by any lack of fit of the model. The experiment can therefore be analysed as were the data on the desorption of carbon monoxide that led to Table 8.1. There the lack of fit sum of squares, on four degrees of freedom, was so small compared to the error mean square, that it was clear there was no systematic lack of fit. If the lack of fit sum of squares had been larger, more powerful tests of model adequacy would have come from breaking the sum of squares into individual degrees of freedom corresponding to quadratic and, perhaps, cubic polynomial terms. Where the degrees of freedom permit, similarly powerful tests for departure can be found for T-optimum designs by fitting combined models such as that of §20.9.1 for linear models. If there are not sufficient degrees of freedom available for fitting such larger models, the lack of fit sum of squares in the analysis of variance already provides the most powerful test available from the data. This would be the case for designs such as (20.28) with four support points for discriminating between two three-parameter models. Power calculations for this four-point design use the non-central chi-squared distribution on one degree of freedom with non-centrality parameter the right-hand side of (20.22) divided by σ^2. Although sequential designs may have appreciably more points of support due to sampling variability in the values of the parameters of the models, approximate replicates can be obtained by grouping observations that are close in $\mathcal{X}$. These can then be treated as true replicates providing a virtually model-free estimate of the error variance. As a consequence the degrees of freedom of the lack of fit sum of squares are reduced, with a subsequent increase in power.

A final comment is on the power of the simulated designs in Figure 20.6. For these simulations the values of σ^2 were chosen to illustrate a variety of different designs. Calculations of the kind mentioned above for the χ_1^2 test with size 5% show that all designs, apart from that with $\sigma^2 = 0$, have power hardly greater than 5% even when $N = 40$. In order to be able to discriminate between these two models for moderate N, the error variance will have to be so small that sequential designs will be indistinguishable from that of Panel (a) in Figure 20.6.

20.12 Further Reading

Uciński and Bogacka (2005) extend T-optimality to multi-response models. Designs for heteroscedastic models are presented by López-Fidalgo, Dette, and Zhu (2005). The further extension to multi-response heteroscedastic dynamic models is in Uciński and Bogacka (2005). References to T-optimality for generalized linear models are given in §22.7.

In a series of papers Dette overcomes local optimality by reducing model testing for several parameters at once to a series of tests of single parameters, when the results of §20.9.1 on the relationship between T- and $\mathrm{D_S}$-optimality remove the dependence on the values of unknown parameters (Dette and Kwiecien 2004). If the models are not naturally nested in this way, there may be some arbitrariness in the order in which it is decided, at the design stage, that the terms will be tested (Dette and Kwiecien 2005).

21

COMPOUND DESIGN CRITERIA

21.1 Introduction

An experimenter rarely has one purpose or model in mind when planning an experiment. For example, the results of a second-order response surface design will typically be used to test the terms in the model for significance as well as being used to test for goodness of fit of the model. The significant terms will then be included in a model, the estimated response from which may be used for prediction at one or several points not necessarily within $\mathcal{X}$. In previous chapters we have found designs optimum for these individual tasks. However, a design that is optimum for one task may be exceptionally inefficient for another. An extreme example is the designs for aspects of the compartmental model in Chapter 17, such as the area under the curve, which contain too few points of support to allow estimation of the parameters of the model. We return to this example in §21.9.

In this chapter we therefore develop designs in which all aspects of interest are included, with appropriate weights, in the design criterion. The resulting compound design criterion is then maximized in the usual way to give an exact or a continuous design. In §21.2 we introduce efficiencies which are used in §21.3 to define general weighted compound design criteria. We then typically find optimum designs for a series of weights and, where possible, choose a design with good efficiencies for all features of interest.

The main emphasis is on D- and $\mathrm{D_S}$-optimality. In §21.4 we find a compound design for one-factor polynomial models of degrees two to six. In the succeeding sections we consider compound designs for model building, parameter estimation, and for discrimination between two or several models, all using aspects of weighted D-optimality. At the end of the chapter we introduce compound criteria for parameter estimation and other aspects of design. Thus, in §21.8, we explore DT-optimum designs that combine D-optimality with T-optimality for discrimination between models. The last criterion, in §21.9, we call CD-optimality; this combines parameter estimation with c-optimality for a feature of interest. Our example is a compartmental model in which it is required both to estimate the parameters of the model and to find the area under the curve.

21.2 Design Efficiencies

In order to use efficiencies to generate compound designs we need to define the efficiencies in a consistent way over different criteria. In particular, we require that the efficiencies for exact designs change in the same way as N changes whatever the design criterion. In this chapter we will look at compound designs involving c-, D-, and T-optimality. For D-optimality let the optimum design be ξ_D^*. Then the D-efficiency of any other design ξ is

$$\text{Eff}^{\,D}(\xi) = \left\{ \frac{|M(\xi)|}{|M(\xi_D^*)|} \right\}^{1/p}. \tag{21.1}$$

Raising the ratio of determinants to the power $1/p$ has the effect that the efficiency has the dimensions of a ratio of variances; two replicates of a design with an efficiency of 50% will estimate parameters with variances similar to a single replicate of the optimum design.

In §10.4 c-optimum designs were introduced to minimize $\text{var}\, c^{\mathrm{T}}\hat{\beta}$ which is proportional to $c^{\mathrm{T}}M^{-1}(\xi)c$, where c is a $p \times 1$ vector of known constants. The definition is already in terms of variance, so we can let

$$\text{Eff}^{\,C}(\xi) = \frac{c^{\mathrm{T}}M^{-1}(\xi_c^*)c}{c^{\mathrm{T}}M^{-1}(\xi)c}. \tag{21.2}$$

The c-optimum design ξ_c^* minimizes the variance of the linear combination and so appears in the numerator of the expression.

Likewise, in Chapter 20, T-optimality was defined in terms of the standardized non-centrality parameter $\Delta(\xi)$ that had the dimensions of a sum of squares that was to be maximized. Then

$$\text{Eff}^{\,T}(\xi) = \frac{\Delta(\xi)}{\Delta(\xi_T^*)}, \tag{21.3}$$

with ξ_T^* the T-optimum design measure.

21.3 Compound Design Criteria

In §10.8 the compound design criterion $\Psi\{M(\xi)\}$ was introduced as a non-negatively weighted linear combination of h convex design criteria

$$\Psi(\xi) = \sum_{i=1}^{h} a_i \Psi_i\{M_i(\xi)\}, \tag{21.4}$$

that was to be minimized by the choice of ξ. We now define a compound criterion as a weighted product of the efficiencies of §21.2 that is to be maximized. Taking logarithms yields a design criterion of the linear form (21.4). Let

$$\Upsilon^{Q}(\xi) = \prod_{i=1}^{h} \left\{ \mathrm{Eff}_i^{Q(i)} \right\}^{\kappa_i}, \tag{21.5}$$

where, in the examples in this chapter, each of the h efficiencies, $Q(i)$ is one of c, D, or T and the κ_i are specified non-negative weights that control the importance of the various efficiencies in the design criterion. Without loss of generality we can take the κ_i to sum to one. If we take logarithms in (21.5) we obtain the criterion

$$\log \Upsilon^{Q}(\xi) = \sum_{i=1}^{h} \kappa_i \log \left\{ \mathrm{Eff}_i^{Q(i)} \right\}, \tag{21.6}$$

which is to be maximized. The individual efficiencies included in (21.5) are functions both of ξ and of the optimum design $\xi^*_{Q(i)}$. The optimum compound design is found by maximizing over ξ. The $\xi^*_{Q(i)}$ are fixed and so do not affect the design although they do affect the value of the design criterion. So (21.6) becomes

$$\log \Upsilon^{Q}(\xi) = \Phi^{Q}(\xi) - \Phi^{Q}(\xi^*_{Q(1)}, \ldots, \xi^*_{Q(h)}), \tag{21.7}$$

where $\Phi^{Q}(\xi)$ is to be maximized. Specifically, for compound D-optimality (21.7) becomes

$$\log \Upsilon^{D}(\xi) = \sum_{i=1}^{h} \kappa_i/p_i \log |M_i(\xi)| - \sum_{i=1}^{h} \kappa_i/p_i \log |M_i(\xi^*_{D(i)})|$$

and we obtain the compound D-optimum criterion

$$\Phi(\xi) = -\Psi(\xi) = \sum_{i=1}^{h} \kappa_i/p_i \log |M_i(\xi)|, \tag{21.8}$$

for which the derivative function

$$\phi(x, \xi) = \sum_{i=1}^{h} (\kappa_i/p_i) d_i(x, \xi). \tag{21.9}$$

In (21.9) $d_i(x, \xi)$ is the variance function (9.8) for the ith model. For the optimum design ξ^*, the maximum value is

$$\bar{\phi}(x, \xi^*) = \sum_{i=1}^{h} \kappa_i, \tag{21.10}$$

providing the bound in the equivalence theorem for compound D-optimum designs. An important special case of (21.8) is D_S-optimality since from

(10.7),

$$\log |M_{11}(\xi) - M_{12}(\xi)M_{22}^{-1}(\xi)M_{12}^{\mathrm{T}}(\xi)| = \log |M(\xi)| - \log |M_{22}(\xi)|. \quad (21.11)$$

This relationship is of particular importance when we consider compound optimum designs for model checking and parameter estimation in §21.5. In many examples we find optimum designs for a range of values of the weights κ_i and select from these a design that has good efficiency for several aspects of the models. But we start by looking at compound optimum designs for a few specific values of the weights.

21.4 Polynomials in One Factor

In Table 11.3 we listed D-optimum designs for the one-factor polynomial

$$\mathrm{E}(y) = \beta_0 + \sum_{j=1}^{d} \beta_j x^j \quad (21.12)$$

over $\mathcal{X} = [-1, 1]$ when the order d ranged from 2 to 6. The designs have weight $1/(d+1)$ at the $d+1$ design points which are at the roots of an equation involving Legendre polynomials. It is most unlikely that a sixth-order polynomial would be needed in practice. But it is of interest to see how good the D-optimum sixth-order design is when used for fitting models of lower order and to make a comparison with the equally weighted compound D-optimum design. The D-efficiencies of the sixth-order design relative to the optimum designs for lower orders are given in column 2 of Table 21.1. As the order of the polynomial decreases, so does the number of support points of the optimum design for that polynomial. Therefore, it is to be expected that the efficiency of the seven-point design for the sixth-order model decreases as the models to be fitted become simpler. However, the decrease is not very sharp; even for estimation of the quadratic model the efficiency is 79.7%.

If in the compound criterion (21.8) we take all $\kappa_i = 1$ we obtain the equally weighted design minimizing

$$\Psi \{M(\xi)\} = -\sum_{d=2}^{6} \frac{\log |M_d(\xi)|}{d+1}, \quad (21.13)$$

where $M_d(\xi)$ is the information matrix for the dth-order polynomial (21.12) which contains $p_i = d+1$ parameters. The derivative function for this

TABLE 21.1. D-efficiencies of D-optimum and compound D-optimum designs for one-factor polynomials from second to sixth order

	D-efficiency Eff^D (%)	
Order of model	Sixth-order polynomial	Equally weighted (21.13)
2	79.75	83.99
3	84.64	89.11
4	88.74	92.40
5	92.81	94.47
6	100.00	95.55

compound design is

$$\phi(x, \xi) = \sum_{d=2}^{6} \frac{d_d(x, \xi)}{d+1}, \tag{21.14}$$

where $d_d(x, \xi)$ is the derivative function for the dth order polynomial. For this optimum compound design

$$\bar{\phi}(\xi^*) = 5. \tag{21.15}$$

Table 21.1 also gives the D-efficiencies for this compound design for polynomials from the second to the sixth-order. These range from 83.99% to 95.55%. For all models except that for $d = 6$ the compound design has higher efficiency than the optimum design for the sixth-order polynomial.

The two designs do not seem very different. The D-optimum design for the sixth-order polynomial is

$$\left\{ \begin{array}{ccccccc} -1 & -0.8302 & -0.4688 & 0 & 0.4688 & 0.8302 & 1 \\ 0.1429 & 0.1429 & 0.1429 & 0.1429 & 0.1429 & 0.1429 & 0.1429 \end{array} \right\},$$

whereas the compound design is

$$\left\{ \begin{array}{ccccccc} -1 & -0.7891 & -0.4346 & 0 & 0.4346 & 0.7891 & 1 \\ 0.1954 & 0.1152 & 0.1248 & 0.1292 & 0.1248 & 0.1152 & 0.1954 \end{array} \right\}.$$

The main difference seems to be that the compound design has weight 0.1954 at $x = \pm 1$, more than the sixth-order design, that has weight $1/7$ at all design points. Of course, the compound design has weights less than $1/7$ at the other design points.

A more marked effect of the design criterion on the properties of the design is revealed in Figure 21.1 which shows the derivative functions. The

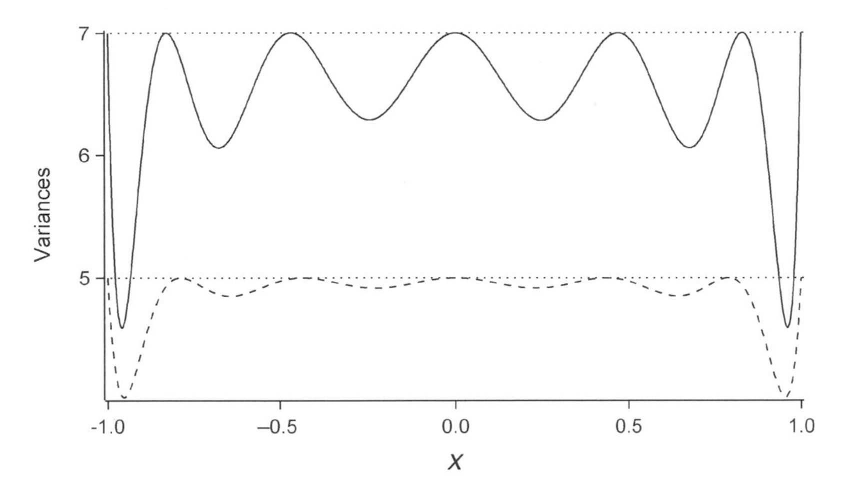

FIG. 21.1. Derivatives $\phi(x, \xi^*)$ for designs for sixth-order polynomial in one factor: continuous, D-optimum design for sixth-order polynomial, $\bar{\phi}(\xi^*) = 7$; dotted, equally weighted design (21.13), $\bar{\phi}(\xi^*) = 5$.

most obvious feature of this figure is that the variance $d(x, \xi)$ for the sixth-order design has a maximum value of seven, whereas from (21.15) that for the compound design is five. However, this feature depends on the scaling of the coefficients for the compound design in (21.13). If we replace $1/(d{+}1)$ by $7/\{5(d{+}1)\}$, the maximum value of (21.14) will also be seven. The important difference is that the curve for optimum design for the sixth-order polynomial fluctuates appreciably more than that for the equally weighted design. As would be expected from the Bayesian D-optimum designs of Chapter 18, averaging the design criterion over five models results in a flatter derivative function at the optimum design.

The comparisons here are for continuous designs. Exact designs for seven or a few more trials can be expected to show a lesser effect of changing the design criterion owing to the difficulty of responding to the slight departures of the weights of the design from the uniform distribution from the D-optimum design for the sixth-order model.

21.5 Model Building and Parameter Estimation

In §20.2.2 we found designs to detect departures from the model

$$\mathrm{E}(y) = F_1\beta_1 \tag{21.16}$$

by extending the model by the addition of further terms to give

$$\mathrm{E}(y) = F_1\beta_1 + F_2\beta_2 = F\beta, \tag{21.17}$$

where β_1 was $r \times 1$, β_2 was $s \times 1$ and β was $p \times 1$ where $p = r + s$. The Bayesian designs of DuMouchel and Jones (1994) did not necessarily require designs with as many as p points of support, so that it was not always possible to estimate all elements of β. However, the $\mathrm{D_S}$-optimum designs of §20.9 do lead to estimation of all parameters, through designs that minimize the generalized variance of the estimates of β_2. A potential disadvantage of this procedure is that the experimental effort is concentrated on checking whether the reduced model is true, rather than on estimating the parameters of that model. In this section compound D-optimality is used to provide dual purpose designs that can be chosen to trade off efficiency of estimation of the parameters β_1 against efficiency for the model checking parameters β_2. We start by considering continuous designs. The information matrix for $\mathrm{D_S}$-optimality is given (21.11). Combining $\mathrm{D_S}$-optimality for β_2 with D-optimality for β_1 gives the compound criterion

$$\begin{aligned}\Phi(\xi) &= \frac{\kappa}{r}\log|M_{11}(\xi)| + \frac{1-\kappa}{s}\{\log|M(\xi)| - \log|M_{11}(\xi)|\}\\ &= \frac{(p\kappa - r)}{rs}\log|M_{11}(\xi)| + \frac{1-\kappa}{s}\log|M(\xi)|.\end{aligned} \tag{21.18}$$

By construction we have D-optimality for β_1 when $\kappa = 1$ and $\mathrm{D_S}$-optimality for β_2 when $\kappa = 0$. However, (21.18) shows that, for $\kappa = r/p$ we have D-optimality for β. This D-optimum design was that with the strongest emphasis on model checking that could be obtained with the procedure of DuMouchel and Jones (1994). Here we obtain a stronger emphasis on model checking with values of $\kappa < r/p$. From (21.9) the derivative function for the criterion (21.18) is

$$\phi(x, \xi) = \frac{(p\kappa - r)}{rs} f_1(x)^{\mathrm{T}} M_{11}^{-1}(\xi) f_1(x) + \frac{1-\kappa}{s} f(x)^{\mathrm{T}} M^{-1}(\xi) f(x). \tag{21.19}$$

For the design ξ^* maximizing (21.18), the maximum value of (21.19) is $\bar{\phi}(x, \xi^*) = 1$. In order to use the criterion a value of κ can be specified that reflects the relative levels of interest in estimation of β_1 and of checking the simple model. A more realistic approach is to find the optimum design for a series of values of κ and to calculate the efficiencies of each design for parameter estimation and model checking. There is then a basis for the choice of design and, implicitly, for the value of κ and the relative importance of the two aspects of the compound design criterion. As a result, exact designs with efficiencies approximately known in advance can be explored

TABLE 21.2. Linear or quadratic regression: D- and D_S-optimum designs for components of the problem

Criterion	κ	w	Eff^D	Eff_s^D
D_S for β_2	0	1/2	0.707	1.000
Equally weighted	0.5	3/5	0.775	0.960
D for β	2/3	2/3	0.816	0.889
D for β_1	1	1	1.000	0.000

for the designated N in the neighbourhood of this value of κ. As examples of continuous designs we take compound criteria for the estimation of first-order models and the detection of departures from them. We start with the simplest case, that of a single factor that was explored in Example 20.7.

Example 21.1. Linear or Quadratic Regression? For linear regression the simple model (21.16) is

$$\mathrm{E}(y_i) = \beta_0 + \beta_1 x_i, \tag{21.20}$$

with the augmented model (21.17)

$$\mathrm{E}(y_i) = \beta_0 + \beta_1 x_i + \beta_2 x_i^2. \tag{21.21}$$

As we have seen, the D-optimum design for (21.20) over the design region $\mathcal{X} = [-1, 1]$ puts weights 1/2 at $x = \pm 1$, whereas the D-optimum design for the augmented model (21.21) puts weight 1/3 at $x = -1, 0$, and 1. The D_S-optimum design for β_2 has the same three support points, but with weights 1/4, 1/2, and 1/4. Because of the simple structure of these designs we do not need to maximize (21.18) over $\mathcal{X}$ for each κ. Rather, we consider only continuous designs of the form

$$\xi_w = \left\{ \begin{array}{ccc} -1 & 0 & 1 \\ w/2 & 1-w & w/2 \end{array} \right\} \quad (1/2 \leq w \leq 1). \tag{21.22}$$

With this structure we can make appreciable analytical progress in finding optimum designs and their properties. In general problems, numerical methods will be needed, but our analytical results illustrate many properties of compound designs from more complicated formulations.

In this example $r = 2$ and $s = 1$, so that D-optimality for β is obtained when $\kappa = r/p = 2/3$. Table 21.2 gives the optimum values of w for four values of κ. To find the optimum w for general κ we note that, for the design (21.22)

$$|M_{11}(\xi_w)| = w \text{ and } |M(\xi_w)| = w^2(1-w),$$

so that (21.18) reduces to maximizing

$$\Phi(\xi) = (1-\kappa/2)\log w + (1-\kappa)\log(1-w) \tag{21.23}$$

for fixed κ. The optimum design weights are therefore given by

$$w^* = \frac{2-\kappa}{4-3\kappa}. \tag{21.24}$$

The D-efficiency of any design ξ_w of the form (21.22) for the first-order model is

$$\text{Eff}^D = \left\{\frac{|M(\xi_w)|}{|M(\xi_D^*)|}\right\}^{1/2} = \sqrt{w}.$$

For D_S-optimality

$$|M(\xi_w)|/|M_{11}(\xi_w)| = w(1-w).$$

The optimum design has $w = 1/2$ so that the D_S-efficiency of ξ_w is

$$\text{Eff}_s^D = 4w(1-w).$$

We can now see how the properties of the optimum design depend on κ. The design itself is represented in the top panel of Figure 21.2 by a plot of w^* against κ. The values rise steadily from 1/2 to one as κ increases. The bottom panel of Figure 21.2 shows the efficiencies as a function of κ. The efficiency Eff^D for the first-order model increases gradually from $\sqrt{2}/2$ to 1 with κ, whereas the efficiency for model checking Eff_s^D decreases with κ. Although it is zero for $\kappa = 1$, it is high over a large range of κ values, so that it is possible to find a design with high efficiency for both aspects. For example, the product of the efficiencies is maximized when $\kappa = 0.5$, that is when $w^* = 3/5$. Then $\text{Eff}^D = 0.775$, whereas Eff_s^D has the surprisingly high value of 0.96. These values are given in Table 21.2 along with the efficiencies for the D-optimum design for β and those for $\kappa = 0$ and 1.

Exact designs can be found by rounding the values of w. For example, the D-optimum design for the second-order model would provide a design with good values of both efficiencies when N is a multiple of 3. However, the equally weighted design with $w = 3/5$ would give an exact design with $N = 10$. If this is too large a design the methods of construction of exact designs should be used. ■

In Example 21.1 the designs for the various criteria were not especially different. We now consider an example with four factors in which the number of support points of the continuous design is appreciably greater than p, the number of parameters. We focus on an exact design for which N is only slightly greater than p.

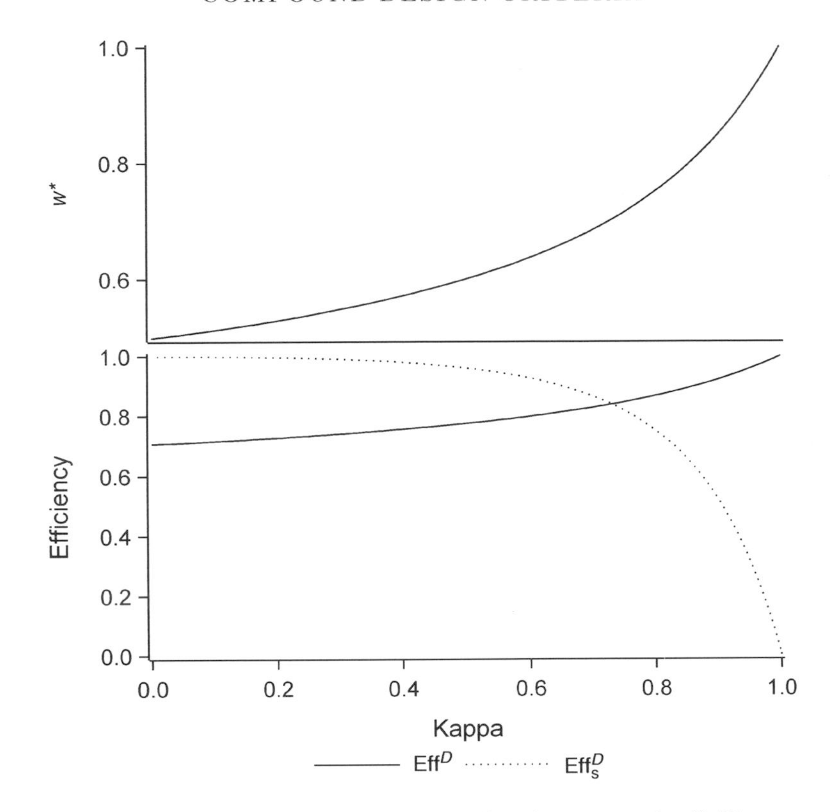

FIG. 21.2. Example 21.1: linear or quadratic regression? Top panel, w^* (combined weight on -1 and 1) against κ; bottom panel, Eff^D and Eff_s^D against κ.

Example 21.2. Response Surface in Four Factors With four factors the results of a 2^4 design can be used to fit a first-order model with interactions. We take as our primary model in (21.17)

$$f_1^{\mathrm{T}}(x)\beta_1 = \beta_0 + \sum_{i=1}^{4} \beta_i x_i + \sum_{i=1}^{3} \sum_{j=i+1}^{4} \beta_{ij} x_i x_j, \tag{21.25}$$

that is a first-order model with two-factor interactions. The secondary terms are

$$f_2^{\mathrm{T}}(x)\beta_2 = \sum_{i=1}^{4} \beta_i x_i^2. \tag{21.26}$$

TABLE 21.3. Example 21.2: response surface in four factors. Efficiencies for exact four-factor designs compound optimum for both parameter estimation and checking the quadratic terms: $\kappa = 0.5$

N	Candidates	Compound	Eff^D	Eff_S^D
15	Grid	0.833	0.589	0.595
	Off-grid	0.858	0.617	0.602
20	Grid	0.944	0.628	0.716
	Off-grid	0.948	0.634	0.715
25	Grid	0.977	0.594	0.811
	Off-grid	0.979	0.597	0.811
Continuous		1.000	0.613	0.824

The numbers of parameters in these models are $r = 11$ and $s = 4$, so that $p = 15$.

The D-optimum continuous design for the full second-order model with 15 parameters is given in Table 11.6. It has support at those points of the 3^4 factorial with 0, 3, and 4 non-zero co-ordinates, in all 49 points of support, more than three times the number of parameters to be estimated. Nonetheless, in Table 11.7 we report a design with $N = 15$ for which the D-efficiency is 87.0%. To calibrate the efficiency of the exact compound designs that we find we also need the D_S-optimum design for the quadratic terms. This has the property that weights w_i are uniform on groups of points that have the same number i of non-zero factor values, with weights given by

$$(w_0, w_1, w_2, w_3, w_4) = (0.065, 0.247, 0.372, 0.257, 0.060).$$

However, we do not need to know these optimum continuous designs when searching for an exact design for specified N, although this knowledge is helpful in comparing designs for different N if the size of the experiment is only approximately specified.

For the four-factor model we obtain the D-optimum design for the second-order model when $\kappa = 11/15$. We take a smaller value $\kappa = 0.5$, so focusing more on model checking. Table 21.3 lists the compound and component efficiencies for exact designs with $N = 15$, 20, and 25 points. For each number of points, the first design listed has points only on the $\{-1, 0, 1\}^4$ grid, while the second design results from improving these points continuously. Apparently, only a small improvement in efficiency is possible by considering points not on the grid. Also listed are the compound and component efficiencies for the optimum continuous design; of course, this design has 100% compound efficiency. Note that at $N = 20$ the compound optimum exact design has nearly 95% compound efficiency and superior D-efficiency, relative to the continuous design. ■

21.6 Non-linear Models

In Example 21.1 we showed how designs for a series of values of the parameter κ could lead to designs with good properties both for estimation of the parameters of a model and for checking that model. In this section we consider balancing the detection of departures from a non-linear model against parameter estimation.

Example 21.3. First-order Growth Linearization of the first-order growth model

$$\eta(t, \theta) = 1 - \exp(-\theta t) \quad (t, \theta \geq 0) \tag{21.27}$$

yields

$$f(\theta, t) = \frac{\partial \eta}{\partial \theta} = t \exp(-\theta t) \tag{21.28}$$

which, apart from a change of sign, is the derivative for the exponential decay model given by (17.9). The value of the sensitivity is zero at $t = 0$, rising to a maximum with increasing t and then declining again to zero as $t \to \infty$. The locally optimum design puts all trials at the maximum point of $f(\theta, t)$, that is at $t = 1/\theta_0$, where θ_0 is the point prior value of θ.

This one-point design provides no information about departures from the model. Following the discussion in §20.4.4 about the detection of departures from non-linear models we apply the second method and add a quadratic term in the parameter sensitivity, obtaining the augmented model

$$\eta(t, \theta) = \beta_1 f(\theta, t) + \beta_2 f(\theta, t)^2.$$

Since f varies between zero and a maximum, the design problem is equivalent to that for the quadratic

$$\mathrm{E}(y) = \beta_1 x + \beta_2 x^2 \quad (0 \leq x \leq 1), \tag{21.29}$$

where the corresponding values of t are found as solutions of

$$x = \theta t \exp(1 - \theta t). \tag{21.30}$$

With this extension of the model, the design for detecting departures from (21.27) is the $\mathrm{D_S}$-optimum design for β_2 in (21.29) which has trials at $\sqrt{2} - 1$

TABLE 21.4. First-order growth model. D_S-optimum design for detecting departures from (21.27) by estimation of a quadratic term in the parameter sensitivity $\partial\eta/\partial\theta$

x	t	w	Eff ^{D}
$\sqrt{2}$ - 1	$\left\{\begin{array}{c}1.83\\29.7\end{array}\right\}$	0.707	
1	10.0	0.293	0.414

and 1 (9.5). The existence of two values of t satisfying (21.30) for all non-negative $x < 1$ allows increased flexibility in the choice of a design. The resulting optimum design for $\theta = 0.1$ is given in Table 21.4. The times of 1.83 and 29.7 both give the same value of x and $\sqrt{2}/2$ of the trials can be arbitrarily divided between them without affecting the efficiency of the design for either parameter estimation or for the detection of departures.

The efficiency of the design of Table 21.4 for the estimation of θ is 0.414. To find designs with greater emphasis on parameter estimation we calculate compound designs for a series of values of κ in a manner similar to that for the linear model of Example 21.1.

Figure 21.3 shows the optimum compound designs. Again $\kappa = 1$ corresponds to the D-optimum design for estimation of θ in the simple model, here (21.27). The top panel of the plot shows the lower solution of (21.30), which together with $t = 10$ forms the support of an optimum design. The middle panel shows the weight on this lower solution. The value decreases gradually from $\sqrt{2} - 1$ to zero as κ increases; the decrease is particularly fast as κ approaches one. Finally, the lower panel in Figure 21.3 shows the efficiencies as a function of κ. The efficiency Eff^D for the first-order model increases gradually from $\sqrt{2}/2$ to 1 with κ, whereas the efficiency for model checking Eff_s^D decreases with κ. Although it is zero for $\kappa = 1$, this efficiency is high over a large range of κ values, for example 0.761 when $\kappa = 0.6$. Again it is possible to find designs with high efficiency for both aspects.

In this example $r = s = 1$ so that the D-optimum design for the augmented model (21.29) corresponds to $\kappa = 0.5$ and is the design for which the product of the efficiencies is maximized. Since this is a D-optimum design with two support points for a model with $p = 2$, the optimum weights are $w^* = 0.5$. It is then straightforward to show that the corresponding support points of the design are $x = 0.5$ and one. For this design $\text{Eff}^D = 0.625$, whereas Eff_s^D has a value of 0.849.

Since the designs are only locally optimum, spreading the design weight over the two values of t corresponding to a single value of x is a sensible

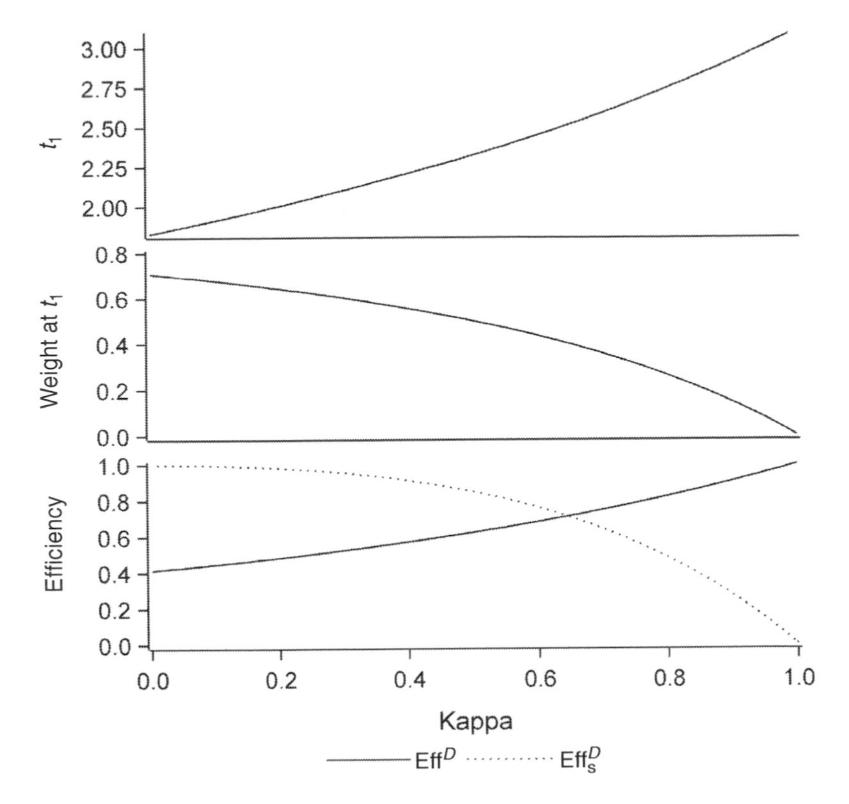

FIG. 21.3. Example 21.3: first-order growth model. Compound optimum designs for parameter estimation and model checking. Top panel, lower design point; middle panel, weight on lower design point; bottom panel, Eff^D (increasing curve) and Eff$_s^D$ against κ for augmented non-linear model.

precaution against a poor choice of θ_0. Despite the wide range of the values of t that this procedure produces, the efficiencies for parameter estimation, if θ_0 holds, are surprisingly high. ■

This non-linear example shares with Example 21.1 the feature that a trade-off between efficiencies can be established by inspecting a range of potential designs. One additional feature of the non-linear design is the choice of experimental conditions from solving (21.30). A second additional feature is that the designs given here are only locally optimum. If a prior distribution is available for θ, rather than a single value, the Bayesian methods of Chapter 18 can be used to provide designs optimum over a distribution of parameter values.

21.7 Discrimination Between Models

21.7.1 Two Models

As a final example of the properties of compound D-optimum designs we return to the problem of discrimination between two models introduced in §20.6 where T-optimum designs were defined. In this section we discuss compound D-optimum designs for the joint purposes of model discrimination and parameter estimation.

We continue with the case when there are two linear models. Then, as in §20.9.1, the combined model can be written

$$\mathrm{E}\,(y) = F\beta = F_1\beta_1 + \tilde{F}_2\tilde{\beta}_2 = \tilde{F}_1\tilde{\beta}_1 + F_2\beta_2,$$

where $\tilde{F}_j\tilde{\beta}_j$ represents the complement of model not j in the combined model $F\beta$. The parameter β_j has r_j elements and the dimension of β is $p \times 1$. So, for example, the dimension of $\tilde{\beta}_2$ is $p - r_1 \times 1$. An alternative to the T-optimum design for model discrimination when model 2 is fitted to the data is the $\mathrm{D_S}$-optimum design for $\tilde{\beta}_1$ mentioned in mentioned in §20.9.2. Provided $\tilde{\beta}_1$ is a vector, these designs, unlike the T-optimum designs, do not depend on the value of the parameter. Such designs are useful in the absence of prior information about parameters when sequential experiments are not possible.

We now derive compound designs that include parameter estimation as well as model discrimination. From (21.18) the compound D-optimum design for estimation of β_1 combined with the detection of departures in the direction of $\tilde{\beta}_2$ maximizes

$$\Phi_1(\xi) = \frac{(p\kappa_1 - r_1)}{r_1(p - r_1)} \log|M_{11}(\xi)| + \frac{1 - \kappa_1}{p - r_1} \log|M(\xi)|.$$

Similarly, when model 2 is fitted and departures are to be detected in the direction of model 1

$$\Phi_2(\xi) = \frac{(p\kappa_2 - r_2)}{r_2(p - r_2)} \log|M_{22}(\xi)| + \frac{1 - \kappa_2}{p - r_2} \log|M(\xi)|,$$

is to be maximized. A variety of compound design criteria can be found by combination of $\Phi_1(\xi)$ and $\Phi_2(\xi)$. We explore only the simplest in which there is the same interest in combining parameter estimation and model discrimination for each model, so that there is a single value of κ. If there is also the same interest in each model we use an unweighted combination and seek designs to maximize

$$\Phi(\xi) = \frac{(p\kappa - r_1)}{r_1(p - r_1)} \log|M_{11}(\xi)| + \frac{(p\kappa - r_2)}{r_2(p - r_2)} \log|M_{22}(\xi)| + (1 - \kappa)\left(\frac{1}{p - r_1} + \frac{1}{p - r_2}\right) \log|M(\xi)|, \qquad (21.31)$$

a criterion that depends on the information matrices of the two-component models and on that of the combined model. A difference from the compound design of §21.5 for a single model is that, unless $r_1 = r_2$, there is no longer a value of κ for which we obtain D-optimality for the combined model.

It is convenient to write (21.31) as

$$\Phi(\xi) = \nu_1 \log|M_{11}(\xi)| + \nu_2 \log|M_{22}(\xi)| + \nu_3 \log|M(\xi)|. \qquad (21.32)$$

Then, from (21.9) the derivative function is

$$\phi(x, \xi) = \nu_1 f_1^{\mathrm{T}}(x) M_{11}^{-1}(\xi) f_1(x) + \nu_2 f_2^{\mathrm{T}}(x) M_{22}^{-1}(\xi) f_2(x) + \nu_3 f^{\mathrm{T}}(x) M^{-1}(\xi) f(x), \qquad (21.33)$$

so that, from (21.10), the optimum design ξ^* is such that $\bar{\phi}(x, \xi^*) = 2$.

Example 21.4. Two Linear Models To exemplify these ideas we return to Example 20.6 in §20.6 where we found the T-optimum design for discriminating between the two models

$$\eta_1(x, \beta_1) = \beta_{10} + \beta_{11} e^{x} + \beta_{12} e^{-x}$$
$$\eta_2(x, \beta_2) = \beta_{20} + \beta_{21} x + \beta_{22} x^2,$$

when the first model is true with parameter values given by the model (20.27). For the compound D-optimum designs of this section, the true values of the parameters do not affect the design. However, the efficiency of the design for model discrimination will depend on the parameter values.

There are several non-compound designs that may be of interest. Each model has three parameters so $r_1 = r_2 = 3$. The T-optimum design in (20.28) maximizing $\Delta_2(\xi)$ has four points of support. The component D-optimum designs for each model are the same, putting weights 1/3 at $x = -1$, 0, and 1. Since the two models share only a single term, $p = 5$ and the D-optimum design maximizing $\log|M(\xi)|$ will have five points of support. Since $r_1 = r_2$ it follows from (21.31) that this is the optimum compound design when $\kappa = 3/5$. We compare this design with the design for $\kappa = 0.5$ that maximizes the product of the efficiencies for parameter estimation and for departure from the individual models.

TABLE 21.5. Example 21.4: two linear models. Individual model efficiencies and T-efficiencies for designs for $\kappa = 0.6$ and 0.5

κ	$x = \{$ −1.000	−0.659	0.000	0.659	1.000$\}$	Eff_1^D	Eff_2^D	Eff^T
0.5	$w = \{$ 0.192	0.208	0.199	0.208	0.192$\}$	0.775	0.787	0.735
0.6	$w = \{$ 0.200	0.200	0.200	0.200	0.200$\}$	0.818	0.827	0.714

Table 21.5 gives the optimum compound designs for $\kappa = 0.6$ and 0.5, together with some efficiencies. The two designs are similar in support points and weights and their properties are similar too. The table gives the efficiencies of the designs for the two individual models and for the T-optimum design of (20.28). To stress that this T-optimum design requires knowledge of the parameters of the true model, we repeat Figure 20.5 as Figure 21.4 with the addition of the efficiency of the design for $\kappa = 0.5$. As σ^2 increases, the variability of the simulations increases and more sequential trials are required for the expected value of the efficiency of the T-optimum design to be equal to that of the compound design found here. ■

Compound D-optimum designs of the kind illustrated here will usually have more points of support than the T-optimum designs and the D-optimum designs for the parameters of individual models. They will therefore be less efficient than the D-optimum designs and also the T-optimum designs for the true, but unknown, parameter values and models. However, as the above example shows, they may be more efficient than the T-optimum designs in the presence of appreciable experimental error, which may mislead the sequential generation of the designs. The D-optimum designs may also often be efficient alternatives to the Bayesian T-optimum designs of §20.8.2 when there is appreciable prior uncertainty about the parameter values.

21.7.2 Three Models

If, as in §20.8.1, there are $v \geq 3$ models, each may be embedded in a more general model and compound designs found for checking and parameter estimation for a weighted criterion including each of the models. Three possible schemes for the model discrimination part of the criterion are:

1. Form a general model by combining all the linearly independent terms in the v regression models. To detect departure from model j in the direction of the other models, the $\mathrm{D_S}$-optimum design would be used for the complement of the jth model in the combined model.

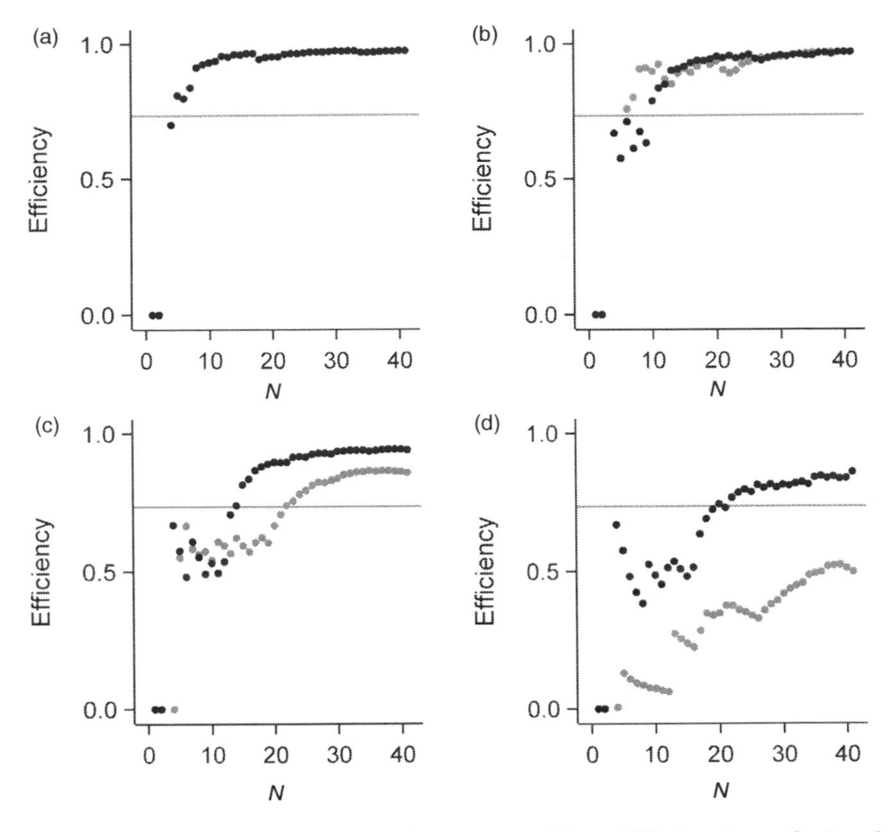

FIG. 21.4. Example 21.4: two linear models. Efficiencies of simulated sequential T-optimum designs for increasing error variance. Reference line marks the efficiency (73.5%) of the compound D-optimum design with $\kappa = 0.5$.

2. Form v general models of the type described in §20.4.4, each specific for departures from one of the models.

3. Form the $v(v-1)/2$ combined models for all pairs of possible models. As in 1 and 2 a composite D-optimum criterion can be used, with appropriate weightings for model discrimination and estimation of the parameters of the v models.

A disadvantage of 1 is that, with several models, the combined model may contain so many terms that the resulting optimum design, with many points of support, may be far from optimum for parameter estimation and for detection of departures for a single model.

Example 21.5. Three Growth Models In §21.6 we used the first-order growth model, Example 21.3, to illustrate the use of a quadratic in the parameter sensitivities in the design of experiments for model checking. We now briefly discuss discrimination between this model and two other models for responses increasing from zero to one. The three models are

$$\begin{array}{rcl} \eta(t,\theta) & = & 1-\exp(-\theta t) \qquad\qquad\qquad (t,\theta\geq 0) \\ \eta(t,\phi) & = & \frac{\phi t}{1+\phi t} \qquad\qquad\qquad\qquad (t,\phi\geq 0) \\ \eta(t,\psi) & = & \left\{\begin{array}{l} \psi t \\ 1 \end{array}\right. \qquad (t,\psi\geq 0), \left\{\begin{array}{l} (t\leq 1/\psi) \\ (t>1/\psi) \end{array}\right. \end{array} \tag{21.34}$$

In the third model the response increases linearly from zero to one at $t = 1/\psi$ and is one thereafter. The model is of a 'bent-stick' form.

Designs for discrimination between these three one-parameter models combined with estimation of the parameters can be found using the methods for model augmentation listed above. However, the three models are all special cases of the model for the reaction $A \rightarrow B$ described by the kinetic differential equation

$$\frac{d[B]}{dt} = \theta[A]^{\lambda}, \tag{21.35}$$

when the initial concentration of A is one and that of B is zero. The models in (21.34) correspond respectively to $\lambda = 1$, 2, and 0. The models can therefore all be combined in a single model and the parameter λ estimated. With this special structure it is not necessary to use the more cumbersome linear combination of models such as that in §21.7 which forms the basis of many of the methods exhibited in this chapter. ■

21.8 DT-Optimum Designs

21.8.1 Two Models

In the last section D-optimum designs were found for the dual purposes of model discrimination and parameter estimation. In this section we explore designs that combine T-optimality for model discrimination with D-optimality for parameter estimation. An advantage is that the designs will have fewer points of support than those derived from D-optimality where the combined model has to be estimable; the DT-optimum designs will therefore have greater efficiency. However, the inclusion of T-optimality in the criterion does mean that the designs depend on the values of unknown parameters and so will be only locally optimum.

The specific compound design criterion will depend upon the particular problem. We assume that there are two linear models. Model 1 is believed true, so we want to estimate its parameters. But we also want to generate evidence that model 2 is false. From (21.6) we required designs to maximize

$$\kappa \log \text{Eff}^{\,D} + (1-\kappa) \log \text{Eff}^{\,T}.$$

Since, as before, the optimum design does not depend on the individual optimum designs, we find a design to maximize

$$\Phi(\xi) = \kappa \log |M_1(\xi)|/p_1 + (1-\kappa) \log \Delta_2(\xi). \tag{21.36}$$

Designs maximizing (21.36) are called DT-optimum and are denoted ξ^*_{DT}. The derivative function for (21.36) is

$$\phi^{(DT)}(x,\xi) = (1-\kappa)\psi_2(x,\xi)/\Delta_2(\xi) + (\kappa/p_1)d_1(x,\xi)$$

$$= (1-\kappa)\{\eta_t(x) - \eta_2(x,\hat{\theta}_{t2})\}^2/\Delta_2(\xi) + (\kappa/p_1)f_1^{\mathrm{T}}(x)M_1^{-1}(\xi)f_1(x). \tag{21.37}$$

The upper bound of $\phi^{(DT)}(x,\xi^*_{DT})$ over $x \in \mathcal{X}$ is one, achieved at the points of the optimum design.

The first term in (21.37) is the derivative of $\log \Delta_2(\xi)$, the logarithm of the criterion for T-optimality. This derivative function $\psi_2(x,\xi)$ is given, for the optimum design, in (20.24). Since

$$\frac{\partial \log \Delta_2(\xi)}{\partial \alpha} = \frac{1}{\Delta_2(\xi)} \frac{\partial \Delta_2(\xi)}{\partial \alpha},$$

we obtain the first term. The second term in $\psi^{(DT)}(x,\xi)$ is that from D-optimality.

Example 21.6 Constant or Quadratic Model? To illustrate these results we return to the example of choice between a constant and a quadratic model introduced in §20.6. The two models are

$$\eta_t(x) = \eta_1(x) = \beta_{1,0} + \beta_{1,1}x + \beta_{1,2}x^2, \quad \eta_2(x,\beta) = \beta_2. \tag{21.38}$$

The simpler model 2 is thus nested within the quadratic. The T-optimum design maximizing $\Delta_2(\xi)$ provides maximum power for testing that $\beta_{1,1}$ and $\beta_{1,2}$ are both zero. The D-optimum design provides estimates of the three parameters in model 1, $\eta_t(x)$. For the parameter values $\beta_{1,0} = 1$, $\beta_{1,1} = 0.5$, and $\beta_{1,2} = 0.8660$, with $\mathcal{X} =$ [-1, 1] the two component optimum designs are

$$\xi^*_T = \left\{ \begin{array}{cc} -0.2887 & 1 \\ \frac{1}{2} & \frac{1}{2} \end{array} \right\}, \quad \xi^*_D = \left\{ \begin{array}{ccc} -1 & 0 & 1 \\ \frac{1}{3} & \frac{1}{3} & \frac{1}{3} \end{array} \right\}. \tag{21.39}$$

The two-point T-optimum design depends solely on the ratio $\beta_{1,2}/\beta_{1,1}$, here $\sqrt{3}$. It has zero efficiency for estimating the parameters of the model. On

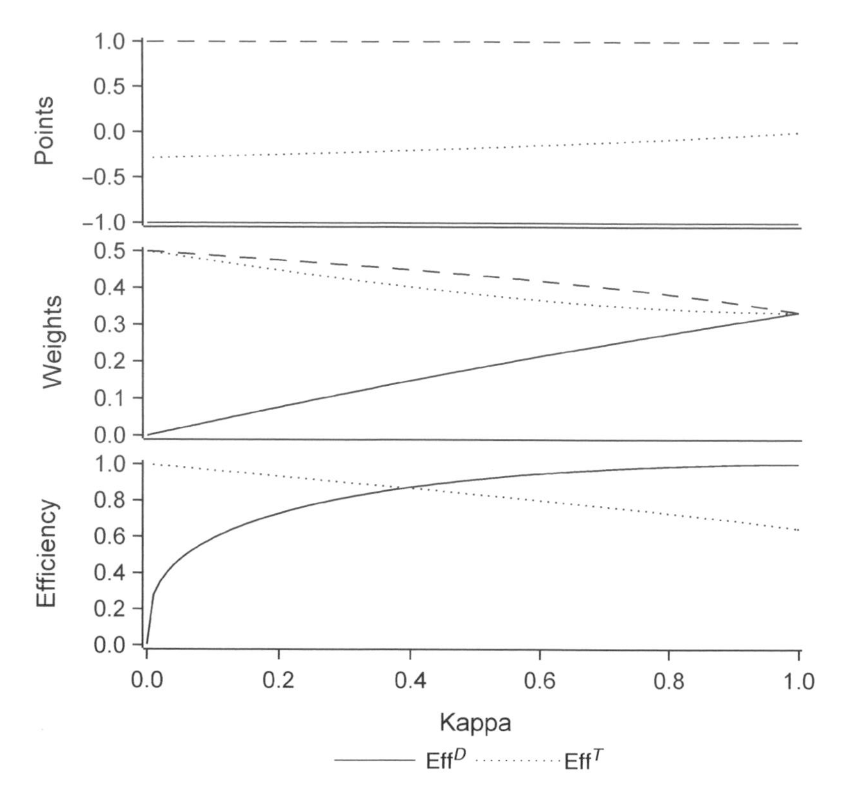

FIG. 21.5. Example 21.6: constant or quadratic model? Structure of DT-optimum designs as κ varies. Top panel, support points x_i; middle panel, design weights w_i; bottom panel, component efficiencies (Eff ^{D} increasing curve). The same line pattern is used in the top two panels for the same value of i.

the other hand, the efficiency of the D-optimum design for testing, Eff ^{T}, is a respectable 64.46%.

To explore the properties of the proposed designs, DT-optimum designs were found for a series of values of κ between zero and one. The resulting designs are plotted in Figure 21.5. All designs were checked for optimality using the equivalence condition based on (21.37). The top panel shows the design points; for $\kappa > 0$ there are three at -1, 1 and an intermediate value which tends to 0 as $\kappa \rightarrow 1$. The weights in the middle panel change in a similarly smooth way from 0, 0.5, and 0.5 to one-third at all design points when $\kappa = 1$.

Also in Figure 21.5, the bottom panel shows that the efficiencies likewise change in a smooth manner. The T-efficiency decreases almost linearly as κ increases, whereas the D-efficiency increases rapidly away from zero as the design weight at $x = -1$ becomes non-negligible. The product of these efficiencies is, as it must be from (21.36), a maximum at $\kappa = 0.5$. The T-efficiency for this value is 83.38%, with the D-efficiency equal to 91.63%. We have found a single design that is highly efficient both for parameter estimation and model testing. ■

21.8.2 Several Models and Non-centrality Parameters

In the simple version of DT-optimality in the previous section we assumed that there was one non-centrality parameter of interest and one set of parameters. This is appropriate when one model is nested within the other. But, even with two non-nested models, such as Example 21.4 with two polynomial models, there are two non-centrality parameters that may be of interest, as well as the two sets of parameters for the two models. The design criterion generalizes in a straightforward manner.

Let there be c standardized non-centrality parameters $\Delta_j(\xi)$ and m sets of model parameters p_k; in some cases only some subsets of the parameters may be of interest. The criterion (21.36) expands to maximization of

$$\Phi^{(GDT)}(\xi) = \sum_{j=1}^{c} a_j \log \Delta_j(\xi) - \sum_{k=1}^{m} (b_k/s_k) \log |A_k^{\mathrm{T}} M_k^{-1}(\xi) A_k|, \qquad (21.40)$$

where the a_j and b_k are sets of non-negative coefficients reflecting the importance of the parts of the design criterion. The matrices of coefficients A_k are $p_k \times s_k$, defining the linear combinations of the p_k parameters in model k that are of interest. For D-optimality A_k is the identity matrix of dimension p_k; for $\mathrm{D_S}$-optimality $s_k < p_k$ and A_k is the $s_k \times s_k$ identity matrix adjoined with $p_k - s_k$ rows of zeroes. The negative sign for the second term on the right-hand side of (21.40) arises because the covariance matrix of the estimates is minimized.

The equivalence theorem states that

$$\phi^{(GDT)}(x, \xi^*_{GDT}) \leq \sum_{j=1}^{c} a_j + \sum_{k=1}^{m} b_k \qquad x \in \mathcal{X}, \text{ where}$$

$$\psi^{(GDT)}(x,\xi) = \sum_{j=1}^{c} a_j \psi_j(x,\xi)/\Delta_j(\xi) + \sum_{k=1}^{m} (b_k/s_k) d_k^A(x,\xi) \text{ and}$$

$$d_k^A(x,\xi) = f_k^{\mathrm{T}} M_k^{-1}(\xi) A_k \{A_k^{\mathrm{T}} M_k^{-1}(\xi) A_k\}^{-1} A_k^{\mathrm{T}} M_k^{-1}(\xi) f_k.$$

The form of $d_k^A(x, \xi)$ comes from the equivalence theorem for D_A-optimality, introduced in §10.2.

21.9 CD-Optimum Designs

As a last example of a compound design criterion we consider the construction of designs that are efficient both for estimating the parameters of a model as well as for estimating features of interest. In §17.5 we found c-optimum designs for three features of a three-parameter compartmental model for the concentration of theophylline in the blood of a horse. The c-optimum designs, for example that for estimation of the area under the curve, all had either two or one points of support and so provided no information on the values of the parameters in the model.

We initially consider design for estimation of the parameters and one feature of interest. The development parallels that of §21.8.1 except that now we find designs to maximize

$$\kappa \log \mathrm{Eff}^{\,D} + (1-\kappa) \log \mathrm{Eff}^{\,C},$$

where $\mathrm{Eff}^{\,C}(\xi)$ is the efficiency for estimation of the linear combination $c^{\mathrm{T}}\beta$. That is, the design should maximize

$$\Phi(\xi) = \kappa \log |M(\xi)|/p - (1-\kappa) \log c^{\mathrm{T}} M^{-1}(\xi) c. \qquad (21.41)$$

Designs maximizing (21.41) are called CD-optimum and are denoted ξ^*_{CD}. The derivative function for (21.41) is

$$\phi^{(CD)}(x, \xi) = (\kappa/p) f^{\mathrm{T}}(x) M^{-1}(\xi) f(x) + (1-\kappa) \frac{\{f^{\mathrm{T}}(x) M^{-1}(\xi) c\}^2}{c^{\mathrm{T}} M^{-1}(\xi) c}. \qquad (21.42)$$

Because of the way the terms in (21.42) have been scaled, the upper bound of $\phi^{(CD)}(x, \xi^*_{DT})$ over $x \in \mathcal{X}$ is one, achieved at the points of the optimum design. The second term in (21.42) is the derivative of $\log c^{\mathrm{T}} M^{-1}(\xi) c$, the logarithm of the criterion for c-optimality. The equivalence theorem for c-optimum designs is in §10.4.

Example 21.7 A Compartmental Model As an example we find CD-optimum designs for the compartmental model of §17.5 that are efficient both for parameter estimation and for estimation of the area under the curve (AUC). For the parameter values $\theta_1^0 = 4.29$, $\theta_2^0 = 0.0589$, and $\theta_3^0 = 21.80$

the two component optimum designs are

$$\xi^*_{AUC} = \left\{ \begin{array}{cc} 0.2331 & 17.6322 \\ 0.0135 & 0.9865 \end{array} \right\}, \; \xi^*_D = \left\{ \begin{array}{ccc} 0.2292 & 1.3907 & 18.4164 \\ \frac{1}{3} & \frac{1}{3} & \frac{1}{3} \end{array} \right\}. \tag{21.43}$$

The weights in the two designs are very different. Not only does ξ^*_{AUC} have two support points, so that it does not provide estimates of the three parameters, but the two design weights are far from equal. As a consequence small exact designs for the AUC may have low efficiency. However, the two support points of ξ^*_{AUC} are very close to the upper and lower support points of ξ^*_D.

The CD-optimum designs were found for a series of values of κ between zero and one. These optimum designs, to within round-off error, seemed to have support at the points of ξ^*_D with equal weight on the two lower design points. To check this structure we found optimum designs constrained both to have the support of ξ^*_D and to have $w_1 = w_2$. From (21.41) the CD efficiency of a design ξ is

$$\text{Eff}^{\,CD}(\xi) = \exp\{\Phi(\xi) - \Phi(\xi^*_{CD})\}.$$

We found that over the whole range of κ this efficiency was at least 99.99%, so that the optimum designs can be taken to have this constrained form. The designs are plotted in Figure 21.6. The weight w_3 in the upper panel decreases linearly from 0.9865 when $\kappa = 0$ to one-third when $\kappa = 1$, with $w_1 = w_2 = 0.5(1 - w_3)$. The efficiencies in the lower panel likewise change in a smooth manner, beginning with 0% D-efficiency and 100% c-efficiency at $\kappa = 0$, with D rising and c falling as κ increases.

This continuous design is only locally optimum. If a prior distribution is available for β it can be sampled as in §18.5 and a Bayesian version of the compound design criterion used. ■

Exact designs, either locally optimum or using the prior distribution, can be found using SAS. Finally, as in §21.8.2, we can take weighted sums of the two component parts of the design criterion (21.41) to find designs for the parameters of several models as well as several features of interest. For the compartmental model we could find a design efficient for parameter estimation and for the three features of interest in §10.4, namely the time to maximum concentration and the maximum concentration as well as the area under the curve. A final extension would be to also include some model testing terms of the form discussed in §21.8.

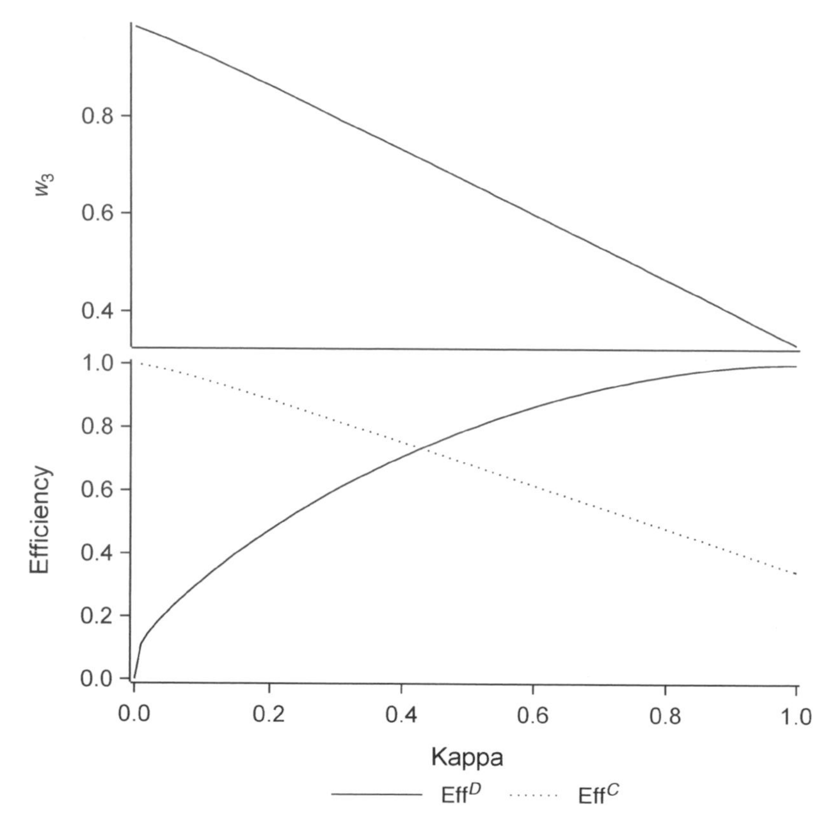

FIG. 21.6. Example 21.7: a compartmental model. Structure of CD-optimum designs as κ varies. For all κ except 0 the support points are indistinguishable from those of the three-point D-optimum design in (21.43). Upper panel, design weight w_3; lower panel, component efficiencies.

21.10 Optimizing Compound Design Criteria in SAS

Most designs discussed in this chapter can be constructed with SAS tools discussed in previous chapters. In particular, the non-linear optimization capabilities of SAS/IML, discussed in §9.6 and §13.4, can find both continuous optimum designs and exact designs over continuous candidate regions, with suitably defined functions to optimize. For example, for the designs suitable for polynomials up to degree 6 discussed in §21.4, the following IML code defines a function `DCrit` to compute the compound criterion for the equally weighted design, 21.13.

```
start llog(x);                /* A safe log() function    */
   if (x > 0) then return(log(x));
   else            return(-9999);
finish;

start MakeZ(x,k);             /* Create the design matrix */
   free F;                    /* for a polynomial of      */
   term = j(nrow(x),1);       /* degree k                 */
   do d = 0 to k;
      F = F || term;
      term = term#x;
      end;
   return(F);
finish;

start DCrit(w,x);             /* Sum the scaled logs of   */
   dd = 0;                    /* the determinants.        */
   do d = 2 to 6;
      Fd = MakeZ(x,d);
      dd = dd + llog(det(Fd`*diag(w)*Fd))/(d+1);
      end;
   return(dd);
finish;
```

In order to optimize such a function in IML, it must be wrapped inside another function with only one argument, which are the parameters being optimized. Thus, when the candidate points are fixed and defined by the vector `x`, the following function can be used to optimize the weights, using the normalized square transformation (9.30) to make the problem unconstrained.

```
start DCritW(w) global(x);
   return(DCrit((w##2)/sum(w##2),x));
finish;
```

A similar function can be used to optimize the design support points `x` for given weights `w`.

> **SAS Task 21.1.** Use the non-linear optimization features of PROC IML to find the continuous compound optimum design for up to a sixth-order polynomial, confirming the results of §21.4.

> **SAS Task 21.2.** Use the non-linear optimization features of PROC IML to find the continuous compound D/D_S optimum design for four factors, Example 21.2.

There are no specialized facilities in SAS to find exact compound optimum designs; in particular, compound optimality is not a feature of PROC OPTEX, which we have used for most exact designs discussed so far. The only recourse we can suggest is to program a specialized exchange algorithm in IML, using the methods of Chapter 12. This is what was done for the designs listed in Table 21.3, using a simple exchange algorithm in that case. The coding is not difficult, but it is too lengthy to discuss in full here. The key is evaluating a current exact design with the addition or deletion of a single point. Thus, if the current design is stored in a matrix **xAdd** and the candidate points to be added in a matrix **xCan**, the following code evaluates the current critical function **DCritX** when the ith candidate is added to the design.

```
xAdd = xCurr // xCan[i,];
dAdd = DCritX(xAdd);
```

Similarly, the following code evaluates the design after deleting the jth point.

```
xDel = xCurr[loc((1:nRow(xCurr)) ^= j),];
dDel = DCritX(xDel);
```

In order to construct a complete exchange algorithm, these bits of code must first be wrapped in loops that evaluate all candidate points and all design points for addition and deletion, respectively, and keeps track of the best ones. Then these loops are wrapped in another loop that updates the current design and iterates until it ceases to change.

> **SAS Task 21.3.** (Advanced). Devise an exchange algorithm to find optimum exact compound D/D$_S$ optimum design for four factors over a $\{-1, 0, 1\}^4$ grid, as given by the appropriate rows of Table 21.3.

21.11 Further Reading

Examples of designs for discriminating between three growth models, combined with parameter estimation, in which the 'bent-stick' model of (21.34) is replaced by a polynomial are given in Table 21.9 of Atkinson and Donev (1992). Atkinson and Bogacka (1997) describe compound optimum designs for determining the order of chemical reactions. Biedermann, Dette, and Zhu (2005) use compound c-optimality to estimate the doses corresponding to several probabilities in dose response models. They work with the sum of the variances of the estimated dose levels. Since these can be very different, they are scaled by the variance from the individual optimum designs. As we saw in §21.3, this scaling would be avoided if the product of variances were used rather than the sum.

Cook and Wong (1994) explore the relationship between constrained and compound optimum designs. They explain their results using graphs similar to Figures 21.5 and 21.6. For example, suppose that for the compartmental model of Example 21.7, the D-optimum design is required subject to a minimum c-efficiency of Eff_*^C. The equivalence with compound designs can then be illustrated with our results on CD-optimality. Let the CD-optimum design in Figure 21.6 with this efficiency arise from a value κ^* of κ. Then the D-optimum design satisfying this constraint is the CD-optimum design for κ^*.

22

GENERALIZED LINEAR MODELS

22.1 Introduction

The methods of experimental design described so far are appropriate if the response, perhaps after transformation, has independent errors with a variance that is constant over $\mathcal{X}$. This is the customary assumption that is made in applications of the normal theory linear model. However, it is not a characteristic of the Poisson and binomial distributions, where there is a strong relationship between mean and variance. The main emphasis of this chapter is on optimum designs for these and other generalized linear models.

The assumptions of normality and constancy of variance for regression models enter the criteria of optimum design through the form of the information matrix $F^{\mathrm{T}}F$. Other forms of information matrix arise from other distributions. Given the appropriate information matrix, the principles of optimum design are the same as those described in earlier chapters. In designs for generalized linear models (McCullagh and Nelder 1989) the asymptotic covariance matrix of the parameters of the linear model is of the form $F^{\mathrm{T}}WF$, where the $N \times N$ diagonal matrix of weights W depends on the parameters of the linear model, on the error distribution and on the link between them. The dependence of the designs on unknown parameters means that, in general, designs for generalized linear models require the specification of prior information: point prior information will lead to locally optimum designs analogous to those for non-linear models in Chapter 17; full specification of a prior distribution leads to Bayesian optimum designs similar to those of Chapter 18.

Because the structure of the information matrix $F^{\mathrm{T}}WF$ is that for weighted least squares, we begin the chapter with a short illustration of design for non-constant variance normal theory linear models. Generalized linear models are introduced in §22.3. The major series of examples concerns experiments with binomial responses. We start in §22.4.1 with models for binomial data and then find D-optimum designs when there are one or two explanatory variables; §22.4.6 explores designs for a full second-order model in two variables. Gamma models are the subject of §22.5. The chapter concludes with references and suggestions for further reading.

22.2 Weighted Least Squares

For the linear regression model $\mathrm{E}(y) = X\beta$ let the variance of an observation at x_i be $\sigma^2/w(x_i)$ $(i = 1, \ldots, N)$, where the $w(x_i)$ are a set of known weights. It is still assumed that the observations are independent. One special case is when the observation at x_i is $\bar{y}_i$, the mean of n_i observations. Then $\mathrm{var}(\bar{y}_i) = \sigma^2/n_i$, so that $w(x_i) = n_i$. If we let $W = \mathrm{diag}\{w(x_i)\}$, the weighted least squares estimator of the parameter β is

$$\hat{\beta} = (F^{\mathrm{T}}WF)^{-1}F^{\mathrm{T}}Wy, \tag{22.1}$$

with

$$\mathrm{var}\hat{\beta} = \sigma^2(F^{\mathrm{T}}WF)^{-1}. \tag{22.2}$$

The design criteria of Chapter 10 can be applied to the information matrix $F^{\mathrm{T}}WF$ to yield optimum designs. In particular, the D-optimum exact design maximizes $|F^{\mathrm{T}}WF|$. For continuous designs the information matrix is

$$M(w, \xi) = \int w(x)f(x)f^{\mathrm{T}}(x)\xi(dx), \tag{22.3}$$

with the D-optimum design maximizing $|M(w, \xi)|$. The equivalence theorem for D-optimality follows by letting

$$\bar{d}(w, \xi) = \max_{x \in \mathcal{X}} w(x)f^{\mathrm{T}}(x)M^{-1}(w, \xi)f(x), \tag{22.4}$$

with $\bar{d}(w, \xi^*) = p$.

In earlier chapters, for example in (9.1), the measure ξ put weight w_i at the support point x_i. In this chapter, to avoid confusion with the weights $w(x_i)$, we denote the design weights by q_i so that

$$\xi = \left\{ \begin{array}{ll} x_1 & x_2 \ldots x_n \\ q_1 & q_2 \ldots q_n \end{array} \right\}.$$

Example 22.1. Weighted Regression Through the Origin The model is $\mathrm{E}(y_i) = \beta x_i$, a straight line through the origin with $x \geq 0$. If the variance of the errors is constant, the D-optimum design puts all trials at the maximum value of x. Suppose, however, that $\mathrm{Var}(y_i) = \sigma^2 e^x$, so that the variance increases exponentially with x. In the notation of this chapter the weight $w(x) = e^{-x}$.

The D-optimum design for this one-parameter model concentrates all mass at one point. Since $f(x) = x$, the information matrix is

$$M(w, \xi) = x^2 e^{-x},$$

which is a maximum when $x = 2$. Then

$$d(x, w, \xi^*) = e^{-x} x M^{-1}(w, \xi^*) x = x^2 \frac{e^{-x}}{4e^{-2}},$$

which takes the value 1 when $x = 2$, which is the maximum over $\mathcal{X}$. Thus the D-optimum design has indeed been found. The rapid increase of variance as x increases leads to a design that does not span the experimental region, provided that the maximum value of x is greater than 2. If the maximum value is $x_u \leq 2$, the optimum design consists of putting all trials at x_u. ■

This example illustrates the effect of the weight $w(x)$. Otherwise the method of finding the design is the same as that for regression with constant variance. In fact, defining $f^*(x) = \sqrt{w(x)} f(x)$, in general the theory and methods for finding optimal unweighted designs for f^* apply to finding optimal weighted designs for f.

22.3 Generalized Linear Models

The family of generalized linear models extends normal theory regression to any distribution belonging to the one-parameter exponential family. As well as the normal, this includes the gamma, Poisson, and binomial distributions, all of which are important in the analysis of data.

The linear multiple regression model can be written as

$$\mathrm{E}(y) = \mu = \eta = \beta^{\mathrm{T}} f(x), \tag{22.5}$$

where μ, the mean of y, is equal to the *linear predictor* η. The extension to the generalized linear model requires the introduction of a *link function* $g(\mu) = \eta$, relating the mean and the linear predictor. In (22.5) since $\mu = \eta$, $g(\mu)$ is the identity. With Poisson data we must have $\mu \geq 0$. A frequently used form for such data is the log link, that is $\log(\mu) = \eta$, or equivalently, $\mu = e^{\eta}$, so that the constraint on μ is satisfied for all η. For the binomial data of §22.4 the link function is such that, however the values of x and β vary, the mean μ satisfies the physically meaningful constraint that $0 \leq \mu \leq 1$. Five such links are mentioned.

The distribution of y determines the relationship between the mean and the variance of the observations. The variance is of the form

$$\text{var}(y) = \phi V(\mu), \tag{22.6}$$

where ϕ is the dispersion parameter, equal to σ^2 for the normal distribution and one for the binomial. The variance function $V(\mu)$ is specific to the error distribution.

The weighted form of the information matrix for generalized linear models (22.3) arises because maximum likelihood estimation of the parameters β of the linear predictor reduces to iterative weighted least squares. The weights for individual observations are given by

$$w = V^{-1}(\mu)\left(\frac{d\mu}{d\eta}\right)^2. \tag{22.7}$$

These weights depend both on the distribution of y and on the link function.

Because the observations are independent the information matrix is of the weighted form explored in §22.2. If the effects of the factors are slight, the means μ_i for all observations will be similar and so will the weights w_i. The weighted information matrix $F^{\mathrm{T}}WF$ will then, apart from a scaling factor, be close to the unweighted information matrix $F^{\mathrm{T}}F$ and the optimum designs will be close. Since the designs maximizing functions of $F^{\mathrm{T}}F$ are the optimum regression designs, these will be optimum, or close to optimum, for generalized linear models with small effects (Cox 1988). We explore the efficiency of regression designs for gamma models in §22.5.2.

The seminal work on generalized linear models is McCullagh and Nelder (1989), with an introduction in Dobson (2001). Atkinson and Riani (2000, Chapter 6) presents a succinct overview. Myers, Montgomery, and Vining (2001) emphasizes data analysis.

22.4 Models and Designs for Binomial Data

22.4.1 Models

For the binomial distribution the variance function (22.6) is

$$V(\mu) = \mu(1-\mu). \tag{22.8}$$

In models for binomial data the response y_i is defined to be R_i/n_i. Sensible models therefore have $0 \leq \mu_i \leq 1$. We list four link functions that have been found useful in the analysis of data.

1. **Logistic.**
$$\eta = \log\left(\frac{\mu}{1-\mu}\right). \tag{22.9}$$
The ratio $\mu/(1-\mu)$ is the odds that $y = 1$. In the logistic model the log odds is therefore equal to the linear predictor. Apart from a change of sign, the model is unaltered if 'success' is replaced with 'failure'.

2. **Probit.**
$$\eta = \Phi^{-1}(\mu), \tag{22.10}$$
where Φ is the normal cumulative distribution function. This link has very similar properties to those of the logistic link.

3. **Complementary log log.**
$$\eta = \log\{-\log(1-\mu)\}, \tag{22.11}$$
which is not symmetrical in 'success' and 'failure'. Interchanging these two gives the

4. **Log–log** link.
$$\eta = \log(-\log \mu). \tag{22.12}$$

Atkinson and Riani (2000, §6.18) describe a fifth link, the arcsine link, which has some desirable robustness properties for binary data.

22.4.2 One-Variable Logistic Regression

The model for logistic regression with one explanatory variable can be written
$$\log\{\mu/(1-\mu)\} = \eta = \alpha + \beta x, \tag{22.13}$$
which is often used to analyse data such as Example 1.6, Bliss's beetle data, in which the variable was the dose of an insecticide.

To calculate W (22.7) requires the derivative $d\mu/d\eta$ or, equivalently from differentiation of the logistic link (22.9),
$$\frac{d\eta}{d\mu} = \frac{1}{\mu(1-\mu)}, \tag{22.14}$$
leading, in combination with (22.8) to the simple form
$$W = \mu(1-\mu). \tag{22.15}$$
Thus, as might be expected, experiments with μ near to zero or one are uninformative; a set of nearly all successes or failures does not provide good parameter estimates.

The optimum design will depend on the values of α and β in (22.13). If we take the *canonical form* $\alpha = 0$ and $\beta = 1$ the D-optimum design for a sufficiently large design region $\mathcal{X}$ is

$$\xi^* = \left\{ \begin{array}{cc} -1.5434 & 1.5434 \\ 0.5 & 0.5 \end{array} \right\}, \tag{22.16}$$

that is equal weights at two points symmetrical about $x = 0$. At these support points $\mu = 0.176$ and $1 - 0.176 = 0.824$.

Although designs for other values of the parameters can likewise be found numerically, design problems for a single x can often be solved in a canonical form, yielding a structure for the designs independent of the particular parameter values (Ford, Torsney, and Wu 1992). The translation into the experimental variable x for other parameter values depends on the particular α and β.

For the upper design point in (22.16) the linear predictor $\eta = 0 + 1 \times x$ has the value 1.5434, which is the value we need for the optimum design whatever the parameterization. If we solve (22.13) for the η giving this value, x_0^*, the upper support point of the design is given by

$$x_0^* = (1.5434 - \alpha)/\beta.$$

Note that as $\beta \rightarrow 0$, the value of x_0^* increases without limit. This is an example of the result of Cox (1988) that, for small effects of the variables, the design tends to that for homoscedastic regression. Here as β decreases the design points go to $\pm\infty$. In practice the design region will not be unlimited and the optimum design for $\beta = 0$ will put equal weight on the boundary points of $\mathcal{X}$.

For Bliss's data $\hat{\alpha} = 0.8893$ and $\hat{\beta} = 2.3937$ so that the support points of the D-optimum design are 53.24 and 65.62. Figure 22.1 shows the data, the fitted logistic model with independent variable dose and the two design points, symmetrical about the estimated LD_{50}, the dose for which $\mu = 0.5$.

22.4.3 One-Variable Regression with the Complementary Log–Log Link

The fitted logistic model in Figure 22.1 appears to describe the data well. We now compare this fitted model and the resulting optimum design with those for the complementary log–log link (22.11).

Differentiation of (22.11) yields

$$\frac{d\eta}{d\mu} = \frac{-1}{(1-\mu)\log(1-\mu)},$$

so that the weights are given by

$$w(\mu) = \frac{1-\mu}{\mu}\left[\log(1-\mu)\right]^2, \tag{22.17}$$

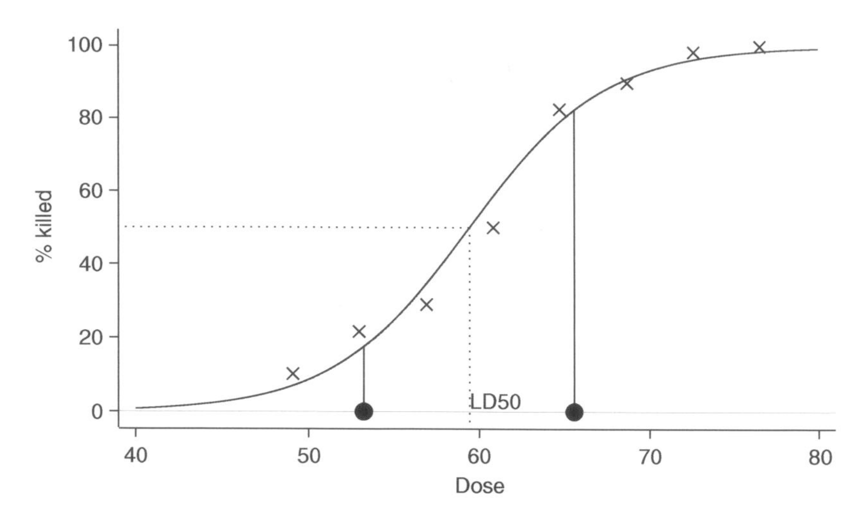

FIG. 22.1. Bliss's data, fitted logistic model, and D-optimum design.

a more complicated form than the logistic weights (22.15). To find the locally D-optimum design we again take the canonical parameter values $\alpha = 0$ and $\beta = 1$. However, (22.11), unlike (22.9), is not symmetrical about $\mu = 0.5$. The canonical locally D-optimum design found by numerical maximization of $|M(w, \xi)|$ is

$$\xi^* = \left\{ \begin{array}{cc} -1.338 & 0.9796 \\ 0.5 & 0.5 \end{array} \right\} \qquad (22.18)$$

at which the values of μ are 0.2308 and 0.9303. Both values of μ are higher than those for the optimum design for the logistic model. In particular, the trials at $x = 0.9796$ yield a probability of success appreciably closer to unity than the 0.8240 for the higher level of x in (22.16).

For Bliss's data the new parameter estimates are $\hat{\alpha} = 0.0368$ and $\hat{\beta} = 1.4928$ so that the support points of the D-optimum design for the complementary log–log link are 54.16 and 69.06. Figure 22.2 repeats Figure 22.1 for this new link. The fit for the lower doses has improved slightly.

Atkinson and Riani (2000, p. 234) use log dose, rather than dose, as the explanatory variable. This model has the advantage of a natural interpretation of zero dose as having no effect. They show that neither the logistic nor the probit link fit these data adequately when combined with the linear predictor (22.13) with $\log x$ rather than x as the independent variable. However, the first-order model is adequate when combined with the complementary log–log link (22.11) and the logarithm of dose. Their formal comparisons could be repeated for the model fitted here with dose as the explanatory

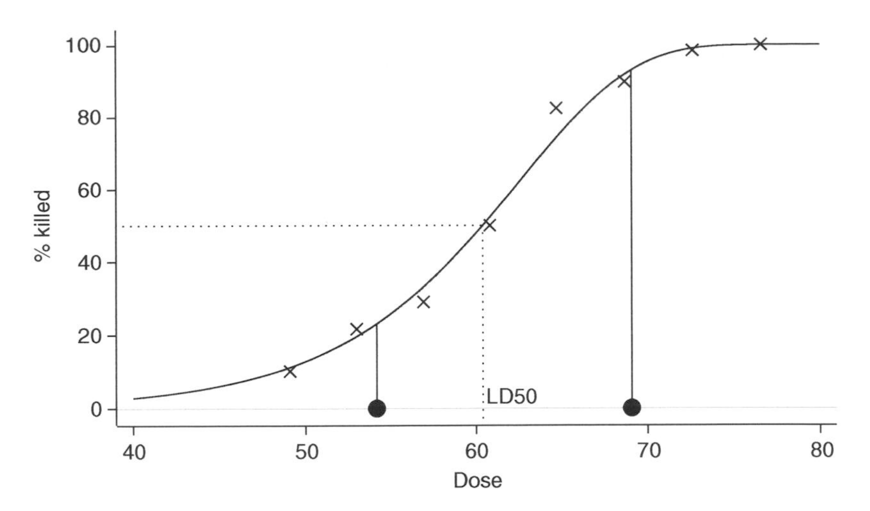

FIG. 22.2. Bliss's data, fitted complementary log–log model, and D-optimum design.

variable. But our interest is in the dependence of the design on the link function.

Although the designs for the two links are somewhat different the design for the logistic link has an efficiency of 86.1% for the complementary log–log link. The reverse efficiency is 93.6%. The results of the next section indicate that incorrect specification of the link is unlikely to be a major source of design inefficiency compared with poor prior knowledge of the parameters of the linear predictor.

22.4.4 Two-Variable Logistic Regression

The properties of designs for response surface models, that is with two or more continuous explanatory variables, depend much more on the experimental region than those where there is only one factor.

Although it was assumed in the previous section that the experimental region $\mathcal{X}$ was effectively unbounded, the design was constrained by the weight w to lie in a region in which μ was not too close to zero or one. But with more than one explanatory variable constraints on the region are necessary. For example, for the two-variable first-order model

$$\log\{\mu/(1-\mu)\} = \eta = \beta_0 + \beta_1 x_1 + \beta_2 x_2, \tag{22.19}$$

TABLE 22.1. D-optimum designs for four binomial models: parameter values for linear predictors, first-order in two variables. Logistic link

Design	β_0	β_1	β_2
B1	0	1	1
B2	0	2	2
B3	2	2	2
B4	2.5	2	2

TABLE 22.2. D-optimum designs for binomial models with the parameter sets B1 and B2 of Table 22.1; q_i design weights

	Design B1				Design B2				
i	x_{1i}	x_{2i}	q_i	μ_i	x_{1i}	x_{2i}	q_i^1	q_i^2	μ_i
1	−1	−1	0.204	0.119	0.1178	−1.0000		0.240	0.146
2					1.0000	−0.1178		0.240	0.854
3	1	−1	0.296	0.500	1.0000	−1.0000	0.327	0.193	0.500
4	−1	1	0.296	0.500	−1.0000	1.0000	0.193	0.327	0.500
5					−1.0000	0.1178	0.240		0.146
6	1	1	0.204	0.881	−0.1178	1.0000	0.240		0.854

with $\beta^T = (0, \gamma, \gamma)$, all points for which $x_1 + x_2 = 0$ yield a value of 0.5 for μ, however extreme the values of x. We now explore designs for the linear predictor (22.19) with the logistic link for a variety of parameter values.

Four sets of parameter values are given in Table 22.1. In all cases we take $\mathcal{X}$ as the square with vertices ± 1. D-optimum designs for the sets B1 and B2 are listed in Table 22.2. The parameter values of B1 $(0, 1, 1)$ are smallest and the table shows that the design has support at the points of the 2^2 factorial, although the design weights are not quite equal, as they would be for the normal theory model and as they become for the logistic model as β_1 and $\beta_2 \to 0$. At those factorial points for which $x_1 + x_2 = 0, \mu = 0.5$ since $\beta_1 = \beta_2$. At the other design points $\mu = 0.119$ and 0.881, slightly more extreme values than the values of 0.176 and 0.824 for the experiment with a single x.

An interesting feature of our example is that the number of support points of the design depends upon the values of the parameters β. From

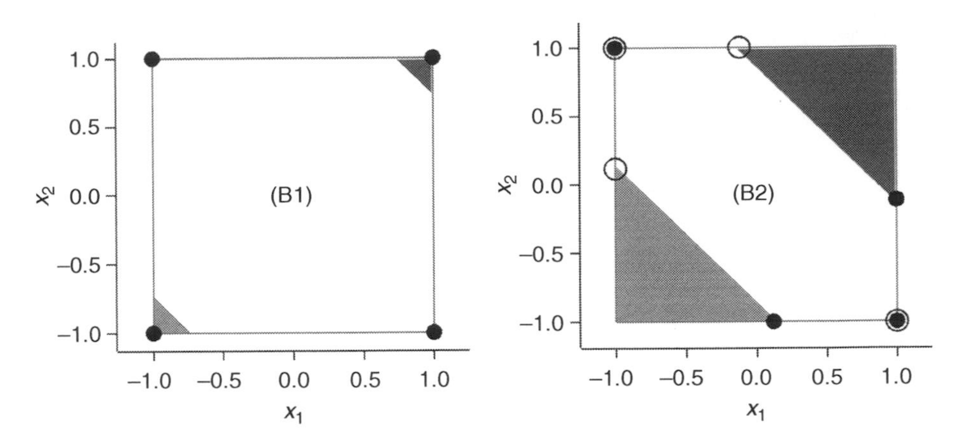

FIG. 22.3. Support points for D-optimum designs for binomial models B1 and B2 in Table 22.1. In the lightly shaded area $\mu \leq 0.15$, whereas, in the darker region, $\mu \geq 0.85$. The two distinct four-point optima for B2 are depicted by distinct markers.

Carathéodory's Theorem discussed in Chapter 9, the maximum number of support points required by an optimum design is $p(p+1)/2$. Our second set of parameters, B2 in which $\beta^T = (0, 2, 2)$, gives two distinct four-point optimum designs, with weights given by q_i^1 and q_i^2 in Table 22.2 and support points where $\mu = 0.146$, 0.5, and 0.854. Any convex combination of these two designs, $\alpha q_i^1 + (1-\alpha)q_i^2$ with $0 \leq \alpha \leq 1$, will also be optimal, and will have six support points, which is the value of the Carathéodory bound when $p = 3$.

The relationship between the support points of the design and the values of μ is highlighted in Figure 22.3 where the pale areas are regions in which $\mu \leq 0.15$, with the dark regions the complementary ones where $\mu \geq 0.85$. Apart from the design points where $\mu = 0.5$, all other design points are close to those boundaries of these regions where μ is around 0.15 and 0.85.

The D-optimum designs for the two remaining sets of parameters in Table 22.1 are given in Table 22.3. These designs have respectively 4 and 3 points of support. When $\beta^T = (2, 2, 2)$, the design points are where $\mu = 0.182$ and 0.818. For $\beta^T = (2.5, 2, 2)$ the values are 0.182 and 0.832. For this three-point design for a three-parameter model, the design weights $q_i = 1/3$.

The relationship between the design points and the values of μ are shown, for these designs, in Figure 22.4. For $\beta^T = (2, 2, 2)$ the design points lie

TABLE 22.3. D-optimum designs for binomial models with the parameter sets B3 and B4 of Table 22.1; q_i design weights

	Design B1				Design B2			
i	x_{1i}	x_{2i}	q_i	μ_i	x_{1i}	x_{2i}	q_i	μ_i
1	−1.0000	−0.7370	0.169	0.186	−1.0000	0.5309	0.333	0.827
2	−1.0000	0.7370	0.331	0.814	−1.0000	−1.0000	0.333	0.182
3	−0.7370	−1.0000	0.169	0.186	0.5309	−1.0000	0.333	0.827
4	0.7370	−1.0000	0.331	0.814				

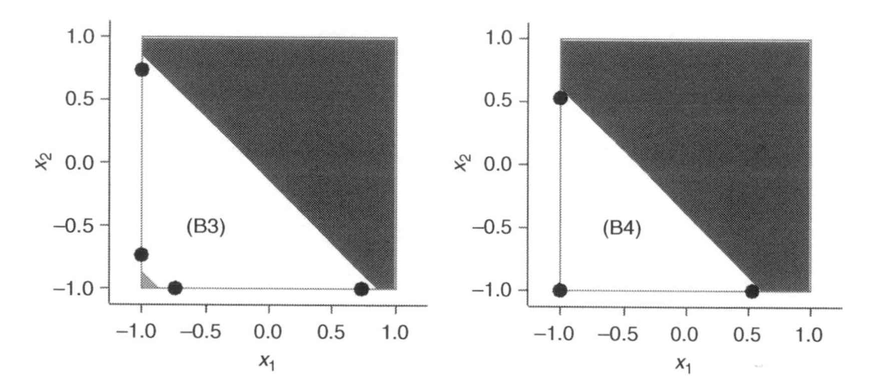

FIG. 22.4. Support points for D-optimum designs for binomial models B3 and B4 in Table 22.1. In the lightly shaded area $\mu \leq 0.15$, whereas, in the darker region, $\mu \geq 0.85$.

slightly away from the boundaries of the regions of high and low values of μ, as they do to a lesser extent in the right-hand panel of the figure. With $\beta^T = (2.5, 2, 2)$ the minimum value of μ, 0.182 at $(-1, -1)$, is sufficiently high that there are no experimental conditions for which $\mu = 0.15$: the corresponding panel of the figure contains only one shaded area.

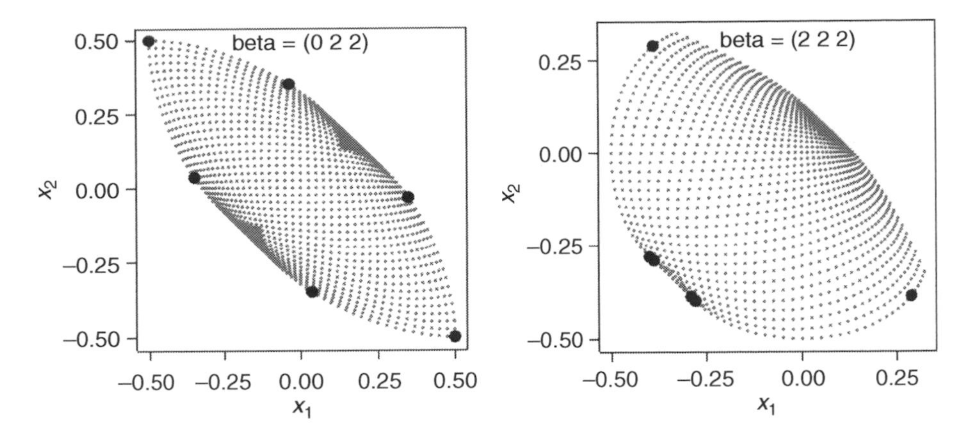

FIG. 22.5. Support points for D-optimum designs for binomial models B2 and B3 in Table 22.1 in the induced design region $\mathcal{Z}$.

22.4.5 Induced Design Region

The properties of designs for generalized linear models are related to the estimation procedure. Because weighted least squares is used, design for the two-variable logistic model (22.19) is equivalent to design for the linear model

$$\eta = \beta_0\sqrt{w} + \beta_1\sqrt{w}x_1 + \beta_2\sqrt{w}x_2 = \beta_0 z_0 + \beta_1 z_1 + \beta_2 z_2. \tag{22.20}$$

The design region $\mathcal{X}$ is then replaced by the induced design region $\mathcal{Z}$, the space in which the values of z can fall as x varies. Since, for this model, $p = 3$, the induced design space $\mathcal{Z}$ is of dimension three. Two examples, projected onto z_1 and z_2 and so ignoring $z_0 = \sqrt{w}$, are given in Figure 22.5 for $\mathcal{X}$ the unit square. In the left-hand panel of the figure $\beta^T = (0, 2, 2)$ so that at the corner of $\mathcal{X}$ for which $x_1 = x_2 = 1$, $\eta = 4$, and $\mu = 0.982$. This is well beyond the range for informative experiments and the projection of the induced design space appears to be folded over. As a consequence, experiments at extreme positions in $\mathcal{Z}$ are not at extreme points in $\mathcal{X}$. The results in the other panel for $\beta^T = (2, 2, 2)$ are similar, but more extreme. For both sets of parameter values the design points lie, as they should, on the boundary of the design region.

These examples show the importance of both the design region and the value of μ in determining the optimum design. In order to reveal the structure of the designs as clearly as possible, the designs considered have all had $\beta_1 = \beta_2$, and so are symmetrical in x_1 and x_2. Both the design region and the values of μ are equally important in the asymmetric designs when the two parameter values are not equal. Asymmetric designs also arise with the log–log and complementary log–log links, since these links are not symmetrical.

22.4.6 A Second-order Response Surface

This section extends the results of earlier sections to the second-order response surface model, again with two factors and again with the logistic link. The D-optimum designs are found, as before, by maximizing $|M(w, \xi)|$ or its exact counterpart. The purpose of the section is to show the relationship with, and differences from, designs for regression models.

To explore how the design changes with the parameters of the model we look at a series of designs for the family of linear predictors

$$\eta = \beta_0 + \gamma(\beta_1 x_1 + \beta_2 x_2 + \beta_{12} x_1 x_2 + \beta_{11} x_1^2 + \beta_{22} x_2^2) \text{ with } \gamma \geq 0 \quad (22.21)$$

and design region the unit square with $-1 \leq x_1 \leq 1$ and $-1 \leq x_2 \leq 1$. When $\gamma = 0$ the result of (Cox 1988) shows that the design is the D-optimum design for the second-order regression model, the unequally weighted 3^2 factorial given in (12.2).

For numerical exploration we take $\beta_0 = 1$, $\beta_1 = 2$, $\beta_2 = 2$, $\beta_{12} = -1$, $\beta_{11} = -1.5$, and $\beta_{22} = 1.5$. As γ varies from 0 to 2, the shape of the response surface becomes increasingly complicated.

Figure 22.6 shows the support points of the D-optimum designs as γ increases from zero in steps of 0.1. The design points are labelled, for $\gamma = 0$, in standard order for the 3^2 factorial, with x_1 changing more frequently. The figure shows how all but one of the design points stay on the boundary of the design region; the circles and black dots are the support points for $\gamma = 1$ and 2, respectively, with the grey dots indicating intermediate values. There is little change in the location of the centre point, point 5, during these changes. Initially the design has nine points, but the weight on point 8 decreases to zero when $\gamma = 0.3$. Thereafter, the design has eight support points until $\gamma = 1.5$ when the weight on observation 6 becomes zero.

The relationship between the support points of the design and the values of μ is highlighted in Figure 22.7 where, as in Figure 22.3, the pale areas are regions in which $\mu \leq 0.15$, with the dark regions the complementary ones where $\mu \geq 0.85$. The left-hand panel of Figure 22.7, for $\gamma = 1$, shows that the eight-point design is a distortion of a standard response surface design, with most points in the white area, several on the boundary of the design

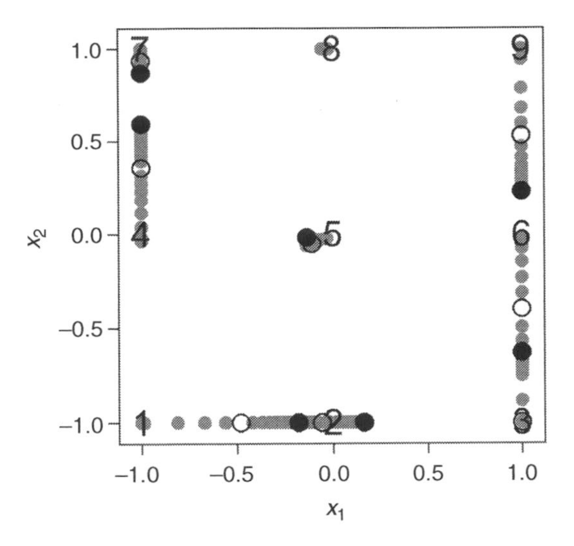

FIG. 22.6. D-optimum designs for binomial model when $0 \leq \gamma \leq 2$. Support points: numbers, $\gamma = 0$; circles, $\gamma = 1$; black dots $\gamma = 2$; and grey dots, intermediate values.

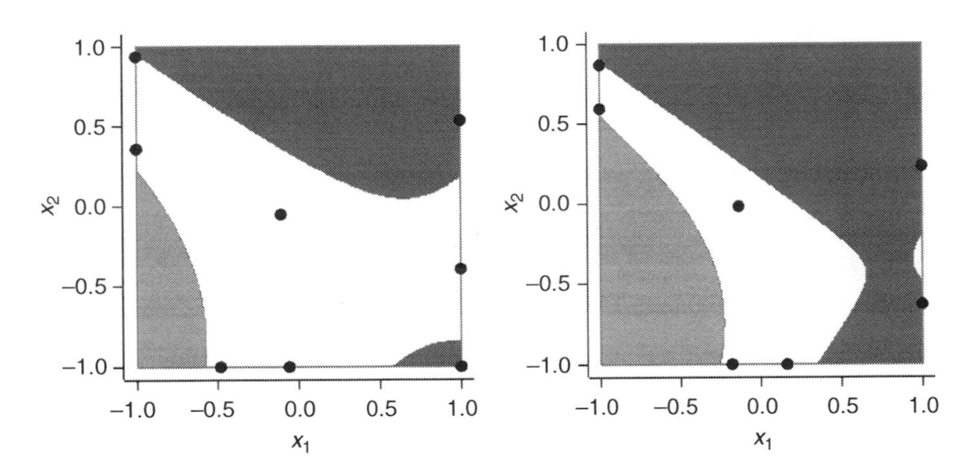

FIG. 22.7. Support points for D-optimum designs for binomial models. Left-hand panel $\gamma = 1$, right-hand panel $\gamma = 2$. In the lightly shaded area $\mu \leq 0.15$, whereas, in the darker region, $\mu \geq 0.85$.

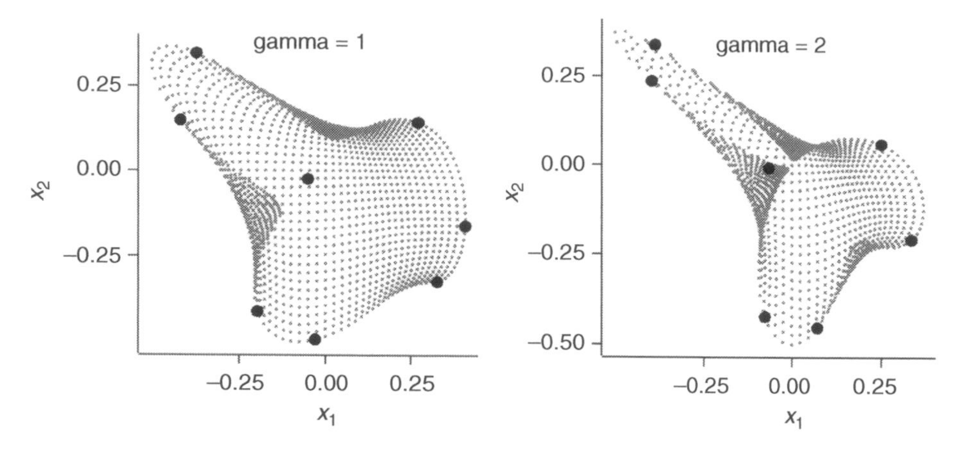

FIG. 22.8. Support points for D-optimum designs for binomial models in the induced design region $\mathcal{Z}$.

region and close to the contours of $\mu = 0.15$ or 0.85. In the numbering of Figure 22.7, points 2 and 6 are on the edge of the design region where μ is close to 0.5. Points 3 and 9 are at higher values of μ.

A similar pattern is clear in the seven-point design for $\gamma = 2$ in the right-hand panel of the figure; four of the seven points are on the edges of the white region, one is in the centre and only points 3 and 9 are at more extreme values of μ.

The two panels of Figure 22.7 taken together explain the trajectories of the points in Figure 22.6 as γ varies. For example, points 1 and 4 move away from $(-1, -1)$ as the response at that point decreases, point 3 remains at $(1, -1)$ until γ is close to one and point 8 is rapidly eliminated from the design as the response near $(0, 1)$ increases with γ.

Further insight into the structure of the designs can be obtained from consideration of the induced design region introduced in §22.4.5. However, the extension of the procedure based on (22.20) to second-order models such as (22.21) is not obvious. The difficulty is the way in which the weights enter in the transformation from $\mathcal{X}$ to $\mathcal{Z}$. If, as in (22.20), $z_j = \sqrt{w}x_j$, then, for example, the interaction term in the linear predictor $\sqrt{w}x_jx_k$ is not equal to z_jz_k.

It is however informative to plot the designs in $\mathcal{Z}$ space. The left-hand panel of Figure 22.8 shows the eight-point design for $\gamma = 1$ plotted against z_1 and z_2; seven points lie on the edge of this region, well spaced and far from the centre, where the eighth point is. The right-hand panel for $\gamma = 2$ shows six points similarly on the edge of $\mathcal{Z}$; the centre point is hidden under the seemingly folded-over region near the origin.

In the induced design region these designs are reminiscent of response surface designs, with a support point at the centre of the region and others at remote points. However the form of $\mathcal{Z}$ depends on the unknown parameters of the linear predictor, so this description is not helpful in constructing designs. In the original space $\mathcal{X}$ we have described the designs for this second-order model as a series of progressive distortions of designs with support at the points of the 3^2 factorial. For small values of γ the unweighted 3^2 factorial provides an efficient design, with a D-efficiency of 97.4% when $\gamma = 0$. However, the efficiency of this design declines steadily with γ, being 74.2% for $\gamma = 1$ and a low 38.0% when $\gamma = 2$. If appreciable effects of the factors are expected, the special experimental design methods of this section need to be used.

22.5 Optimum Design for Gamma Models

The gamma model is often an alternative to response transformation. In particular, with a log link, it is may be hard to distinguish the gamma from a linear regression model with logged response. A discussion is in §§8.1 and 8.3.4 of McCullagh and Nelder (1989) with examples of data analyses in §7.5 of Myers *et al.* (2001).

The gamma family is one in which the correct link is often in doubt. We use the Box and Cox link in our examples, which is generally equivalent to the power link.

22.5.1 Box and Cox Link

A useful, flexible family of links is the Box and Cox family, in which

$$g(\mu) = \begin{cases} (\mu^\lambda - 1)/\lambda & (\lambda \neq 0) \\ \log \mu & (\lambda = 0). \end{cases} \tag{22.22}$$

This is seemingly equivalent to the power family of links $g(\mu) = \mu^\lambda$ but is continuous as $\lambda \to 0$. Differentiation of (22.22) yields

$$\frac{d\eta}{d\mu} = \mu^{\lambda - 1}. \tag{22.23}$$

Since the variance function for the gamma distribution is

$$V(\mu) = \mu^2,$$

the combination of (22.7) and (22.23) shows that the weights for the gamma distribution with this link family are

$$w = V^{-1}(\mu)\left(\frac{d\mu}{d\eta}\right)^2 = \mu^{-2\lambda}. \tag{22.24}$$

When $\lambda = 0$, that is for the log link, (22.24) shows that the weights are equal to one. It therefore follows that optimum designs for gamma models with this link are identical to optimum designs for regression models with the same linear predictors. Unlike designs for binomial generalized linear models, the designs do not depend on the parameters β.

To find designs that illustrate the difference between regression and the gamma GLM requires a value of $\lambda \neq 0$. Here we investigate designs for an example when the linear predictor is second-order and $\lambda \neq 0$.

Atkinson and Riani (2000, §6.9) use a gamma model to analyse data from Nelson (1981) on the degradation of insulation due to elevated temperature at a series of times. The data do not yield a particularly clean model as there seem to be some identifiable subsets of observations which do not completely agree with the fitted response-surface model. However, for our purposes, a second-order model is required in the two continuous variables and the gamma model fits best with a power link with $\lambda = 0.5$. A theoretical difficulty with such a value of λ is that μ must be > 0, while η is, in principle, unconstrained.

We scale the variables so that the design region $\mathcal{X}$ is the unit square with vertices $(-1, -1)$, $(-1, 1)$, $(1, -1)$, and $(1, 1)$. The linear predictor is the quadratic

$$\eta = \beta_0 + \beta_1 x_1 + \beta_2 x_2 + \beta_{11} x_1^2 + \beta_{22} x_2^2 + \beta_{12} x_1 x_2, \tag{22.25}$$

that is the same as (22.21) with $\gamma = 1$. Then the standard D-optimum design for the normal theory regression model, given in (12.2), has unequally weighted support at the points of the 3^2 factorial. This design is, from what was said above, also optimum for the gamma model with log link.

For other links the design will depend on the actual values of the parameters β in (22.25). Any design found will therefore be locally optimum. With $\lambda = 0.5$, it follows from (22.24) that the weights

$$w_i = 1/\mu_i.$$

We take β to have the values given in Table 22.4, G1 being rounded from an analysis of Nelson's data.

The exact optimum nine-point design for G1, found by searching over a grid of candidates with steps of 0.01 in x_1 and x_2, are in Table 22.5. This

TABLE 22.4. D-optimum designs for two gamma models: parameter values for linear predictors, second-order in two variables. Power link, $\lambda = 0.5$

Design	β_0	β_1	β_2	β_{11}	β_{22}	β_{12}
G1	3.7	−0.46	−0.65	−0.19	−0.45	−0.57
G2	3.7	−0.23	−0.325	−0.095	−0.225	−0.285

TABLE 22.5. D-optimum designs for the parameter sets G1 and G2 of Table 22.4

	Design G1				Design G2			
i	x_{1i}	x_{2i}	n_i	μ_i	x_{1i}	x_{2i}	n_i	μ_i
1	−1.00	−1.00	1	12.96	−1.00	−1.00	1	13.32
2	−1.00	1.00	2	11.83	−1.00	1.00	1	12.74
3	1.00	−1.00	2	14.59	1.00	−1.00	1	14.14
4	1.00	1.00	1	1.90	1.00	1.00	1	6.45
5	0.11	0.15	1	12.46	−1.00	0.00	1	14.71
6	0.26	1.00	1	5.38	−0.01	− 1.00	1	14.44
7	1.00	0.29	1	7.07	0.07	0.09	1	13.33
8					0.08	1.00	1	9.66
9					1.00	0.09	1	11.01

shows that, at the points of the design, the minimum value of μ is 1.90 and the maximum 14.59. If these were normal responses that had to be non-negative, a range of this order would indicate a power transformation. As Table 22.5 and Figure 22.9 show, the support points of the design are a slight, and nearly symmetrical, distortion of those of the 3^2 factorial.

We have already seen, for instance in design B1, that designs for binomial models tend towards those for regression models as the effects decrease. To illustrate this point for gamma models we found the D-optimum nine-point design for the set of parameter values G2 in Table 22.4 in which all parameters, other than β_0, have half the values they have for design G1. As Table 22.5 shows, the range of means at the design points is now 6.45 to 14.71, an appreciable reduction in the ratio of largest to smallest response. The support points of the design for G2 are shown in Figure 22.9

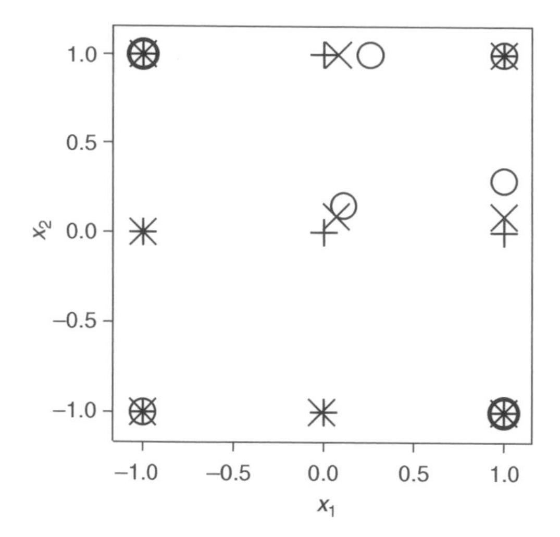

FIG. 22.9. Points for D-optimum nine-point designs for gamma models in Table 22.5: +, the points of the 3^2 factorial; ∘, G1 and ×, G2. Points for G1 which are replicated twice are darker.

by the symbol ×. All are moved in the direction of the factorial design when compared to the points of support of G1.

22.5.2 Efficient Standard Designs for Gamma Models

The designs for second-order gamma models with the parameter sets G1 and G2 of Table 22.4 are both slight distortions of the 3^2 factorial. As the values of the parameters, apart from β_0, tend to zero, the design tends towards the D-optimum design for the second-order regression model which has unequal support at the points of the 3^2 factorial. The simplest design with this support is the 3^2 factorial in which all weights are equal to 1/9. We now explore how well such regression designs perform for gamma models.

We compare two designs for their D-efficiency relative to the D-optimum design for G1 shown in Table 22.5, that is the design for the more extreme parameter set G1 of Table 22.4. The D-optimum design for the less extreme parameter set G2, also given in Table 22.5, has efficiency 97.32%, while the equi-replicated 3^2 factorial has efficiency 96.35%. The main feature of these designs is how efficient they are for the gamma model, both efficiencies being greater than 90%. The design for parameter G2 is for a model with smaller effects than G1, so that the design is between that for G1 and the equi-weighted design, the optimum design for the normal theory model.

An indication of the example with a gamma model is that standard designs may be satisfactory for gamma models. The same conclusion does not hold for our binomial example in §22.4.6.

22.6 Designs for Generalized Linear Models in SAS

Several procedures in SAS/STAT software can be used to fit generalized linear models. The LOGISTIC procedure fits linear logistic regression models for discrete response data, with many specialized analysis features for such models. The more general GENMOD procedure fits generalized linear models for a wide variety of distributions and links. Finally, the GLIMMIX procedure fits generalized linear *mixed* models. The basic syntax for all of these procedures is similar—mainly, a MODEL statement that names the response to be fitted and the independent effects, and if necessary, the distribution and link functions. Refer to the SAS/STAT documentation (SAS Institute Inc. 2007*d*) for details and examples of usage.

> **SAS Task 22.1.** Use PROC LOGISTIC to analyse Example 1.6, Bliss's beetle data.

> **SAS Task 22.2.** Use PROC GENMOD to analyse Example 1.6, Bliss's beetle data, using a complementary log–log link.

Considering the construction of optimum designs for generalized linear models, while the design tools in SAS are not specifically designed for this, it is fairly easy to coax them to be so. Recall that an optimum design for weighted regression with mean $E(y) = f(x, \beta)$ and weight function $w(x)$ is equivalent to an optimal design for *unweighted* regression with mean $E(y) = \sqrt{w(x)}f(x, \beta)$. Thus, simply multiplying the mean function by the square root of the weight function allows us to reuse the computational techniques for optimum designs for unweighted regression discussed in Chapter 12.

For example, recall that if candidate doses are stored in a data set named `Candidates`, then the following statements will direct PROC OPTEX to find an optimum design for a simple linear model, storing it in a data set named `Design`.

```
data Candidates;
   do Dose = -2 to 2 by 0.01;
      output;
      end;
proc optex data=Candidates coding=orthcan;
   model Dose;
   generate n=2 method=m_fedorov niter=1000 keep=10;
   output out=Design;
run;
```

For a generalized linear model for binomial data the weight function is given in terms of the mean μ by (22.8). For the logistic link

$$\mu = \frac{e^{\eta}}{e^{\eta}+1}.$$

Thus, the following code uses this weight function to transform the candidate points for the values $\alpha = 0$ and $\beta = 1$, then uses OPTEX with the transformed candidates to find an optimum design for the logistic model.

```
data GLMCandidates; set Candidates;
   eta = 0 + 1*Dose;
   mu = exp(eta)/(1+exp(eta));
   W = mu*(1-mu);
   J1 = sqrt(W)*1;
   J2 = sqrt(W)*Dose;
proc optex data=GLMCandidates coding=orthcan;
   model J1 J2 / noint;
   generate n=2 method=m_fedorov niter=1000 keep=10;
   id Dose;
   output out=GLMDesign;
proc print data=GLMDesign;
run;
```

The DATA step in the code above computes the columns of the Jacobian, and the only change in the OPTEX code is in the MODEL statement, where the two columns of the Jacobian are specified and the NOINT option specifies that the constant column corresponding to the (unweighted) intercept is not to be added. The resulting exact design corresponds to the continuous design of (22.16).

SAS Task 22.3. Use PROC OPTEX to find exact optimum designs for one-variable linear binomial regression over the range $0 < x \leq 80$ using the parameter estimates from Example 1.6, Bliss's beetle data, and

- a logistic link, and
- a complementary log–log link.

For each link, use the parameter estimates from the appropriate fit of the original data.

Finally, continuous designs can be constructed with SAS/IML matrix programming using the techniques discussed in Chapter 13, but redefining the D-optimality criterion. An IML function to compute the criterion for unweighted regression assuming a previously defined function `MakeZ()` that computes the design matrix Z might be the following.

```
start DCrit(w,x);
   Z = MakeZ(x,1);
   return(det(Z'*diag(w)*Z));
finish;
```

The simple modification of this required for a logistic model for binomial data simply involves using the design matrix and prior parameter values `b` to compute the weights and again premultiplying by the square root of the weights.

```
start DCrit(w,x) global(b);
   Z = MakeZ(x,1);

   eta = Z*b;
   mu  = exp(eta)/(1 + exp(eta));
   wgt = mu#(1-mu);
   F = sqrt(diag(wgt))*Z;

   return(det(F'*diag(w)*F));
finish;
```

> **SAS Task 22.4.** (Advanced). Use SAS/IML to find the continuous optimum design for logistic regression shown in (22.16).

22.7 Further Reading

Chaloner and Larntz (1989) find Bayesian optimum designs for the one-variable logistic model. As the results of §22.4.4 suggest, increasing the number of explanatory variables greatly increases the complexity of the structure of the optimum design for logistic models. Sitter and Torsney (1995*a*) explore designs for two-variable models and Torsney, in a series of papers, for example Sitter and Torsney (1995*b*) and Torsney and Gunduz (2001), explores designs for higher dimensions. Burridge and Sebastiani (1994) find optimum designs for the gamma model of §22.5. They establish conditions on the values of the parameters β in the first-order linear predictor under which two-level factorial designs are optimum; under other conditions the optimum designs have only one factor at a time not at the lower level.

Sitter (1992) and King and Wong (2000) find minimax designs for the one variable logistic model that reduce the harmful effect on the design of poor initial guesses of the parameters. Woods *et al.* (2006) calculate robust designs for models with several explanatory variables by using compound D-optimality to average over parameters and over link functions.

The approach of Khuri and Mukhopadhyay (2006) to robust design uses an extension of the variance–dispersion graphs of §6.4 to compare designs.

The emphasis in this chapter has been on parameter estimation and D-optimum designs. The T-optimum designs for model discrimination of §20.6 were extended by Ponce de Leon and Atkinson (1992) to generalized linear models. Waterhouse *et al.* (2007) refine T-optimality for nested generalized linear models. They compare the designs from their new criterion with D- T-, and D_S-optimality for binomial models with linear predictors that are second-order polynomials in one or two factors.

23

RESPONSE TRANSFORMATION AND STRUCTURED VARIANCES

23.1 Introduction

In the regression models that are the major subject of this book it is assumed that the error variance is constant. In §22.2 this assumption was relaxed to allow symmetrical errors with non-constant variances of the form $\sigma^2/w(x_i)$, where the $w(x_i)$ were a set of known weights. Weighted least squares was the appropriate form of analysis.

We start this chapter with design when the mean–variance relationship is such that the response may be transformed to approximate normality. The untransformed observations accordingly have a skewed distribution. The mean–variance relationship for power transformations of the response is introduced in §23.2.1 and the Box and Cox parametric family of power transformations is describe in §23.2.2.

The D-optimum designs for estimation of the parameters for the Box-Cox transformation of the response in regression data, including those of the linear model, are derived in §23.3. The effect of transformation on design for the parameters of non-linear models is indicated in §23.4.1. Such designs require transformation of both sides of the model, introduced in §23.4.2. The remainder of §23.4 is devoted to illustrating these methods through developing designs for the exponential decay model, including robust designs that are efficient for a variety of transformations. The chapter concludes in §23.6.1 with designs for symmetrical errors, extending the methods of optimum design to normal-theory response models with parameterized variance functions. A special case is when the variance is a function of the mean, but the error distribution is approximately normal, rather than skewed as it is for data that may be transformed to normality.

23.2 Transformation of the Response

23.2.1 Empirical Evidence for Transformations

The analysis of the data on the viscosity of elastomer blends in §8.3 led to a model in which the response was the log of viscosity rather than viscosity itself, a special case of the Box–Cox power transformation. In general, power transformation of the response is helpful when the variance of y increases as a power of the expected value $\mathrm{E}(y)$ of y. If

$$\mathrm{var}(y) \propto \{\mathrm{E}(y)\}^{2(1-\lambda)}, \tag{23.1}$$

Taylor series expansion shows that the variance is approximately stabilized by using as the response

$$\begin{array}{ll} y^{\lambda} & \lambda \neq 0 \\ \log y & \lambda = 0. \end{array} \tag{23.2}$$

So, for $\lambda = 1$, the variance is independent of the mean and no transformation is necessary. When $\lambda = 0.5$, the variance is proportional to the mean and the square root transformation is indicated, whereas, when $\lambda = 0$, the standard deviation is proportional to the mean and the logarithmic transformation provides approximately constant variance.

If the power law (23.1) holds with $\lambda < 1$, large observations will have larger standard deviations than small ones. Taking logarithms of the square root of both sides of this relationship yields

$$\log(\mathrm{s.d.}y) = \gamma_0 + (1-\lambda)\log\{\mathrm{E}(y)\}, \tag{23.3}$$

where s.d. is the standard deviation of y. If replicate observations are available, a plot of log standard deviation against log mean will indicate whether the power law holds. See Atkinson (2003*b*) for such a plot.

A linear relationship between log standard deviation and log mean is well established in analytical chemistry, where it is known as Horwitz's rule, an empirical relationship between the variability of chemical measurements and the concentration of the analyte. Lischer (1999) states that, even then, it was supported by the results of studies involving almost 10,000 individual data sets; the average transformation is to the power 0.14. This specific value goes against the standard statistical advice of using values with simple physical interpretations, such as the square root or the one-third power for volumes. However the evidence for this rule is overwhelming.

23.2.2 Box–Cox Transformation

For the usual linear regression model $\mathrm{E}(y) = f(x)^{\mathrm{T}}\beta$, Box and Cox (1964) analyse the power transformation

$$y(\lambda) = \begin{cases} (y^{\lambda} - 1)/\lambda & \lambda \neq 0 \\ \log y & \lambda = 0, \end{cases} \tag{23.4}$$

which is continuous at $\lambda = 0$. We have already used this function in the analysis of the viscosity data, where it was given as (8.5), and as the link of the gamma generalized linear model in §22.5.1.

The model to be fitted is

$$y(\lambda) = f^{\mathrm{T}}(x)\beta + \epsilon \tag{23.5}$$

for some lambda for which the ϵ are independent and, at least approximately, normally distributed with constant variance. If the required transformation is known, that is it is known that $\lambda = \lambda_0$, the design problem is the usual one of providing good data for estimating the parameters β of a linear model. Only if it also desired to estimate λ do new considerations arise.

For inference about the value of λ, Box and Cox (1964) use the likelihood ratio test. A computationally simpler alternative test that fits naturally with optimum design theory for regression models is the approximate score statistic derived by Taylor series expansion of (23.4) as

$$y(\lambda) \doteq y(\lambda_0) + (\lambda - \lambda_0)v(\lambda_0), \tag{23.6}$$

where

$$v(\lambda_0) = \frac{\partial y(\lambda)}{\partial \lambda}|_{\lambda=\lambda_0}.$$

The combination of (23.6) and the regression model (23.5) yields the model

$$\begin{aligned} y(\lambda_0) &= f^{\mathrm{T}}(x)\beta - (\lambda - \lambda_0)v(\lambda_0) + \epsilon \\ &= f^{\mathrm{T}}(x)\beta + \gamma\, v(\lambda_0) + \epsilon. \end{aligned} \tag{23.7}$$

Because (23.7) is again a regression model with an extra variable $v(\lambda_0)$ derived from the transformation, the new variable is called the *constructed variable* for the transformation. The approximate score statistic for testing the transformation is the t statistic for regression on $v(\lambda_0)$ in (23.7). Our design problem is to provide information on the value of γ, the coefficient of this constructed variable (or, equivalently, on λ) as well, perhaps, as providing information about the parameters β of the linear model and the variance σ^2.

23.3 Design for a Response Transformation

23.3.1 Information Matrices

The transformation (23.4) clearly only applies to $y > 0$. The purpose of the transformation is to provide homogeneous data for which the proposed linear model holds and for which the error distribution, after transformation, is not far from normal. Evidence about the transformation can come both from the mean and from the change of variance with the mean. Data that cover several cycles typically provide the strongest evidence about transformations. For example, in the wool data analysed by Box and Cox (1964) where the non-negative response is the number of cycles to failure of a test specimen, the smallest observation is 90 and the largest 3,636, a ratio of over 40. There is very strong evidence of the need for a transformation.

Atkinson and Cook (1997) find D-optimum designs for simultaneous estimation of the parameters β, σ^2 and λ in (23.5) as well as D_S-optimum designs for the subsets β and λ. The resulting D-optimum designs can be interpreted using the results on multivariate D-optimality in §10.10. However, obtaining the design criterion is complicated by the non-linear nature of the successive derivatives of $y(\lambda)$ with λ, the expectations of which are needed to calculate the expected information matrix.

We assume that the density of $y(\lambda)$ is well approximated by a normal distribution

$$l(y,\theta) = \frac{1}{\sigma\sqrt{(2\pi)}} J(\lambda) \exp\left[-\frac{\{y(\lambda) - f^{\mathrm{T}}(x)\beta\}^2}{2\sigma^2}\right], \tag{23.8}$$

where

$$J(\lambda) = \left|\frac{\partial y(\lambda)}{\partial y}\right|$$

is the Jacobian for the transformation, allowing for the change in the value of the response $y(\lambda)$ on transformation. The expected information per observation is, from (23.8)

$$M(\theta) = M(\beta, \sigma^2, \lambda) = -\mathrm{E}\left\{\frac{\partial^2}{\partial\theta^2} l(y,\theta)\right\}. \tag{23.9}$$

From (23.8)

$$M(\theta) = \begin{bmatrix} f(x)f^{\mathrm{T}}(x)/\sigma^2 & 0 & -f(x)\mathrm{E}\dot{y}(\lambda)\sigma^2 \\ & 1/2\sigma^4 & -\mathrm{E}\{\epsilon\dot{y}(\lambda)\}/\sigma^4 \\ & & [\mathrm{E}\{\epsilon\ddot{y}(\lambda)\} + \mathrm{E}\{\dot{y}(\lambda)^2\}]/\sigma^2 \end{bmatrix} \tag{23.10}$$

where $\epsilon = y(\lambda) - f^{\mathrm{T}}(x)\beta$, the expectations are taken with respect to (23.8) and the single and double dots indicate first and second derivatives with respect to λ. In particular

$$\dot{y}(\lambda) = v(\lambda) = \frac{y^\lambda \log y^\lambda - y^\lambda + 1}{\lambda^2},$$

with a lengthier expression for $\ddot{y}(\lambda)$. Apart from the term in $\ddot{y}(\lambda)$, (23.10) is the $(p+2) \times (p+2)$ information matrix that would be obtained from the model including the constructed variable $v(\lambda)$ (23.7) with σ^2 unknown. The $(p+1) \times (p+1)$ upper left-hand submatrix of 23.10) is the usual information matrix for the regression model without transformation.

Atkinson and Cook use Taylor series expansions to obtain approximations to the expectations in (23.10). For the value of λ for which the transformation holds, let

$$\mu(x, \theta) = \mathrm{E}(y^\lambda) = \lambda f^{\mathrm{T}}(x)\beta + 1. \tag{23.11}$$

The approximate information matrix per observation is then

$$M_a(\theta) = \begin{bmatrix} f(x)f^{\mathrm{T}}(x)/\sigma^2 & 0 & -f(x)\mathrm{E}\dot{y}(\lambda)/\sigma^2 \\ & 1/(2\sigma^4) & -\log\mu(x,\theta)/(\lambda\sigma^2) \\ & & 2\log^2\mu(x,\theta)/\lambda^2 + \{\mathrm{E}\dot{y}(\lambda)\}^2/\sigma^2 \end{bmatrix}, \tag{23.12}$$

where

$$\mathrm{E}\dot{y}(\lambda) = \mathrm{E}v(\lambda) \doteq \{\mu(x,\theta)\log\mu(x,\theta) - \mu(x,\theta) + 1\}/\lambda^2.$$

This information matrix can be written as the sum of two information matrices, that is

$$M_a(\theta) = KK^{\mathrm{T}} + LL^{\mathrm{T}}, \tag{23.13}$$

which is the form of the information matrix for independent bivariate responses.

23.3.2 D-optimum Designs

We begin by recalling a simplified form of the results of §10.10 on D-optimum designs for multivariate observations. These provide a general equivalence framework for our designs.

Let h independent responses be measured, the values of which are functions of the vector explanatory variable u which may include both x and z. Further, let the variance of all observations be unity. The information matrix per observation is then the sum of h rank-one matrices

$$M(u,\theta) = \sum_{j=1}^{h} f_j(u,\theta) f_j^{\mathrm{T}}(u,\theta), \tag{23.14}$$

where θ is a $r \times 1$ vector of parameters. The information matrix for the experimental design ξ is, as usual,

$$M(\xi,\theta) = \int_{\mathcal{X}} M(u,\theta)\xi(dx). \tag{23.15}$$

The notation allows for the possibility of locally optimum designs.

The standardized variance of prediction for response j can be written

$$d_j(u,\xi,\theta) = f_j^{\mathrm{T}}(u,\theta) M^{-1}(\xi,\theta) f_j(u,\xi,\theta), \tag{23.16}$$

with $M(\xi,\theta)$ given by (23.15). The equivalence theorem for D-optimum designs then applies to

$$d(u,\xi,\theta) = \sum_{j=1}^{h} d_j(u,\xi,\theta).$$

The, perhaps locally, D-optimum design ξ_D^* maximizing $|M(\xi,\theta)|$ is such that $d(u,\xi_D^*,\theta) \le r$ for $u \in \mathcal{X}$.

In the nomenclature of (23.14) the variables in the information matrix (23.13) are

$$\begin{array}{lcllr} f_1^{\mathrm{T}}(x,\theta) & = & (f(x)/\sigma & 0 & -\mathrm{E}v(\lambda)/\sigma), \\ f_2^{\mathrm{T}}(x,\theta) & = & (0 & 1/(\sqrt{2}\sigma^2) & -\sqrt{2}\log\mu(x,\theta)/\lambda). \end{array}$$

When $\lambda = 0$ these variables become

$$\begin{array}{lcllr} f_1^{\mathrm{T}}(x,\theta) & = & (f(x)/\sigma & 0 & -(f^{\mathrm{T}}(x)\beta)^2/2\sigma), \\ f_2^{\mathrm{T}}(x,\theta) & = & (0 & 1/(\sqrt{2}\sigma^2) & -\sqrt{2}f^{\mathrm{T}}(x)\beta). \end{array} \tag{23.17}$$

These variables exhibit the two sources of information about the transformation. One comes from the constructed variable $v(\lambda)$ in $f_1(x,\theta)$. The transformation information in the variance function comes from the logarithm of the regression function $\log\mu(x,\theta)$. The sum of squares of $\log(\mu)$ over the design enters into $M_a(\theta)$ through the last term of $f_2(x,\theta)$, indicating a preference for designs with relatively large changes in the variance. The relative importance of these two terms depends upon the value of σ^2. As (23.12) shows, for small σ^2 the term in $v(\lambda)$ dominates. The design then becomes that based on regression including a constructed variable (23.7).

Example 23.1. Box and Cox Wool Data Box and Cox (1964) give the number of cycles to failure of a wool (worsted) yarn under cycles of repeated loading. The results are from a single 3^3 factorial experiment. The three factors and their levels are:

x_1: length of test specimen (25, 30, 35 cm)
x_2: amplitude of loading cycle (8, 9, 10 mm)
x_3: load (40, 45, 50 g).

The number of cycles to failure ranges from 90, for the shortest specimen subject to the most severe conditions, to 3,636 for the longest specimen subjected to the mildest conditions. In their analysis Box and Cox (1964) recommend that the data be fitted after the log transformation of y, a conclusion supported by the analysis of Atkinson and Riani (2000) using the forward search.

When the data are logged, that is $\lambda = 0$, a first-order model is adequate. The parameter estimates are $\hat{\beta} = (6.335,\ 0.832,\ -0.631,\ -0.392)$ and $\hat{\sigma} = 0.1856$; these estimates are computed from regressing log (cycles to failure) on a linear model in the scaled values x_1, x_2, and x_3 of the factors. We take these as the values of β and σ for our design. As discussed earlier, the D-optimum design maximizes

$$\begin{aligned} M(\xi, \theta) &= \int_{\mathcal{X}} M(u, \theta)\xi(dx) \\ &= \int_{\mathcal{X}} f_1(u, \theta) f_1^{\mathrm{T}}(u, \theta)\xi(dx) + \int_{\mathcal{X}} f_2(u, \theta) f_2^{\mathrm{T}}(u, \theta)\xi(dx) \\ &= M_1(\xi, \theta) + M_2(\xi, \theta), \end{aligned} \tag{23.18}$$

where f_1 and f_2 are given by (23.17). The resulting design is shown in Table 23.1, together with the values of $\beta^{\mathrm{T}} f(x)$ for these support points.

This design was constructed by first building up the support points and their approximately optimum weights using sequential design augmentation over a grid on $[-1, 1]^3$ with increment 0.1. The weights for these support points were then refined by non-linear optimization.

Several features of the design of Table 23.1 are notable.

- The design has the same support as the D-optimum design for an untransformed response, which is the 2^3 factorial, but the weights are different.
- The strength of evidence for a transformation depends in part on the range of values of $\mathrm{E}(y) = \beta^{\mathrm{T}} f(x)$. The design conditions in Table 23.1 are ordered by these values. The greatest weights are at the minimum, central and maximum values of $\beta^{\mathrm{T}} f(x)$.

TABLE 23.1. Example 23.1 Box and Cox Wool Data. D-optimum design for fitting a linear model and testing $\lambda = 0$ in the Box–Cox transformation

Weight	x_1	x_2	x_3	$\beta^{\mathrm{T}} f(x)$
0.1842	−1	1	1	4.480
0.0875	−1	1	−1	5.264
0.1064	−1	−1	1	5.742
0.1219	1	1	1	6.144
0.1219	−1	−1	−1	6.526
0.1064	1	1	−1	6.928
0.0875	1	−1	1	7.406
0.1842	1	−1	−1	8.190

- The weights are orthogonal to the factor values for the support points. Equivalently, the weights are balanced with respect to each of the factors. Consequently the upper 4 ×4 submatrix of $M_1(\xi, \theta)$ has zeroes in the first row and column apart from element (1,1).

Optimum designs for estimating both λ and β are often not so close to the optimum design for estimation of β as that of Table 23.1. For example, the designs for second-order response surfaces in two factors given by Atkinson and Cook (1997) have up to four more support points than the nine of the D-optimum design for the linear model. We can obtain designs with similar properties for the first-order model for the wool data by changing the parameter values. If we replace $\beta_1 = 0.832$ by 10 times that value we obtain the 10-point design of Table 23.2.

The support points of the designs are again those of the 2^3 factorial, but now augmented with two points at the central value of x_1, the factor with the greatest effect now that the value of β_1 has been modified. Otherwise many of the features of the design are similar to those of Table 23.1, for example that there is appreciable design weight at the minimum, central and maximum values of the linear predictor $\beta^{\mathrm{T}} f(x)$. The design weights are also orthogonal to the factors so that $M_1(\xi, \theta)$ again has the block diagonal structure noted in the results of Table 23.1.

The observations in Table 23.2 are ordered by the value of the linear predictor, the minimum values of which are negative. For the power transformation (23.4) to be applicable, all observations have to be positive. This imposes the restriction that the right-hand side of (23.11) must also be positive. However, for the log transformation there is no restriction on the

TABLE 23.2. Example 23.1 Box and Cox Wool Data with modified parameters. D-optimum design for fitting a linear model and testing $\lambda = 0$ in the Box–Cox transformation; $\beta = (6.335,\ 10\times0.832,\ -0.631,\ -0.392)^{\mathrm{T}}$

Weight	x_1	x_2	x_3	$\beta^{\mathrm{T}} f(x)$
0.1581	−1	1	1	−3.008
0.0405	−1	1	−1	−2.224
0.0311	−1	−1	1	−1.746
0.1097	−1	−1	−1	−0.962
0.1606	0	1	−1	6.096
0.1606	0	−1	1	6.574
0.1097	1	1	1	13.632
0.0311	1	1	−1	14.416
0.0405	1	−1	1	14.894
0.1581	1	−1	−1	15.678

value of the linear predictor; the negative values in Table 23.2 correspond to positive values of μ.

The ordering of the design points by the value of the linear predictor in both tables reveals a symmetry of the design weights over values of the linear predictor. However, the important point for the theory of these designs is that, as the range of values of the linear predictor increases with σ remaining fixed, the number of support points of the design becomes greater than that for the D-optimum design for β alone. ■

23.4 Response Transformations in Non-linear Models

23.4.1 Transformations and Optimum Design for Exponential Decay

Provided the value of λ is known, transformation of the response in linear models has no effect on the optimum design; the response is straightforwardly transformed, with the D-optimum design for the parameters β independent of the transformation. However, if the model is non-linear, transforming the response often does affect the design, even for a known transformation. For instance, if a kinetic model is such that the concentrations of the chemical components sum to one, the sum of the power-transformed components will not be one. The model has also to be transformed for this constraint to be satisfied.

A simple example of this effect of transformation of the response on experimental design comes from the non-linear response model resulting from first-order decay, Example 17.1, in which the concentration of chemical A at time t is given by the non-linear function

$$[A] = \eta_A(t, \theta) = e^{-\theta t} \quad (\theta, t \geq 0), \tag{23.19}$$

if it is assumed that the initial concentration of A is 1. The simple statistical model of the observations assumed in Chapter 17 is

$$y = \eta_A(t, \theta) + \epsilon,$$

where the errors ϵ are independently distributed with zero mean and constant variance. The variance of the least squares estimator $\hat{\theta}$ then depends on the parameter sensitivity

$$f(t, \theta) = \frac{d\eta_A(t, \theta)}{d\theta} = -t\exp(-\theta t). \tag{23.20}$$

As we showed in §17.2, the locally D-optimum design minimising the variance of $\hat{\theta}$ takes all measurements where $f(t, \theta)$ is a maximum, that is at the time $t^* = 1/\theta$.

Now suppose that the model needs to be transformed to give constant variance. If the log transformation is appropriate and $[A]$ is measured, taking logarithms of both sides of (23.19), combined with additive errors, yields the statistical model

$$\log y = \log\{\eta_A(t, \theta)\} + \epsilon = -\theta t + \epsilon. \tag{23.21}$$

The log transformation thus results in a linear statistical model with response $\log y$, for which the parameter sensitivity is just the time t. The optimum design puts all observations at the maximum possible time, when the concentration is as small as possible, an apparently absurd answer. Thus a seemingly slight assumption about the error distribution can have a huge effect on the optimum experimental design.

Rocke and Lorenzato (1995) do question the model in which error variance becomes negligible as concentration decreases. They suggest an alternative with two error components for which, although standard deviation decreases with concentration, it does not go to zero. Such an error model would give less extreme designs than those found in later sections for our Example 23.2 when λ has very small positive values. A transformation when the two error components are respectively normal and lognormal is presented by McLachlan, Do, and Ambroise (2004).

23.4.2 Transforming Both Sides of a Non-linear Model

The simple example of §23.4.1 for exponential decay shows the dependence of design for non-linear models on the transformation, even when λ is known. We now find simple expressions for the parameter sensitivities when the response is transformed.

When, for example, $\eta(x, \theta)$ is a mechanistic model based on chemical kinetics, the relationship between the response and the concentrations of the other reactants needs to be preserved after transformation. This is achieved by transformation of both sides of the model, as described in Chapter 4 of Carroll and Ruppert (1988). For fixed $\lambda \neq 0$, estimation of the parameters θ after transformation does not depend on whether the response is $y(\lambda)$ (23.4) or straightforwardly y^{λ}. If the response is multivariate, each response will need to be transformed, we assume with the same value of λ. Simplification of the model and the introduction of observational error on this transformed scale leads to the statistical model

$$y_u^{\lambda} = \{\eta_u(x, \theta)\}^{\lambda} + \epsilon_u, \tag{23.22}$$

for the uth response, $u = 1, \ldots, h$.

The notation for the parameter sensitivities has to be extended to accommodate transformation. For response u in the absence of transformation let

$$f_{uj}(1; x, \theta) = \frac{\partial \eta(x, \theta)}{\partial \theta_j}. \tag{23.23}$$

The parameter sensitivities for the multivariate version of the transformation model (23.22) are found by differentiation to be

$$\frac{\partial \{\eta_u(t, \theta)\}^{\lambda}}{\partial \theta_j} = \lambda \{\eta_u(t, \theta)\}^{\lambda - 1} \frac{\partial \eta_u(t, \theta)}{\partial \theta_j} = \lambda \{\eta_u(t, \theta)\}^{\lambda - 1} f_{uj}(1; t, \theta). \tag{23.24}$$

For fixed λ, multiplication by λ in (23.24) does not change the optimum design, so the sensitivities have the easily calculated form

$$f_{uj}(\lambda; t, \theta) = \{\eta_u(t, \theta)\}^{\lambda - 1} f_{uj}(1; t, \theta) = f_{uj}(1; t, \theta) / \{\eta_u(t, \theta)\}^{1 - \lambda}. \tag{23.25}$$

If $\lambda < 1$, the variance of the observations increases with the value of $\eta_u(t, \theta)$. Thus transformation of both sides for such values of λ will increase the relative value of the sensitivities for times where the response is small. We can expect that designs for $\lambda < 1$ will include observations at lower concentrations than those when no transformation is needed.

Example 23.2. Exponential Decay As a simple example of the effect of transformation of the response in a non-linear model on experimental design we continue with the model for exponential decay.

As we saw in §23.4.1, the optimum design in the absence of transformation, that is for $\lambda = 1$, puts all trials at $t^* = 1/\theta$. At the other extreme the design for the log transformed model (23.21), that is for $\lambda = 0$, puts all observations at the maximum possible time, when the concentration is as small as possible. We now find less extreme designs for values of λ between zero and one.

Since no material is lost during the reaction, the concentrations in the absence of error obey the relationship $[A] + [B] = 1$. From (23.19) the concentration of B at time t is therefore

$$[B] = \eta_B(t, \theta) = 1 - e^{-\theta t} \quad (\theta, t \geq 0).$$

If $[A]$ is measured in the absence of transformation the parameter sensitivity is

$$f_A(1; t, \theta) = -t \exp(-\theta t), \tag{23.26}$$

whereas if $[B]$ is measured

$$f_B(1; t, \theta) = t \exp(-\theta t),$$

both of which have their extreme value at the time $t^* = 1/\theta$. Therefore, if the purpose of the experiment is to estimate θ with minimum variance, all readings should be taken at this one value of time. The result holds not only if $[A]$ or $[B]$ are measured on their own, but also if both $[A]$ and $[B]$ are measured.

Now suppose that the model needs to be transformed to give constant variance. From (23.24) the parameter sensitivity for the power transformation λ when $[A]$ is measured is

$$f_A(\lambda; t, \theta) = \{\eta_A(t, \theta)\}^{\lambda-1} f_A(1; t, \theta) = -t \exp(-\lambda \theta t). \tag{23.27}$$

The optimum design is therefore at a time of $1/(\lambda\theta)$. As λ decreases, the time for the optimum design increases reaching, as we have seen, infinity when $\lambda = 0$, the log transformation.

The analysis when $[B]$ is measured is similar, but does not yield an explicit value for the optimum time. The sensitivity is now

$$f_B(\lambda; t, \theta) = \{\eta_B(t, \theta)\}^{\lambda-1} f_B(1; t, \theta) = t \exp(-\theta t)\{1 - \exp(-\theta t)\}^{\lambda-1}, \tag{23.28}$$

which is maximized by the optimum time. As $\lambda \to 0$, the optimum time does likewise; when $\lambda = 0$, $t = 0$.

Figure 23.1 shows the optimum time at which the reading of the concentration of A or B should be taken as a function of λ, when $\theta = 0.2$ as well as for the multivariate experiment in which both $[A]$ and $[B]$ are measured.

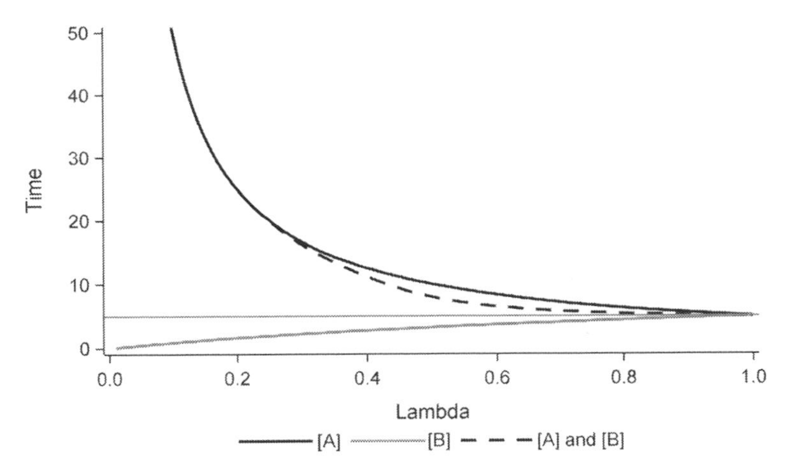

FIG. 23.1. Exponential decay: optimal design points as a function of λ when only $[A]$ is measured, when only $[B]$ is measured, and when both $[A]$ and $[B]$ are measured.

We assume that the errors in the two responses are independent with the same variance. Then the figure shows this design is similar to that when only $[A]$ is measured. For the Horwitz value of 0.14 for λ, mentioned in §23.2.1 as typical in analytical chemistry, the optimum times are 35.7 when $[A]$ or both $[A]$ and $[B]$ are measured and 1.19 when only $[B]$ is measured. The concentration of B is small at the beginning of the experiment and that of A is small for large times. The figure shows that, as λ decreases and a stronger transformation is needed, so the design points when only one response is measured move to regions of lower concentration. ■

23.4.3 Efficiencies of Designs

If the optimum designs vary with λ, as they do in Example 23.2, it is likely that a design for the wrong λ will be inefficient. To quantify this effect let the optimum design for a specified λ_0 be ξ_0^* and for some other λ be ξ_λ^*. The value of the information matrix depends not only on ξ and θ but also on the parameter λ. When the value of the transformation parameter is λ the information matrix for the design ξ_0^* can be written as $M(\xi_0^*, \theta, \lambda)$. Then, as in (21.1), the D-efficiency of the design ξ_0^* for some λ is the ratio of determinants

$$\text{Eff}^{\,D}(\xi_0^*, \lambda) = \{|M(\xi_0^*, \theta, \lambda)|/|M(\xi_\lambda^*, \theta, \lambda)|\}^{1/p}, \tag{23.29}$$

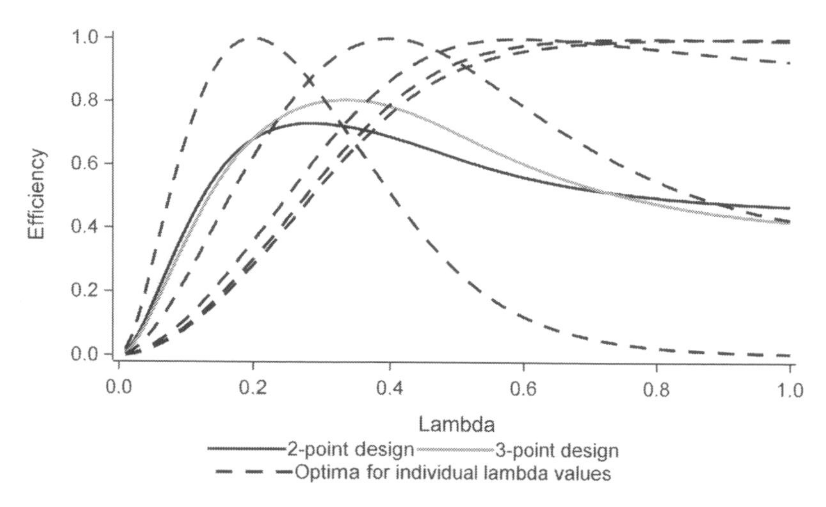

FIG. 23.2. Exponential decay: efficiencies of single and multipoint designs when both [A] and [B] are measured. Values of λ for individual optima are 0.2, 0.4, 0.6, 0.8, and 1.0.

where θ is $p \times 1$. The value of $\text{Eff}^{D}(\xi_0^*, \lambda_0)$ is, by definition, 100%.

Example 23.2. Exponential Decay continued Figure 23.2 shows the efficiencies defined in (23.29) of the D-optimum designs for five values of λ_0: 0.2, 0.4, 0.6, 0.8, and 1. These are plotted over a range of values of λ between 0.01 and one. The maximum value of each efficiency is 100% when $\lambda = \lambda_0$. What is particularly noticeable is that all designs are inefficient for low values of λ, a result of the rapidly increasing value of the optimum time as $\lambda \to 0$. The design for $\lambda_0 = 0.2$ is the most efficient of those shown for low values of λ, but is inefficient for high values of λ. Thus none of these one-point designs will be efficient for estimating θ if there is virtually no information about the true value of λ. ■

23.5 Robust and Compound Designs

Robust Designs. The optimum designs of Figure 23.1 for exponential decay all have one point of support. An alternative is to investigate multi-point designs that possibly have better properties over a range of λ. These robust designs will, of course, have a reduced efficiency for any specific λ_0.

TABLE 23.3. Exponential decay when both [A] and [B] are measured. Multipoint robust designs

	t, w	Design		
Two-point	t_i	6.462	24.96	
	w_i	0.5	0.5	
Three-point	t_i	6.25	12.5	25.0
	w_i	0.333	0.333	0.333

We first illustrate the properties of two arbitrary multipoint designs for exponential decay and then indicate how optimum designs can be found using compound D-optimality.

Example 23.2. Exponential Decay continued When $\lambda_0 = 0.6$ in the exponential decay model, the optimum time for measurement is 6.462, whereas when $\lambda_0 = 0.2$ it is 24.96. Figure 23.2 shows that these two designs are efficient over different ranges of λ. We form a multipoint design from the linear combination of these two designs with weights 0.5. For comparison a three-point design with times of 25, 12.5, and 6.25 is also included. Such designs, with constant spacing in log time or log concentration, are frequent in medical and pharmaceutical experiments: an example is in Downing, Fedorov, and Leonov (2001). The two designs are in Table 23.3.

The efficiencies of these two multipoint designs are also plotted in Figure 23.2. The maximum efficiency for the two-point design is 73%, whereas that for the three-point design is just over 80%, both occurring in the range of 0.3 to 0.4 for λ. Outside these values the two designs have similar efficiencies, with that for the two-point design being slightly less peaked. The efficiencies of these designs are however not otherwise strikingly different from those of the optimum one-point designs plotted in the same figure. ■

Compound Designs. The two hopefully robust multipoint designs for the exponential decay model were found by guessing the distribution of times of observation that might lead to an efficient design. A preferable alternative is to consider other design criteria. A natural criterion is to maximize the product of efficiencies in (23.29) for l values of λ. The argument of §21.3 shows that this leads to the form of compound D-optimality in which

$$\Phi(\xi) = \sum_{i=1}^{l} \kappa_i \log |M(\xi, \theta, \lambda_i)|, \tag{23.30}$$

is maximized. A distinction from (21.8) is that all models have the same number of parameters, so we do not have to adjust the individual terms by a factor $1/p_i$. An example of such robust designs for a three-parameter non-linear model is given by Atkinson (2005).

23.6 Structured Mean–Variance Relationships

23.6.1 Models

In the earlier sections of the chapter the emphasis was on response transformation, which arose from the attempt to normalize asymmetrical response distributions in which the variance is a function of the mean. For the remainder of the chapter we concentrate instead on heteroscedastic responses. In order to make progress we assume that the responses are normal, so that their distribution is symmetrical. In this section we outline the design consequences of a parameterized variance function. The details are given by Atkinson and Cook (1995).

Statistical models in which both means and variances are functions of explanatory variables have become increasingly important in quality control (Nelder and Lee 1991; Box 1993; Lee and Nelder 1999; Box 2006) although the design consequences have been less explored. The possibility of additive heteroscedastic errors, known up to a constant of proportionality, is routinely considered by, for example, Fedorov (1972). Here the model has the more general form

$$y = f^{\mathrm{T}}(x)\beta + \sigma[\tau\{g^{\mathrm{T}}(z)\alpha\}]^{1/2}\epsilon, \tag{23.31}$$

where x and z are design vectors of dimension p and q with $f(x)$ and $g(z)$ respectively $p \times 1$ and $q \times 1$ vectors of linearly independent continuous functions. The error term ϵ is standardized to have expectation zero and unit variance. In order to derive information matrices it will, in addition, be taken to have a normal distribution. The unknown parameters are α, β and $\sigma > 0$. It follows from (23.31) that, at the point (x, y), $\mathrm{E}(y) = f^{\mathrm{T}}(x)\beta$ and $\mathrm{var}(y) = \sigma^2[\tau\{g^{\mathrm{T}}(z)\alpha\}]$. Thus we have the standard linear model for the mean with the variance a function of another linear predictor.

For applications it is often useful to take τ to be the exponential function and then to work with a linear model for the logarithm of the variance

$$\log\{\mathrm{var}(y)\} = \log\sigma^2 + g^{\mathrm{T}}(z)\alpha. \tag{23.32}$$

Atkinson and Cook (1995) identify two special cases of (23.31) that deserve attention. One is when the design variables influencing the mean are the same as those influencing the variance, that is $x = z$, so that (23.31) becomes

$$y = f^{\mathrm{T}}(x)\beta + \sigma[\tau\{g^{\mathrm{T}}(x)\alpha\}]^{1/2}\epsilon. \tag{23.33}$$

A further specialization is when the variance depends on x only through the mean so that

$$y = f^{\mathrm{T}}(x)\beta + \sigma[\tau\{\nu f^{\mathrm{T}}(x)\beta\}]^{1/2}\epsilon, \tag{23.34}$$

where ν is an unknown real-valued parameter that allows for the strength of dependence of the variance function on the mean.

23.6.2 Information Matrices

We return to the model (23.31) with general variance function τ. The structure of the information matrices reflects the contributions to the estimation of the parameters by information coming from the mean and from the variance. When $\alpha = \alpha_0$ and $\sigma^2 = \sigma_0^2$ are known the information per observation on β in (23.31) has the form

$$M_\mu(x, z|\beta, \alpha_0, \sigma_0^2) = \frac{f(x)f^{\mathrm{T}}(x)}{\sigma_0^2\tau\{g^{\mathrm{T}}(z)\alpha_0\}}, \tag{23.35}$$

leading to estimation by (non-iterative) weighted least squares treated in §22.2. The information on (α, σ^2) for known $\beta = \beta_0$ can also be found and is

$$M_\sigma(z|\alpha, \sigma^2) = JJ^{\mathrm{T}}, \tag{23.36}$$

where $J^{\mathrm{T}}(z|\alpha, \sigma^2) = \{g^{\mathrm{T}}(z)r(z|\alpha), \sigma^{-2}\}/\sqrt{2}$ with

$$r(z|\alpha) = \frac{\dot{\tau}\{g^{\mathrm{T}}(z)\alpha\}}{\tau\{g^{\mathrm{T}}(z)\alpha\}},$$

and the dot above τ indicates the first derivative. As the notation implies, $M_\sigma(z|\alpha, \sigma^2)$ does not depend on β_0. The function r measures the rate of change in the log variance. When τ is the exponential function, $r(z|\alpha) = 1$. See model IV of §23.6.3.

The information matrix per observation for $(\beta, \alpha, \sigma^2)$ in model (23.31) can now be represented as

$$M(x, z|\beta, \alpha, \sigma^2) = \begin{bmatrix} M_\mu(x, z|\beta, \alpha, \sigma^2) & 0 \\ 0 & M_\sigma(z|\alpha, \sigma^2) \end{bmatrix}. \tag{23.37}$$

The $(p + q + 1) \times (p + q + 1)$ information matrix for all the parameters is therefore block diagonal.

The information matrix for model (23.33) is obtained by simply setting $z = x$ in (23.37) and writing the information matrix as a function of x alone.

For (23.34), in which the variance is a function of the mean, there are $p+2$ parameters. The information matrix for one observation can be written in the additive form $KK^{\mathrm{T}} + LL^{\mathrm{T}}$ of (23.13) with

$$K^{\mathrm{T}}(x|\beta,\nu,\sigma^2) = \{\nu f^{\mathrm{T}}(x) r(x|\nu\beta), f^{\mathrm{T}}(x)\beta r(x|\nu\beta), \sigma^{-2}\}/\sqrt{2}$$

$$L^{\mathrm{T}}(x|\beta,\nu,\sigma^2) = [f^{\mathrm{T}}(x)\tau^{-1/2}\{\nu f^{\mathrm{T}}(x)\beta\}, 0, 0]/\sigma$$

and

$$r(z|\nu\beta) = \frac{\dot{\tau}\{\nu f^{\mathrm{T}}(z)\beta\}}{\tau\{\nu f^{\mathrm{T}}(z)\beta\}}.$$

Comparison with (23.35) shows the extra precision that can be obtained when information about β comes both from the structure of the mean and of the variance.

The block diagonal form of (23.37) and the additive form (23.13) are both helpful in the construction of optimum designs.

23.6.3 D-optimum Designs

We now use the structure of the multivariate D-optimum designs of §23.3.2 to explore the properties of designs for the information matrices of §23.6.2. We first consider the general model (23.31) in which there are non-overlapping models for the mean and variance. In sequence we look at the special cases when the parameters of the variance function and then of the mean are known, before looking at the product information matrix (23.37) when neither set of parameters is known. We conclude with design when the variance is a function of the mean.

*I. General Model with Known Variance Structure*Fixing $\alpha = \alpha_0$, the information for (β, σ^2) is block diagonal. The total information for the mean is

$$M_\mu(\beta|\alpha_0,\sigma^2) = \int M_\mu(x,z|\beta,\alpha_0,\sigma^2)\xi(dx,dz), \qquad (23.38)$$

where $M_\mu(x,z|\beta,\alpha_0,\sigma^2)$ is given by (23.35). Because σ^2 enters $M_\mu(\beta|\alpha_0,\sigma^2)$ as a proportionality constant, the design will not depend on its value. This then is essentially the standard situation with known efficiency function. It is obtained as a special case of the results on multivariate D-optimality by setting $\theta = \beta$, $u^{\mathrm{T}} = (x^{\mathrm{T}}, z^{\mathrm{T}})$, $h = 1$ and $f_1(u,\theta) = f(x)/[\tau\{g^{\mathrm{T}}(z)\alpha_0\}]^{1/2}$. To construct optimal designs, methods discussed in previous chapters for both continuous and exact designs can be applied to $f_1(u,\theta)$.

When x and z are structurally unrelated the design region can be written as $\mathcal{X} = \mathcal{X}_x \times \mathcal{X}_z$ so that, for each element of one design space, we have all elements of the other space. Let ξ_z denote an arbitrary marginal design on $\mathcal{X}_z$ and let $\xi^*_{D(x)}$ denote the D-optimum design on $\mathcal{X}_x$ for the homoscedastic model $y = f^{\mathrm{T}}(x)\beta + \epsilon$. The overall design is then $\xi = \xi^*_{D(x)} \times \xi_z$. The global D-optimum design maximizing $|M_\mu(\beta|\alpha_0, \sigma^2)|$ is then $\xi^*_D = \xi^*_{D(x)} \times \xi^*_z$, where ξ^*_z places mass 1 at the value of z that minimizes τ.

II. General Model with Known Mean Structure When β is assumed known, the information per observation is given in (23.36). The total information $M_\sigma(\alpha, \sigma^2)$ follows by integration over the design measure ξ, analogously to the derivation in (23.38). The determinant of the total information depends only on σ^2 through a proportionality constant; consequently the D-optimum design will, at most, depend on the value of α. The information $M_\sigma(\alpha, \sigma^2)$ is proportional to the total information (23.15) when we put $\theta^{\mathrm{T}} = (\beta^{\mathrm{T}}, \sigma^2)$, $u = z$, $h = 1$ and $f_1(u, \theta) = \{1, g^{\mathrm{T}}(z) r(z|\alpha)\}$. Again, methods of constructing optimal designs discussed in previous chapters can be applied to $f_1(u, \theta)$.

When τ is specified as the exponential function, $M_\sigma(\alpha, \sigma^2)$ is independent of α and is proportional to the total information on γ in the homoscedastic model $y = \gamma_0 + g^{\mathrm{T}}(z)\gamma_1 + \epsilon$. Consequently, under the exponential variance function, optimum designs for variance parameters when β is known can be constructed using standard algorithms.

III. General Model with Mean and Variance Structures Both Unknown The information matrix is given in (23.37). Because of the block-diagonal structure, the determinant of the total information $M(\beta, \alpha, \sigma^2)$ can be expressed as

$$|M(\beta, \alpha, \sigma^2) = |M_\mu(\beta|\alpha, \sigma^2)| \times |M_\sigma(\alpha, \sigma^2)|, \qquad (23.39)$$

so that the criterion function is the product of those considered in the two preceding special cases. The determinant of the total information again depends on σ^2 only through a constant of proportionality, so that the D-optimum design will depend, at most, on the value of α.

To apply (23.14) and the other results on multivariate D-optimality in this case set $\theta^{\mathrm{T}} = (\beta^{\mathrm{T}}, \alpha^{\mathrm{T}}, \sigma^2)$, $u^{\mathrm{T}} = (x^{\mathrm{T}}, z^{\mathrm{T}})$, $h = 2$,

$$f_1^{\mathrm{T}}(u, \theta) = (f^{\mathrm{T}}(x)/[\tau\{g^{\mathrm{T}}(z)\alpha\}]^{1/2}, 0^{\mathrm{T}}, 0),$$

and

$$f_2^{\mathrm{T}}(u, \theta) = \{0^{\mathrm{T}}, g^{\mathrm{T}}(z) r(x|\alpha), 1\}/\sqrt{2}.$$

Then the total information in the general formulation (23.15) is proportional to (23.39). The total variance $d(u, \xi, \theta)$ in this case is just the sum of the variances in the two preceding special cases I and II.

IV. General Model with Exponential Variance Structure In the absence of strong prior knowledge about the structure of the variances, it seems natural to take τ to be the exponential function. Then r is unity and the two parts of the design criterion in III become

$$f_1^{\mathrm{T}}(u,\theta) = (f^{\mathrm{T}}(x)/[\exp\{g^{\mathrm{T}}(z)\alpha\}/2], 0^{\mathrm{T}}, 0),$$

and

$$f_2^{\mathrm{T}}(u,\theta) = \{0^{\mathrm{T}}, g^{\mathrm{T}}(z), 1\}/\sqrt{2}.$$

If, in addition, it can be assumed that the variance does not vary strongly over the experimental region, we can design as if $\alpha = 0$. Then

$$f_1^{\mathrm{T}}(u,\theta) = (f^{\mathrm{T}}(x), 0^{\mathrm{T}}, 0).$$

Because of the block diagonal structure of the information matrix (23.39), the design criterion becomes a simple form of compound D-optimality. When x and z are structurally unrelated so that $\mathcal{X} = \mathcal{X}_x \times \mathcal{X}_z$ we can use a criterion of the form

$$\kappa \log M_\mu(x, z|\beta, \alpha = 0, \sigma^2) + (1-\kappa)\log M_\sigma(z|\gamma, \sigma^2). \tag{23.40}$$

The two parts of (23.40) are standard information matrices for linear models, one that for the mean and the other that for the variances as in II. The value of κ can be chosen to reflect interest in the parameters β of the mean and the parameters α of the variance. If $\alpha \neq 0$, some optimality will be lost because the design for the mean parameters assumes homoscedasticity. However the design for the variance over $\mathcal{X}_z$ does not depend on the value of β.

V. Variance a Function of the Mean When the variance is a function of the mean, the information per observation for model (23.34) is again in the general form (23.14). Now we let $\theta^{\mathrm{T}} = (\beta^{\mathrm{T}}, \nu, \sigma^2), u = x, \mathcal{X} = \mathcal{X}_x, h = 2, f_1 = K$, and $f_2 = L$, where K and L are given in (23.13). In this case the total information depends on the values of all parameters, including σ^2. The design criterion is thus, perhaps surprisingly, more complicated; for example, Bayesian designs would require a prior distribution on σ^2.

When the variance has the exponential form (23.32)

$$f_1^{\mathrm{T}}(x|\beta,\nu,\sigma^2) = \{\nu f^{\mathrm{T}}(x), f^{\mathrm{T}}(x)\beta, \sigma^{-2}\}/\sqrt{2}$$

and

$$f_2^{\mathrm{T}}(x|\beta,\nu,\sigma^2) = (f^{\mathrm{T}}(x)\exp[-\{\nu f^{\mathrm{T}}(x)\beta\}/2], 0, 0\}/\sigma.$$

Comparison with (23.35) shows again, but for this special case, the extra precision that can be obtained when information about β comes from the structure of both the mean and of the variance. As with all the designs discussed in this section, standard methods of construction for D-optimum designs apply.

Atkinson and Cook (1995) give examples of designs for a two-factor response surface model. Downing, Fedorov, and Leonov (2001) and Fedorov, Gagnon, and Leonov (2002) give examples of designs for linear and non-linear models and discuss both estimation and inference.

24

TIME-DEPENDENT MODELS WITH CORRELATED OBSERVATIONS

24.1 Introduction

Observations that occur as a time series may not be independent. In this chapter we see how the methods of optimum design have to be adapted if there is indeed correlation between observations at different time points.

The nonlinear examples of Chapter 17 illustrate the general point. In Example 17.3 with two consecutive first-order reactions, the concentrations of chemicals evolved over time. On the assumption that the errors were independent we found the D-optimum design for the two parameters, which consisted of taking observations at just two time points. This design is optimum if individual measurements are taken on different independent runs of the experiment. But, if several readings are taken on one run, the observations will have some correlation. In §24.3 we see how the presence of correlation alters the design for this example.

The effect of correlation on the optimum design is not trivial. Although it is straightforward to write down the extension of the design criteria to correlated errors, it is difficult to build an efficient algorithm for finding designs. In our numerical example, with covariance kernel (24.9), replicate observations are completely correlated. The optimum design accordingly provides a spread of times at which readings are to be taken, the spread depending on the rate at which correlation decreases. In addition, the design is exact, depending on the number of observations to be taken. The result is a list of N distinct times at which readings are to be taken.

Although a list of precise times at which measurements are to be taken may be a feasible design in a technological experiment, such designs may not be possible when measurements are taken on animal or human subjects. For example, patients attending a clinic may receive an injection or other treatment on, or shortly after, arrival. Measurements can then be taken during the rest of the day, or during succeeding days, but it may not be possible to take measurements outside ordinary working hours. There may also be restrictions on the number of times it is possible to take measurements on an individual patient. In §24.4 we find optimum designs when the choice is between a finite number of possible distinct measurement schedules. If measurements on different patients are independent, we show that the problem can be reformulated as a standard one in design optimality, even though the measurements on a single individual are correlated.

The chapter concludes with a discussion of a variety of other design problems that arise with correlated observations, for example the design of experiments where observations are taken in space.

24.2 The Statistical Model

To emphasize the importance of the error process we write the model for the univariate response as

$$y(t) = \eta(t, \theta) + \epsilon(t) \quad t \in \mathcal{T}, \tag{24.1}$$

where θ is a $p \times 1$ vector of unknown parameters. The error process $\epsilon(t)$ is the crucial difference between the designs considered in this chapter and those in the other chapters of the book.

In (24.1) the error term $\epsilon(t)$ is a stochastic process with zero mean and known continuous covariance kernel

$$\mathrm{E}[\epsilon(t)\epsilon(s)] = c(t, s) \ \text{ on } \mathcal{T}^2. \tag{24.2}$$

The generalized least-squares estimator of θ from the N experimental observations is given by

$$\hat{\theta} = \arg\min_{\theta} \sum_{i=1}^{N} \sum_{j=1}^{N} w_{ij}[y(t_i) - \eta(t_i, \theta)][y(t_j) - \eta(t_j, \theta)]. \tag{24.3}$$

The weights w_{ij} are the elements of the matrix

$$W = \begin{bmatrix} w_{11} & \cdots & w_{1N} \\ \vdots & \ddots & \vdots \\ w_{N1} & \cdots & w_{NN} \end{bmatrix} = \begin{bmatrix} c(t_1, t_1) & \cdots & c(t_1, t_N) \\ \vdots & \ddots & \vdots \\ c(t_N, t_1) & \cdots & c(t_N, t_N) \end{bmatrix}^{-1}. \tag{24.4}$$

The information matrix from this N-trial design can be written, as before, as

$$M(\theta) = F^{\mathrm{T}} W F, \tag{24.5}$$

where F is the $N \times p$ matrix of parameter sensitivities (17.12). When W is the identity matrix the observations are independent and the errors are homoscedastic, as they are in nearly all chapters of this book. Then (24.2) has the special form

$$E[\epsilon(t)\epsilon(s)] = \begin{cases} \sigma^2 & \text{if } s = t, \\ 0 & \text{otherwise.} \end{cases} \tag{24.6}$$

When $W = \text{diag}\, w_i$ the model becomes that for which weighted least squares (22.1) is the appropriate form of analysis. A consequence is that, in (24.3), there is only a single summation. Here, with correlated observations, W contains, at least some, non-zero off-diagonal elements.

Optimum designs can, as before, in principle be found by application of the design criteria of Chapter 10. We restrict attention to D-optimum designs, that is we find designs maximizing $\log|M(\theta)|$ in (24.5). This is however very different from previous applications of D-optimality. Since all measurements are to be taken on one run of the process (24.1), replications are not allowed. We are then restricted to N-point exact designs

$$\xi_{NO} = \left\{ \begin{array}{ccc} t_1 & \ldots & t_N \\ 1/N & \ldots & 1/N \end{array} \right\}. \tag{24.7}$$

We use ξ_{NO} instead of ξ_N to emphasize that the designs have no replication, that is we require $t_i \neq t_j$ for $i, j = 1, \ldots, N$ and $i \neq j$. This requirement has an effect not only on the designs, but also on the algorithms used to construct them. It also means that the design cannot be represented by a continuous measure ξ to which an equivalence theorem applies.

24.3 Numerical Example

Example 24.1 Two Consecutive First-order Reactions We return to the model for two consecutive first-order reactions introduced as Example 17.3 in which the concentration of B at time t is

$$\eta(t, \theta) = [B] = \frac{\theta_1}{\theta_1 - \theta_2}\{\exp(-\theta_2 t) - \exp(-\theta_1 t)\} \quad (t \geq 0), \tag{24.8}$$

provided that $\theta_1 > \theta_2 > 0$ and that the initial concentration of A is one.

The parameters θ_1 and θ_2 are to be estimated based on measurements made at N different time moments. For independent errors (24.6) we know that the D-optimum design does not depend on the value of the error variance σ^2. The same is true for correlated observations so, without loss of generality, the measurements can be assumed to be corrupted by zero-mean correlated noise with covariance kernel

$$c(t, s) = \exp(-\tau|t - s|). \tag{24.9}$$

The covariances thus tends to one as $|t - s| \to 0$. Two observations at the same time point therefore provide no more information than one. Observations that are far apart have low correlation, the correlation dying off more quickly as τ becomes larger.

The nominal parameter values $\theta_1^0 = 1.0$ and $\theta_2^0 = 0.5$ were used. For independent errors, we saw in Chapter 17 that the optimum design then puts equal numbers of observations at $t = 0.78$ and 3.43.

The results obtained for various numbers of measurements N are presented in Figure 24.1. The influence of the correlation between observations was tested by varying the coefficient τ in (24.9). In particular, the τ values of 1, 5, and 50 were chosen as representative ones for considerable, medium, and small correlations, respectively; the covariance kernel (24.9) implies that interactions between different points are certainly negligible at distances of 5, 1, and 0.1, respectively.

Two main conclusions can be drawn from our results. First, the greater the correlation, the more the optimal observations are spread over the domain where measurements can be taken. If the correlation is less strong, observations tend to form two clusters with approximately equal numbers of elements. From Panel (c) of the figure it is clear that the design points are clustering around the times 0.78 and 3.43.

Note that the higher the correlation, the lower is the value of the D-optimality criterion. This phenomenon occurs because higher correlations impose stronger relationships between noise realizations at different points, so that the information in the experiment is reduced.

The designs of Figure 24.1 were found by direct numerical optimization of $|M(\theta)|$ using the N-trial information matrix defined in (24.5). For examples with more parameters there may be advantages in using the exchange-type algorithm of Uciński and Atkinson (2004) with the disadvantage of more complicated programming.

24.4 Multiple Independent Series

24.4.1 Background

In the preceding sections we assumed that observations were taken on a single time series with correlated observations. The design problem was to find the optimum set of times $t_1, \ldots, t_N$ at which to take measurements. We now consider observations on several series of correlated observations,

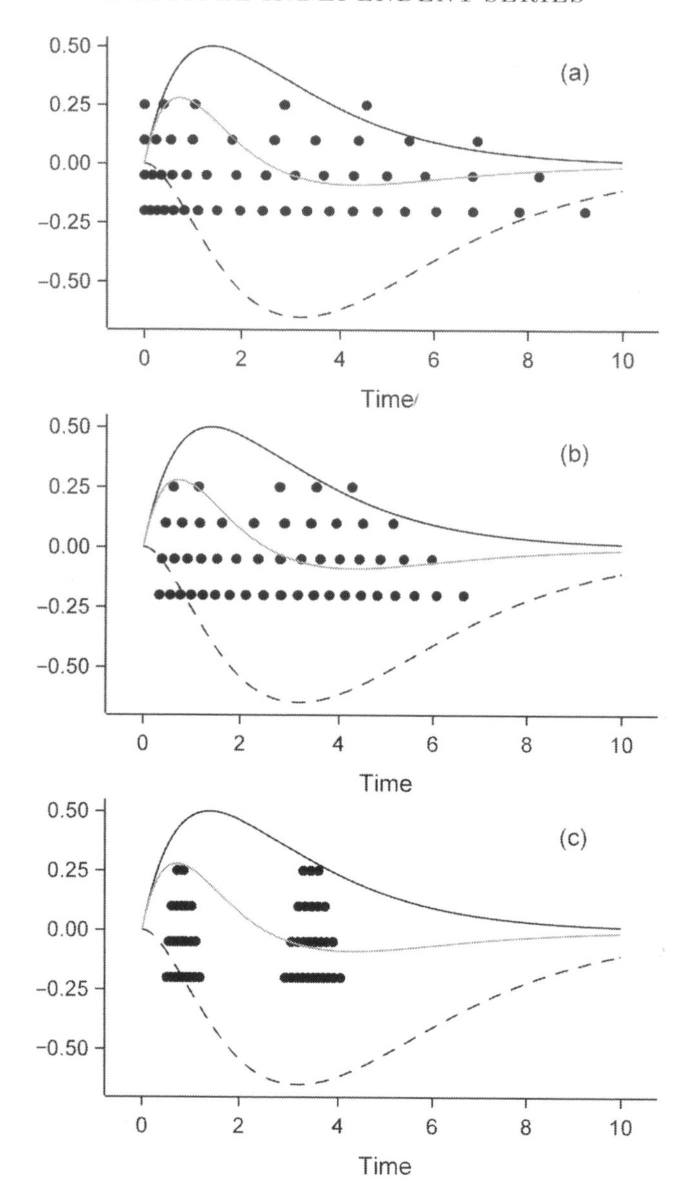

FIG. 24.1. Time evolution of the reactant concentration $[B]$ (solid line) and its sensitivities to parameters θ_1 and θ_2. Circles denote D-optimum measurement moments for $N = 5, 10, 15$, and 20, from top to bottom. (a) considerable correlation ($\tau = 1$); (b) medium correlation ($\tau = 5$); (c) small correlation ($\tau = 50$).

when observations on different series are independent of each other. We allow different series of time points for the different realizations of the series. These s series of observations, or measurement profiles, are specified in advance and we want to choose the optimum combination of them. They thus form the candidate set for our design. If profile i consists of measurements at $N_i = N(i)$ time points $t_{1(i)}, \ldots, t_{N(i)(i)}$, the connection with the standard model for independent observations is emphasized by writing

$$x_i = \left\{t_{1(i)} \quad t_{2(i)} \cdots t_{N(i)(i)}\right\}. \tag{24.10}$$

With correlated observations we take one observation at each time point. Thus the measure for the non-replicate exact design $\xi_{NO(i)}$ puts weight $1/N_i$ at each point of x_i.

The profiles may be subject to physical constraints which prevent any of them from being individually optimum for the series of correlated observations. A trivial example is when none of the profiles on its own yields a non-singular information matrix.

Because the problem is that of choosing between known measurement profiles for independent realizations of the process we avoid optimizations involving correlated observations. In §24.4.2 we given the additive structure of the information matrices, which arises from the independence of the realizations of the series. We then outline the properties of D-optimum designs, which are a simple extension of those for independent observations. Finally, in §24.4.4, we present an example in which both the optimum series of observations and their number depend on the strength of the correlation between observations.

24.4.2 Information Matrices

Let there be s different measurement schemes or profiles. The profiles may have different numbers of observation N_i; for the ith profile the design measure is $\xi_{NO(i)}$. Since the profiles are defined by the $\xi_{NO(i)}$, all of the design points in a profile must be distinct. The information matrix for profile i can, from (24.5), be written as

$$M_i(\theta) = F_i^{\mathrm{T}} W_i F_i = F(x_i)^{\mathrm{T}} W(x_i) F(x_i) = N_i M_i\{\xi_{NO(i)}, \theta\}, \tag{24.11}$$

where F_i and W_i are determined by the design $\xi_{NO(i)}$. We write

$$\Upsilon(x) = F^{\mathrm{T}}(x) W(x) F(x) \text{ and } \Upsilon^1(x) = M\{\xi_{NO}, \theta\}.$$

Let there be R_i replications of profile i. The total number of profiles observed is

$$P = \sum_{i=1}^{s} R_i$$

and the total number of observations is

$$O = \sum_{i=1}^{s} R_i N_i.$$

Since observations on different profiles are independent, the information matrix for the whole experiment is

$$M(\theta) = \sum_{i=1}^{s} R_i \Upsilon(x_i). \tag{24.12}$$

This is of the standard additive form for information matrices and we can apply the methods of optimum design for independent observations. Now the design region $\mathcal{X}$ contains s points x_i, the ith of which gives the information matrix $\Upsilon(x_i)$. If in (24.12) we write $q_i = R_i/P$ the information matrix becomes

$$M(\theta) = P \sum_{i=1}^{s} q_i \Upsilon(x_i). \tag{24.13}$$

We can then consider exact designs $\xi_n = \{x_i, q_i\}$, which give non-zero weights R_i/P to n of the s profiles in $\mathcal{X}$. Likewise we can find continuous designs ξ that distribute weight over the s profiles with (24.13) replaced in the standard way by

$$M(\xi, \theta) = \int_{\mathcal{X}} \Upsilon(x) \xi(dx).$$

If all $N_i = N$, the interpretation of the design criterion is clear; changing weight on various profiles does not alter the total number of observations taken. But, if the N_i are not all equal, profiles may be chosen by a design algorithm for which the N_i are relatively large, since these might provide more information in total than profiles with fewer observations that are, however, more efficient per observation. If we rewrite the information matrix as a function of the number of observations (24.13) becomes

$$M(\theta) = O \sum_{i=1}^{s} q_i^1 \Upsilon^1(x_i), \tag{24.14}$$

where the weights $q_i^1 = R_i N_i / O$. If we now divide by O we obtain the normalized information matrix

$$M^1(\theta) = \sum_{i=1}^{s} q_i^1 \Upsilon^1(x_i). \tag{24.15}$$

Now the design region can be called $\mathcal{X}^1$ in which the s matrices are the information matrices per observation $\Upsilon^1(x_i)$. Optimum designs can be found over either design region.

To overcome variation in the N_i we have defined the information matrices per observation as

$$N_i \Upsilon^1(x_i) = F_i^{\mathrm{T}} W_i F_i = F_i^{\mathrm{T}} C_i^{-1} F_i, \qquad (24.16)$$

where C_i is the covariance matrix with elements $c(t, s)$ (24.4) for profile i. The interpretation of (24.16) when $C = I$ is that for the standard theory with independent observations. However, Fedorov and Hackl (1997, p. 70) comment that caution is necessary when using this standardization with highly correlated processes, including long memory processes, when $\Upsilon^1(x_i) \to 0$ as $N_i \to \infty$.

24.4.3 Optimum Designs

We shall only be concerned with D-optimum designs maximizing $|M(\xi, \theta)|$ or their exact versions. Since we have a design problem with an additive information matrix, many of the properties are the same as those for independent observations. For example, the optimum design will have $n \leq p(p+1)/2$ support points. However, in contrast to standard results for single-response models, the lower bound on the number of support points is now unity, and not p. This arises because the matrix terms $\Upsilon(x_i)$ in the weighted sum 24.12 are non-singular when the matrices $F(x)$ have full column rank. A design with one support point can therefore give a non-singular information matrix for the whole experiment, even if the one-point design is not optimum.

For the standard case of independent observations the General Equivalence Theorem for D-optimum designs stated that maximization of $|M(\xi, \theta)|$ imposed a minimax condition on the standardized variance

$$d(x, \xi, \theta) = f^{\mathrm{T}}(x) M^{-1}(\xi, \theta) f(x) = \mathrm{tr}\, \{M^{-1}(\xi, \theta) f(x) f^{\mathrm{T}}(x)\}, \qquad (24.17)$$

where $f(x) f^{\mathrm{T}}(x)$ is the information matrix from an observation at x. For the measurement trajectories that replace individual observation in this section, the contribution of an individual trajectory is $\Upsilon(x)$ and we can write

$$d(x, \xi, \theta) = \mathrm{tr}\, \{M^{-1}(\xi, \theta) \Upsilon(x)\}. \qquad (24.18)$$

The theorem then says that the following conditions are equivalent:

1. The design ξ^* maximizes $|M(\xi, \theta)|$.
2. The design ξ^* minimizes $\max_{x \in \mathcal{X}} \mathrm{tr}\, \{M^{-1}(\xi) \Upsilon(x), \}$.
3. $\max_{x \in \mathcal{X}} \mathrm{tr}\, \{M^{-1}(\xi^*) \Upsilon(x)\} = p$.

For designs with the normalized information matrices the theorem applies to $M^1(\xi, \theta)$ (24.15) and $\Upsilon^1(x)$.

24.4.4 Numerical Example

Example 24.2 A Compartmental Model continued To illustrate optimum multiple series design, consider Example 1.5, the experiment on the concentration of theophylline in a horse's bloodstream, now with the stipulation that each horse can have its blood sampled only twice at times t_1 and t_2 15 seconds apart. Assume that the two measurements on each horse are correlated as in (24.9), and that measurements on different horses are independent. Then the design has the form of multiple independent series.

Optimum multiple series designs can be computed as in §13.4, where non-linear optimization is employed to find the optimum weights for given candidates; the only difference is in how the D-criterion is computed. Table 24.1 shows the support of the optimum design for several representative values of τ in $[0.1, 100]$. The support points are, in general, near the optimum times for uncorrelated, unconstrained measurements of 0:14, 1:24, and 18:24 (min:sec).

For all values of τ the weight on the upper support point is one-third, the measurement series starting at values ranging from 17:45 to 18:30. The lower support points show much more variation. For the most highly correlated series ($\tau = 0.5$), the support is divided between series starting at 0:00 and at 0:15, so that two-thirds of the horses have a measurement at the overlapping point 0:15. For $\tau = 1$ there are three series at 0:00, 1:15 and 18:15, each with weight one-third. For lower correlations the first time of measurement is 0:15, with the second series starting at 1:15 or 1:30. In the case of $\tau = 8$ the series are split between these two starting values.

The efficiency of the optimum multiple series design for uncorrelated measurements ($\tau = \infty$) relative to the optimum designs for particular values of τ is shown in Figure 24.2. When the measurements are highly correlated, this design has less than 50% efficiency; as the correlation decreases, it becomes fully efficient, as expected. Notice that the slope of the efficiency is non-smooth at several points. These correspond to values of τ where the support of the optimal design is transitioning between those depicted in Table 24.1.

24.5 Discussion and Further Reading

Correlated errors have been considered for treatment allocation designs in which t treatments have to be allocated along a line or over a plane. The main requirement is that of neighbour balance (Williams 1952). Kiefer and Wynn (1984) explore the relationship with coding theory. The introduction

TABLE 24.1. Example 24.2: optimum multiple series designs for theophylline sampling. Each support point consists of a series of two samples at times t_1 and t_2 separated by 15 seconds. Times are given as [min:sec]

τ		Support points (two times) and weights				
0.5	t_1	0 : 00	0 : 15			17 : 45
	t_2	0 : 15	0 : 30			18 : 00
	w	0.239	0.428			0.333
1	t_1	0 : 00		1 : 15		18 : 15
	t_2	0 : 15		1 : 30		18 : 30
	w	0.333		0.333		0.333
4	t_1		0 : 15	1 : 15		18 : 15
	t_2		0 : 30	1 : 30		18 : 30
	w		0.400	0.267		0.333
8	t_1		0 : 15	1 : 15	1 : 30	18 : 15
	t_2		0 : 30	1 : 30	1 : 45	18 : 30
	w		0.373	0.106	0.188	0.333
16	t_1		0 : 15		1 : 30	18 : 30
	t_2		0 : 30		1 : 45	18 : 45
	w		0.364		0.303	0.333

to Azzalini and Giovagnoli (1987) discusses this literature and extends the results to repeated measurement designs with nuisance covariates.

The theory of experimental designs for regression problems with correlated errors was studied in Parzen (1961), Sacks and Ylvisaker (1966, 1968, 1970), and Wahba (1971). The problem considered there differed from that of this chapter in that an optimal number of support points was sought in addition to their co-ordinates. The main difficulty in such a formulation stems from the fact that every new observation gives a new piece of information

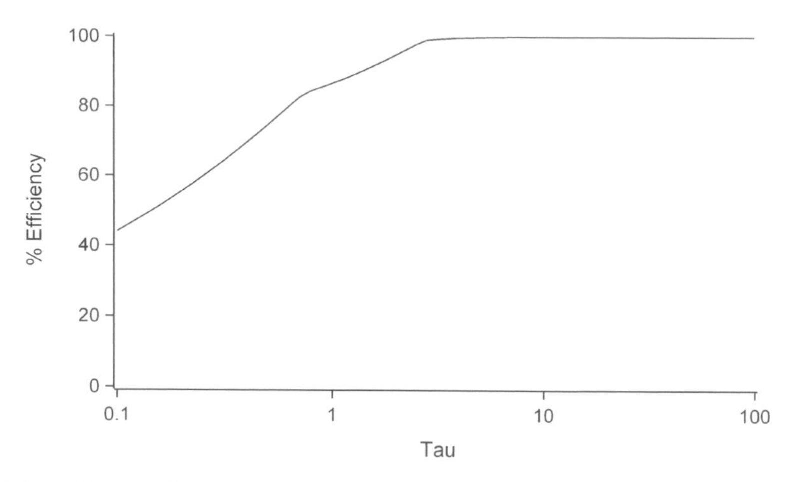

FIG. 24.2. Example 24.2: efficiency of optimum design for uncorrelated measurements at different values of τ.

about the parameters, so usually a solution with a finite number of support points does not exist. An exception is when the basis functions multiplying the estimated parameters in the linear regression model can be represented in the form $f(t) = \sum_{i=1}^{N} a_i k(t, t_i)$ for some finite N and fixed numbers a_i and t_i (more generally, they are elements of the reproducing kernel Hilbert space associated with the kernel $k(t, s)$). But this assumption is too strong to be satisfied in most practical situations. Some relaxation of this condition comes from introducing the notion of the asymptotic optimality for a sequence of designs.

Brimkulov, Krug, and Savanov (1980) introduced an exchange algorithm for correlated observations. The booklength treatment of Brimkulov, Krug, and Savanov (1986) considers design for correlated errors in detail. Their examples however are for simple linear models. The algorithm of Müller and Pázman (2003) adds a second, independent, error component to the model (24.1) with correlated errors. The optimum design measure is found by searching over a fine grid of many more than N design points. The variance for this second component increases as the measure decreases in such a way that the measure finally does have N points of support. Patan and Bogacka (2007) extend the exchange algorithm to multivariate observations that are serially correlated and apply their algorithm to the construction of designs for the two consecutive first-order reactions of Example 24.1. The structure of the resulting designs, such as those of Figure 24.1, is not straightforward. Several authors (Stehlík 2005; Dette, Kunert, and Pepelyshev 2007; Pepelyshev 2007) have shown that, for fixed N, the points of support of the design are discontinuous with respect to τ.

Correlated observations constitute a fundamental problem in spatial statistics, although the emphasis is rather more on prediction than on the parameter estimation central to the approach of this chapter. For a book-length treatment of design for spatial processes see Müller (2007), with some introductory material in Chapter 5 of Fedorov and Hackl (1997).

The measurement profiles of §24.4 arise naturally in the designs for population parameters discussed in §25.5. The fixed effects models of most of this book are there replaced by random effects; in Example 24.2 each horse would have its own set of randomly distributed parameters. Interest would be in the population mean of each parameter.

25

FURTHER TOPICS

25.1 Introduction

Although the topics covered in this chapter receive relatively brief treatment, this is not because they are of little practical importance.

The crossover designs that are presented in §25.2 are often used in medical research for the comparison of several treatments on individual subjects; as we note, their use in other areas is increasing. In §25.3 we continue with designs again useful in medical research, but consider instead clinical trials in which patients arrive sequentially. The problem is to provide an allocation of treatments that achieves some balance over prognostic factors whilst also including sufficient randomness to avoid biases. The following section extends the methods to include adaptive designs that respond to the responses of earlier patients in the trial. An extra design objective is to reduce the number of patients receiving inferior treatments. The population designs of §25.5 apply when subject to subject variation is sufficiently large that it needs to be modelled. We describe designs for linear mixed models in which the parameters for each individual have a normal distribution about a mean vector that is the parameter of interest. Designs for non-linear mixed models are shown to be harder to calculate.

Many of our experiments involve taking observations over time. In addition to the correlation structure that this may involve, that was the subject of Chapter 24, experiments may have factors that vary in a smooth way over time. We indicate in §25.6 how splines may be used for the parsimonious modelling of such profiles and so yield a tractable design problem. In §25.7 we describe the selection of scenarios for training neural nets. In the final section we briefly mention designs robust against mis-specification of the model and the design of computer simulation experiments in which the observations are expensive, but, unlike the observations considered in the rest of the book, come without error.

25.2 Crossover Designs

Crossover designs are often used in medical research. Customarily their main objective is to find precise estimates of the differences in the effects of t treatments. Unlike parallel group designs where every subject receives a single treatment, every subject enrolled in a crossover trial receives sequentially a number of treatments over several periods of time. Therefore, the use of crossover designs is limited to cases when such a sequential administration of treatments is possible. However, their structure allows for the comparisons of interest to be made 'within subjects' and possible substantial variation in the response of different subjects would not obscure the differences of interest. A disadvantage of using these designs is that there are many nuisance effects, such as carry-over effects, period effects, subject effects, even interactions between some of them that may affect the response. Consequently more assumptions have to be made about the parameters of the model describing the data than if a parallel group design is used.

A simple linear model for an observation of the response y_{ij}, taken on subject i in period j of a crossover trial, is

$$y_{ij} = \mu + \tau_{d[i,j]} + \lambda_{d[i,j-1]} + \pi_j + s_i + \epsilon_{ij}, \tag{25.1}$$

where μ is the general mean, $d[i, j]$ denotes the treatment applied to subject i in period j ($i = 1, 2, \ldots, s$ and $j = 1, 2, \ldots, p$), $\tau_{d[i,j]}$ is the direct effect of treatment $d[i, j]$, $\lambda_{d[i,j-1]}$ is the carry-over effect of treatment $d[i, j-1]$ observed in period j for subject i and $\lambda_{d[i,0]} = 0$, π_j is the effect of period j and s_i is the effect of subject i. In some applications the errors ϵ_{ij} are assumed to be normally distributed and independent with zero mean and variance σ^2. However, the observations taken on a subject can be correlated. Usually an equal number of subjects, say r, is allocated to each of a number of pre-specified treatment sequences. Therefore, if all subjects are available for p periods, the number of the observations is equal to $N = rsp$. Other model forms can also be used (Jones and Donev 1996). However, it is important that a model that adequately represents the experimental situation is chosen. Jones and Kenward (2003) and Senn (1993) summarize the main results in this area and give practical examples and an extensive list of references.

Crossover designs that are optimum with respect to a chosen criterion can be constructed using optimum design theory, though the optimality of the experimental designs for such studies will depend on the unknown correlation structure of the observations. The search is made more complicated because the exchange of a treatment with another treatment also changes the carry-over effect for the subsequent period. As interest is usually in estimating the differences between the treatments, an often-used criterion is to minimize their variances. For example, Donev (1997) presented a group-exchange algorithm for the construction of A-optimum crossover designs for which

$$A = \sum_{i=1}^{t-1} \sum_{j=i+1}^{t} \mathrm{var}(\hat{\tau}_i - \hat{\tau}_j) \tag{25.2}$$

is minimum. John, Russell, and Whitaker (2004) also propose an algorithm for constructing crossover designs. Donev and Jones (1995) show that when model (25.1) is reparameterized by setting the sum of the parameter values for the treatments, the carry-over effects, the period effects, and the group effects to zero, equation (25.2) simplifies to

$$A = 2t \sum_{i=1}^{t-1} \sum_{j=i}^{t-1} m^{ij},$$

where m^{ij} is the i, jth element of the inverse of the information matrix M^{-1}.

Example 25.1 Crossover Design for Two Treatments As an illustration of the advantages and the limitations of the methods for constructing crossover designs let us consider the case when the comparison of two treatments (A and B) is required and the subjects are available for four periods, that is, $t = 2$ and $p = 4$. Table 25.1 shows three possible designs. Design D1 divides the subjects in four treatment groups, while for designs D2 and D3 the number of the groups is two. For example, the second group in design D1 receives the sequence of treatments B A A and B, possibly with a wash-out period between the treatments, and the response in the last period may be explained by the effect of treatment B used in this period, the carry-over effect of treatment A used in the previous period, the effect for period 4, and the effect of the subject receiving this treatment sequence. In this example the A-value in equation (25.2) reduces to the variance of the estimate for the difference between the two treatments, that is, $A = \mathrm{var}(\hat{\tau}_A - \hat{\tau}_B)$. When

TABLE 25.1. Crossover designs for two treatments, four periods and four or two treatment sequences

D1				D2				D3			
A	B	B	A	A	A	B	B	A	B	B	A
B	A	A	B	B	B	A	A	B	A	A	B
A	A	B	B								
B	B	A	A								

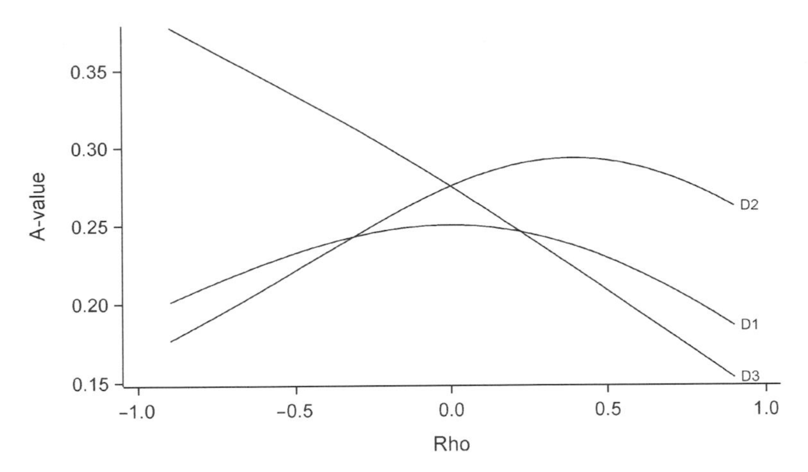

FIG. 25.1. Plots of the A-criterion of optimality for the crossover designs given in Table 25.1: (a) D1, (b) D2, (c) D3 (equal group sizes).

the sizes of the groups of subjects receiving the same treatment sequence is the same, design D1 has a smaller A-value than Design B and Design C and is therefore better.

Design D1 is not the best design to use when the observations taken on the same subject are correlated. Suppose the correlation structure is explained by the first-order autoregressive model, that is, the covariance between the observations in periods j and k on subject i is assumed to be

$$\mathrm{cov}(\epsilon_{ij}, \epsilon_{ik}) = \sigma^2 \rho^{|j-k|}/(1-\rho^2).$$

Figure 25.1 shows the A-values of the designs of Table 25.1 for different values of the correlation coefficient ρ and equal group sizes. As ρ increases, the A-value for Design D3 decreases and this design becomes better than Design D1 for moderate or large values of ρ. Similarly, Design D2 is to be preferred for moderate or high negative values of ρ. These observations agree with Matthews (1987) who shows that there are potential benefits to be gained if the group sizes are different. Donev (1998) uses a Bayesian algorithmic approach to design crossover trials taking into account the dependence between the observations and allows for optimum groups sizes to be found numerically. Practical considerations are also discussed. ■

It was assumed in (25.1) that the carry-over effects $\lambda_{d[i,j-1]}$ were additive. Bailey and Kunert (2006) find optimum designs when these effects are proportional to the treatment effects. The premise is that treatments with appreciable effects can be expected to have larger carry-overs than treatments with small effects. Hedayat, Stufken, and Yang (2006) consider design when the subject effects are random. The implicit framework in these papers, as in our discussion, is subjects receiving medical treatments. However, crossover designs are also used in non-medical areas of research. For instance, examples from the food industry are given in Jones and Wang (1999) and Deppe *et al.* (2001).

25.3 Biased-coin Designs for Clinical Trials

25.3.1 Introduction

There is a vast literature on clinical trials. Books include Piantadosi (2005), Whitehead (1997), Matthews (2006) and Rosenberger and Lachin (2002). Here we consider sequential clinical trials in which some randomization is required in the allocation of treatments. The appreciable literature on the resulting 'biased-coin' designs is reviewed by Atkinson (2002). This section shows how optimum design can be used to provide efficient designs of known properties when it is desired to adjust the responses for covariates such as body weight, cholesterol level, or previous medical history; much of the arbitrariness of other procedures is avoided.

In this section we find sequential designs for one patient at a time in the absence of knowledge of the responses of earlier patients. Adaptive designs, in which knowledge of the responses of earlier patients is used to guide the allocation, are the subject of §25.4.

Patients for a clinical trial arrive sequentially and are each to be allocated one of t treatments. All treatment allocations are to be made on the basis of patients' prognostic factors or covariates before any outcomes are available. The data, perhaps after transformation, are to be analysed adjusting for the covariates using a linear model, which, as in (5.16) we write as

$$\mathrm{E}\,(y) = X\gamma = Z\alpha + F\beta. \tag{25.3}$$

In (25.3) the matrix Z, of dimension $N \times t$, consists of the indicator variables for the treatments whereas F is the $N \times (q-1)$ matrix of covariates, including powers and products if appropriate. To ensure that the model is full rank, F does not include a constant term. The treatment parameters α are of interest, whereas the coefficients β of the covariates in F are nuisance parameters.

In the customary sequential generation of $\mathrm{D_S}$-optimum designs we would select vectors z_{N+1} and f_{N+1} to minimize the generalized variance of the estimates $\hat{\alpha}$. There are however several important differences here from standard design construction:

1. Only some linear combinations $c_i^{\mathrm{T}}\alpha$ may be of interest.
2. The vector of covariates f_{N+1} is known, rather than chosen by the design.
3. There should be some element of randomness in the treatment allocation.

With two treatments the treatment difference is customarily of interest, with the mean response level a nuisance parameter. The parameter of interest can be written $\delta = \alpha_1 - \alpha_2$. Then, in point 1 above, $c^{\mathrm{T}} = (-1 \quad 1)$.

For general t, interest is in linear combinations of the α, the mean level of response again being a nuisance parameter making q nuisance parameters in all. Let C^T be a $(t-1) \times t$ matrix of contrasts orthogonal to the mean. An example is given by Atkinson (1982). Since the volume of the normal theory confidence ellipsoid for least squares estimates of the contrasts is unaffected by non-singular linear transformations of the contrasts, the exact form of C is unimportant, provided the contrasts span the $t-1$ dimensional space orthogonal to the overall mean. Because the β in (25.3) are nuisance parameters, the combinations need augmenting by a $(t-1) \times (q-1)$ matrix of zeroes

$$A^T = (C^T \quad 0)$$

to reflect interest solely in contrasts in the treatment parameters. If only $s < t-1$ specific combinations are of interest, C can be modified accordingly.

From the allocations for the first N patients, the covariance matrix of the linear combinations is

$$\mathrm{var}\,\{A^T\hat{\gamma}\} = \sigma^2 A^T (X_N^T X_N)^{-1} A. \tag{25.4}$$

The generalized variance of these combinations is minimized by finding the D_A-optimum design to minimize

$$\Delta_N = |A^T (X_{N+1}^T X_{N+1})^{-1} A|. \tag{25.5}$$

If treatment j is allocated, X_{N+1} is formed from X_N by addition of the row

$$x_{N+1}^{\mathrm{T}}(j) = (z^{\mathrm{T}}(j) \quad f_{N+1}^{\mathrm{T}}). \tag{25.6}$$

In (25.6) f_{N+1} is not at the choice of the experimenter; $z(j)$ has element $j = 1$ and all other elements zero. The design region $\mathcal{X}$ therefore contains just t elements, each corresponding to the allocation of one treatment. In the sequential construction of these designs the allocation is made for which the variance

$$d_A(j, f_{N+1}) = x_{N+1}^T(j)(X_N^T X_N)^{-1} A\{A^T (X_N^T X_N)^{-1} A\}^{-1} \times A^T (X_N^T X_N)^{-1} x_{N+1}(j) \quad (j = 1, \ldots, t), \tag{25.7}$$

is a maximum over $\mathcal{X}$.

25.3.2 Randomization

Since the trial is sequential, it is not known exactly how many patients there will be, so the number of patients over whom the treatments are to be allocated is uncertain. If recruitment of patients ceases when the trial is unbalanced, the variance of the estimated treatment effects will be larger than if the trial were balanced, even after adjustment for the prognostic factors. The $\mathrm{D_A}$-optimum allocation rule given by sequentially allocating at the maximum of (25.7) provides the most balanced design, given the particular sequence of prognostic factors f_N with the patients present, and so the parameter estimates with the smallest generalized variances. However, this rule needs expanding to allow for some element of randomness in allocation.

There are many reasons for partially randomizing the allocation, including the avoidance of bias due to secular trends and the avoidance of selection bias, measured as the ability of the clinician to guess which treatment will be allocated next. The design with least bias is completely random allocation of treatments with probabilities equal to the design weights in the $\mathrm{D_A}$-optimum design for the model $\mathrm{E}(y) = Z\alpha$, that is ignoring the covariates.

In order to provide a randomized form of the sequential construction (25.7), Atkinson (1982) suggests allocating treatment j with probability

$$\pi_A(j|f_{N+1}) = \frac{d_A(j, f_{N+1})}{\sum_{i=1}^{t} d_A(i, f_{N+1})}. \tag{25.8}$$

In (25.8) the variances $d_A(.)$ could be replaced by any monotone function $\psi\{d_A(.)\}$. In Atkinson (1999) it is shown that the sequential version of the general Bayesian biased-coin procedure of Ball, Smith, and Verdinelli (1993) which uses D_A-optimality leads to

$$\psi(u) = (1 + u)^{1/\gamma}, \tag{25.9}$$

with γ a parameter to be elucidated from the experimenter. This rule is a special case of that for adaptive designs derived in §25.4.

25.3.3 Efficiencies

To compare designs we need measures of performance. We begin with those related to the variances of parameter estimates.

The variance–covariance matrix of the set of s estimated linear combinations for some design X_N is given by (25.4) with generalized variance (25.5). The D_A-efficiency of the design, relative to the optimum design X_N^* is

$$E_N = \frac{|A^{\mathrm{T}}(X_N^{*\mathrm{T}} X_N^*)^{-1} A|^{1/s}}{|A^{\mathrm{T}}(X_N^{\mathrm{T}} X_N)^{-1} A|^{1/s}}. \tag{25.10}$$

Optimum designs, which minimize (25.5), are balanced over the covariates in (25.3). That is, $Z^T F = 0$ and the covariates can be ignored when considering the properties of the optimum design.

In general the optimum design is a function of A that will have to be found numerically. However, when only one linear combination of the t treatment parameters is of interest, the design criterion is that of c-optimality and some analytical progress can be made. In particular, the estimate of the linear combination $c^{\mathrm{T}}\alpha$ has variance

$$\operatorname{var} c^{\mathrm{T}}\hat{\alpha} = \sigma^2 a^T (X_N^T X_N)^{-1} a. \tag{25.11}$$

If N_j patients receive treatment j

$$\operatorname{var} c^{\mathrm{T}}\hat{\alpha} = \sigma^2 \sum_{i=1}^{t} c_i^2 / N_i,$$

which is minimized when the allocation uses the optimum numbers

$$N_j^* = \frac{N|c_j|}{\sum_{i=1}^{t} |c_i|}.$$

Then, for the optimum design,

$$\operatorname{var} c^{\mathrm{T}}\hat{\alpha} = \frac{\sigma^2}{N} \left(\sum_{i=1}^{t} |c_i| \right)^2.$$

For example, when $t = 2$ and the quantity of interest is the treatment difference $\delta = \alpha_1 - \alpha_2$, $\operatorname{var} c^{\mathrm{T}}\hat{\alpha} = 4\sigma^2/N$.

It is convenient, and useful in the discussion of adaptive designs in §25.4, to let

$$p_j = |c_j| / \sum_{i=1}^{t} |c_i|, \tag{25.12}$$

when the optimum design can be written simply as

$$N_j^* = N p_j.$$

For this optimum design

$$\operatorname{var} p^{\mathrm{T}}\hat{\alpha} = \sigma^2/N.$$

The p_j are the proportions of observations allocated to each treatment in the optimum design.

Comparisons of designs can use either the efficiency E_N, (25.10) or the loss introduced by Burman (1996), which leads to a more subtle understanding of the properties of designs. Since the transformation (25.12) from c to p does not affect the optimum design, we can find the variance of $p^{\mathrm{T}}\hat{\alpha}$ from (25.11). The variance from the optimum design is σ^2/N and

$$E_N = 1/\left\{Np^T(X_N^T X_N)^{-1}p\right\}.$$

The variance of $p^{\mathrm{T}}\hat{\alpha}$ for a non-optimum design is greater than σ^2/N. The loss L_N is defined by writing the variance (25.11) as

$$\mathrm{var}\,\{p^T\hat{\alpha}\} = \frac{\sigma^2}{N - L_N}. \qquad (25.13)$$

The loss can be interpreted as the number of patients on whom information is unavailable due to the lack of optimality of the design. For a general design for linear combinations A and efficiency given by (25.10)

$$L_N = N(1 - E_N). \qquad (25.14)$$

The loss L_N is a random variable, depending upon the particular trial and pattern of covariates and also on the randomness in the allocation rule. Let $E(L_N) = \mathcal{L}_N$. The results of Burman (1996) cover the allocation of two treatments when the treatment difference is of interest. With random allocation, as N increases $\mathcal{L}_N \rightarrow q$, the number of nuisance parameters. Other designs that force more balance have lower values of $\mathcal{L}_N$. For the sequential allocation of $\mathrm{D_A}$-optimality $\mathcal{L}_N \rightarrow 0$. All reasonable rules have values of $\mathcal{L}_N$ within this range. An example of an 'unreasonable' rule would be one which deliberately sought imbalance, for example by persistently allocating to small values of $d_A(j, f_{N+1})$ in (25.7).

Atkinson (2002) uses simulation to exhibit the average small sample properties of 11 rules for unskewed treatment allocation for two treatments. These include rules for five values of the parameter γ in (25.9). The results show that one advantage of loss as a measure of design performance is that it approaches the informative asymptotic value relatively quickly. Since the loss tends to a constant value as N increases, it follows from (25.14) that the efficiencies of all designs asymptotically tend to one.

The comparisons show that designs with little randomness in the allocation have small loss but large selection bias; given knowledge of previous allocations and of f_{N+1} there is a high probability of correctly guessing the next treatment allocation. Conversely, near random allocation has virtually zero selection bias. Atkinson (2003*a*) uses simulation to explore the properties of designs for individual trials.

25.4 Adaptive Designs for Clinical Trials

25.4.1 Introduction

We now extend the applications of optimum design theory to include adaptive designs for clinical trials. The designs are intended to provide information about treatment comparisons whilst minimizing the number of patients receiving inferior treatments. This minimization uses information on the responses of earlier patients. In the previous section we found designs that allocated treatments to minimize the variance of the linear combination $p^{\mathrm{T}}\hat{\alpha}$. We now show how to choose the vector p for each patient to reflect information on the responses and covariates of previous patients as well as the utility function of the clinician. We further show how these utilities can be reduced to the specification of the asymptotic probabilities p_j^* of allocating the treatment ranked j.

The resulting designs give specified probabilities of allocating the ordered treatments. Since the ordering is determined by the results of previous patients it will typically change as the trial progresses. One advantage of the procedure is that the properties of the designs can be very simply specified. Another advantage is that the ordering of treatments may depend in a complicated way on many objective and subjective factors. All that is required is that the treatments be ordered in desirability. The procedure applies to the comparison of as many treatments as are required, adjusted for an arbitrary number of covariates.

25.4.2 Utility

We start with a utility function introduced by Ball *et al.* (1993) to balance randomization and parameter estimation in non-sequential designs. The designs maximize the utility

$$U = U_V - \gamma U_R, \tag{25.15}$$

where the contribution of U_V is to provide estimates with low variance, whereas U_R provides randomness. The parameter γ provides a balance between the two.

With π_j the probability of allocating treatment j, let

$$U_V = \sum_{j=1}^{t} \pi_j \phi_j,$$

where ϕ_j is a measure of the information from applying treatment j. In §25.4.3 we define information in terms of D_A-optimality.

To combine randomness with greater allocation to the better treatments we introduce a set of gains $G_1, \ldots, G_t$, with $G_1 > G_2 \geq \cdots \geq G_t$ when

$$U_R = \sum_{j=1}^{t} \pi_j\{-G_{R(j)} + \log \pi_j\}. \qquad (25.16)$$

In (25.16) $R(j)$ is the rank of treatment j. When all G_j are equal, minimization of U_R leads to random allocation with equal probabilities. The G_j skew allocation towards the better treatments.

To maximize the utility (25.15) subject to the constraint $\sum_{j=1}^{t} \pi_j = 1$ we introduce the Lagrange multiplier λ and maximize

$$U = \sum_{j=1}^{t} \pi_j\phi_j - \gamma \sum_{j=1}^{t} \pi_j\{-G_{R(j)} + \log \pi_j\} + \lambda \left(\sum_{j=1}^{t} \pi_j - 1 \right). \qquad (25.17)$$

Since the G_j occur in U with a positive coefficient, maximization of U gives large values of π_j for treatments with larger $G_{R(j)}$. Differentiation of (25.17) with respect to π_j leads to the t relationships

$$\phi_j - \gamma\{-G_{R(j)} + 1 + \log \pi_j\} + \lambda = 0,$$

so that all quantities

$$\phi_j/\gamma + G_{R(j)} - \log \pi_j$$

must be constant. Since $\sum_{j=1}^{t} \pi_j = 1$, we obtain

$$\pi_j = [\exp\{\phi_j/\gamma + G_{R(j)}\}]/S = \{\exp(\psi_j/\gamma)\}/S, \qquad (25.18)$$

where

$$\psi_j = \phi_j + \gamma G_{R(j)}$$

and

$$S = \sum_{j=1}^{t} \{\exp(\phi_j/\gamma) + G_{R(j)}\} = \sum_{j=1}^{t} \exp(\psi_j/\gamma).$$

25.4.3 Optimum and Sequential Designs

The probabilities of allocation π_j (25.18) depend on the information measure ϕ_j. For this measure we again use D_A-optimality to minimize the logarithm of the determinant of the covariance matrix (25.5). Thus we find designs to maximize the information measure

$$\phi_j = -\log |A^T (F_{n+1,j}^T F_{n+1,j})^{-1} A| = -\log \Delta_j.$$

Substitution of this expression for ϕ_j in (25.18) yields

$$\pi_j = \Delta_j^{-1/\gamma} \exp\{G_{R(j)}\}/S. \tag{25.19}$$

In the sequential construction of adaptive designs we again exploit the results on sequential generation of D_A-optimum designs (25.7) to obtain

$$\pi(j|f_{N+1}) = \frac{\{1 + d_A(j, f_{N+1})\}^{1/\gamma} \exp\{G_{R(j)}\}}{\sum_{i=1}^{t} \{1 + d_A(i, f_{N+1})\}^{1/\gamma} \exp\{G_{R(i)}\}}. \tag{25.20}$$

25.4.4 Gain and Allocation Probabilities

The allocation probabilities (25.19) and (25.20) depend on γ, on the gains $G_{R(j)}$ and on the matrix of linear combinations A, here a vector. The parameter γ determines the balance between randomization and parameter estimation, small values giving less randomness. When all $G_{R(j)}$ are equal, (25.20) reduces to the non-adaptive criterion (25.9). The dependence of these sequential, non-adaptive designs on the value of γ has been extensively explored by Atkinson (2002). Initially, small values of γ force balance and so efficient parameter estimation. As N increases the emphasis on balance decreases. We now relate the values of the $G_{R(j)}$ to the coefficients p_j of (25.12).

At the optimum design, that is when there is balance across all covariates, all $d_A(j, f_{N+1})$ are equal and the treatments are correctly ordered. Then, from (25.20)

$$\pi(j|_{N+1}) = p^*_{R(j)} = \frac{\exp\{G_{R(j)}\}}{\sum_{i=1}^{t} \exp\{G_{R(i)}\}}. \tag{25.21}$$

This relationship is crucial to the interpretation of our method, providing the relationship between the asymptotic proportion of patients receiving each of the ordered treatments and the gains G_j. In designing the trial, the p_j^* are the fundamental quantities which are to be specified, rather than the gains G_j.

The probabilities of allocation in (25.20) and (25.21) are unaltered if we replace $G_{R(j)}$ with

$$G^a_{R(j)} = G_{R(j)} + a.$$

We choose a so that $\sum_{i=1}^{t} \exp(G^a_{R(i)}) = 1$. Then (25.21) becomes

$$G^a_{R(j)} = \log p^*_{R(j)}$$

and the allocation probabilities (25.20) have the simple form

$$\pi(j|f_{N+1}) = \frac{\{1 + d_A(j, f_{N+1})\}^{1/\gamma} p^*_{R(j)}}{\sum_{i=1}^{t} \{1 + d_A(i, f_{nN+1})\}^{1/\gamma} p^*_{R(i)}}, \quad (25.22)$$

provided the ranking of the treatments is known.

25.4.5 Adaptive Allocations

The asymptotic proportion of patients receiving the jth best treatment is specified by the clinician as p^*_j. Since the correct ordering is not known, adaptive designs use estimated ranks $\hat{r}(j)$ and the set of coefficients

$$p_j = p^*_{\hat{r}(j)},$$

instead of the $p^*_{R(j)}$ in (25.22). The allocation probabilities then depend on the, perhaps incorrect, experimental ordering of the treatments when patient $N + 1$ arrives. Atkinson and Biswas (2006) explore the properties of designs for several values of p^* with treatments ordered by the magnitude of the estimates $\hat{\alpha}_1, \ldots, \hat{\alpha}_t$. They study the loss L_N together with the proportion of patients receiving treatment j and the rate at which this converges to p^*_j.

In this scheme the probability of allocating the treatments depends on p^* and on the ordering of the α_j, but not on the differences between them. Suppose there are two treatments. Then, if $\alpha_1 > \alpha_2$, treatment 1 will eventually be allocated in a proportion p^*_1 of the trials regardless of the value of $\delta = \alpha_1 - \alpha_2$. Of course, if δ is small relative to the measurement error, in many of the initial trials, $\hat{\alpha}_1 < \hat{\alpha}_2$ and it will seem that treatment 2 is better. Then the allocation will be skewed in favour of treatment 2 with proportion p^*_1, that is $p_{\hat{r}(2)}$. When $\hat{a}_1 > \hat{\alpha}_2$, treatment 1 will be preferred.

25.4.6 Regularization

As the experiment progresses, the values of the $d_A(.)$ in (25.20) decrease with N, provided all treatments continue to be allocated. Atkinson and Biswas (2006) report the empirical advantages of regularization in ensuring that all treatments continue to be allocated, although decreasingly often, whatever their performance. For two-treatment trials they allocate 5 of the first 10 patients to treatment 1 and the other 5 to treatment 2. Thereafter, if the number allocated to either treatment is below $\sqrt{N}$, that treatment is allocated when N is an integer squared. For an 800 trial design the first regularization could occur when $N = 36$ and the last when $N = 784$. There is nothing special about $\sqrt{N}$: all that is required is a bounding sequence that avoids very extreme allocations.

25.4.7 Discussion

Although the ordering of the treatments does not depend on the assumption of normality, the variance $d_A(\cdot)$ used in calculating the probabilities in (25.20) does assume that the linear model (25.3) with errors of constant variance is appropriate for the analysis of the data. But, as the results of Chapter 22 show, the design construction is also appropriate for other models including generalized linear models where the treatment effects are sufficiently small that the effect on the design of the iterative weights can be ignored and for gamma models with the log link, when the effects do not have to be small.

Recent books on randomization and adaptive design in clinical trials include Matthews (2000), Rosenberger and Lachin (2002), and Hu and Rosenberger (2006), which is concerned with adaptive design and the ensuing inferences. Virtually all of the reported work on adaptive designs is for binary responses in the absence of prognostic factors, with designs generated from urn models. Atkinson and Biswas (2005*a*,*b*) report other approaches to the use of optimum design theory in the construction of adaptive designs.

25.5 Population Designs

In pharmacokinetics and other medical studies there is often appreciable subject to subject variation around unknown population values. This section discusses the design of experiments when, in particular, interest is in the parameters describing the overall population.

As an example, in Chapter 17 we found optimum designs for the non-linear compartmental model (17.6) describing the time trace of the concentration of theophylline in the blood of a single horse. Davidian and Giltinan (1995) present similar data for the time traces of 12 subjects who have each been given a single dose, again of theophylline. The individual curves are similar to those of Figure 1.2, but there is appreciable variability in the time to maximum concentration and an over 50% variation in that maximum concentration, although all curves seem to decline at much the same rate. The design question, of at what times to take measurements on which individual, depends both on the purpose of the design and the model to be fitted. Davidian and Giltinan (2003) give a careful discussion of such non-linear mixed models and the purposes of analysis. Inference for non-linear models, let alone design, is computationally demanding. As a consequence, the structure of the designs seems appreciably more obscure than those for linear models. Accordingly we start with the simpler linear model.

We assume that the curves for individuals are generated by a common functional structure, with random parameters for each individual and the usual normal theory error structure. There is a wide literature for such linear mixed models under a variety of terms, for example random coefficient regression models, latent variable regression, hierarchical linear models, multi-level models, and empirical Bayes regression. See, for example, Smith (1973), Pinheiro and Bates (2000), or Verbeke and Molenberghs (2000). There are S subjects. For the jth measurement on subject i the linear model is

$$y_{ij} = b_i^{\mathrm{T}} f(x_{ij}) + \epsilon_{ij}, \tag{25.23}$$

$j = 1, \ldots, N_i$ and $i = 1, \ldots, S$. The regression function $f(x_{ij})$ is known and the same for all individuals, whereas b_i is the vector of parameters for the curve associated with individual i. The experimental variables can be quite general—for example, Davidian and Giltinan (2003) discuss the possibility that the parameters b_i may be functions of some concomitant variables z_{ij} such as body weight or past medical history. For conciseness we assume that the experimental variables are solely times of measurement.

The individuals' parameter vectors b_i are most simply assumed to be sampled from a homogeneous population with mean $\mathrm{E}\, b_i = \beta$ and covariance matrix $\sigma^2 D$. Further all distributions are assumed normal, with b_i independent of $b_{i'}, i \neq i'$ and of the observational errors. Then

$$b_i \sim \mathcal{N}(\beta, \sigma^2 D) \quad \text{and} \quad \epsilon_{ij} \sim \mathcal{N}(0, \sigma^2). \tag{25.24}$$

We consequently obtain the linear mixed model

$$y_{ij} = \beta^{\mathrm{T}} f(x_{ij}) + (b_i - \beta)^{\mathrm{T}} f(x_{ij}) + \epsilon_{ij}. \tag{25.25}$$

With the model written in this form we see not only, as expected, that $\mathrm{E}\,(y_{ij}) = \beta^{\mathrm{T}} f(x_{ij})$, but that

$$\operatorname{var}(y_{ij}) = v_{ij} = \sigma^2\{1 + f^{\mathrm{T}}(x_{ij}) D f(x_{ij})\}. \tag{25.26}$$

If we write the N_i observations for subject i as Y_i, with $N_i \times p$ design matrix F_i, the correlation between the Y_i is expressed as

$$\operatorname{var} Y_i = \sigma^2 V_i, \quad \text{where } V_i = I_{N(i)} + F_i D F_i^{\mathrm{T}} \tag{25.27}$$

and $I_{N(i)}$ is the $N_i \times N_i$ identity matrix.

The model for all $N = \sum_{i=1}^{S} N_i$ observations can be written by forming a vector Y of the S vectors Y_i and the $N \times p$ extended design matrix F, formed from the individual matrices F_i. The covariance matrix of Y is block diagonal with diagonal blocks $\sigma^2 V_i$. The details are in Entholzner *et al.* (2005) which this description follows. The population parameter β is estimated by least squares and has variance

$$\text{var}\,\hat{\beta} = \sigma^2 \left\{ \sum_{i=1}^{S} F_i^{\mathrm{T}} V_i^{-1} F_i \right\}^{-1}. \tag{25.28}$$

If some of the N patients follow the same measurement schedule, then there will be repeated entries in the summand of (25.28).

In the case of continuous designs, the results of Schmelter (2005) show that only one measurement schedule is necessary for the optimum design. However, for exact designs, more than one schedule may be necessary if none are close to the optimum. The situation is similar to that for the independent series of correlated observations in §24.4.

If there is just one measurement schedule, explicit results can be obtained for the estimator $\hat{\beta}$ and for some properties of optimum designs. Let the extended design matrix for the common measurement schedule be F_1. The variance of the N_1 observations Y_1 is, from (25.27), $\sigma^2(I_1 + F_1 D F_1^{\mathrm{T}})$, where I_1 is the $N_1 \times N_1$ identity matrix. Then, from (22.1),

$$\text{var}\,\hat{\beta} = \sigma^2 \{F_1^{\mathrm{T}}(I_1 + F_1 D F_1^{\mathrm{T}})^{-1} F_1\}^{-1} = \sigma^2\{(F_1^{\mathrm{T}} F_1)^{-1} + D\}. \tag{25.29}$$

The simple form on the right-hand side of (25.29) follows from standard results on matrix inverses, for example Smith (1973) or Rao (1973, p. 33). Optimum designs will then minimize appropriate functions of the covariance matrix $(F_1^{\mathrm{T}} F_1)^{-1} + D$. Since there is a common design for all S subjects, the exact design does not depend on the total number of observations $N = SN_1$.

For the moment we stay with exact designs. Entholzner *et al.* (2005) consider two design criteria. If interest is in the variances of the s linear combinations $\hat{\psi} = A^{\mathrm{T}}\hat{\beta}$, where A is a $p \times s$ matrix of constants, it follows from (25.29) that the sum of the variances is minimized by the linear optimum design (10.17) minimizing

$$\begin{aligned} \text{var} \sum_{i=1}^{p} \hat{\psi}_i &\propto \text{tr}\,\left[A^{\mathrm{T}}\{(F_1^{\mathrm{T}} F_1)^{-1} + D\}A\right] \\ &= \text{tr}\,\left[A^{\mathrm{T}}\{(F_1^{\mathrm{T}} F_1)^{-1}\}A\right] + \text{tr}\,A^{\mathrm{T}} D A. \end{aligned} \tag{25.30}$$

Since the second term on the right-hand side does not depend on the design, it can be ignored in calculating the optimum design which therefore does not depend on the variance D of the β_i. Thus, for a linear criterion function, the optimum design for the fixed effects model $\mathrm{E}\,(y) = \beta^{\mathrm{T}} f(x)$ in which $D = 0$ is also optimum for the mixed effects model, that is when $D \neq 0$, provided there is only one measurement schedule. Special cases of the linear criterion to which this result applies include A-optimality when $A = I$ (§10.1); c-optimality, in which one linear combination of the parameters is of interest and I or V optimality in which the integrated mean squared error of the predicted response is minimized (§10.6). However, it is clear from (25.28) that the equivalence between mixed and fixed effects models does not necessarily hold when the design requires two or more measurement schedules.

We now consider D-optimality. For the exact N_1-trial design in which $|(F_1^{\mathrm{T}}F_1)^{-1} + D|$ is to be minimized, the optimum design now depends on the value of the covariance matrix D, except in the trivial case when $p = 1$. Some properties of the D-optimum design can be established for the continuous case with design measure ξ. However, these designs still depend on N_1, the total number of observations per subject. We write

$$N_1 M(\xi) = F_1^{\mathrm{T}} F_1, \tag{25.31}$$

so that the, not necessarily integer, number of observations at support point i is $N_1 w_i$. With N_1 fixed the D-optimum design for the population parameter β will minimize

$$|\{N_1 M(\xi)\}^{-1} + D| \propto |M^{-1}(\xi) + N_1 D|. \tag{25.32}$$

Increasing N_1 thus increases the relative effect of the covariance D. Fedorov and Hackl (1997, §5.2) discuss designs not only for the estimation of β, but also for the estimation of the individual parameters β_i and for estimation of the variances σ^2 and D. If the D-optimum design for estimation of β is ξ^*, that is ξ^* minimizes (25.32), the equivalence theorem states that, over $\mathcal{X}$

$$\begin{aligned} &f^{\mathrm{T}}(x) M^{-1}(\xi^*)\{M^{-1}(\xi^*) + N_1 D\}^{-1} M^{-1}(\xi^*) f(x) \\ &\qquad \leq p - \mathrm{tr}\,\{M^{-1}(\xi^*) + N_1 D\}^{-1} M(\xi^*), \end{aligned} \tag{25.33}$$

a form similar to that for design augmentation in §19.2.2. Standard matrix results such as those already used in this section provide the alternative form

$$d(x, \xi^*) - d_{\mathrm{tot}}(x, \xi^*) \leq p - \mathrm{tr}\,\{M(\xi^*) + D^{-1}/N_1\} M^{-1}(\xi^*), \tag{25.34}$$

where

$$\begin{aligned} d(x, \xi) &= f^{\mathrm{T}}(x) M^{-1}(\xi) f(x) \quad \text{and} \\ d_{\mathrm{tot}}(x, \xi) &= f^{\mathrm{T}}(x)\{M(\xi) + D^{-1}/N_1\}^{-1} f(x). \end{aligned} \tag{25.35}$$

In (25.34) $d(x, \xi)$ is the variance in the absence of random effects while $d_{\mathrm{tot}}(x, \xi)$ incorporates prior information about D. Entholzner *et al.* (2005) find exact G-optimum designs minimizing the maximum over $\mathcal{X}$ of $d(x, \xi)$.

These can be very different from the exact D-optimum designs, the continuous versions of which would satisfy (25.34).

Although we have written the information matrix in (25.31) as $M(\xi)$, it might be better written as $M_1(\xi)$, since the complete design problem as formulated here concerns the allocation of N measurements among S subjects with $N = SN_1$. If N_1 is not specified, calculation of the optimum design requires the introduction of the costs of each measurement on a subject and of recruitment of new subjects, leading to an optimum balance between the number of subjects and the number of measurements on each. In practice, the number of measurements per patient will often be determined by practical constraints; optimum designs will then minimize functions of the exact-design covariance matrix (25.29).

It is very much more difficult to find and characterize optimum designs for non-linear models. In place of (25.25) for the response of subject i we now have the non-linear mixed model

$$y_{ij} = \eta(x_{ij}, \theta_i) + \epsilon_{ij}, \tag{25.36}$$

where the random parameter θ_i has a distribution $p(\theta_i, D)$. With $\mathrm{E}\,\epsilon_{ij} = 0$, the marginal mean of y_{ij} is

$$\mathrm{E}\,(y_{ij}) = \int \eta(x_{ij}, \theta_i) p(\theta_i, D) d\theta_i. \tag{25.37}$$

Even with $\theta_i \sim \mathcal{N}(0, D)$ this integral will be intractable and will have to be evaluated numerically. The information matrix will likewise contain entries that require numerical evaluation. Since numerical integration will have to be employed, the use of distributions other than the normal does not appreciably complicate the calculation of optimum designs. For example, the parameters in many non-linear models have interpretations as rates of reaction, which cannot be negative. In these circumstances a more plausible assumption about the parameters is that they have lognormal distributions; that is, $\log\,\theta_i$ can be taken to be normally distributed. Details of fitting non-linear mixed models are given by Pinheiro and Bates (2000).

There are two main approaches to finding population designs for non-linear models. The first (Mentré, Mallet, and Baccar 1997; Fedorov, Gagnon, and Leonov 2002) is to expand the model in a Taylor series, as we did in Chapter 17 for non-linear models with fixed effects. Here the expansion leads to linear random effects models similar to (25.25). An alternative is the use of Bayesian methods to incorporate the random effects. Stroud, Müller, and Rosner (2001) find optimum sampling times for a sequential design in which the results from previous patients are available before sampling times are assigned for the current patient. Han and Chaloner (2004) assume different priors for design and analysis. All four studies include sampling costs. Because of the computational requirements it may be hard to explore the whole design region. Han and Chaloner (2004) compare eight designs; Mentré, Mallet, and Baccar (1997) find optimum designs, including some that require two separate measurement schedules. Han and Chaloner (2004) comment that the linear approximation to the non-linear mixed-effects model is inaccurate unless the between-subject variability is small or the model nearly linear. Although a poor linear approximation need not result in a poor design, investigation of the quality of designs is hampered by the numerical complications flowing from the integral in (25.37).

25.6 Designs Over Time

Many of our experimental designs have been for measurements over time. In Chapter 17 the observational errors were assumed independent, whereas in Chapter 24 there was correlation between the observations. But, whatever the error structure, we have assumed that the spacing of the design points is solely determined by the design algorithm responding to the parameter sensitivities and the correlation structure. However, there are situations in which there is physically a minimum time between observations, for example because a sampling instrument is blocked for a fixed time period until the sample or analysis have been transmitted. Some examples are given by Bohachevsky, Johnson, and Stein (1986) for independent errors. Inclusion of such constraints in our algorithm for correlated observations is theoretically possible through use of a correlation function that is zero if the observations are sufficiently far apart and one if they are not.

There are however situations in which the time points are not an issue, measurements for example being taken at regular intervals. Then the experimental variables are the starting conditions and the profiles of other variables during the run. For example, Uciński and Bogacka (2004) find optimum designs for a kinetic experiment in which the reaction rates are a function of temperature. The experimental problem is then to choose a temperature profile that maximizes the design criterion, in their case T-optimality although the same strategy would apply for other criteria such as D-optimality. The n measurement points are given and the design should specify the n temperatures at these times, together with the initial concentrations of the reactants. A smooth temperature profile is achieved by using B-splines (Press *et al.* 1992, §3.3) to interpolate the temperatures given by the design, the resulting temperature profile then being used in the integration of the kinetic differential equation defining the model. In this way definition of a continuous profile is reduced to the selection of n times from a design region $\mathcal{T}$. In the example of Uciński and Bogacka (2004) $\mathcal{T}$ specifies maximum and minimum temperatures. The parameterization of time profiles for experimental variables is also discussed by Fedorov and Hackl (1997, p. 72).

In the kinetic examples described in this book a reaction starts with specific concentration of reactants, runs for a certain time, and is then considered finished. Many industrial processes are not of this batch kind but are continuous; there may be a stirred reactor to which components are added and from which product is withdrawn. Then the experimental design will also consist of time profiles of, for example, flows, temperatures, and concentrations, together perhaps with time points of observation. An example is Bauer *et al.* (2000).

25.7 Neural Networks

Ideas of the optimum design theory can be extended to organize an efficient data collection required to build neural networks. A commonly used type of neural network is the feed-forward network. The data are used to estimate the parameters of non-linear regression or classification models. For example, the model

$$\mathrm{E}(y) = g\left\{ w_0 + \sum_{k=1}^{u} w_k f\left(v_{k_0} + \sum_{j=1}^{m} v_{k_j} x_j \right) \right\} \tag{25.38}$$

may describe a categorical response for different values of the variables, usually called inputs or covariates, x_j. The 'activating' functions g and f describe the architecture of the network. For instance, $f(a)$ could be the logistic function $\exp(a)/\{1+\exp(a)\}$. In (25.38) w_k and v_{k_j} are called 'weights' and their estimation is required. When the values of the inputs can be selected, standard results presented in Chapter 17 can be used to ensure efficient network training. Sequential methods are usually used. Examples are given in Haines (1998). Cohn (1996) points out that in high-dimensional problems the existence of multiple optima can lead to finding suboptimum solutions.

The model describing the neural network may not be known in advance and may be refined as the data are collected. However, the ideas presented in Chapters 18, 19, 22, and 23 can be extended to address this issue. For example, the Bayesian approach has been used by MacKay (1992) and Haines (1998). Titterington (2000) provides a survey of the recent literature in this area.

25.8 In Brief

Optimum experimental designs depend upon the assumed model, including assumptions about the error distribution, and on the design criterion. In the case of non-linear models they will also depend on the values of the parameters. Some or all of these aspects may be uncertain or poorly specified at the time of design.

In several places in our book we have suggested the use of compound designs to reduce the effect of these uncertainties. In §21.4 we compared compound D-optimum designs for a one-factor polynomial with designs for individual polynomials. More generally, many of the compound design criteria of Chapter 21 provide designs that are robust to assumption or, like the CD-optimum designs of §21.9, robust in the sense of giving good designs for a variety of criteria. In §22.7 we refer to the use of compound D-optimum designs to average over parameters and link functions for a binomial model. An alterative to these averaging designs are the minimax designs, such as those mentioned in §17.3.3, where the best design is found for the most extreme departure from assumptions.

A series of papers by Wiens and co-workers explores robust designs for inadequate models in a more systematic way. Wiens (1998) and Fang and Wiens (2000) find, respectively, continuous and exact designs when there is a bound on the departure from the assumed model and when the customary assumption of homoscedasticity may be violated. The minimax solutions provide a scatter of design points. Fang and Wiens (2000, §7) conclude that 'the designs that protect against these very general forms of model bias and error heteroscedasticity may be approximated by taking the (homogeneous) variance-minimizing designs, which typically have replicates at p sites, and replacing these replicates by clusters of points at nearby but distinct sites'. The extension to sequential designs for approximate non-linear models is given by Sinha and Wiens (2002).

Finally, we consider designs for simulations that do not involve error. Although some simulations are stochastic, many are deterministic. Sacks *et al.* (1989) describe several computer models arising, for example, in studies of combustion. The controllable inputs are the parameters of the system, such as chemical rate constants and mixing rates. The deterministic response is often hard to calculate, as it involves the numerical solution of sets of simultaneous partial differential equations. Because the response is deterministic, replication of a design point yields an identical result. Yet the choice of input settings for each simulation is a problem of experimental design. Sacks *et al.* (1989) identify several design objectives, but concentrate on the prediction of the response at an untried point. By viewing the deterministic response that is to be approximated as a realization of a stochastic process, they provide a statistical basis for the design of experiments.

A more recent general discussion of computer experiments is in Bates *et al.* (1996), with a booklength treatment given by Santner *et al.* (2003). Many of the proposed designs are Latin hypercubes that fill space whilst guaranteeing uniform samples for the marginal distribution of each experimental input. Further goals, such as higher-dimensional marginal uniformity and orthogonality between first- and second-order terms are considered by Steinberg and Lin (2006).

26

EXERCISES

The exercises in this chapter are designed to help readers consolidate their understanding of the basic ideas of optimum experimental design. Unlike the SAS tasks of earlier chapters, the use of computers is not necessary, except for data analysis.

1. Set up a scenario for a new response surface investigation. It could either be a hypothetical problem or a problem that is related to your work. Determine what response variables need to be measured and which factors could influence the results. Which factors could be regarded as nuisance variables? Is it possible to control the variation due to these variables? Identify which of the factors will be varied and which of them will be kept fixed. Define the number of levels of each of the discrete factors, as well as the intervals within which the continuous factors will vary. Write down the latter in the original form and coded between -1 and 1. Identify the design region and create the analogue of Figure 2.1 for your experiment giving the names of the variables. State what model is likely to explain the relationship between the measured responses and the factors, as well as the assumptions under which use of the model would be appropriate. (Chapters 1–4, 6, 7, 10)

2. Give an example of an experiment where both qualitative and quantitative factors have to be studied. How does such an experiment differ from those in which some of the qualitative factors are regarded as fixed blocking factors, while others are regarded as random blocking variables? (Chapters 1, 14, 15)

3. Consider the experimental design given in Table 26.1. Is this design orthogonal, for a first-or second-order polynomial model in x if the blocking variables are fixed or if they are random?

4. Reanalyse the data of Table 1.2 separately for Filler A and for Filler C. Based on the results: are the designs orthogonal; if so why? A new investigation has to be carried out. Only one filler, chosen at the previous stage, will be included. Using the results of your statistical analysis, consider the appropriateness of using a complete or fractional

TABLE 26.1. A 18-trial design for one factor and two blocking variables

Blocking variable 1	Blocking variable 2	N_i	x		
1	1	2	-1	0	1
1	2	2	-1	0	1
2	1	1	-1	0	1
2	2	1	-1	0	1

factorial design; a composite design and an algorithmically constructed design. Propose a design. Which of the criteria for a good experiment listed in Chapter 6 are satisfied; which are not and why? (Chapters 1, 4–8)

5. A good design should make it possible to detect lack of fit of the fitted model. Give an example when a complete or a fractional factorial design would fail to do that. (Chapters 4, 7)

6. The effect of eight factors $x_1, \ldots, x_8$ on a response variable y is to be studied. Due to various constraints, the experimenter can afford to make only 16 observations. In the cases below write down the alias structure of designs that you consider.
(a). A model which includes all main effects for the factors is assumed. In addition, there is a possibility that the two-factor interactions between x_1, x_2, and x_3 are significant. Construct an appropriate fractional factorial design.
(b). Investigate whether the design obtained in (a) allows in addition to what was required in (a) for the two-factor interactions of x_8 with x_4, x_5, x_6, and x_7 to be estimated. What other interactions can be estimated?
(c). Suppose that the experiment has to be carried out over two days and there is a possibility that the results may be affected by this new (blocking) variable. Construct a fractional factorial design in two blocks and decide whether its use will be appropriate if the above model assumptions are correct.
(d). Consider the case when the experiment has to be carried out over four days. What factorial design would be needed in order to be able to estimate all main effects, the two-factor interactions between x_1, x_2, and x_3, and the two-factor interactions of x_8 with x_4, x_5, x_6, and x_7? (Chapter 7)

7. Consider again the scenario of Exercise 6. Suppose that two more observations can be collected. How will the experimental designs change and why? (Chapters 7, 12)

8. The experimenter has resources to carry out a statistical investigation with 14 observations. Construct a central composite design ($\alpha = 1$) for three explanatory variables. Describe the range of models that can be fitted to data from such an experiment. (Chapters 4, 7)

9. Prove that all equi-replicated complete or fractional two-level factorial designs are orthogonal and both D- and G-optimum for all estimable polynomial models. (Chapters 7, 9–11)

10. An experimenter has collected data on a health benefit (measured by the continuous variable y) of patients receiving different doses (D mg per body weight) of a new drug. As the benefit is expected to increase with the dose within the studied range of doses, a simple regression model is likely to explain the data. The experimental design that has been used consists of four measurement of y at $D = 1$; 1.4; 1.6; and 2.0 mg/bw.

    (a). Scale the values of the doses used in the experimental design to vary between -1 and 1.
    (b). Derive the standardized variance of the prediction using this design.
    (c). Investigate the optimality of the design used by the experimenter in terms of the D- and G-optimality criteria and of any other feature that might be useful.
    (d). The experimenter would like to take a new observation. What dose level would you advise her to use if D-optimality is required and the range of values of D cannot be extended? That is, she must have $1 \leq D \leq 2$. What design should be used if G-optimality is required?
    (e) Now, instead of (d), consider the case where the experimenter would like to extend the design region to 3 mg/bw. She intends to continue the experiment by taking one new observation at a time so that each time the new observation is chosen using the D-optimality criterion. What dose should be used next? Obtain the expression for the standardized variance of the prediction when your (new) design is used and explain how this can be used to find the next new dose. (Chapters 2, 9–12)

11. Consider a design with three observations $x = -1, 1$, and a and assume a simple regression model. Find the value of a for which the design is D-optimum when $\mathcal{X} = [-1, 1]$. Derive an expression for the standardized variance of the prediction $d(\xi, x)$, evaluate it for several values of a

and sketch it. What do the results suggest about the optimality of the designs? (Chapters 9–12)

12. Check whether the design including the points $(-1,-1)$, $(-1,1)$, $(1,-1)$, and $(1,1)$ is both D- and G-optimum for the model:

$$\mathrm{E}(y) = \beta_1 x_1 + \beta_2 x_2.$$

Using the points of the first design as a candidate list, find the D- and G-optimum designs with two observations for the same model. Would you recommend this design in practice? (Chapters 6, 9–12)

13. Investigate whether or not the following designs are both D- and G-optimum if a simple regression model has to be estimated and $\mathcal{X} = [-1, 1]$:

(a) $x = -1, 1$;
(b) $x = -1, -1, 1, 1$;
(c) $x = -1, 0, 1$;
(d) $x = -1, -1/3, 0, 1/3, 1$;
(e) $x = -1, 1, 1, 1$.

Which point has to be added to each of the designs in order to improve them most with respect to D-optimality? (Chapters 9–12)

14. Give an example of a design which is both D-optimum and A-optimum. Can you conjecture a general sufficient condition for this to be true? (Chapters 9, 10)

15. Show that with orthogonal coding, an A-optimum design minimizes the average standardized variance of prediction over the candidate points. In §10.6 this design criterion is called *I*- or *V-optimality*. (Chapters 9, 10)

16. $\mathrm{D_S}$-optimality may be required when only a subset of the model parameters are of interest. Why are the remaining parameters not simply omitted from the model? (Chapters 5, 10)

17. Construct a simplex-centroid design and a simplex lattice design for a four-component mixture and a third-order Scheffé polynomial model. Compare the designs. (Chapter 16)

18. Consider the case when a three-component mixture with no constraints has to be studied.

(a). Show that the Lattice Design is D-optimum if a second-order Scheffé polynomial has to be fitted.
(b). Derive the estimators for the parameters of the second-order Scheffé polynomial.

TABLE 26.2. Reaction velocity versus substrate concentration with enzyme treated with Promycin

Substrate concentration (ppm)	Velocity (counts/min^2)
0.02	76
0.02	47
0.06	97
0.06	107
0.11	123
0.11	139
0.22	159
0.22	152
0.56	191
0.56	201
1.10	207
1.10	200

(c). Add a couple of observations at the blend when the first two components are used in equal proportions to form the mixture and investigate the properties of prediction with the fitted model. In particular, compare the variances of prediction at the design points before and after this design augmentation. Show that the standardized variance of the predictions at the design points is equal to NN_i^{-1}, where N_i is the number of replications at the ith design point. (Chapter 16)

19. Consider other cases where the result in 16(d) holds. (Chapters 9, 10)

20. Bates and Watts (1988, p. 269) explore data on the dependence of enzymatic reaction velocity (y, counts/min^2) on the substrate concentration (x, in parts per million). The data are listed in Table 26.2. The model (Michaelis and Menten 1913)

$$E(y) = \frac{\alpha x}{\beta + x} \tag{26.1}$$

is used, assuming independent additive errors with a homogeneous variance.

(a). Analyse the data and verify that the estimates for the model parameter are $\hat{\alpha} = 0.2127 \times 10^3$ and $\hat{\beta} = 0.6412 \times 10^{-1}$.

(b). Write down the matrix F for a design with observations at x_1 and x_2.

(c). Show that $F^{\mathrm{T}}F =$

$$\left\{ \begin{array}{cc} x_1^2(\beta + x_1)^{-2} + x_2^2(\beta + x_2)^{-2} & -\alpha \left[x_1^2(\beta + x_1)^{-3} + x_2^2(\beta + x_2)^{-3}\right] \\ -\alpha \left[x_1^2(\beta + x_1)^{-3} + x_2^2(\beta + x_2)^{-3}\right] & \alpha^2 \left[x_1^2(\beta + x_1)^{-4} + x_2^2(\beta + x_2)^{-4}\right] \end{array} \right\}.$$

(d). Consider the experimental design with $x_1 = \beta$ and $x_2 = \infty$. Evaluate the asymptotic covariance matrix $(F^{\mathrm{T}}F)^{-1}$ for this design using the estimates of the model parameters given in (a) as prior values for α and β. Derive also an expression for the standardized variance $d(x)$ and sketch it. Show that the design is locally D-optimum.

(e). Interpret the value of the response when $x = \beta$.

(f). Suppose now that the largest value that x can take in the experiment is x_{max}. An experimental design with

$$x_1 = \frac{\beta}{1 + 2\beta x_{\text{max}}^{-1}}$$

and $x_2 = x_{\text{max}}$ is locally D-optimum. Check this assertion for $x_{\text{max}} = 1$ and the value of β given in (a). (Chapter 17)

21. The model (Hill 1913)

$$\mathrm{E}(y) = \delta + \frac{\alpha - \delta}{1 + (\beta/x)^{\gamma}}, \tag{26.2}$$

with additive independent errors of constant variance, has frequently been used to describe data collected in bioassay.

(a) Show that it is a generalization of model (26.1) which is obtained when $\delta = 0$ and $\gamma = 1$.

(b) Consider the case when it is known that $\delta = 0$ and $\alpha = 1$. Suppose now that prior values for γ and β are 1 and 2.5119, respectively. Show that the values of x_1 and x_2 for which the design is locally D-optimum are 0.885 and 7.132. Interpret the value of β. Obtain an expression for the standardized variance for this design and sketch it. (Chapter 17)

BIBLIOGRAPHY

Abramowitz, M. and Stegun, I. A. (1965). *Handbook of Mathematical Functions.* Dover Publications, New York.

Aitchison, J. (2004). *The Statistical Analysis of Compositional Data.* Blackburn Press, Caldwell, New Jersey.

Atkinson, A. C. (1982). Optimum biased coin designs for sequential clinical trials with prognostic factors. *Biometrika*, **69**, 61–67.

Atkinson, A. C. (1985). *Plots, Transformations, and Regression.* Oxford University Press, Oxford.

Atkinson, A. C. (1988). Recent developments in the methods of optimum and related experimental designs. *International Statistical Review*, **56**, 99–115.

Atkinson, A. C. (1992). A segmented algorithm for simulated annealing. *Statistics and Computing*, **2**, 221–230.

Atkinson, A. C. (1999). Bayesian and other biased-coin designs for sequential clinical trials. *Tatra Mountains Mathematical Publications*, **17**, 133–139.

Atkinson, A. C. (2002). The comparison of designs for sequential clinical trials with covariate information. *Journal of the Royal Statistical Society, Series A*, **165**, 349–373.

Atkinson, A. C. (2003*a*). The distribution of loss in two-treatment biased-coin designs. *Biostatistics*, **4**, 179–193.

Atkinson, A. C. (2003*b*). Horwitz's rule, transforming both sides and the design of experiments for mechanistic models. *Applied Statistics*, **52**, 261–278.

Atkinson, A. C. (2005). Robust optimum designs for transformation of the response in a multivariate chemical kinetic model. *Technometrics*, **47**, 478–487.

Atkinson, A. C. and Bailey, R. A. (2001). One hundred years of the design of experiments on and off the pages of *Biometrika*. *Biometrika*, **88**, 53–97.

Atkinson, A. C. and Biswas, A. (2005*a*). Bayesian adaptive biased-coin designs for clinical trials with normal responses. *Biometrics*, **61**, 118–125.

Atkinson, A. C. and Biswas, A. (2005*b*). Optimum design theory and adaptive-biased coin designs for skewing the allocation proportion in clinical trials. *Statistics in Medicine*, **24**, 2477–2492.

Atkinson, A. C. and Biswas, A. (2006). Adaptive designs for clinical trials that maximize utility. Technical report, London School of Economics.

Atkinson, A. C. and Bogacka, B. (1997). Compound, D- and D_s-optimum designs for determining the order of a chemical reaction. *Technometrics*, **39**, 347–356.

Atkinson, A. C., Bogacka, B., and Zhigljavsky, A. (eds) (2001). *Optimal Design 2000*, Kluwer, Dordrecht.

Atkinson, A. C., Bogacka, B., and Zocchi, S. S. (2000). Equivalence theory for design augmentation and parsimonious model checking: response surfaces and yield density models. *Listy Biometryczne—Biometrical Letters*, **37**, 67–95.

Atkinson, A. C., Chaloner, K., Herzberg, A. M., and Juritz, J. (1993). Optimum experimental designs for properties of a compartmental model. *Biometrics*, **49**, 325–337.

Atkinson, A. C. and Cook, R. D. (1995). D-optimum designs for heteroscedastic linear models. *Journal of the American Statistical Association*, **90**, 204–212.

Atkinson, A. C. and Cook, R. D. (1997). Designing for a response transformation parameter. *Journal of the Royal Statistical Society, Series B*, **59**, 111–124.

Atkinson, A. C. and Cox, D. R. (1974). Planning experiments for discriminating between models (with discussion). *Journal of the Royal Statistical Society, Series B*, **36**, 321–348.

Atkinson, A. C., Demetrio, C. G. B., and Zocchi, S. S. (1995). Optimum dose levels when males and females differ in response. *Applied Statistics*, **44**, 213–226.

Atkinson, A. C. and Donev, A. N. (1989). The construction of exact D–optimum experimental designs with application to blocking response surface designs. *Biometrika*, **76**, 515–526.

Atkinson, A. C. and Donev, A. N. (1992). *Optimum Experimental Designs.* Oxford University Press, Oxford.

Atkinson, A. C. and Donev, A. N. (1996). Experimental designs optimally balanced for trend. *Technometrics*, **38**, 333–341.

Atkinson, A. C. and Fedorov, V. V. (1975*a*). The design of experiments for discriminating between two rival models. *Biometrika*, **62**, 57–70.

Atkinson, A. C. and Fedorov, V. V. (1975*b*). Optimal design: experiments for discriminating between several models. *Biometrika*, **62**, 289–303.

Atkinson, A. C., Hackl, P., and Müller, W. G. (eds) (2001). *MODA 6—Advances in Model-Oriented Design and Analysis.* Physica–Verlag, Heidelberg.

Atkinson, A. C., Pronzato, L., and Wynn, H. P. (eds) (1998). *MODA 5—Advances in Model-Oriented Data Analysis and Experimental Design.* Physica–Verlag, Heidelberg.

Atkinson, A. C. and Riani, M. (2000). *Robust Diagnostic Regression Analysis.* Springer–Verlag, New York.

Atkinson, A. C., Riani, M., and Cerioli, A. (2004). *Exploring Multivariate Data with the Forward Search.* Springer–Verlag, New York.

Atkinson, A. C. and Zocchi, S. S. (1998). Parsimonious designs for detecting departures from nonlinear and generalized linear models. Technical report, London School of Economics.

Azzalini, A. and Giovagnoli, A. (1987). Some optimal designs for repeated measurements with autoregressive errors. *Biometrika*, **74**, 725–734.

Bailey, R. A. (1991). Strata for randomized experiments (with discussion). *Journal of the Royal Statistical Society, Series B*, **53**, 27–78.

Bailey, R. A. (2004). *Association Schemes: Designed Experiments, Algebra and Combinatorics.* Cambridge University Press, Cambridge.

Bailey, R. A. (2006). *Design of Comparative Experiments.* http://www.maths.qmul.ac.uk/~rab/DOEbook/.

Bailey, R. A. and Kunert, J. (2006). On optimal crossover designs when carryover effects are proportional to direct effects. *Biometrika*, **93**, 613–625.

Ball, F. G., Smith, A. F. M., and Verdinelli, I. (1993). Biased coin designs with a Bayesian bias. *Journal of Statistical Planning and Inference*, **34**, 403–421.

Bandemer, H., Bellmann, A., Jung, W., and Richter, K. (1973). *Optimale Versuchsplanung.* Akademie Verlag, Berlin.

Bandemer, H., Bellmann, A., Jung, W., Son, L. A., Nagel, S., Näther, W., Pilz, J., and Richter, K. (1977). *Theorie und Anwendung der optimalen Versuchsplanung: I Handbuch zur Theorie.* Akademie Verlag, Berlin.

Bandemer, H. and Näther, W. (1980). *Theorie und Anwendung der optimalen Versuchsplanung: II Handbuch zur Anwendung.* Akademie Verlag, Berlin.

Barnard, G. A., Box, G. E. P., Cox, D. R., Seheult, A. H., and Silverman, B. W. (eds) (1989). *Industrial Quality and Reliability.* Royal Society, London.

Bates, D. M. and Watts, D. G. (1980). Relative curvature measures of nonlinearity (with discussion). *Journal of the Royal Statistical Society, Series B*, **42**, 1–25.

Bates, D. M. and Watts, D. G. (1988). *Nonlinear Regression Analysis and Its Applications.* Wiley, New York.

Bates, R. A., Buck, R. J., Riccomagno, E., and Wynn, H. P. (1996). Experimental design and observation for large systems (with discussion). *Journal of the Royal Statistical Society, Series B*, **58**, 77–94.

Battiti, R. and Tecchiolli, G. (1992). The reactive tabu search. *ORSA Journal on Computing*, **6**, 126–140.

Bauer, I., Bock, H. G., Körkel, S., and Schlöder, J. P. (2000). Numerical methods for optimum experimental design in DAE systems. *Journal of Computational and Applied Mathematics*, **120**, 1–25.

Baumert, L., Golomb, S. W., and Hall, M. (1962). Discovery of an Hadamard matrix of order 92. *American Mathematical Society Bulletin*, **68**, 237–238.

Beale, E. M. L. (1960). Confidence regions in nonlinear estimation (with discussion). *Journal of the Royal Statistical Society, Series B*, **22**, 41–88.

Becker, N. G. (1968). Models for the response of a mixture. *Journal of the Royal Statistical Society, Series B*, **30**, 349–358.

Becker, N. G. (1969). Regression problems when the predictor variables are proportions. *Journal of the Royal Statistical Society, Series B*, **31**, 107–112.

Becker, N. G. (1970). Mixture designs for a model linear in the proportions. *Biometrika*, **57**, 329–338.

Bendell, A., Disney, J., and Pridmore, W. A. (eds) (1989). *Taguchi Methods: Applications in World Industry.* IFS Publications, Bedford, UK.

Berger, M. and Wong, W.K. (eds) (2005). *Applied Optimal Designs*, Wiley, New York.

Biedermann, S., Dette, H., and Pepelyshev, A. (2004). Maximin optimal designs for a compartmental model. In *MODA 7—Advances in Model-Oriented Design and Analysis* (eds A. Di Bucchianico, H. Läuter, and H. P. Wynn), pp. 41–49. Physica-Verlag, Heidelberg.

Biedermann, S., Dette, H., and Zhu, W. (2005). Compound optimal designs for percentile estimation in dose–response models with restricted design intervals. In *Proceedings of the 5th St Petersburg Workshop on Simulation* (eds S. Ermakov, V. Melas, and A. Pepelyshev), pp. 143–148. NII Chemistry University Publishers, St Petersburg.

Bingham, D. R. and Sitter, R. R. (2001). Design issues in fractional factorial split-plot experiments. *Journal of Quality Technology*, **33**, 2–15.

Bliss, C. I. (1935). The calculation of the dosage-mortality curve. *Annals of Applied Biology*, **22**, 134–167.

Bogacka, B., Johnson, P., Jones, B., and Volkov, O. (2007). D-efficient window experimental designs. *Journal of Statistical Planning and Inference*. (In press).

Bohachevsky, I. O., Johnson, M. E., and Stein, M. L. (1986). Generalized simulated annealing for function optimization. *Technometrics*, **28**, 209–217.

Box, G., Bisgaard, S., and Fung, C. (1989). An explanation and critique of Taguchi's contribution to quality engineering. In *Taguchi Methods: Applications in World Industry* (eds A. Bendel, J. Disney, and W. A. Pridmore), pp. 359–383. IFS Publications, Bedford, UK.

Box, G. E. P. (1952). Multi-factor designs of first order. *Biometrika*, **39**, 49–57.

Box, G. E. P. (1993). Quality improvement—the new industrial revolution. *International Statistical Review*, **61**, 1–19.

Box, G. E. P. (2006). *Improving Almost Anything, 2nd Edition.* Wiley, New York.

Box, G. E. P. and Behnken, D. W. (1960). Some new 3 level designs for the study of quantitative variables. *Technometrics*, **2**, 455–475.

Box, G. E. P. and Cox, D. R. (1964). An analysis of transformations (with discussion). *Journal of the Royal Statistical Society, Series B*, **26**, 211–246.

Box, G. E. P. and Draper, N. R. (1959). A basis for the selection of a response surface design. *Journal of the American Statistical Association*, **54**, 622–654.

Box, G. E. P. and Draper, N. R. (1963). The choice of a second order rotatable design. *Biometrika*, **50**, 335–352.

Box, G. E. P. and Draper, N. R. (1975). Robust designs. *Biometrika*, **62**, 347–352.

Box, G. E. P. and Draper, N. R. (1987). *Empirical Model-Building and Response Surfaces.* Wiley, New York.

Box, G. E. P. and Hunter, J. S. (1961*a*). The 2^{k-p} fractional factorial designs Part I. *Technometrics*, **3**, 311–351.

Box, G. E. P. and Hunter, J. S. (1961*b*). The 2^{k-p} fractional factorial designs Part II. *Technometrics*, **3**, 449–458.

Box, G. E. P. and Hunter, W. G. (1965). Sequential design of experiments for nonlinear models. In *Proceedings IBM Scientific Computing Symposium: Statistics*, pp. 113–137. IBM, New York.

Box, G. E. P., Hunter, W. G., and Hunter, J. S. (2005). *Statistics for Experimenters: Design, Innovation, and Discovery, 2nd Edition.* Wiley, New York.

Box, G. E. P. and Lucas, H. L. (1959). Design of experiments in nonlinear situations. *Biometrika*, **46**, 77–90.

Box, G. E. P. and Meyer, D. R. (1986). An analysis for unreplicated fractional factorials. *Technometrics*, **28**, 11–18.

Box, G. E. P. and Wilson, K. B. (1952). On the experimental attainment of optimum conditions (with discussion). *Journal of the Royal Statistical Society, Series B*, **13**, 1–45.

Box, M. J. and Draper, N. R. (1971). Factorial designs, the $|F^T F|$ criterion and some related matters. *Technometrics*, **13**, 731–742: Correction, **14**, 511 (1972); **15**, 430 (1973).

Brenneman, W. A. and Nair, V. J. (2001). Methods for identifying dispersion effects in unreplicated factorial experiments: a critical analysis and proposed strategies. *Technometrics*, **43**, 388–405.

Brimkulov, U. N., Krug, G. K., and Savanov, V. L. (1980). Numerical construction of exact experimental designs when the measurements are correlated. *Zavodskaya Laboratoria (Industrial Laboratory)*, **36**, 435–442. (In Russian).

Brimkulov, U. N., Krug, G. K., and Savanov, V. L. (1986). *Design of Experiments in Investigating Random Fields and Processes.* Nauka, Moscow. (In Russian).

Brownlee, K. A. (1965). *Statistical Theory and Methodology in Science and Engineering, 2nd Edition*, Wiley, New York.

Burman, C.-F. (1996). *On Sequential Treatment Allocations in Clinical Trials.* Department of Mathematics, Göteborg.

Burridge, J. and Sebastiani, P. (1994). D–optimal designs for generalised linear models with variance proportional to the square of the mean. *Biometrika*, **81**, 295–304.

Calinski, T. and Kageyama, S. (2000). *Block Designs: A Randomization Approach. Volume I: Analysis.* Lecture Notes in Statistics 150. Springer-Verlag, New York.

Calinski, T. and Kageyama, S. (2003). *Block Designs: A Randomization Approach. Volume II: Design.* Lecture Notes in Statistics 170. Springer-Verlag, New York.

Carroll, R. J. and Ruppert, D. (1988). *Transformation and Weighting in Regression.* Chapman and Hall, London.

Chaloner, K. (1988). An approach to experimental design for generalized linear models. In *Model-oriented Data Analysis* (eds V. V. Fedorov and H. Läuter). Springer, Berlin.

Chaloner, K. (1993). A note on optimal Bayesian design for nonlinear problems. *Journal of Statistical Planning and Inference*, **37**, 229–235.

Chaloner, K. and Larntz, K. (1989). Optimal Bayesian design applied to logistic regression experiments. *Journal of Statistical Planning and Inference*, **21**, 191–208.

Chaloner, K. and Verdinelli, I. (1995). Bayesian experimental design: a review. *Statistical Science*, **10**, 273–304.

Cheng, C. S. and Tang, B. (2005). A general theory of minimum aberration and its applications. *Annals of Statistics*, **33**, 944–958.

Chernoff, H. (1953). Locally optimal designs for estimating parameters. *Annals of Mathematical Statististics*, **24**, 586–602.

Claringbold, P. J. (1955). Use of the simplex design in the study of joint action of related hormones. *Biometrics*, **11**, 174–185.

Cobb, G. W. (1998). *Introduction to the Design and Analysis of Experiments.* Springer-Verlag, New York.

Cochran, W. G. and Cox, G. M. (1957). *Experimental Designs, 2nd Edition.* Wiley, New York.

Cohn, D. A. (1996). Neural network exploration using optimum experimental designs. *Neural Networks*, **9**, 1071–1083.

Cook, R. D. and Fedorov, V. V. (1995). Constrained optimization of experimental designs. *Statistics*, **26**, 129–178.

Cook, R. D. and Nachtsheim, C. J. (1980). A comparison of algorithms for constructing exact D-optimal designs. *Technometrics*, **22**, 315–324.

Cook, R. D. and Nachtsheim, C. J. (1982). Model robust, linear-optimal designs. *Technometrics*, **24**, 49–54.

Cook, R. D. and Weisberg, S. (1990). Confidence curves in nonlinear regression. *Journal of the American Statistical Association*, **85**, 544–551.

Cook, R. D. and Weisberg, S. (1999). *Applied Regression Including Computing and Graphics.* Wiley, New York.

Cook, R. D. and Wong, W. K. (1994). On the equivalence between constrained and compound optimal designs. *Journal of the American Statistical Association*, **89**, 687–692.

Coombes, N. E., Payne, R. W., and Lisboa, P. (2002). Comparison of nested simulated annealing and reactive tabu search for efficient experimental designs with correlated data. In *COMPSTAT 2002 Proceedings in Computational Statistics* (eds W. W. Härdle and B. Rönz), pp. 249–254. Physica-Verlag, Heidelberg.

Cornell, J. A. (1988). Analyzing data from mixture experiments containing process variables: A split-plot approach. *Journal of Quality Technology*, **20**, 2–23.

Cornell, J. A. (2002). *Experiments with Mixtures: Designs, Models, and the Analysis of Mixture Data, 3rd Edition.* Wiley, New York.

Cox, D. R. (1958). *Planning of Experiments.* Wiley, New York.

Cox, D. R. (1971). A note on polynomial response functions for mixtures. *Biometrika*, **58**, 155–159.

Cox, D. R. (1988). A note on design when response has an exponential family distribution. *Biometrika*, **75**, 161–164.

Cox, D. R. and Reid, R. (2000). *The Theory of the Design of Experiments.* Chapman and Hall/CRC Press, Boca Raton.

Crosier, R. B. (1984). Mixture experiments: geometry and pseudocomponents. *Technometrics*, **26**, 209–216.

Davidian, M. and Giltinan, D. M. (1995). *Nonlinear Models for Repeated Measurement Data.* Chapman and Hall/CRC Press, Boca Raton.

Davidian, M. and Giltinan, D. M. (2003). Nonlinear models for repeated measurements: an overview and update. *Journal of Agricultural, Biological, and Environmental Statstics*, **8**, 387–419.

Dean, A. M. and Lewis, S. M. (Eds) (2006). *Screening Methods for Experimentation in Industry, Drug Discovery, and Genetics.* Springer-Verlag, New York.

Dean, A. M. and Voss, D. (2003). *Design and Analysis of Experiments.* Springer-Verlag, New York.

Dehnad, K. (ed.) (1989). *Quality Control, Robust Design, and the Taguchi Method.* Wadsworth & Brooks/Cole, Pacific Grove, CA.

Deppe, C., Carpenter, R., and Jones, B. (2001). Nested incomplete block designs in sensory testing: construction strategies. *Food Quality and Preference*, **12**, 281–290.

Derringer, G. C. (1974). An empirical model for viscosity of filled and plasticized elastomer compounds. *Journal of Applied Polymer Science*, **18**, 1083–1101.

Dette, H. (1996). A note on Bayesian c- and D-optimal designs in nonlinear regression models. *Annals of Statistics*, **24**, 1225–1234.

Dette, H. and Biedermann, S. (2003). Robust and efficient designs for the Michaelis–Menten model. *Journal of the American Statistical Association*, **98**, 679–686.

Dette, H. and Haines, L. M. (1994). E-optimal designs for linear and nonlinear models with two parameters. *Biometrika*, **81**, 739–754.

Dette, H., Kunert, J., and Pepelyshev, A. (2007). Exact optimal designs for weighted least squares analysis with correlated errors. *Statistica Sinica*, **17**. (To appear).

Dette, H. and Kwiecien, R. (2004). A comparison of sequential and non-sequential designs for discrimination between nested regression models. *Biometrika*, **91**, 165–176.

Dette, H. and Kwiecien, R. (2005). Finite sample performance of sequential designs for model identification. *Journal of Statistical Computing and Simulation*, **75**, 477–495.

Dette, H., Melas, V. B., and Pepelyshev, A. (2003). Standardized maximin E-optimal designs for the Michaelis–Menten model. *Statistica Sinica*, **4**, 1147–1163.

Dette, H., Melas, V. B., and Pepelyshev, A. (2004). Optimal designs for a class of nonlinear regression models. *Annals of Statistics*, **32**, 2142–2167.

Dette, H., Melas, V. B., and Wong, W. K. (2005). Optimal design for goodness-of-fit of the Michaelis–Menten enzyme kinetic function. *Journal of the American Statistical Association*, **100**, 1370–1381.

Dette, H. and Neugebauer, H. M. (1997). Bayesian D-optimal designs for exponential regression models. *Biometrika*, **60**, 331–349.

Dette, H. and O'Brien, T. E. (1999). Optimality criteria for regression models based on predicted variance. *Biometrika*, **86**, 93–106.

Dette, H. and Sahm, M. (1998). Minimax optimal designs in nonlinear regression models. *Statistica Sinica*, **8**, 1249–1264.

Di Bucchianico, A., Läuter, H., and Wynn, H. P. (eds) (2004). *MODA 7—Advances in Model-Oriented Design and Analysis*, Physica-Verlag, Heidelberg.

Dobson, A. (2001). *An Introduction to Generalized Linear Models, 2nd Edition.* Chapman and Hall, London.

Donev, A. N. (1988). *The Construction of D-optimum Experimental Designs.* PhD Thesis, University of London.

Donev, A. N. (1989). Design of experiments with both mixture and qualitative factors. *Journal of the Royal Statistical Society, Series B*, **51**, 297–302.

Donev, A. N. (1997). Algorithm AS 313: an algorithm for the construction of crossover trials. *Applied Statistics*, **46**, 288–295.

Donev, A. N. (1998). Crossover designs with correlated observations. *Journal of Biopharmaceutical Statistics*, **8**, 249–262.

Donev, A. N. (2004). Design of experiments in the presence of errors in factor level. *Journal of Statistical Planning and Inference*, **126**, 569–585.

Donev, A. N. and Atkinson, A. C. (1988). An adjustment algorithm for the construction of exact D-optimum experimental designs. *Technometrics*, **30**, 429–433.

Donev, A. N. and Jones, B. (1995). Construction of A–optimum cross-over designs. In *MODA 4*—Advances in Model-Oriented Data Analysis (eds C. P. Kitsos and W. G. Müller), pp. 165–171. Physica-Verlag, Heidelberg.

Downing, D., Fedorov, V. V., and Leonov, S. (2001). Extracting information from the variance function: optimal design. In *MODA 6—Advances in Model-Oriented Design and Analysis* (eds A. C. Atkinson, P. Hackl, and W. G. Müller), pp. 45–52. Physica-Verlag, Heidelberg.

Draper, N. R. and Hunter, W. G. (1966). Design of experiments for parameter estimation in multiresponse situations. *Biometrika*, **53**, 596–599.

Draper, N. R. and St. John, R. C. (1977). A mixtures model with inverse terms. *Technometrics*, **19**, 37–46.

Draper, N. R. and Lin, D. K. J. (1990). Small response surface designs. *Technometrics*, **32**, 187–194.

Draper, N. R., Prescott, P., Lewis, S. M., Dean, A. M., John, P. W. M., and Tuck, M. G. (1993). Mixture designs for four components in orthogonal blocks. *Technometrics*, **35**, 268–276. Correction (1994). *Technometrics* **36**, 234.

Draper, N. R. and Smith, H. (1998). *Applied Regression Analysis, 3rd Edition.* Wiley, New York.

Dubov, E. L. (1971). D–optimal designs for nonlinear models under the Bayesian approach (in Russian). In *Regression Experiments* (ed. V. V. Fedorov). Moscow University Press, Moscow.

DuMouchel, W. and Jones, B. (1994). A simple Bayesian modification of D–optimal designs to reduce dependence on an assumed model. *Technometrics*, **36**, 37–47.
Dykstra, O. (1971*a*). The augmentation of experimental data to maximize $|X^T X|$. *Technometrics*, **13**, 682–688.
Dykstra, O. (1971*b*). Addendum to 'The augmentation of experimental data to maximize $|X^T X|$'. *Technometrics*, **13**, 927.
Elfving, G. (1952). Optimum allocation in linear regression theory. *Annals of Mathematical Statistics*, **23**, 255–262.
Entholzner, M., Benda, M., Schmelter, T., and Schwabe, R. (2005). A note on designs for estimating population parameters. *Listy Biometryczne–Biometrical Letters*, **42**, 25–41.
Ermakov, S. M. (ed.) (1983). *The Mathematical Theory of Planning Experiments.* Moscow, Nauka. (In Russian).
Ermakov, S. M. and Zhiglijavsky, A. A. (1987). *The Mathematical Theory of Optimum Experiments.* Nauka, Moscow. (In Russian).
Evans, J. W. (1979). The computer augmentation of experimental designs to maximize $|X^T X|$. *Technometrics*, **21**, 321–330.
Fang, Z. and Wiens, D. P. (2000). Integer-valued, minimax robust designs for estimation and extrapolation in heteroscedastic, approximately linear models. *Journal of the American Statistical Association*, **95**, 807–818.
Farrell, R. H., Kiefer, J., and Walbran, A. (1968). Optimum multivariate designs. In *Proceedings of the 5th Berkeley Symposium*, Volume 1, pp. 113–138. University of California Press, Berkeley, CA.
Fedorov, V. V. (1972). *Theory of Optimal Experiments.* Academic Press, New York.
Fedorov, V. V. (1981). Active regression experiments. In *Mathematical Methods of Experimental Design* (ed. V. B. Penenko). Nauka, Novosibirsk. (In Russian).
Fedorov, V. V. and Atkinson, A. C. (1988). The optimum design of experiments in the presence of uncontrolled variability and prior information. In *Optimal Design and Analysis of Experiments* (eds Y. Dodge, V. V. Fedorov, and H. P. Wynn), pp. 327–344. North-Holland, Amsterdam.
Fedorov, V. V., Gagnon, R. C., and Leonov, S. L. (2002). Design of experiments with unknown parameters in the variance. *Applied Stochastic Models in Business and Industry*, **18**, 207–218.
Fedorov, V. V. and Hackl, P. (1997). *Model-Oriented Design of Experiments.* Lecture Notes in Statistics 125. Springer–Verlag, New York.
Fedorov, V. V. and Leonov, S. L. (2005). Response-driven designs in drug development. In *Applied Optimal Designs* (eds M. Berger and W.-K. Wong), Chapter 5, pp. 103–136. Wiley, New York.

Finney, D. J. (1945). The fractional replication of factorial arrangements. *Annals of Eugenics*, **12**, 291–301.

Firth, D. and Hinde, J. P. (1997). On Bayesian D-optimum design criteria and the equivalence theorem in non-linear models. *Journal of the Royal Statistical Society, Series B*, **59**, 793–797.

Fisher, R. A. (1960). *The Design of Experiments, 7th Edition*. Oliver and Boyd, Edinburgh.

Flury, B. (1997). *A First Course in Multivariate Statistics*. Springer-Verlag, New York.

Ford, I., Titterington, D. M., and Kitsos, C. P. (1989). Recent advances in nonlinear experimental design. *Technometrics*, **31**, 49–60.

Ford, I., Torsney, B., and Wu, C. F. J. (1992). The use of a canonical form in the construction of locally optimal designs for non-linear problems. *Journal of the Royal Statistical Society, Series B*, **54**, 569–583.

Franklin, M. F. and Bailey, R. A. (1985). Selecting defining contrasts and confounded effects in p^{n-m} factorial experiments. *Technometrics*, **27**, 165–172.

Fresen, J. (1984). Aspects of bioavailability studies. M.Sc. Dissertation, Department of Mathematical Statistics, University of Capetown.

Galil, Z. and Kiefer, J. (1980). Time-and space-saving computer methods, related to Mitchell's DETMAX, for finding D-optimum designs. *Technometrics*, **21**, 301–313.

Garvanska, P., Lekova, V., Donev, A. N., and Dencheva, R. (1992). Mikrowellenabsorbierende, electrisch zeitende Polymerpigmente für Textilmaterialien: Teil I. Optmierung der Herstellung von mikrowellenabsorbierenden Polymerpigmenten. Technical report, University of Chemical Technology, Sofia, Bulgaria.

Giovagnoli, A. and Wynn, H. P. (1985). Group invariant orderings with applications to matrix orderings. *Linear Algebra and its Applications*, **67**, 111–135.

Giovagnoli, A. and Wynn, H. P. (1996). Cyclic majorization and smoothing operators. *Linear Algebra and its Applications*, **239**, 215–225.

Goos, P. (2002). *The Optimal Design of Blocked and Split-plot Experiments*. New York: Springer.

Goos, P. and Donev, A. N. (2006*a*). Blocking response surface designs. *Journal of Computational Statistics and Data Analysis*, **15**, 1075–1088.

Goos, P. and Donev, A. N. (2006*b*). The D-optimal design of blocked experiments with mixture components. *Journal of Quality Technology*, **38**, 319–332.

Goos, P. and Donev, A. N. (2007*a*). D-optimal minimum support mixture designs in blocks. *Metrika*, **65**, 53–68.

Goos, P. and Donev, A. N. (2007*b*). Tailor-made split-plot designs for mixture and process variables. Journal of Quality Technology (In press).

Goos, P. and Vandebroek, M. (2001*a*). D-optimal response surface designs in the presence of random block effects. *Computational Statistics and Data Analysis*, **37**, 433–453.

Goos, P. and Vanderbroek, M. (2001*b*). Optimal split-plot designs. *Journal of Quality Technology*, **33**, 436–450.

Goos, P. and Vandebroek, M. (2003). D-optimal split-plot designs with given numbers and sizes of whole plots. *Technometrics*, **45**, 235–245.

Guest, P. (1958). The spacing of observations in polynomial regression. *Annals of the Mathematical Statistics*, **29**, 294–299.

Haines, L. (1993). Optimal design for nonlinear regression models. *Communications in Statistics—Theory and Methods*, **22**, 1613–1627.

Haines, L. M. (1987). The application of the annealing algorithm to the construction of exact optimal designs for linear-regression models. *Technometrics*, **29**, 439–447.

Haines, L. M. (1998). Optimal designs for neural networks. In *New Developments and Applications in Experimental Design* (eds N. Flournoy, W. F. Rosenberger, and W. K. Wong), Volume 34, pp. 123–124. IMS Lecture Notes-Monograph Series, IMS, Hayward, CA.

Hald, A. (1998). *A History of Mathematical Statistics from 1750 to 1930.* Wiley, New York.

Hamilton, D. C. and Watts, D. G. (1985). A quadratic design criterion for precise estimation in nonlinear regression models. *Technometrics*, **27**, 241–250.

Han, C. and Chaloner, K. (2003). D- and c-optimal designs for exponential regression models used in viral dynamics and other applications. *Journal of Statistical Planning and Inference*, **115**, 585–601.

Han, C. and Chaloner, K. (2004). Bayesian experimental design for nonlinear mixed-effects models with application to HIV dynamics. *Biometrics*, **60**, 25–33.

Hardin, R. H. and Sloane, N. J. A. (1993). A new approach to the construction of optimal designs. *Journal of Statistical Planning and Inference*, **37**, 339–369.

Hartley, H. O. (1959). Smallest composite designs for quadratic response surfaces. *Biometrics*, **15**, 611–624.

Harville, D. (1974). Nearly optimal allocation of experimental units using observed covariate values. *Technometrics*, **16**, 589–599.

Harville, D. (1975). Computing optimum designs for covariate models. In *A Survey of Statistical Design and Linear Models* (ed. J. N. Srivastava), pp. 209–228. North-Holland, Amsterdam.

Hedayat, A. S., Stufken, J., and Yang, M. (2006). Optimal and efficient crossover designs when subject effects are random. *Journal of the American Statistical Association*, **101**, 1031–1038.

Hedayat, A. S., Zhong, J., and Nie, L. (2003). Optimal and efficient designs for 2-parameter nonlinear models. *Journal of Statistical Planning and Inference*, **124**, 205–217.

Heiberger, R. M., Bhaumik, D. K., and Holland, B. (1993). Optimal data augmentation strategies for additive models. *Journal of the American Statistical Association*, **88**, 926–938.

Hill, A. V. (1913). The combinations of haemoglobin with oxygen and with carbon monoxide. *Biochemical Journal*, **7**, 471–480.

Hoel, P. G. (1958). Efficiency problems in polynomial estimation. *Annals of Mathematical Statistics*, **29**, 1134–1145.

Hotelling, H. (1944). On some improvements in weighing and other experimental techniques. *Annals of Mathematical Statistics*, **15**, 297–306.

Hu, F. and Rosenberger, W. F. (2006). *The Theory of Response-Adaptive Randomization in Clinical Trials.* Wiley, New York.

Huang, P., Chen, D., and Voelkel, J. O. (1998). Minimum-aberration two-level split-plot designs. *Technometrics*, **40**, 314–326.

John, J. A., Russell, K. G., and Whitaker, D. (2004). Crossover: an algorithm for the construction of efficient cross-over designs. *Statistics in Medicine*, **23**, 2645–2658.

John, P. W. M. (1984). Experiments with mixtures involving process variables. Technical Report 8, Center for Statistical Sciences, University of Texas, Austin, Texas.

Johnson, M. E. and Nachtsheim, C. J. (1983). Some guidelines for constructing exact D-optimal designs on convex design spaces. *Technometrics*, **25**, 271–277.

Jones, B. (1976). An algorithm for deriving optimal block designs. *Technometrics*, **18**, 451–458.

Jones, B. and Donev, A. N. (1996). Modelling and design of cross-over trials. *Statistics in Medicine*, **15**, 1435–1446.

Jones, B. and Eccleston, J. A. (1980). Exchange and interchange procedures to search for optimal designs. *Journal of the Royal Statistical Society, Series, B*, **42**, 238–243.

Jones, B. and Kenward, M. G. (2003). *Design and Analysis of Cross-Over Trials, 2nd Edition.* Chapman and Hall/CRC Press, Florida.

Jones, B. and Wang, J. (1999). The analysis of repeated measurements in sensory and consumer studies. *Food Quality and Preference*, **11**, 35–41.

Kenworthy, O. O. (1963). Fractional experiments with mixtures using ratios. *Industrial Quality Control*, **19**, 24–26.

Khuri, A. I. (1984). A note on D–optimal designs for partially nonlinear regression models. *Technometrics*, **26**, 59–61.
Khuri, A. I. (ed.) (2006). *Response Surface Methodology and Related Topics.* World Scientific, Singapore.
Khuri, A. I. and Cornell, J. A. (1996). *Response Surfaces, 2nd Edition.* Marcel Dekker, New York.
Khuri, A. I. and Mukhopadhyay, S. (2006). GLM designs: the dependence on unknown parameters dilemma. In *Response Surface Methodology and Related Topics* (ed. A. I. Khuri), pp. 203–223. World Scientific, Singapore.
Kiefer, J. (1959). Optimum experimental designs (with discussion). *Journal of the Royal Statistical Society, Series B*, **21**, 272–319.
Kiefer, J (1961). Optimum designs in regression problems II. *Annals of Mathematical Statistics*, **32**, 298–325.
Kiefer, J. (1975). Optimal design: Variation in structure and performance under change of criterion. *Biometrika*, **62**, 277–288.
Kiefer, J. (1985). *Jack Carl Kiefer Collected Papers III.* Springer, New York.
Kiefer, J. and Wolfowitz, J. (1959). Optimum designs in regression problems. *Annals of Mathematical Statistics*, **30**, 271–294.
Kiefer, J. and Wolfowitz, J. (1960). The equivalence of two extremum problems. *Canadian Journal of Mathematics*, **12**, 363–366.
Kiefer, J. and Wynn, H. P. (1984). Optimum and minimax exact treatment designs for one-dimensional autoregressive error processes. *Annals of Statistics*, **12**, 414–450.
King, J. and Wong, W. K. (2000). Minimax D-optimal designs for the logistic model. *Biometrics*, **56**, 1263–1267.
Kitsos, C. P., Titterington, D. M., and Torsney, B. (1988). An optimum design problem in rhythmometry. *Biometrics*, **44**, 657–671.
Kôno, K. (1962). Optimum designs for quadratic regression on k-cube. *Memoirs of the Faculty of Science, Kyushu University, Series A*, **16**, 114–122.
Kowalski, S., Cornell, J. A., and Vining, G. G. (2000). A new model and class of designs for mixture experiments with process variables. *Communications in Statistics: Theory and Methods*, **29**, 2255–2280.
Kowalski, S. M., Cornell, J. A., and Vining, G. G. (2002). Split-plot designs and estimation methods for mixture experiments with process variables. *Technometrics*, **44**, 72–79.
Kuhfeld, W. F. and Tobias, R. D. (2005). Large factorial designs for product engineering and marketing research applications. *Technometrics*, **47**, 132–141.
Kurotschka, V. G. (1981). A general approach to optimum design of experiments with qualitative and quantitative factors. In *Proceedings of the Indian Statistical Institute Golden Jubilee International Conference*

on Statistics: Applications and New Directions (eds J. Ghosh and J. Roy), pp. 353–368. Indian Statistical Institute, Calcutta.

Kuroturi, I. S. (1966). Experiments with mixtures of components having lower bounds. *Industrial Quality Control*, **22**, 592–596.

Läuter, E. (1974). Experimental design in a class of models. *Mathematische Operationsforschung und Statistik, Serie Statistik*, **5**, 379–398.

Läuter, E. (1976). Optimum multiresponse designs for regression models. *Mathematische Operationsforschung und Statistik, Serie Statistik*, **7**, 51–68.

Lee, Y. and Nelder, J. A. (1999). Generalized linear models for the analysis of quality-improvement experiments. *Quality Control and Applied Statistics*, **44**, 323–326.

Lenth, R. V. (1989). Quick and easy analysis of unreplicated factorials. *Technometrics*, **31**, 469–473.

Lewis, S. M., Dean, A., Draper, N. R., and Prescott, P. (1994). Mixture designs for q components in orthogonal blocks. *Journal of the Royal Statistical Society, Series B*, **56**, 457–467.

Lim, Y. B., Studden, W. J., and Wynn, H. P. (1988). A general approach to optimum design of experiments with qualitative and quantitative factors. In *Statistical Decision Theory and Related Topics IV* (eds J. O. Burger and S. S. Gupta), Volume 2, pp. 353–368. Springer, New York.

Lin, C. S. and Morse, P. M. (1975). A compact design for spacing experiments. *Biometrics*, **31**, 661–671.

Lindsey, J. K. (1999). *Models for Repeated Measurements, 2nd Edition.* Oxford University Press, Oxford.

Lischer, P. (1999). Good statistical practice in analytical chemistry. In *Probability Theory and Mathematical Statistics* (ed. B. Grigelionis), pp. 1–12. VSP, Dordrecht.

Littell, R., Stroup, W., and Freund, R. (2002). *SAS for Linear Models, 4th Edition.* SAS Press, Cary, NC.

Logothetis, N. and Wynn, H. P. (1989). *Quality Through Design.* Clarendon Press, Oxford.

López-Fidalgo, J., Dette, H., and Zhu, W. (2005). Optimal designs for discriminating between heteroscedastic models. In *Proceedings of the 5th St Petersburg Workshop on Simulation* (eds S. Ermakov, V. Melas, and A. Pepelyshev), pp. 429–436. NII Chemistry University Publishers, St Petersburg.

MacKay, D. A. (1992). Information-based objective functions for active data selection. *Neural Computation*, **4**, 590–604.

Martin, R. J., Bursnall, M. C., and Stillman, E. C. (2001). Further results on optimal and efficient designs for constrained mixture experiments. In *Optimal Design 2000* (eds A. C. Atkinson, B. Bogacka, and A. Zhigljavsky), pp. 225–239. Kluwer, Dordrecht.

Matthews, J. N. S. (1987). Optimal crossover designs for the comparison of two treatments in the presence of carry over effects and autocorrelated errors. *Biometrika*, **74**, 311–320.

Matthews, J. N. S. (2000). *An Introduction to Randomized Controlled Clinical Trials*. Edward Arnold, London.

Matthews, J. N. S. (2006). *An Introduction to Randomized Controlled Clinical Trials, 2nd Edition*. Edward Arnold, London.

McCullagh, P. and Nelder, J. A. (1989). *Generalized Linear Models, 2nd Edition*. Chapman and Hall, London.

McGrath, R. M. and Lin, D. K. J. (2001). Testing multiple dispersion effects in unreplicated fractional factorial designs. *Technometrics*, **43**, 406–414.

McLachlan, G., Do, K.-A., and Ambroise, C. (2004). *Analyzing Microarray Gene Expression Data*. Wiley, New York.

McLean, R. A. and Anderson, V. L. (1966). Extreme vertices design of mixture experiments. *Technometrics*, **8**, 447–454.

Melas, V. (2006). *Functional Approach to Optimal Experimental Design*. Lecture Notes in Statistics 184. Springer–Verlag, New York.

Melas, V. B. (2005). On the functional approach to optimal designs for nonlinear models. *Journal of Statistical Planning and Inference*, **132**, 93–116.

Mentré, F., Mallet, A., and Baccar, D. (1997). Optimal design in random-effects regression models. *Biometrika*, **84**, 429–442.

Meyer, R. K. and Nachtsheim, C. J. (1995). The coordinate exchange algorithm for constructing exact optimal experimental designs. *Technometrics*, **37**, 60–69.

Michaelis, L. and Menten, M. L. (1913). Die Kinetik der Invertinwirkung. *Biochemische Zeitschrift*, **49**, 333–369.

Miller, A. J. (2002). *Subset Selection in Regression, 2nd Edition*. Chapman and Hall/CRC Press, Boca Raton.

Miller, A. J. and Nguyen, N.-K. (1994). Algorithm AS 295. A Fedorov exchange algorithm for D-optimal design. *Applied Statistics*, **43**, 669–677.

Mitchell, T. J. (1974). An algorithm for the construction of 'D-optimum' experimental designs. *Technometrics*, **16**, 203–210.

Mitchell, T. J. and Miller, F. L. (1970). Use of "design repair" to construct designs for special linear models. In *Report ORNL-4661*, pp. 130–131. Oak Ridge National Laboratory, Oak Ridge, TN.

Montgomery, D. C. (2000). *Design and Analysis of Experiments, 5th Edition*. Wiley, New York.

Mood, A. M. (1946). On Hotelling's weighing problem. *Annals of Mathematical Statistics*, **17**, 432–446.

Mukerjee, R. and Wu, C. F. J. (2006). *A Modern Theory of Factorial Design*. Springer–Verlag, New York.

Mukhopadhyay, S. and Haines, L. M. (1995). Bayesian D-optimal designs for the exponential growth model. *Journal of Statistical Planning and Inference*, **44**, 385–397.

Müller, W. G. (2007). *Collecting Spatial Data, 3rd Edition*. Springer-Verlag, Berlin.

Müller, W. G. and Pázman, A. (2003). Measures for designs in experiments with correlated errors. *Biometrika*, **90**, 423–434.

Myers, R. H. and Montgomery, D. C. (2002). *Response Surface Methodology, 2nd Edition*. Wiley, New York.

Myers, R. H., Montgomery, D. C., and Vining, G. G. (2001). *Generalized Linear Models: with Applications in Engineering and the Sciences*. Wiley, New York.

Myers, R. H., Montgomery, D. C., Vining, G. G., Borror, C. M., and Kowalski, S. M. (2004). Response surface methodology: A retrospective and literature survey. *Journal of Quality Technology*, **36**, 53–77.

Nachtsheim, C. J. (1989). On the design of experiments in the presence of fixed covariates. *Journal of Statistical Planning and Inference*, **22**, 203–212.

Nelder, J. A. and Lee, Y. (1991). Generalized linear models for the analysis of Taguchi-type experiments. *Applied Stochastic Models and Data Analysis*, **7**, 107–120.

Nelson, W. (1981). The analysis of performance-degradation data. *IEEE Transactions on Reliability*, **R-30**, 149–155.

Nigam, A. K. (1976). Corrections to blocking conditions for mixture experiments. *Annals of Statistics*, **47**, 1294–1295.

O'Brien, T. E. (1992). A note on quadratic designs for nonlinear regression models. *Biometrika*, **79**, 847–849.

O'Brien, T. E. and Rawlings, J. O. (1996). A nonsequential design procedure for parameter estimation and model discrimination in nonlinear regression models. *Journal of Statistical Planning and Inference*, **55**, 77–93.

Parzen, E. (1961). An approach to time series analysis. *Annals of Mathematical Statistics*, **32**, 951–989.

Patan, M. and Bogacka, B. (2007). Optimum designs for dynamic systems in the presence of correlated errors. (Submitted).

Patterson, H. D. and Thompson, R. (1971). Recovery of inter-block information when block sizes are unequal. *Biometrika*, **58**, 545–554.

Payne, R. W., Coombes, N. E., and Lisboa, P. (2001). Algorithms, generators and improved optimization methods for design of experiments. In *A Estatistica em Movimento: Actas do VIII Congresso Anual da Sociedade Portuguesa de Estatistica* (eds M. M. Neves, J. Cadima, M. J. Martins, and F. Rosado), pp. 95–103. Sociedade Portugesa de Estatistica, Lisbon.

Pázman, A. (1986). *Foundations of Optimum Experimental Design.* Reidel, Dordrecht.

Pázman, A. (1993). *Nonlinear Statistical Models.* Reidel, Dordrecht.

Pázman, A. and Pronzato, L. (1992). Nonlinear experimental design based on the distribution of estimators. *Journal of Statistical Planning and Inference*, **33**, 385–402.

Pepelyshev, A. (2007). Optimal design for the exponential model with correlated observations. In *MODA 8—Advances in Model-Oriented Design and Analysis* (eds J. Lopez-Fidalgo, J. M. Rodriguez-Diaz, and B. Torsney). Physica-Verlag, Heidelberg. (To appear).

Petkova, E., Shkodrova, V., Vassilev, H., and Donev, A. N. (1987). Optimization of the purification of nickel sulphate solution in the joint presence of iron (II), copper (II) and zinc (II). *Metallurgia*, **6**, 12–17. (In Bulgarian).

Piantadosi, S. (2005). *Clinical Trials, 2nd Edition.* Wiley, New York.

Piepel, G. F. (2006). 50 years of mixture experiment research: 1955–2004. In *Response Surface Methodology and Related Topics* (ed. A. I. Khuri), pp. 283–327. World Scientific, Singapore.

Piepel, G. F. and Cornell, J. A. (1983). Models for mixture experiments when the response depends on the total amount. *Technometrics*, **27**, 219–227.

Pilz, J. (1983). *Bayesian Estimation and Experimental Design in Linear Regression Models.* Teubner, Leipzig.

Pilz, J. (1991). *Bayesian Estimation and Experimental Design in Linear Regression Models.* Wiley, New York.

Pinheiro, J. C. and Bates, D. M. (2000). *Mixed-Effects Models in S and S-Plus.* Springer-Verlag, New York.

Plackett, R. L. and Burman, J.P. (1946). The design of optimum multifactorial experiments. *Biometrika*, **33**, 305–325.

Ponce de Leon, A. M. and Atkinson, A. C. (1992). The design of experiments to discriminate between two rival generalized linear models. In *Advances in GLIM and Statistical Modelling: Proceedings of the GLIM92 Conference, Munich* (eds L. Fahrmeir, B. Francis, R. Gilchrist, and G. Tutz), pp. 159–164. Springer, New York.

Prescott, P. (2000). Projection designs for mixture experiments in orthogonal blocks. *Communications in Statistics: Theory and Methods*, **29**, 2229–2253.

Press, W. H., Teukolsky, S. A., Vetterling, W. T., and Flannery, B. P. (1992). *Numerical Recipes in Fortran, 2nd Edition.* Cambridge University Press, Cambridge, England.

Pronzato, L. and Pázman, A. (2001). Using densities of estimators to compare pharmacokinetic experiments. *Computers in Biology and Medicine*, **31**, 179–195.

Pronzato, L. and Walter, E. (1985). Robust experimental design via stochastic approximations. *Mathematical Biosciences*, **75**, 103–120.

Pukelsheim, F. (1993). *Optimal Design of Experiments.* Wiley, New York.

Pukelsheim, F. and Rieder, S. (1992). Efficient rounding of approximate designs. *Biometrika*, **79**, 763–770.

Rafajłowicz, E. (2005). *Optymalizacja Eksperymentu z Zastosowaniami w Monitorowaniu Jakości Produkcji.* Oficyna Wydawnicza Politechniki Wrocławskiej, Wrocław.

Rao, C. R. (1973). *Linear Statistical Inference and its Applications, 2nd Edition.* Wiley, New York.

Rasch, D. and Herrendörfer, G. (1982). *Statistische Versuchsplanung.* Deutscher Verlag der Wissenschaften, Berlin.

Ratkowsky, D. A. (1989). *Nonlinear Regression Modeling.* Marcel Dekker, New York.

Ratkowsky, D. A. (1990). *Handbook of Nonlinear Regression Models.* Marcel Dekker, New York.

Rocke, D. M. and Lorenzato, S. (1995). A two-component model for measurement error in analytical chemistry. *Technometrics*, **37**, 176–184.

Roquemore, K. G. (1976). Hybrid designs for quadratic response surfaces. *Technometrics*, **18**, 419–423.

Rosenberger, W. F. and Lachin, J. L. (2002). *Randomization in Clinical Trials: Theory and Practice.* Wiley, New York.

Ryan, T. P. (1997). *Modern Regression Methods.* Wiley, New York.

Sacks, J., Welch, W. J., Mitchell, T. J., and Wynn, H. P. (1989). Design and analysis of computer experiments. *Statistical Science*, **4**, 409–435.

Sacks, J. and Ylvisaker, D. (1966). Designs for regression problems with correlated errors. *Annals of Mathematical Statistics*, **37**, 66–89.

Sacks, J. and Ylvisaker, D. (1968). Designs for regression problems with correlated errors: Many parameters. *Annals of Mathematical Statistics*, **39**, 49–69.

Sacks, J. and Ylvisaker, D. (1970). Designs for regression problems with correlated errors III. *Annals of Mathematical Statistics*, **41**, 2057–2074.

Sams, D. A. and Shadman, F. (1986). Mechanism of potassium-catalysed carbon/CO_2 reaction. *AIChE Journal*, **32**, 1132–1137.

Santner, T. J., Williams, B. J., and Notz, W. (2003). *The Design and Analysis of Computer Experiments.* Springer–Verlag, New York.

SAS Institute Inc. (2007*a*). *SAS/ETS User's Guide, Version 9.2.* SAS Institute Inc., Cary, NC.

SAS Institute Inc. (2007*b*). *SAS/IML User's Guide, Version 9.2.* SAS Institute Inc., Cary, NC.

SAS Institute Inc. (2007*c*). *SAS/QC User's Guide, Version 9.2.* SAS Institute Inc., Cary, NC.

SAS Institute Inc. (2007*d*). *SAS/STAT User's Guide, Version 9.2.* SAS Institute Inc., Cary, NC.

Savova, I., Donev, T. N., Tepavicharova, I., and Alexandrova, T. (1989). Comparative studies on the storage of freeze-dried yeast strains of the genus saccharomyces. In *Proceedings of the 4th International School on Cryobiology and Freeze-drying, 29 July 6 August 1989, Borovets, Bulgaria*, pp. 32–33. Bulgarian Academy of Sciences Press, Sofia.

Saxena, S. K. and Nigam, A. K. (1973). Symmetric-simplex block designs for mixtures. *Journal of the Royal Statistical Society, Series B*, **35**, 466–472.

Scheffé, H. (1958). Experiments with mixtures. *Journal of the Royal Statistical Society, Series B*, **20**, 344–360.

Schmelter, T. (2005). On the optimality of single-group designs in linear mixed models. Preprint 02/2005, Otto von Guericke Universität, Fakultät für Mathematik, Magdeburg.

Schwabe, R. (1996). *Optimum Designs for Multi-Factor Models.* Lecture Notes in Statistics 113. Springer-Verlag, Heidelberg.

Seber, G. A. F. (1977). *Linear Regression Analysis.* Wiley, New York.

Seber, G. A. F. and Wild, C. J. (1989). *Nonlinear Regression.* Wiley, New York.

Senn, S. J. (1993). *Cross-over Trials in Clinical Research.* Wiley, Chichester.

Shah, K. R. and Sinha, B. K. (1980). *Theory of Optimal Design.* Lecture Notes in Statistics 54. Springer-Verlag, Berlin.

Shinozaki, K. and Kira, T. (1956). Intraspecific competition among higher plants. VII. Logistic theory of the C–D effect. *Journal of the Institute of Polytechnics, Osaka City University*, **D7**, 35–72.

Sibson, R. (1974). D_A-optimality and duality. In *Progress in Statistics, Vol.2 – Proceedings of the 9th European Meeting of Statisticians, Budapest* (eds J. Gani, K. Sarkadi, and I. Vincze). North-Holland, Amsterdam.

Silvey, S. D. (1980). *Optimum Design.* Chapman and Hall, London.

Silvey, S. D. and Titterington, D. M. (1973). A geometric approach to optimal design theory. *Biometrika*, **60**, 15–19.

Silvey, S. D., Titterington, D. M., and Torsney, B. (1978). An algorithm for optimal designs on a finite design space. *Communications in Statistics A – Theory and Methods*, **14**, 1379–1389.

Sinha, S. and Wiens, D. P. (2002). Robust sequential designs for nonlinear regression. *Canadian Journal of Statistics*, **30**, 601–618.

Sitter, R. R. (1992). Robust designs for binary data. *Biometrics*, **48**, 1145–1155.

Sitter, R. R. and Torsney, B. (1995*a*). Optimal designs for binary response experiments with two variables. *Statistica Sinica*, **5**, 405–419.

Sitter, R. S. and Torsney, B. (1995*b*). D-optimal designs for generalized linear models. In *MODA 4—Advances in Model-Oriented Data Analysis* (eds C. P. Kitsos and W. G. Müller), pp. 87–102. Physica-Verlag, Heidelberg.

Smith, A. F. M. (1973). A general Bayesian linear model. *Journal of the Royal Statistical Society, Series B*, **35**, 67–75.

Smith, K. (1916). On the 'best' values of the constants in frequency distributions. *Biometrika*, **11**, 262–276.

Smith, K. (1918). On the standard deviations of adjusted and interpolated values of an observed polynomial function and its constants and the guidance they give towards a proper choice of the distribution of observations. *Biometrika*, **12**, 1–85.

Snee, R. D. and Marquardt, D. W. (1974). Extreme vertices designs for linear mixture models. *Technometrics*, **16**, 391–408.

Stehlík, M. (2005). Covariance related properties of D-optimal correlated designs. In *Proceedings of the 5th St Petersburg Workshop on Simulation* (eds S. Ermakov, V. Melas, and A. Pepelyshev), pp. 645–652. NII Chemistry University Publishers, St Petersburg.

Steinberg, D. M. and Lin, D. K. J. (2006). A construction method for orthogonal Latin hypercube designs. *Biometrika*, **93**, 279–288.

Street, A. P. and Street, D. J. (1987). *Combinatorics of Experimental Design.* Oxford University Press, Oxford.

Stromberg, A. (1993). Computation of high breakdown nonlinear regression parameters. *Journal of the American Statistical Association*, **88**, 237–244.

Stroud, J. R., Müller, P., and Rosner, G. L. (2001). Optimal sampling times in population pharmacokinetic studies. *Applied Statistics*, **50**, 345–359.

Studden, W. J. (1977). Optimal designs for integrated variance in polynomial regression. In *Statistical Decision Theory and Related Topics II* (eds S. S. Gupta and D. S. Moore), pp. 411–420. Academic Press, New York.

Tack, L. and Vandebroek, M. (2004). Budget constrained run orders in optimum design. *Journal of Statistical Planning and Inference*, **124**, 231–249.

Taguchi, G. (1987). *Systems of Experimental Designs (Vols 1 and 2, 1976 and 1977, with 1987 translation)*. UNIPUB, Langham, MD.

Titterington, D. M. (1975). Optimal design: Some geometrical aspects of D-optimality. *Biometrika*, **62**, 313–320.

Titterington, D. M. (2000). Optimal design in flexible models, including feed-forward networks and nonparametric regression (with discussion). In *Optimum Design 2000* (eds A. C. Atkinson, B. Bogacka, and A. Zhigljavsky), pp. 261–273. Kluwer, Dordrecht.

Torsney, B. and Gunduz, N. (2001). On optimal designs for high dimensional binary regression models. In *Optimal Design 2000* (eds A. C. Atkinson, B. Bogacka, and A. Zhigljavsky), pp. 275–285. Kluwer, Dordrecht.

Torsney, B. and Mandal, S (2004). Multiplicative algorithms for constructing optimizing distributions : Further developments. In *MODA 7—Advances in Model-Oriented Design and Analysis* (eds A. Di Bucchianico, H. Läuter, and H. P. Wynn), pp. 163–171. Physica-Verlag, Heidelberg.

Trinca, L. A. and Gilmour, S. G. (2000). An algorithm for arranging response surface designs in small blocks. *Computational Statistics and Data Analysis*, **33**, 25–43.

Trinca, L. A. and Gilmour, S. G. (2001). Multi-stratum response surface designs. *Technometrics*, **43**, 25–33.

Tsai, P.-W., Gilmour, S. G., and Mead, R. (2000). Projective three-level main effects designs robust to model uncertainty. *Biometrika*, **87**, 467–475.

Uciński, D. (1999). *Measurement Optimization for Parameter Estimation in Distributed Systems*. Technical University Press, Zielona Góra.

Uciński, D. (2005). *Optimal Measurement Methods for Distributed Parameter System Identification*. CRC Press, Boca Raton.

Uciński, D. and Atkinson, A. C. (2004). Experimental design for processes over time. *Studies in Nonlinear Dynamics and Econometrics*, **8**(2). (Article 13). http://www.bepress.com/snde/vol8/iss2/art13.

Uciński, D. and Bogacka, B. (2004). T-optimum designs for multiresponse dynamic heteroscedastic models. In *MODA 7—Advances in Model-Oriented Design and Analysis* (eds A. Di Bucchianico, H. Läuter, and H. P. Wynn), pp. 191–199. Physica-Verlag, Heidelberg.

Uciński, D. and Bogacka, B. (2005). T-optimum designs for discrimination between two multiresponse dynamic models. *Journal of the Royal Statistical Society, Series B*, **67**, 3–18.

Valko, P. and Vajda, S. (1984). An extended ODE solver for sensitivity calculations. *Computers and Chemistry*, **8**, 255–271.

Van Schalkwyk, D. J. (1971). *On the Design of Mixture Experiments*. PhD Thesis, University of London.

Verbeke, G. and Molenberghs, G. (2000). *Linear Mixed Models for Longitudinal Data.* Springer–Verlag, New York.

Vuchkov, I. N. (1977). A ridge-type procedure for design of experiments. *Biometrika*, **64**, 147–150.

Vuchkov, I. N. (1982). Sequentially generated designs. *Biometrical Journal*, **24**, 751–763.

Vuchkov, I. N. and Boyadjieva, L. N. (2001). *Quality Improvement with Design of Experiments: A Response Surface Approach.* Kluwer, Dordrecht.

Vuchkov, I. N., Damgaliev, D. L., and Yontchev, C. A. (1981). Sequentially generated second order quasi D-optimal designs for experiments with mixture and process variables. *Technometrics*, **23**, 233–238.

Wahba, G. (1971). On the regression design problem of Sacks and Ylvisaker. *Annals of Mathematical Statistics*, **42**, 1035–1053.

Wald, A. (1943). On the efficient design of statistical investigations. *Annals of Mathematical Statistics*, **14**, 134–140.

Walter, E. and Pronzato, L. (1997). *Identification of Parametric Models from Experimental Data.* Springer-Verlag, New York.

Waterhouse, T. H., Woods, D. C., Eccleston, J. A., and Lewis, S. M. (2007). Design selection criteria for discrimination between nested models for binomial data. *Journal of Statistical Planning and Inference*. (In press).

Weisberg, S. (2005). *Applied Linear Regression, 3rd Edition.* Wiley, New York.

Welch, W. J. (1982). Branch and bound search for experimental designs based on D-optimality and other criteria. *Technometrics*, **24**, 41–48.

Welch, W. J. (1984). Computer-aided design of experiments for response estimation. *Technometrics*, **26**, 217–224.

Whitehead, J. (1997). *The Design and Analysis of Sequential Clinical Trials, 2nd Edition.* Wiley, Chichester.

Whittle, P. (1973). Some general points in the theory of optimal experimental design. *Journal of the Royal Statistical Society, Series B*, **35**, 123–130.

Wiens, D. P. (1998). Minimax robust designs and weights for approximately specified regression models with heteroscedastic errors. *Journal of the American Statistical Association*, **93**, 1440–1450.

Wierich, W. (1986). On optimal designs and complete class theorems for experiments with continuous and discrete factors of influence. *Journal of Statistical Planning and Inference*, **15**, 19–27.

Williams, R. M. (1952). Experimental designs for serially correlated observations. *Biometrika*, **39**, 151–167.

Woods, D. C., Lewis, S. M., Eccleston, J. A., and Russell, K. G. (2006). Designs for generalized linear models with several variables and model uncertainty. *Technometrics*, **48**, 284–292.

Wu, C.-F. and Wynn, H. P. (1978). General step-length algorithms for optimal design criteria. *Annals of Statistics*, **6**, 1273–1285.

Wu, C. F. J. and Hamada, M. (2000). *Experiments: Planning, Analysis, and Parameter Design Optimization.* Wiley, New York.

Wynn, H. P. (1970). The sequential generation of D-optimal experimental designs. *Annals of Mathematical Statistics*, **41**, 1055–1064.

Wynn, H. P. (1972). Results in the theory and construction of D-optimum experimental designs. *Journal of the Royal Statistical Society, Series B*, **34**, 133–147.

Wynn, H. P. (1985). Jack Kiefer's contributions to experimental design. In *Jack Carl Kiefer Collected Papers III* (eds L. D. Brown, I. Olkin, J. Sacks, and H. P. Wynn), pp. xvii–xxiv. Wiley, New York.

Zellner, A. (1962). An efficient method of estimating seemingly unrelated regressions and tests of aggregation bias. *Journal of the American Statistical Association*, **57**, 348–368.

AUTHOR INDEX

SUBJECT INDEX

Made in the USA
Lexington, KY
25 April 2014